Mating Systems and Strategies

MONOGRAPHS IN
BEHAVIOR AND ECOLOGY

Edited by John R. Krebs and
Tim Clutton-Brock

Foraging Theory, by David W. Stephens and John R. Krebs

Dynamic Modeling in Behavioral Ecology, by Marc Mangel and Colin W. Clark

The Evolution of Parental Care, by T. H. Clutton-Brock

Parasitoids: Behavioral and Evolutionary Ecology, by H.C.J. Godfray

Sexual Selection, by Malte Andersson

Polygyny and Sexual Selection in Red-Winged Blackbirds, by William A. Searcy and Ken Yasukawa

Leks, by Jacob Höglund and Rauno V. Alatalo

Social Evolution in Ants, by Andrew F. G. Bourke and Nigel R. Franks

Female Control: Sexual Selection by Cryptic Female Choice, by William G. Eberhard

Sex, Color, and Mate Choice in Guppies, by Anne E. Houde

Foundations of Social Evolution, by Steven A. Frank

Parasites in Social Insects, by Paul Schmid-Hempel

Levels of Selection in Evolution, edited by Laurent Keller

Social Foraging Theory, by Luc-Alain Giraldeau and Thomas Caraco

Model Systems in Behavioral Ecology: Integrating Conceptual, Theoretical, and Empirical Approaches, edited by Lee Alan Dugatkin

Sperm Competition and Its Evolutionary Consequences in the Insects, by Leigh W. Simmons

The African Wild Dog: Behavior, Ecology, and Conservation, by Scott Creel and Nancy Marusha Creel

Mating Systems and Strategies, by Stephen M. Shuster and Michael J. Wade

Mating Systems and Strategies

STEPHEN M. SHUSTER
& MICHAEL J. WADE

Princeton University Press
Princeton and Oxford

Published by Princeton University Press, 41 William Street, Princeton, New Jersey 08540
In the United Kingdom: Princeton University Press, 3 Market Place, Woodstock, Oxfordshire OX20 1SY

Library of Congress Cataloging-in-Publication Data

Shuster, Stephen M., 1954–
Mating systems and strategies / Stephen M. Shuster, Michael J. Wade.
p. cm. — (Monographs in behavior and ecology)
Includes bibliographical references (p.).
ISBN 0-691-04930-0 (cl : alk. paper) —
ISBN 0-691-04931-9 (pb : alk. paper)
1. Sexual behavior in animals. I. Wade, Michael John, 1949– II. Title. III. Series.

QL761 .S54 2003
591.56′3—dc21 2002072855

British Library Cataloging-in-Publication Data is available

This book has been composed in Times Roman

Printed on acid-free paper. ∞

www.pupress.princeton.edu

Printed in the United States of America

10 9 8 7 6 5 4 3 2 1

Contents

Preface

We have written this book to provide a comparative, conceptual, and statistical framework for studying mating systems and alternative mating strategies in natural populations. When we began our discussions on these topics in 1988 over coffee and M&Ms, we were struck by the explosive growth of verbal, evolutionary models "explaining" male and female reproductive behavior, many unconstrained by the principles of evolutionary genetics. It seemed as though every human failing or inadequacy was being elevated by clever argument to the level of an adaptive strategy, and added to the repertoire of animal behavior by natural or, most often, sexual selection. We have attempted to make the fundamental relationships between the mean, variance, and covariance of male and female reproductive selection explicit and thereby establish a more solid theoretical foundation for future studies on mating systems and strategies.

In chapter 1, we explore the *Quantitative Paradox of Sexual Selection*, namely, the fact that the microevolutionary process of sexual selection and the macroevolutionary pattern it produces pose a paradox. On the one hand, the taxonomic pattern of extreme phenotypic diversity between males of closely related species suggests the action of a very strong and rapid evolutionary force. In contrast, selection restricted to one sex, to a single component of fitness within that sex, and opposed in the other sex, should be a very weak microevolutionary force, producing change slowly if at all.

Using a statistical approach for measuring the strength of selection developed by James F. Crow (Crow 1958, 1962), we show that the *sex difference in the opportunity for selection* is the key to resolving the quantitative paradox between microevolutionary process and macroevolutionary pattern. We illustrate how and why sexual selection, although restricted to one sex and opposed in the other, can be, as the taxonomic pattern indicates, one of the strongest and fastest of all evolutionary forces. Focusing on the sex difference in the opportunity for selection makes it clear that the spatial and temporal clustering of *female receptivity* as well as female reproductive life history are fundamental to understanding the more conspicuous male mating strategies now used for mating system classification. Our statistical framework encompasses the roles of both sexes in mating system evolution.

We apply our approach in chapters 2 and 3, respectively, to quantify the relationships between the *spatial* and *temporal* distributions of receptive females and the opportunity for sexual selection. We show how the spatial aggregation of female mating receptivity enhances the sex difference in the opportunity for sexual selection, while temporal aggregation of female receptivity (i.e., reproductive synchrony) diminishes it and, thereby, limits sexual selection. We show how the opportunity for sexual selection, caused by the spatial and temporal clustering of female receptivity, is related to the ecological concept of mean crowding (*sensu* Lloyd 1967) applied directly to the spatial and temporal aggregation of receptive females.

Sexual selection based on the spatial clustering of reproduction can lead to a runaway process that results in an overcrowding of females at resources. However, the trade-off between mate guarding and mate seeking by males can bring this runaway to a halt *without increased male mortality*. Similarly, we show that a temporal runaway process can favor reproductive asynchrony of females and that it can be halted by a trade-off between extremely early (and extremely late) receptivity and reduced fecundity. This is a runaway process that is halted by selection on females as opposed to the classic runaway Fisherian process, which is halted by male mortality. In fact, we identify more than ten possible runaway processes between male and female traits, most of which are limited by factors other than male mortality. In this context, we suggest that the high frequency of hidden or stolen copulations, now being revealed by molecular paternity analysis, is the expected equilibrium signature of both the spatial runaway process based on female aggregation and the temporal runaway process based on female reproductive asynchrony.

In chapter 4, we explain how female promiscuity (both mating-number and mate-number promiscuity) diminishes the sex difference in the opportunity for selection by limiting sexual selection on males. We combine male and female influences on the opportunity for sexual selection to predict overall patterns of male and female mating behavior.

In chapter 5, we explore the effects of female reproductive life history (multiple mating and multiple reproductive episodes) on the variance in offspring numbers among females. We show how the variance in apportionment of paternity by multiple mating in females affects sexual selection on males. We also show how female reproductive life history affects the sex difference in the opportunity for selection, ΔI, and how to measure ΔI in natural and experimental populations. We suggest that male genetic quality is *least* important to females in species with strong sexual selection and large values of ΔI, and *most* important to females in species in which sexual selection is negligible and ΔI is small. However, mating systems in which ΔI is small may be evolutionarily unstable.

In chapter 6 we propose a statistical framework for quantifying within- and between-species variation in the values of ΔI. We examine how the "ΔI surface" changes with changes in the spatial and temporal patchiness of receptive females. This approach is similar in concept to earlier methods for quantifying the opportunity for selection in nature (Wade and Arnold 1980; Arnold and Wade 1984a, b; Clutton-Brock 1991), and it provides us with a quantitative framework for classifying mating systems.

In chapter 7, we discuss what we consider to be difficulties with and contradictions in current theories of behavioral evolution, especially those applied to the evolution of mating behaviors. Much of existing mating system theory is based on parental investment theory (Trivers 1972), as well as on patterns of paternal care (Clutton-Brock 1994). We suggest that studies of mating systems that focus on mate numbers and factors affecting ΔI, rather than patterns of parental investment, can lead to greater insights into how and when sexual selection operates and the mating systems and strategies that develop as a result.

In chapter 8 we describe the behaviors of males and females most likely to influence mating system evolution, and in chapter 9 we combine the methods and concepts introduced in the first seven chapters to generate a predictive scheme for classifying mating systems. We combine the spatial and temporal distributions of female receptivity with female reproductive life histories to examine their aggregate effects on the opportunity for sexual selection. We generate specific predictions about the occurrence of sperm competition, mate guarding, sexual dimorphism, sexual conflict, and alternative mating strategies. We illustrate our scheme using selected animal and plant taxa and we modify the nomenclature introduced by Emlen and Oring (1977) for mating system classification to address previously unrecognized mating systems.

In chapter 10, we discuss Darwin's perspective on alternative male mating strategies. Darwin clearly recognized male polymorphism as well as the necessary conditions for new male forms to invade and persist in populations. However, we show that he overlooked "female mimicry," one of the most common male mating strategies, in his emphasis on the timing of expression of sex-limited mutations.

In chapter 11, we discuss key conceptual developments in the study of alternative mating strategies since Darwin, as well as gaps and inconsistencies in existing hypotheses regarding alternative mating strategies. We identify the conditions under which genetic polymorphism in male mating behavior may persist and we argue that such polymorphism is probably more common than presently recognized.

In chapter 12, we describe patterns of alternative mating strategies in nature, including male polymorphisms, sequential hermaphroditism, and parental manipulation of family sex ratio. We also discuss gaps and inconsistencies in existing data, notably the apparent ubiquity of condition-dependent mating strategies. We describe factors likely to contribute to the evolution of alternative mating strategies, and we provide a statistical framework, based on parameters introduced in earlier chapters, for measuring selection on alternative mating strategies. We provide a new classification scheme for alternative mating strategies based, not on the relative degree to which phenotypes are condition dependent, but instead on the ecological conditions in which they are likely to arise. Lastly, we present our conclusions on alternative mating strategies operating within animal mating systems, and we propose directions for future research.

We are grateful to our families for their support and encouragement. In particular, Michelle M. Pitts and Debra L. Rush-Wade provided insightful comments on content and exposition as well as a loving and rewarding family environment, replete with chocolate, caffeine, and gentle but unmistakable encouragement. We also thank the members of our respective laboratories and discussion groups, whose comments and criticisms of earlier drafts of this book significantly influenced its final form. These include Christie Clark, Jeff Demuth, Steve Freedberg, Heidi Hornstra, Kevin Johnson, Veijo Jormalainen, Jennifer Learned, Tim Linksmayer, Bryce Marshall, Peter Nelson, Emily Omana, Meaghan Saur, Steve Vuturo, and Herb Wildey. Throughout this effort our colleagues, Chris Boake, Butch Brodie, Russ Balda, Lynda Delph, Norman Johnson, Curt Lively, Ellen Ketterson, Bill Rowland, Con Slobodchikoff, and Tad Theimer, generously offered encouragement, insightful commentary, and novel ideas. Carl Schlichting, Peter Price, and Steve Frank shared their book-writing experiences with us and, by anecdote and example, helped us indirectly through some of the rougher patches. We are especially grateful to the four referees, Barry Sinervo, Mart Gross, John Alcock, and Derek Roff, who thoughtfully commented on the first draft of our manuscript. We have seldom received such detailed and constructive advice. Despite our differing opinions on some issues, we thank these gentlemen for their sincere support as well as for their rigorous and cogent criticisms. Lastly, we thank our editor, Sam Elworthy, and his assistant, Sarah Harrington, at Princeton University Press, and copyeditor Jennifer Slater for their assistance in producing this book.

S.M.S.
Flagstaff, Arizona
M.J.W.
Bloomington, Indiana

1 The Opportunity for Selection

"When the males and females of any animal have the same general habits of life, but differ in structure, colour, or ornament, such differences have been mainly caused by sexual selection."
—(Darwin 1859, p. 89)

Sexual Selection and the Sex Difference in Variance of Reproductive Success

Darwin recognized two patterns in nature and used them to frame the central questions of sexual selection (Darwin 1859):

1. Why do males and females of the same species differ from one another, with males exhibiting morphological and behavioral phenotypes more exaggerated than those of females?
2. Why do the males of closely related species exhibit much greater differences in morphology and behavior than the females of closely related species?

The first pattern is a microevolutionary one, seen commonly within species of almost all taxa with separate sexes, including plants. It indicates that some kind of selection is working to differentiate the sexes and it is affecting males to a much greater degree than females. The second pattern is a macroevolutionary one, observed across species within genera or families of almost all taxa (e.g., many avian taxa). These large differences in male phenotype among closely related taxa are the signature of a very strong and rapid evolutionary force. Darwin noted that, in many species, the phenotypic differences between the sexes are associated neither with essential reproductive physiology nor with development of the male and female gametes. The exaggerated plumage, coloration, behavior, and morphology of males are correlated with but not necessary to reproduction.

Both of Darwin's patterns are reflected in the language of natural history. In many species, the male is so conspicuously different from the female that the common name of a species describes only the male sex. Only male red-winged blackbirds (*Aegelaius phonecius phonecius*; Searcy 1979; Weatherhead and Robertson 1979), are black with red epaulets on their wings, whereas the females are inconspicuous and dull brown in color (fig. 1.1a). In the bullfrog, *Rana catesbania* (Howard 1984), it is only the male that makes the deep call for which the species gets its common name; female bullfrogs

(a)

(b)

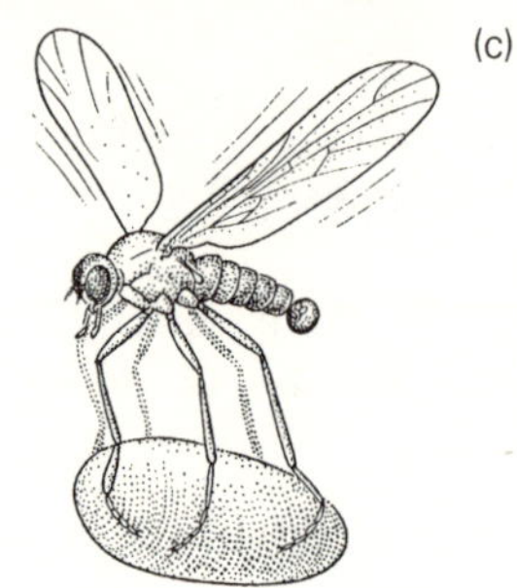

(c)

Figure 1.1. (a) Redwinged blackbird male (photo by Delbert Rust); (b) male bullfrog (photo by J. White) male balloon fly (from Alcock 1975).

are silent (fig. 1.1b). Only males carry balloons of silk as nuptial gifts for females in balloon flies, *Hilara santor* (fig. 1.1c; Kessel 1955). Neither the epaulets of the male blackbird, nor the call of the male bullfrog, nor the bower of the male bowerbird, nor the balloons of male balloon flies, are essential for sperm production or other physiological aspects of reproductive function. These males differ from the females of their own species in a rather arbitrary suite of phenotypic traits when considered across taxa.

Darwin used the term "trivial" to describe many of these exaggerated, male-limited characters because they appeared to have no clear relationship to viability or reproductive fitness. Despite very similar ways of life, closely related species could have males with very different phenotypes. Why should male tail length in one species be greatly elongated while, in another species of the same genus, males might possess a cape of expandable neck feathers and be rather ordinary in tail length (Gilliard 1962; Borgia 1986)? Why would longer tails be adaptive for males of one species but not the other? Furthermore, if these traits were adaptive for males, why were they not also adaptive for females? Darwin saw no obvious functional relationship between the exaggerated traits of males and the physical environment as he did for many other characters. Indeed, the "fit" between certain male phenotypes and the abiotic environment was exceptionally poor; exaggerated male characters might actually lower male viability.

The macroevolutionary pattern of large phenotypic differences between males of closely related species suggests that the selection responsible for these exaggerated male traits was *rapid and strong*. In contrast, we know from microevolutionary theory and empirical studies of artificial selection that selection acting on only one sex is considerably *slower and weaker* than selection acting in both sexes. In fact, selection on one sex but not on the other is only half as effective as selection acting on both, because half of the genes in any generation are derived from each sex in the previous generation (Falconer 1989). Selection restricted to one sex is tantamount to drawing half of the genes at random, and unselected. Selection that acts in *opposing* directions in the two sexes is slower still than selection absent in one sex.

To understand the microevolutionary perspective on single-sex selection, consider an experiment in which a laboratory or captive population is subjected to artificial selection to increase tail length in males. There are several different ways that we might impose artificial selection and these have different effects on the expected rate of response. Consider first artificial selection on both males and females. After measuring tail lengths of all males, those with long tails are chosen as parents and those with shorter tails are discarded and prevented from breeding. We can quantify the strength of this selection using the standardized selection differential experienced by males, S_{males} (fig. 1.2.a). The difference in average tail length between the selected males and the unselected males, divided by the standard deviation of male tail length, equals S_{males}. Similarly, we measure tail lengths of all females and choose those with the longest tails for parents and discard those with shorter tails (fig. 1.2.b). The strength of this selection, S_{females}, is defined just like S_{males}, but relative to the female trait distribution. Total selection on our hypothetical population is the average of these two selection differentials, S_{total} or $(S_{\text{males}} + S_{\text{females}})/2$. If our selection is as strong in males as it is in females, so that S_{males} equals S_{females}, which equals S, then S_{total} also equals S.

Now consider artificial selection *only* on males and not on females (fig. 1.3). As before, we measure tail length of males and select those with the longest tails as breeders (fig. 1.3a). However, we choose female parents at random with regard to tail length (fig. 1.3b). Thus, the selection differential in females, S_{females}, must be zero because the mean tail lengths of the breeding and nonbreeding females are the same. By using the *selected* males and *unselected* females as parents, fully half of the genes of each offspring, namely, those descending through the females, are not subject to any selection at all. Because half of the genetic material affecting tail length in the offspring generation has not been selectively screened but rather has been

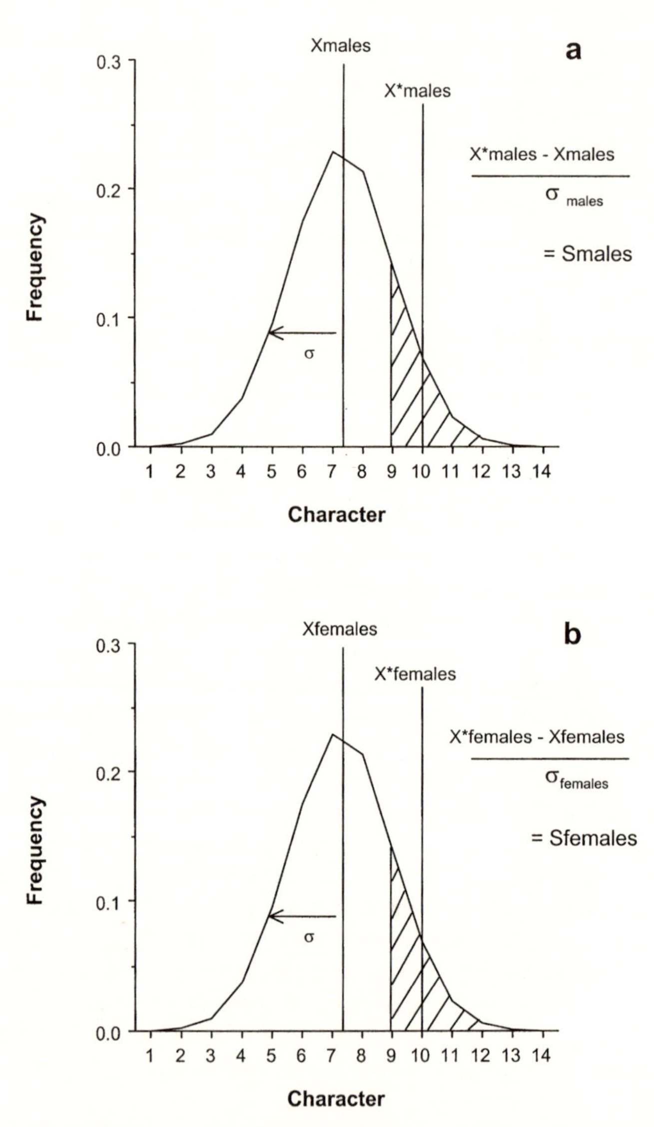

Figure 1.2. Selection on both sexes simultaneously; hatched areas represent breeding individuals; open areas represent nonbreeding individuals.

chosen at random, the total selection differential, averaged across the sexes, equals $S_{\text{males}}/2$ or, if we select on males as strongly as we did above, $S/2$. This makes single-sex selection weaker by half than selection on both sexes. Single-sex artificial selection experiments conducted on a number of species confirm this theoretical expectation (Robertson 1980).

The taxonomic observation of conspicuous male divergence between

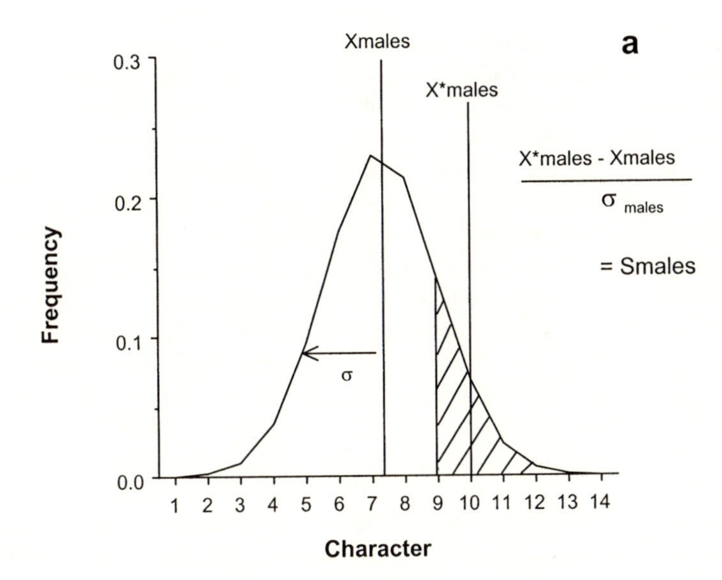

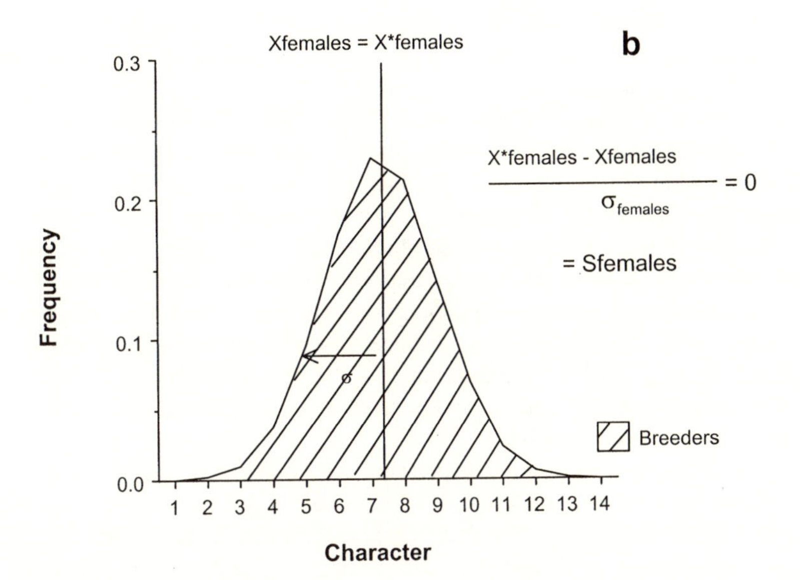

Figure 1.3. Single-sex selection, that is, selection on only one sex; hatched areas represent breeding individuals; open areas represent nonbreeding individuals.

closely related species stands in sharp contrast with the expectation from microevolutionary theory of a slower response to single-sex selection. This contrast between macroscopic pattern and microevolutionary process becomes even starker when we consider the evolution of sex-limited expression of male phenotypes. When genes are carried by both sexes but expressed in only one sex, we say that the expression of the gene is *sex-limited.* Most genes are expressed in both sexes; the limitation of a gene's expression to one sex is itself an evolved property of that gene or the developmental genetic system. Indeed, when artificial selection is practiced on only *one* sex, the focal character responds in *both* sexes, because selection in one sex does not limit the expression of the genes to that sex. For this reason, in our second artificial selection experiment (see above), we expect the tails of the unselected females to increase in length as a result of selection for increased male tail length. In the terminology of phenotypic selection models, the same trait (e.g., tail length) expressed in males and in females can be considered as two distinct but genetically correlated traits, one expressed in males and one in females. Because the genetic correlation across the sexes is positive, direct selection on one sex results in a correlated response in the homologous trait in the other owing to indirect selection.

With sexual selection, however, the selection differential in females is not zero, but actually less than zero (i.e., $S_{females} < 0$). Why is this so? Because, as Wallace (1868) argued, the exaggerated traits favored in males are *selected against* in females. Thus, the sex-specific selection differentials are of opposite sign, $S_{females} < 0 < S_{males}$. It is as though we divided a population's genetic composition into two separate pools of genes: one from females and one from males. In our hypothetical population, in the male gene pool, we select for those genes that increase tail length or body size ($0 < S_{males}$; fig. 1.4a). In the female pool, we select *against* these very same genes ($S_{females} < 0$; fig. 1.4b)! We then combine the two divergently selected pools by mating the selected males and females to create the offspring. The result is an evolutionary process that is even *weaker and slower* than single-sex selection (fig. 1.3) because the average selection differential on the trait $[(S_{males} - S_{females})/2]$ is *less than* that of single-sex selection, $S_{males}/2$. If the selection differentials were equal in the two sexes (but of opposite sign), then S_{total} would equal zero.

How does sex-limited gene expression arise? The evolution of gene expression that exists in one sex and not the other is believed to occur via the evolution of *modifiers*, genetic factors that modify the normal pattern of gene expression during development (Fisher 1928; Altenberg and Feldman 1987). The fitness advantage to the modifier, which is otherwise neutral, accrues when there is selection favoring the phenotype in one sex and opposing it in the opposite sex. That is, the evolution of sex-limited gene expression requires a sex difference in the direction of selection, as occurs with sexually selected traits. As modifier genes spread through the population, the expres-

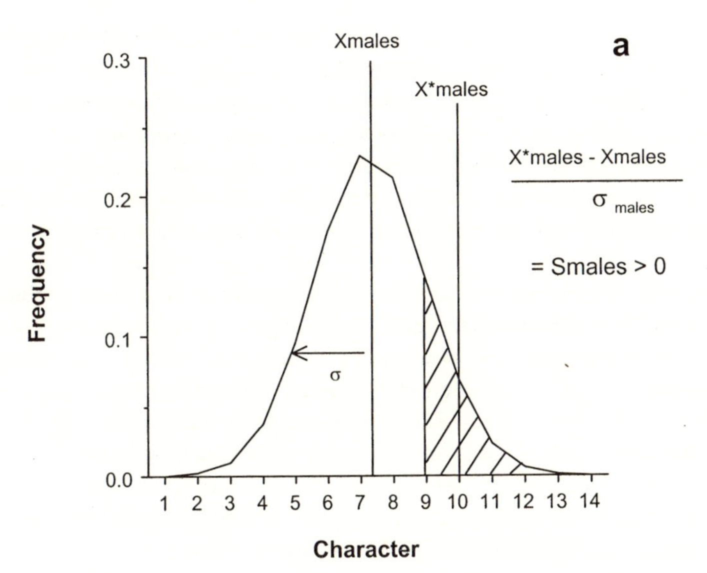

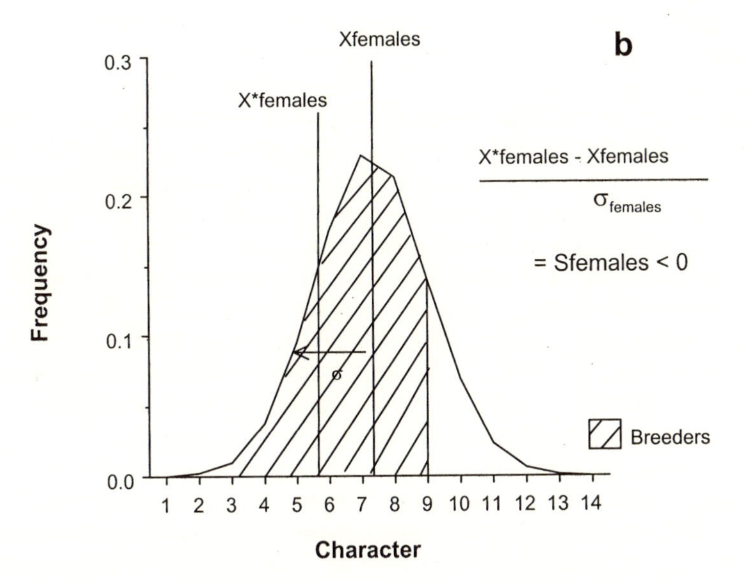

Figure 1.4. Directional selection in opposing directions between the sexes; hatched areas represent breeding individuals; open areas represent nonbreeding individuals.

sion of the genes affecting tail length is diminished in the female sex and the trait can become sex-limited. Such modifiers can be viewed as genes that reduce the genetic correlation between a trait expressed in males and the same trait expressed in females. Reducing the genetic correlation reduces the limitation imposed by the sex difference in the direction of selection. However, the evolution of such modifier genes takes time, so the rate of evolutionary response for genes with initial phenotypic expression in both sexes is slowed.

There is an additional reason for sexual selection to be a slow evolutionary process. Exaggerated male traits do not appear to be adaptive at all life stages even in the male sex. Darwin (1859) reasoned that seasonal patterns in the expression of exaggerated male-limited traits indicate that these phenotypes may not be of selective advantage at other times during the life of males. Wallace (1868) went further, suggesting that such traits were selected against at these other times (see chapter 10). This conflict in the direction of selection at different life stages in the male sex makes total selection on these traits *weaker* and thus makes their evolution *slower*.

Consider again our population subject to artificial selection on males (fig. 1.5). In this population, suppose that selection is imposed on males at two different stages in the life cycle. When males are young and immature, selection may favor small tail size and act against males with larger tails, so that $S_{\text{males}}(\text{early}) < 0$ (fig. 1.5a). Only males with the smallest tails at this stage are permitted to mature. Later, at maturity, selection among the remaining males favors only those with the largest tails becoming parents, so that $S_{\text{males}}(\text{late}) > 0$ (fig. 1.5b). Genes that increase tail length at all ages will experience conflicting selection pressures, and, in particular, they will be culled and discarded at the first episode of selection. The total selection differential in males is the *combination* of these two opposing components, further weakening the overall strength of selection for these traits in the male sex.

Only those genes that fortuitously act later in male life to increase tail size will experience a coherent selection pressure and avoid the opposing juvenile selection. It is possible, and even likely, that there are "modifier" genes that could delay the timing of expression of tail size genes to late in male life. However, age-limited expression, like the sex-limited expression discussed above, represents another derived property of the genetic architecture and it further slows the expected rate of evolution under sexual selection.

Such conflicts between the fitness components of male viability and male reproduction may exist for many exaggerated male traits. It is often hypothesized that exaggerated male traits lower male viability by making males more conspicuous to predators (e.g., bird territorial calls, plumage, and displays) or by imposing high energetic costs on males (e.g., male vocalizations in frogs and toads). In some species, these costs to males have been well documented. For example, only male frogs call, and calling has been shown by ecological physiologists to be extremely costly in energetic terms (Ryan

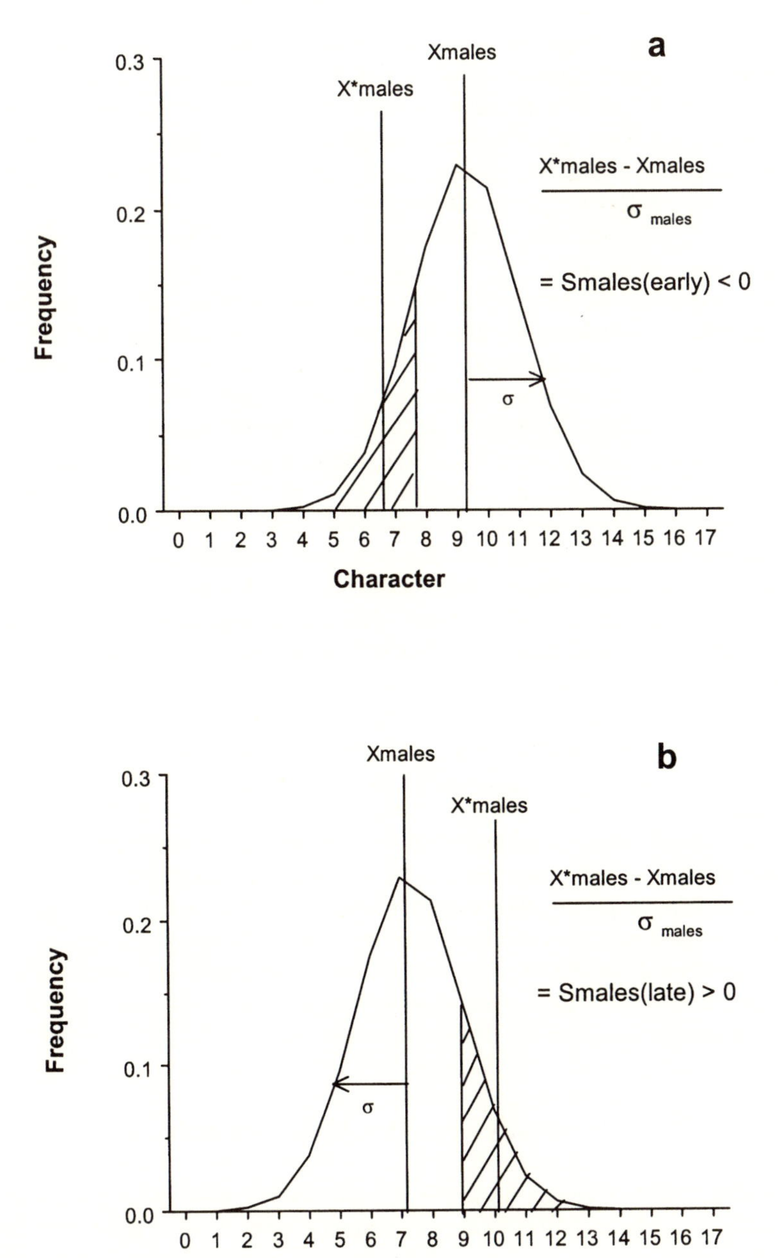

Figure 1.5. Directional selection in opposing directions within one sex over different life stages; hatched areas represent breeding individuals; open areas represent nonbreeding individuals.

and Tuttle 1981). Males of many anuran species expend large amounts of energy calling and, although more than 90% of matings will take place over three or four nights, the males might call every night for two or three months. Also, in at least one species, *Physalemus pustuluosus*, males are victims of frog-eating bats that use the sex-specific calls to locate their prey (Ryan and Tuttle 1981; Ryan 1983b).

There is a similar risk to males in some species of lampyrid beetles or "lightning bugs" (Lloyd 1966). In a typical species, the males fly around and emit a luminous, blinking signal. The males' visual calling elicits a response from females, which are flightless in many species. A signaling male finds a mate when a female responds to his signal with her own light. Not only is the energetic cost of flying borne solely by the males but also the more conspicuous males encounter a sex-specific risk of predation, sometimes from heterospecific females that mimic the response signal of another species and eat the responding males (fig. 1.6; Lloyd 1975). McCauley (1982) has shown that in mating pairs of milkweed beetles, *Tetraopes tetraophthalmous*, males in copula and guarding females succumb to wheel-bug predation more often than females. Male scorpion flies, genus *Panorpa*, forage in spider webs for insect prey that they then use as nuptual gifts to facilitate copulation with females (Thornhill 1981). In general, as a consequence of the expression of sexually selected traits, males suffer greater mortality than females. Indeed, many believe that lack of strong predation pressure probably facilitated the evolution of ground display habits by some birds of paradise as well as bower building by all bowerbirds (S. Pruett-Jones, pers. comm.).

In summary, there is an apparent conflict between the microevolutionary perspective on sexual selection and the macroevolutionary pattern. Whereas the comparative pattern indicates that sexual selection is one of the *fastest and strongest* evolutionary forces, microevolutionary analysis suggests that sexual selection should be *weak and slow* owing to single-sex selection, sex-limited expression, age-limited or stage-limited expression, and conflict between viability and fertility fitness within males.

The Quantitative Paradox of Sexual Selection

The Quantitative Paradox of sexual selection is this: *How can sexual selection be strong enough to counter the combined, opposing forces of male and female viability selection?* In studying the forces of evolution, we want to understand both the mechanisms by which they work, as well as the relative strength of each in relation to the others. Weak natural selection does not always override other evolutionary forces, like random genetic drift or mutation. Many equilibrium states in evolutionary genetics represent a balance between opposing evolutionary forces, such as mutation-selection balance

Figure 1.6. *Photurus* male being eaten by *Photinus* female (photo by J. Lloyd; see also Lloyd 1975).

(Hartl and Clark 1989; Lynch et al. 1998). The evolution of one trait can be limited by opposing natural selection on other genetically correlated traits. Although Darwin provided two mechanisms (see below) that permit us to understand *how* sexual selection operates, he did not address the *quantitative* issue of its *strength* relative to other evolutionary forces. How can sexual selection be such a strong and rapid evolutionary process that it can create large differences in morphology, physiology, and behavior among males of closely related species? How can Darwin's "less rigorous" process of sexual selection (see below) overpower the opposing forces of natural selection in both sexes?

We can address these questions by partitioning the effects of exaggerated male characters into separate components of male and female reproductive and viability fitness (table 1.1). When we do this, we find that only one component is positive and all the rest are negative. The paradox of sexual selection arises because the fitness decrements associated with exaggerated male traits appear to outweigh the fitness increments (see table 1.1). As explained below, Darwin proposed two compelling mechanisms of sexual selection, namely, female choice and male-male combat. However, he did not identify or even discuss this more quantitative issue. We will first discuss Darwin's mechanisms for sexual selection and then address the resolution of what we have called the Quantitative Paradox of sexual selection.

Table 1.1
The sex-specific selection differentials, partitioned into reproductive and viability components of fitness, for an exaggerated male trait.

	Viability	*Reproduction*	*Total Selection Differential*
Selection on Males	$S_{early} < 0$	$S_{late} > 0$	S_{males}
Selection on Females	$S_{early} < 0$	$S_{late} < 0$	$S_{females} < 0$

Note: A selection differential greater than zero indicates that the trait enhances this component of fitness and is favored by selection while a negative selection differential (< 0) indicates the opposite.

The Mechanisms of Sexual Selection

Darwin proposed sexual selection as a special mode of natural selection to explain the evolution of extreme male phenotypes. He defined sexual selection as that selection which occurs within one sex as a result of competition among members of that sex, for reproduction with members of the other sex. From the viewpoint of one sex, generally males, members of the other sex, generally females, are a scarce resource. Thus, males must compete among themselves for access to this scarce resource, that is, for females. Darwin argued that males who won this sexual competition for females obtained more mates (a higher mating success) than males who lost in this competition. He further postulated that the extreme or exaggerated phenotypes of males were beneficial to male fitness *solely in regard to this competition with other males for mates*. Wallace (1868) argued further that these same male phenotypes were selected against in females as well as in males at times other than during the courtship season (see above table 1.1 and chapter 10).

The evidence that these phenotypes were detrimental to viability and selected against in females came from several different lines of reasoning. First, because these "male" traits are lacking or expressed only mildly in females, Wallace reasoned that they could not enhance survival. If such traits were generally good for survival, they would be expressed by both sexes. Second, he observed that many male-limited phenotypes are displayed only during the breeding season. For example, the bright plumage of many male birds, the antlers of some deer and other ungulates, and the calls of male frogs are expressed only during a brief period of the year, the courtship and breeding season. If these traits were advantageous in ways other than competition for mates, Wallace reasoned that they would be displayed year round, especially considering the tremendous investment of energy expended during the growth and development of some male structures. If the antlers of male deer were generally useful for repelling predators, why were they shed after the mating season and not retained year round? Darwin inferred that exaggerated male traits *must* function in mate acquisition. He was not clear whether they were disadvantageous to males and females in other respects as Wallace argued (see Chapter 10). For these reasons, Darwin proposed two mechanisms whereby more ornamented males could achieve a greater mating success than less ornamented males: (1) male-male competition for mates, and (2) female choice of mates.

Male-Male Competition for Mates

By male-male competition for mates, Darwin meant those cases in which males contested directly with one another during the breeding season for access to females. In his own words, sexual selection ". . . depends, not on a struggle for existence, but on a struggle between males for possession of the females; the result is not death of the unsuccessful competitor, but few or no offspring. Sexual selection is, therefore, less rigorous than natural selection" (1859, p. 88). Winning males mate with more females than losing males, and as a result sire more offspring. This quotation from Darwin places the Quantitative Paradox of sexual selection in high relief. If it is less rigorous, how can sexual selection override opposing natural selection in both sexes?

Male-male competition is clearly the mechanism of sexual selection in organisms such as deer, elk, or horned beetles (Eberhard 1979; Clutton-Brock et al. 1982; Emlen 1996). In these species, males, but not females, develop a sexual ornament and intrasexual weapon, antlers, just prior to the breeding season. During the mating season, males establish territories and engage in prolonged head-to-head combat using these antlers. Vanquished males are excluded from access to females and the winning males mate successfully with many does. Antlers in most species are not only a male-specific phenotype but also play a central role in the male-male contests for mates, which are a pivotal component of male reproductive fitness. Clearly, as Darwin suggested, sexual selection has played a role in the evolution of this kind of male-specific trait because its adaptive function is mate acquisition.

Similar male-male contests have been observed in many other species, including many beetles (Eberhard 1979; Emlen 1996). Male stag beetles (*Lucanus cervus*; Price 1996), for example, possess enlarged mandibles (fig. 1.7). Although carried by males throughout adult life, the horns are used only at the time of mating. When a male encounters another male copulating with a female, he uses the horns to pry his rival off the female and sometimes off the tree entirely. The winning male then mounts the female himself if she has remained in the vicinity. In many beetles, males will attack other

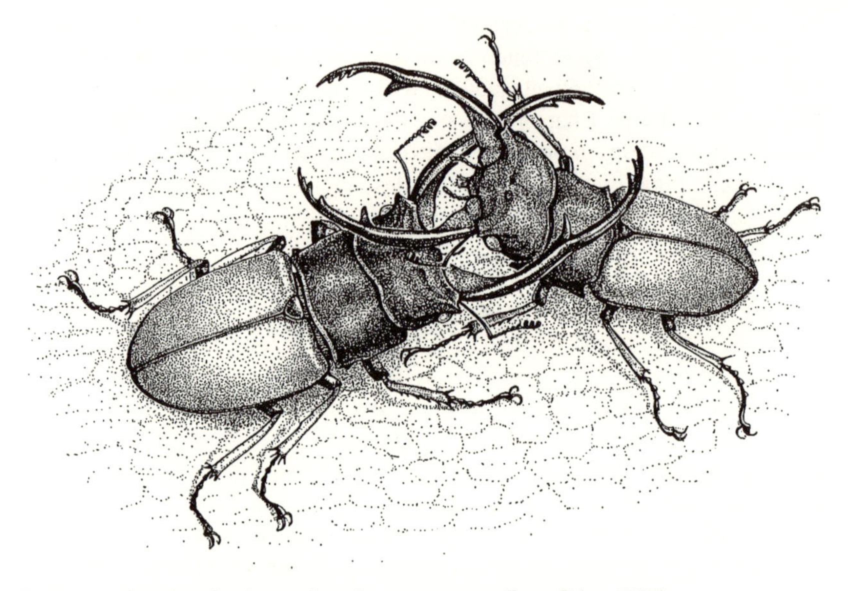

Figure 1.7. Stag beetles in combat, *Lucanus cervus* (from Price 1997).

mating males by biting the legs or antennae, and these attacking males often interrupt copulation. A similar evolutionary enlargement of protuberances on the pronotum of males has occurred in the forked fungus beetle, *Bolitotherus cornutus* (Conner 1988) and in the broad-horned flour beetle, *Gnathocerus cornutus* (fig. 1.8). The former species is found living on shelf fungus growing on dead trees. Females congregate on the bracts of fungi to lay eggs, and the larvae tunnel into and feed on the fungus. Males also aggregate where females oviposit and fight to exclude other males from the fungal bracts. These beetles are so sedentary in their habits that the same individuals can be found on the same bracts of fungus for up to six years! The male defending a bract can mate repeatedly with the females who lay eggs on that bract and sire many offspring.

In many primates, males sport enormously developed fangs, manelike ornamentation, and thick fur around the neck, as in hamadryas baboons (*Papio hamadryas*; fig. 1.9). Single dominant males may have reproductive access to a harem of several females. In this instance, the "fitness" stakes of a single male-male contest are not just the offspring of a single female but that of an entire group of females. With this kind of fitness reward attendant on male-male combat, it is obvious that successful males will have a much higher fitness than other, losing males. Indeed, in a species with a fifty-fifty Mendelian sex ratio at breeding, if one successful male has several mates, several males must necessarily have no mates at all. It is this consequence of sexual selection that establishes the fundamental relationship between the

Figure 1.8. Sexual dimorphism in *Gnathocerus cornutus*; male (right) with enlarged processes on the pronotum; female (left) unmodified (photo by M. J. Wade).

variance in male fitness and the variance in female fitness, which we will use to resolve the Quantitative Paradox of sexual selection (see below).

There are also male structures that function in reproductive combat but are less apparent than antlers, horns, manes, and fangs, because they are not used in face-to-face combat. Copulatory and seminal combat between males also occurs and appears to be commonplace in some taxonomic groups, especially insects (Smith 1984; Birkhead and Moller 1993a,b; Baker and Belis 1995; Eberhard 1996; Howard 1999). For example, males of some species of damselflies have a caudal structure employed at the time of mating whose function is to remove the sperm of other males from the reproductive tract of the female (Waage 1979). Males of some grasshoppers and katydids have an inflatable bulb on the end of the aedeagus, which also has a central groove. By inflating the bulb, a mating male mechanically forces the sperm of previously mating males out of the female reproductive tract back along the groove. Conversely, males of some species transfer not only sperm but also a

Figure 1.9. Male baboon with enlarged neck fur (photo by R. Willey).

"sperm plug" which may impede the copulatory efforts of subsequently mating males (Gwynne 1984).

In insects, the structure of sperm itself is often wildly elaborated and varied among closely related species (fig. 1.10) in much the same way that plumage and behaviors differ among males of related species of birds (Sivinski 1980; Eberhard 1996). It seems likely that postcopulatory but prezygotic

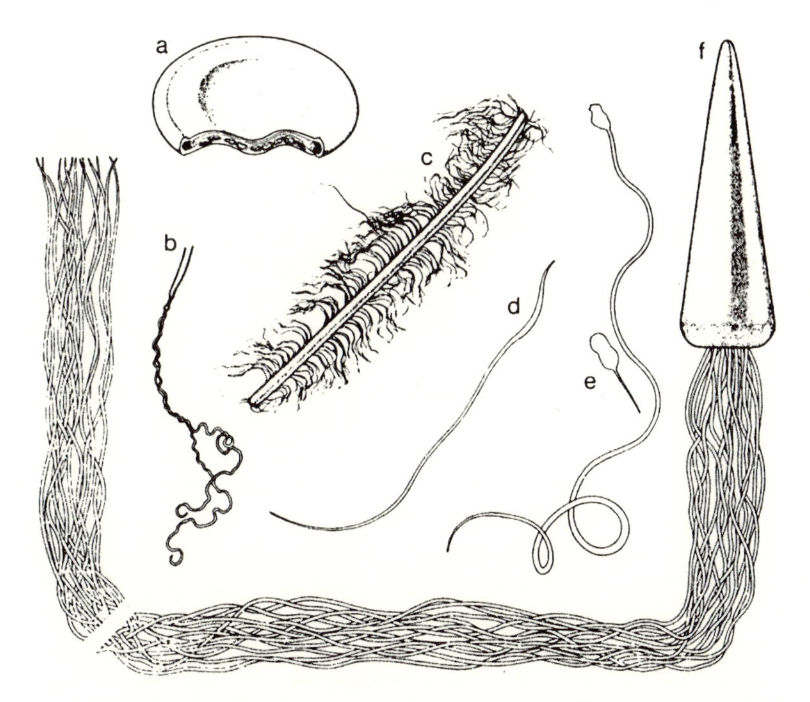

Figure 1.10. Elaborate insect sperm. (a) Proturan sperm, *Eosentonon transitorium*; (b) firebrat sperm, *Thermobia domestica*; (c) spermatostyle and spermatozoa from a gyrinid beetle, *Dineutus* sp.; (d) firefly sperm, *Pyractomena barberi*; (e) symphlan ssperm, *Symphylella vulgaris*; (f) termite sperm, *Mastotermes darwiniensis* (from Sivinski 1980).

competition between sperm of different males within the reproductive tract of multiply inseminated females results in sexual selection on sperm morphology (Pitnick 1996). This is a form of direct gametic competition between males with important ramifications for speciation (Robinson et al. 1994; Howard 1999).

Female Choice of Mates

All sex differences in phenotype cannot be explained by male-male competition alone. In many species, males differ from females in the extreme development of traits that appear to have no direct or plausible connection to male-male combat. Male peacocks do not fight one another with their tails nor do male bowerbirds fight one another directly with their bowers. Darwin believed that male-male combat was inadequate to account for the evolution of these other kinds of male-limited characters, and he postulated sexual selection via female choice of mates.

By female choice of mates, Darwin meant that females exhibited mating preferences for different kinds of males; i.e., they were more willing to accept certain males as mates than other males. Darwin reasoned, "if man can in a short time give elegant carriage and beauty to his bantams, according to his standard of beauty, I can see no good reason to doubt that female birds, by selecting, during thousands of generations, the most melodious or beautiful males, according to their standard of beauty, might produce a marked effect" (1859 p. 89). If female mating preferences were operating, then the traits of preferred males would become exaggerated by this mechanism of sexual selection in the same way and for the same reason that combat-related male traits evolved.

Questions about the existence of female choice, as well as how and why it operates on males, represent active areas of research today. It is only in the past ten years that strong female mating preferences for males with extreme ornamentation have been experimentally demonstrated in natural populations (review in Andersson 1994; Basolo 1990a,b). These studies show unequivocally that males with extreme values of preferred traits achieve more matings than males with less extreme values of these same traits. The number of mates that a male acquires in his lifetime indicates the degree to which he is preferred by females relative to other males. Here, differences among males in reproductive fitness result from the mating preferences exerted by females rather than from male-male combat. "Successful" males may have very high fitness, but the proximate reason for such differential mating success is female choice rather than male combat. Again, for every male who is accepted by and successfully mates with several females, there must be several males who *do not mate at all.* This simultaneous addition of winning males at one end of the mate number distribution and losing males at the other means that sexual selection always increases the *variance* in male mating success. It is this effect on the variance in male reproductive success that creates a sex difference in the variance in fitness. It is this sex difference in fitness variance that is fundamental to resolving the Quantitative Paradox of sexual selection.

The Strength of Sexual Selection Relative to Natural Selection: Resolving the Quantitative Paradox of Sexual Selection

Although the twin mechanisms of male-male combat and female choice explain why certain male traits have adaptive value in mate acquisition, we are still left with a critical quantitative issue (see table 1.1): *Why is the single positive component of male reproductive fitness sufficient to outweigh the totality of the several negative components in the evolution of these exaggerated male traits?* How can sexual selection be one of the strongest evolutionary forces when it affects only one sex, is opposed in the other sex, affects only one fitness component, and is exposed to selection only some of the time in the favored sex? It is not sufficient to say, as Darwin did, that the trait is favored in males by male-male competition for mates or by female choice of mates. Nor is it sufficient to measure the degree to which a particular character is modified within each sex and then attribute its greater modification in males to either of Darwin's mechanisms. Analyses that focus on the mechanisms of sexual selection are no substitute for measurements of the strength of selection. Identifying the mechanism of selection is different from quantifying its evolutionary effect.

We can resolve the Quantitative Paradox of sexual selection by considering, first, the *variance* in fitness and its relationship to the strength of selection and, second, the *sex difference* in variation of fitness between males and females. We are interested in this variation in fitness not only because fitness variance is required for selection, but also because the strength of selection is *proportional* to the variance in fitness: *the greater the variance in fitness, the stronger the force of selection.* Not only is fitness variation necessary for evolution, but also the variance in fitness sets an upper bound on the *rate* of evolution. Large differences in fitness between individuals mean strong natural selection. The absence of fitness differences between individuals (no fitness variance) means that natural selection, and, hence, adaptive evolution, are not possible.

Fitness variation determines the maximum rate (i.e., the upper limit) of evolution by natural selection because it limits the degree to which breeding parents differ from the average individual in the population before selection. The breeding parents are a subset of all individuals in the population and only they contribute genes to the next generation. The resemblance of offspring to their parents will be imperfect, to a degree that varies between zero and 1, a parameter better known as heritability. Because heritabilities are less than 1, the phenotypic mean of the offspring *must* differ from that of the breeding parents and be closer to that of the unselected population. If the average phenotype of the breeding parents is limited by the variation in fitness, the average phenotype of the selected offspring descended from those parents must also be limited. It is in this way that the total variation in fitness sets an upper limit to the rate of evolutionary change.

The Variance in Fitness and the Strength of Selection

These qualitative principles can be explained more precisely with simple mathematical terms and figures. Let the reproductive fitness of an individual with phenotypic value z be $W(z)$, and the frequency of such individuals in a population be $p(z)$ (fig. 1.11). The mean fitness of this population *before selection*, is, by definition,

$$W = \int W(z)p(z)dz. \qquad [1.1]$$

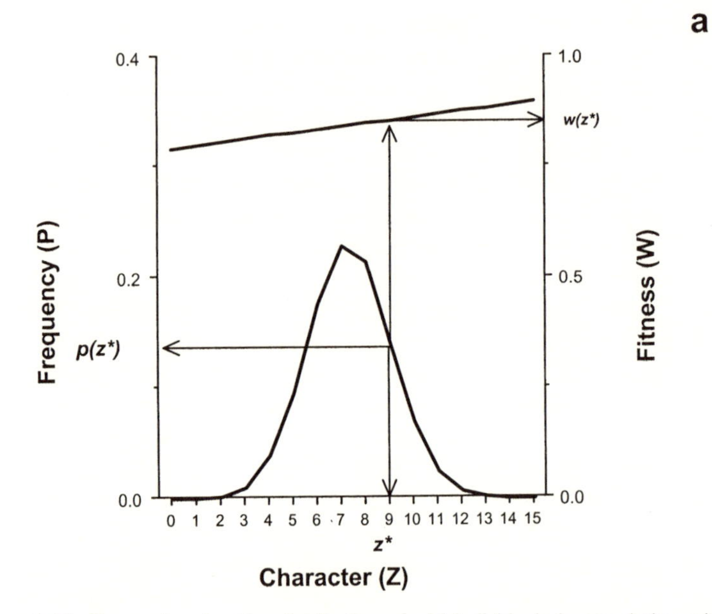

Figure 1.11. Figure showing the distribution of $p(z)$ individuals in population with $W(z)$ fitnesses.

That is, the fraction of the population exhibiting each phenotypic value, $p(z)$, is multiplied by its corresponding fitness $W(z)$, and the resulting products are summed, or integrated, over all values of z. The mean phenotype in the population before selection, Z, is

$$Z = \int zp(z)dz. \qquad [1.2]$$

That is, each value of z in the phenotypic distribution is multiplied by the proportion of the population, $p(z)$, exhibiting that phenotype, and then integrated over all z phenotypes. The relative fitness $w(z)$, for individuals with phenotype z is simply the ratio $W(z)/W$.

As a result of selection, the distribution of phenotypes changes from $p(z)$ in the population *before selection* to $p'(z)$ in the population *after selection*. These two distributions are related multiplicatively by relative fitness so that

$$p'(z) = w(z)p(z). \qquad [1.3]$$

The mean fitness W' of the parents selected to breed equals

$$W' = \int W(z)p'(z)dz = \int [W^2(z)/W]p(z)dz. \qquad [1.4]$$

Thus, the difference in fitness, ΔW, between breeding parents and the unselected parent population before selection is given by the difference between eqs. [1.4] and [1.1] or

$$\Delta W = (W' - W) = \int [W^2(z)/W]p(z)dz - \int W(z)p(z)dz. \qquad [1.5]$$

This expression can be rewritten as

$$\Delta W = [\int W^2(z)p(z)dz - W^2]/[W] = V_W/W, \tag{1.6}$$

where V_W is the variance in fitness of the parent population *before reproductive selection.*

The *relative* change in mean fitness by natural selection, $\Delta W/W$, is thus equal to

$$\Delta W/W = V_W/W^2 = V_w, \tag{1.7}$$

where V_w is the variance in relative fitness. This is the increase in average fitness of the breeding parents *relative* to that of the parent population before selection. This entire change is not *transmitted* across generations to the offspring because not all of the variation in parental fitness is heritable. The offspring mean is given by the product of the parent mean (eq. [1.4]) and heritability. It is also given by adding the product of eq. [1.6] and heritability to the mean before selection. Offspring mean fitness is necessarily less than or equal to that of the breeding parents because heritability is a fraction always equal to or less than 1. Heritability is usually less than 1 because both genetic and environmental factors influence the expression of characters. The more the environment influences variation in phenotypic expression, the smaller heritability will be (Falconer and Mackay 1996).

Thus, the variance in relative fitness, V_w, places an upper bound on the change in mean fitness possible from one generation to the next. In fact, *the variance in relative fitness, V_w, places an upper bound on the change in the mean of any phenotypic trait Z.* The mean phenotype of the breeding parents, Z', is defined as

$$Z' = \int zp'(z)dz. \tag{1.8}$$

We substitute $w(z)p(z)$ for $p'(z)$ to obtain

$$Z' = \int zw(z)p(z)dz. \tag{1.9}$$

The variable Z', is equal to X^*_{males} in figs. 1.3a, 1.4a, and 1.5 or X^*_{females} in figs. 1.3b and 1.4b.

The change in mean phenotype *before* and *after* selection is equal to

$$\Delta Z = (Z' - Z) = \int zw(z)p(z)dz - \int zp(z)dz. \tag{1.10}$$

It is important to note that ΔZ is the numerator of S, the selection differential (cf. figs. 1.3–1.5). Or, differently put, S is ΔZ, divided by the standard deviation of the phenotypic distribution $p(z)$. This conversion of ΔZ into units of standard deviation is important for comparing selection intensities across different populations, different experiments, and different traits, where both the mean and variance vary.

We note that $\int w(z)p(z)dz = 1$, and we can rewrite eq. [1.10] as

$$\Delta Z = \int zw(z)p(z)dz - [\int zp(z)dz]\,[\int w(z)p(z)dz]. \tag{1.11}$$

We can recognize this expression as the covariance between z and $w(z)$, i.e., the covariance between phenotype and relative fitness. Thus, the change in mean phenotype resulting from selection is

$$\Delta Z = \mathrm{Cov}(z,w[z]). \qquad [1.12]$$

As with W, the fraction of ΔZ transmitted across generations depends on the heritability of the phenotype, which is always less than or equal to 1.

It is important to understand that the average fitness as well as the average phenotype may change as a result of selection. However, it is more important to understand that there is a *relationship* between the variance in fitness, V_W and the covariance between phenotype and fitness, $\mathrm{Cov}(z,w[z])$. To understand this relationship, we first must recognize that the ratio $\mathrm{Cov}(z,w[z])/(V_Z\ V_w)^{1/2}$ is the product moment correlation between a phenotype z and its relative fitness $w[z]$. Unless a perfect correlation exists between phenotype and relative fitness, this expression will always be less than 1. Next, note that the ratio $\mathrm{Cov}(w[z],w[z])/V_w$ is the product moment correlation between relative fitness and itself; thus its value is equal to 1. We can use these relationships to establish that

$$\mathrm{Cov}(z,w[z])/(V_z\ V_w)^{1/2} < 1 = \mathrm{Cov}(w[z],w[z])/\ V_w. \qquad [1.13]$$

We can further transform our phenotypic values of z and $w(z)$ to the unit normal scale x, where x equals $(z - Z)/\sigma_z$ and $w(x) = \frac{w(z) - 1}{\sigma_w}$. Thus, eq. [1.13] becomes

$$\mathrm{Cov}(x,w[x]) < V_w. \qquad [1.14]$$

Hence, the variance in relative fitness places an *upper bound* not only on the change in mean fitness itself, but also on the standardized change in the mean of *every other* phenotypic trait. It was for this reason, that Crow (1958, 1962) defined I, the "opportunity for selection," as

$$I = V_W/W^2 = V_w. \qquad [1.15]$$

It is this "opportunity for selection" that sets an upper bound on the rate of evolutionary change in the mean of *all* phenotypes.

The Sex Difference in the Variance in Fitness

We will now examine the opportunity for selection in the male and the female sex. We will see that there is a fundamental algebraic relationship between the opportunity for selection in males, I_{males}, and that in females, I_{females}. Based on the derivation above, note that the opportunity for selection for each sex is equal to the variance in fitness among members of that sex, V_i, divided by the squared average in fitness among members of that sex, X_i^2. Thus, $I_{males} = V_{males}/X^2_{males}$ and $I_{females} = V_{females}/X^2_{females}$. These expressions are linked together through the sex ratio and mean fitness. Furthermore, we will show that the variation among males in the numbers of mates caused by either male-male competition or female choice has two consequences:

1. the opportunity for selection in males exceeds the opportunity for selection in females, i.e., $I_{males} - I_{females} > 0$; and
2. given certain assumptions, the sex difference in the opportunity for selection equals I_{mates}, where I_{mates} is defined as the ratio of V_{mates}/X^2_{mates}, and where X_{mates} is the average number of mates per male. Thus, $I_{males} - I_{females} = I_{mates}$.

Hence, the opportunity for selection that results from the competition among males for mates, I_{mates}, is responsible, in large part, for the sex difference in the strength of selection.

We can now rewrite table 1.1 in terms of the strength of each component of selection (table 1.2). This notation will allow us to show why I_{mates} is often greater than the weighted sum of the other three terms in table 1.2 and thus resolves the Quantitative Paradox of sexual selection. However, to fully understand why this is so, we must first quantify the strength of sexual selection in males relative to the strength of natural selection on males and females to resolve the Quantitative Paradox concerning the evolution of extreme male traits (tables 1.1 and 1.2). Let us begin by considering the variation in male fitness associated with mating success.

The Average and Variance in Mating Success

In most breeding populations, males can be divided into a series of mating classes, k_i. The number of mates males obtain defines each mating class. Thus, k_0 males do not mate, k_1 males mate once, k_2 males mate twice, and so on. The number of males in each mating class, m_i, depends on how variable male are in their mating success. For example, in fig. 1.12a, all of the males

Table 1.2
The opportunities for selection represented by the components of reproductive and viability fitness as they affect the evolution of an exaggerated male trait.

	Reproduction	*Viability*
Male Fitness	I_{mates}	$I_{viability}$
Female Fitness	$I_{fecundity}$	$I_{viability}$

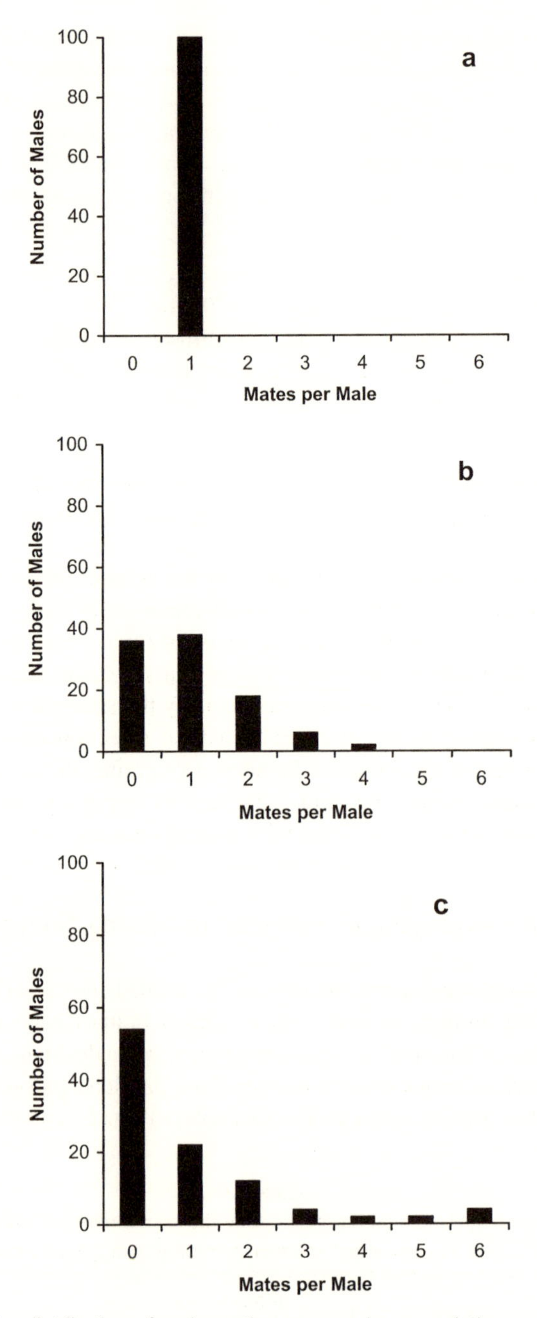

Figure 1.12. The distribution of male mating success in a population consisting of 100 males and 100 females (a) who each mate only once; all males secure a single mate; (b) mating occurs approximately at random; (c) certain males mate with more than one female.

mate once. Thus, all m_i categories equal zero except $m_1 = 100$. The sex ratio R in this or any population can be expressed as the ratio of the number of females to the number of males. In this case, $R = N_{\text{females}}/N_{\text{males}} = 1$ (Wade 1995).

The average mating success per male, M, equals the number of females in the k_ith mating class, multiplied by the number of males in the m_ith class, added up over all i classes, and then divided by the total number of males. Thus, $M = (\Sigma\ k_i\ m_i)/\Sigma\ m_i$, or in fig. 1.12a, $M_a = [(0)(0) + (1)(100) + (2)(0) + (3)(0) + (4)(0) + (5)(0) + (6)(0)]/100 = 1$. Clearly, $R_a = M_a$.

The average harem size H equals the number of mates per *mating* male. H is usually larger than the sex ratio ($H > R$). This is true because, whenever one male secures two or more females, other males are excluded from mating (Darwin 1874). The average mating success per mating male equals the number of females in each mating class, multiplied by the number of males in that class, added up over all classes, and then divided by the sum of the males who *actually mate*. Thus, $H = (\Sigma\ k_i\ m_i)/[(\Sigma m_i) - m_0]$. Since m_0 in fig. 1.12a equals zero, $R_a = M_a = H_a = 1$.

In fig. 1.12b, mating occurs approximately at random. In this situation, $M_b = [(0)(36) + (1)(38) + (2)(18) + (3)(6) + (4)(2) + (5)(0) + (6)(0)]/100 = 1$, and here again, $M_b = R_b$. However, $H_b = [(0)(36) + (1)(38) + (2)(18) + (3)(6) + (4)(2) + (5)(0) + (6)(0)]/[100 - 36] = 1.56$. Thus, with random mating, R_b still equals M_b, but H_b exceeds both values. This condition changes further when some males obtain more mates than expected by chance. In fig. 1.12c, again, $M_c = [(0)(54) + (1)(22) + (2)(12) + (3)(4) + (4)(2) + (5)(2) + (6)(4)]/100 = 1 = R_c$. However, $H_c = [(0)(54) + (1)(22) + (2)(12) + (3)(4) + (4)(2) + (5)(2) + (6)(4)]/[100 - 54] = 2.17$. Thus, in this example, as well as in general, as fewer males obtain more mates, R and M are equivalent and remain unchanged, but H increases still further.

How do such changes in the distribution of females with males affect the variance in mate numbers? The answer is this: As average harem size changes, so does the variance in mate numbers among males. The variance in mate numbers describes the "spread" of the distribution of mates per male around the population average. This variance can be calculated by squaring the value of each mating class, k_i, multiplying each squared value by the number of males in each mating class, m_i, adding up the products, and then subtracting this quantity from the squared average in mate numbers per male. Thus, V_M equals the average of the squared number of mates per male, minus the square of the average number of mates per male (Sokal and Rohlf 1995). That is, $V_M = [(\Sigma\ k_i^2 \text{m}_\text{i})/\Sigma\ m_i] - [\Sigma\ k_i\ m_i/\Sigma\ m_i]^2$.

When all males in the population secure one mate, the variance in mating success among males is zero, that is, $V_{Ma} = \{[(0)^2(0) + (1)^2(100) + (2)^2(0) + (3)^2(0) + (4)^2(0) + (5)^2(0) + (6)^2(0)]/100\} - \{[(0)(0) + (1)(100) + (2)(0) + (3)(0) + (4)(0) + (5)(0) + (6)(0)]/100\}^2 = 0$ (fig. 1.12a). When some males mate more than once *by chance*, other males must still be excluded from mating. Consequently, the distribution of male mating success around the average widens. That is, $V_{Mb} = \{[(0)^2(36) + (1)^2(38) + (2)^2(18) + (3)^2(6) + (4)^2(2) + (5)^2(0) + (6)^2(0)]/100\} - \{[(0)(36) + (1)(38) + (2)(18) + (3)(6) + (4)(2) + (5)(0) + (6)(0)]/100\}^2 = 1$ (fig. 1.12b). The more some males mate with more than one female, the larger the variance in mating success becomes. Thus, $V_{Mc} = [(0)^2(54) + (1)^2(22) + (2)^2(12) + (3)^2(4) + (4)^2(2) + (5)^2(2) + (6)^2(4)]/100\} - \{[(0)(54) + (1)(22) + (2)(12) + (3)(4) + (4)(2) + (5)(2) + (6)(4)]/100\}^2 = 2.32$ (fig. 1.12c). Clearly, as the variance in mating success, V_M, becomes larger, the average mating success of mating males, H, must also increase.

As the example above graphically demonstrates, when there are equal numbers of breeding males and females, and if some males have many mates, then, necessarily, some males will have no mates at all. If a male loses in competition with other males for mates, then he has a reproductive success of zero (0). Conversely, a winning male might mate with one or more different females. As we have shown above, R equals the average number of mates per male. The parameter R also equals the sex ratio, $N_{\text{females}}/N_{\text{males}}$. In future examples, we will let V_{mates} be the variance in mate numbers among males.

Males might win or lose mates owing to either the Darwinian mechanism of sexual selection, direct male-male combat, or female choice. We will consider the relative effects of these mechanisms later (chapters 4 and 5). For now, to allow comparison of the effects of natural and sexual selection on both sexes, we must consider male and female fitness using the same units. Let O be the average number of offspring per female. Different females may produce more offspring or fewer offspring than the "average" female and we represent this variation in offspring numbers among females by V_O, the variance among females in fecundity fitness. The opportunity for fecundity selection in females, I_{females}, equals V_O/O^2. Because half of the genes in the offspring of our hypothetical species come from these females, half of the genes experience natural selection *against* the trait, possibly as strong as I_{females}.

The average male enjoys a reproductive success equal to RO, i.e., R times the reproductive success of the average female (O). When we say that the average number of offspring per male is the product RO, we are assuming that mate numbers and offspring numbers are independent of one another, i.e., not correlated. We will relax this assumption in chapter 4. If the breeding sex ratio is one male to every female, then for every male with k mates, there must be $(k - 1)$ males with 0 mates. As a result, the *least successful* males have a *lower fitness* than the least successful breeding females. In addition, unsuccessful males *outnumber* winning males whenever $k < 1$. Conversely, the fitness of the *most successful* males exceeds that of the most successful females *by a factor greater than R*. Differently put, there are many males with a fitness lower than the lowest in the female distribution and some males with a fitness much, much higher than the highest in the female distribution. The mean fitness is the same for both sexes when R is 1, because each offspring has one mother and one father (Fisher 1930). However, *the variance in fitness is much greater for males than it is for females.*

The average number of mates per male, R, is the total number of reproducing females, N_{females}, divided by the total number of available breeding males, N_{males}, which is also the sex ratio R. When the sex ratio R is expressed in this way it has the advantage that it is equivalent to the average number of mates per male. In many other discussions, the sex ratio is expressed as the reciprocal of R or $N_{\text{males}}/N_{\text{females}}$, as a means of expressing the degree of competition among males. This ratio of the number of mating males to the number of sexually receptive females is called the *operational sex ratio* or *OSR* (Emlen and Oring 1977). We will use the symbol R_O (i.e., $[1/R]$) for the OSR.

The OSR captures the intuitive idea that the greater the excess of males over females at the time of breeding, the greater the intensity of reproductive competition among males for mates. We show below that R_O is only one component, albeit an important one, of the sex difference in the opportunity for selection. Although it is a reasonable idea, R_O is *not* equivalent to the sex difference in the strength of selection (see Reynolds 1996; Kvarnemo and Ahnesjo 1996 for discussions to the contrary). Thus, two mating systems with the *same* value of OSR (our $1/R$ or R_O) can *differ* in the strength of sexual selection. Conversely, two mating systems with *different* OSRs can have the *same* strength of sexual selection. Because the OSR is not always predictive of the variance in male reproductive success, it is not always correlated with the strength of selection affecting male-female dimorphism.

We can divide males into two categories: those that are unsuccessful at mating, and thus have a fitness of zero, and those males that are successful and have one or more mates. We let H be the average number of mates of the successful males. We define p_m as the proportion of mating males, and p_0 as the proportion of nonmating, unsuccessful males. (Remember that $\{p_m + p_0\} = 1$.) These two frequencies are connected through the sex ratio R. The average number of mates per male, R, equals the average number of mates across our two categories of males. We can express this in terms of our definitions as

$$R = (H)p_m + (0)p_0. \qquad [1.16]$$

Writing the frequency of successful males, p_m, as $[1 - p_0]$, we find

$$p_0 = 1 - (R/H) \text{ or} \qquad [1.17]$$

$$p_0 = 1 - (1/[R_O H]). \qquad [1.18]$$

It is clear from eqs. [1.18] that the greater the number of mates per successful male, H, the larger must be the proportion of males with no mates at all,

p_0. As Darwin put it, females are a scarce or limited resource from the perspective of males competing for mates. Success in reproductive competition for some males *necessarily* means failure for others. Thus, whenever we add successful males with many mates to the right-hand side of the distribution of male fitness, we also must add one or more unsuccessful males at zero, the opposite side of the distribution.

To calculate the total variance in male fitness, V_{males}, we need to know the distribution of the number of mates of males. Let p_k represent the fraction of males that have k mates, where k ranges between 0 and a maximum of N_{females}. Thus, p_0 is the frequency of males with no mates at all and p_5 is the frequency of males with five mates, i.e., $k = 5$. The frequency of successfully mating males, p_m, is equal to the sum $\Sigma p_j = (1 - p_0)$. For all categories of mating males, we assume that the family size of each female is a random draw from the distribution of female reproductive success with mean O and variance V_O, which, in this case, is V_{females}. (We later relax this assumption in chapter 4 and permit male mate numbers to affect O.) The population of males can be viewed as consisting of the k categories given in table 1.3.

The total variance in male reproductive success is the sum of two components: (a) the average variance in offspring numbers among males within the categories of table 1.3; and (b) the variance in average in offspring numbers among the categories. Thus, we now have

$$V_{\text{males}} = \Sigma\, p_j(jV_{\text{females}}) + \Sigma\, p_j(jO - RO)^2 \qquad [1.19]$$

$$= RV_{\text{females}} + O^2V_{\text{mates}}. \qquad [1.20]$$

Equation [1.20] illustrates the insight of Bateman (1948): *the fundamental cause of the sex difference in fitness variance is the variation of mate numbers among males.* However, the variance in male mating success is not as useful a comparative measure of relative male mating success as I, the opportunity for selection (Crow 1958, 1962; Wade 1979, 1995; Wade and Arnold 1980).

Table 1.3
The distribution of mates among males.

Number of Mates	*Frequency*	*Mean Number of Offspring*	*Variance in Offspring*
k	p_k	kX	kV_X
0	p_0	X	0
1	p_1	X	$1V_X$
2	p_2	$2X$	$2V_X$
3	p_3	$3X$	$3V_X$
4	p_4	$4X$	$4V_X$
k	p_k	kX	kV_X
Total: N_{females}	1	$N_{\text{females}}X$	$N_{\text{females}}V_{\text{females}}$

The opportunity for selection on males, I_{males}, equals the variance in male fitness, V_{males}, divided by the square of mean male fitness, $(RO)^2$. Dividing eq. [1.20] by $(RO)^2$, we have

$$(V_{males}/[RO]^2) = (RV_{females}/[RO]^2) + (O^2V_{mates}/[RO]^2), \quad [1.21a]$$

$$I_{males} = (1/R)(I_{females}) + I_{mates}, \quad [1.21b]$$

$$I_{males} = (R_O)(I_{females}) + I_{mates}. \quad [1.21c]$$

When R and R_O are equal to 1, this expression reduces to

$$I_{males} = I_{females} + I_{mates}. \quad [1.22]$$

When we say that the opportunity for sexual selection, I_{mates}, is the *only* cause of the sex difference in the strength of selection, we mean that

$$\Delta I = I_{males} - I_{females} = I_{mates}. \quad [1.23]$$

I_{mates} is necessarily greater than zero whenever there is variation among males in the numbers of mates. Thus, I_{mates} provides a *standardized measure of the intensity of sexual selection on males and the sex difference in strength of selection* (Wade 1979; Wade and Arnold 1980; Wade 1995).

We need not assume that R and R_O are equal to 1. More generally, we find that the sex difference in the opportunity for selection can be found by subtracting $I_{females}$ from both sides of eq. 1.21c. This expression becomes

$$I_{males} - I_{females} = (R_O - 1)\, I_{females} + I_{mates} \quad [1.24a]$$

or

$$I_{males} - I_{females} = (1/R - 1)\, I_{females} + I_{mates}. \quad [1.24b]$$

When R and R_O equal 1, eq. [1.24a] and eq. [1.24b] both reduce to eq. [1.23]. From the more general expression (eq. [1.4]), we see that the sex difference in the opportunity for selection is clearly affected by R_O, the OSR, as Emlen and Oring (1977) argued. When the OSR is less than 1, so that there are more females than males competing for them, then the sex difference in the strength of selection is diminished. When the OSR exceeds 1, so that there are more males than there are available females, then the intensity of selection on males exceeds that on females by more than I_{mates}. However, it is also clear from eq. [1.24], that the OSR *does not* estimate the sex difference in the strength of selection, which is important for the evolution of sex dimorphism by sexual selection. The OSR is *only one* component of the sex difference in strength of selection. Whenever I_{mates} is much larger than $I_{females}$, then $(R_O - 1)\, I_{females}$ will be only a small component of the sex difference in selection. On the other hand, when I_{mates} is much smaller than $I_{females}$, then the OSR will explain much of the sex difference in the strength of selection that results from sexual selection. We will provide an example of the influence of OSR on the sex difference in the opportunity for selection in chapter 5.

The Dimensionality of I_{mates}

In his definition of the opportunity for selection, I, Crow (1958) emphasized that the variance in fitness, divided by the square of the average fitness, is the variance in *relative* fitness. Thus, it measures the maximum change in a populational phenotype that can result from a single episode of selection. Note that I_{mates} equals the variance in mate numbers, V_{mates}, divided by the square of the average number of mates per male, R. This expression is the component of male relative fitness that results from reproductive competition among males. Since both the numerator and denominator of I_{mates} are expressed in units of $(mates)^2$, the opportunity for sexual selection associated with differences in mate numbers between the sexes is *dimensionless* (although "I_{mates}" remains a convenient, specific notation).

Other examples of dimensionless numbers include coefficients of variation, pondural indices, and drag and growth coefficients (Vogel 1988; Charnov and Berrigan 1991). In evolutionary biology, examples include the ratio of gene flow distance to spatial change in selection coefficient (Slatkin 1973, 1987; Kirkpatrick 1996), and the measures of population subdivision, G_{ST} and F_{ST} (Hartl and Clark 1989). Such parameters are extremely useful in cross-taxonomic comparisons because they capture the essence of the process independent of scale. Thus, I_{mates} permits comparisons of the strength of sexual selection within and among taxa. In later chapters, we will show how I_{mates} arises from the microspatial and temporal distributions of females. This is an important point. In our attempt to resolve the Quantitative Paradox of sexual selection, we will link microevolutionary processes to macroevolutionary patterns. Such explicit within- and between-taxonomic comparisons of processes and their outcomes require dimensionless parameters.

Potential Difficulties with I_{mates}

Several authors have criticized the use of I_{mates} as a means for identifying the intensity of sexual selection. Grafen (1987) argued that the study of sexual selection is "historical" (emphasizing taxonomic patterns) rather than "empirical" (emphasizing microevolutionary processes), and thus that quantification of the intensity of sexual selection has no intrinsic value. Grafen (1987, p. 222) noted that, "Darwin (1871) discovered almost everything important now known about sexual selection and did so without measurement." Grafen also reasoned that, because the study of sexual selection has proceeded at a vigorous rate without measures of its intensity, ". . . the desirability of precise mathematical modeling does not imply the desirability of measuring the parameters of any of these models in the field." He further concluded that empirical studies of selection intensities seem unlikely to increase our understanding of its process or its outcomes and "quantification has no intrinsic virtues . . . and can set back the cause of science." This argument, while extreme, is not surprising. It is the same one used by molecular geneticists and cell biologists, when biology departments discuss whether or not statistics and calculus should be required courses for biology majors.

Our response to this perspective is that one can seldom understand nature less by studying it more. While Darwin's mechanisms of sexual selection are clear, so too is the Quantitative Paradox of sexual selection. The discrepancy between microevolutionary theory and historical pattern *requires* quantitative measures of the sex difference in the strength of selection in natural populations to be assessed and resolved. We hope readers may share our view by the end of this book.

The fact that I includes all of the variance in fitness, both selected and random, has disturbed other authors. For example, Sutherland (1985) argued that, since chance alone (i.e., random mating) can generate nonzero values for I ($= I_{\text{mates}}$), this estimator is a poor measure of sexual selection. Indeed, Crow (1958, 1962) addressed this very point and showed that, *when random or chance deaths occur, the effectiveness of selective deaths must be diminished.* By chance, bad things happen to good genes and vice versa. If one eliminates the random variation in mortality and reproduction, one *overestimates* the strength of selection. This is another reason why I_{mates} sets an *upper limit* on the response to sexual selection: not all mating is differential with respect to male characters. By chance, some males obtain more mates than others, just as, by chance, some individuals die before expressing their good (or bad) genes for viability. Mutations with deleterious effects late in life accumulate in populations to a greater degree than deleterious mutations expressed early in life for this very reason: random, early mortality interferes with their expression.

Wade (1987) illustrated this same point in an analysis of laboratory data on the difference in mating success of male *Panorpa* scorpion flies with claspers intact and claspers immobilized (Thornhill 1981). Although males without functional claspers obtained no mates, and some males with intact claspers obtained more than one mate, many other males with intact claspers also did not mate. There was variance in mating success within as well as between the two artificial categories of experimental males. Because the variance in mating success between these two types of males was only a small part of the total variance in mating success, the strength of selection on claspers as a male trait (present or absent) was actually quite small. This was true despite the fact that the difference in average mating success between the two male groups was large and, statistically speaking, clear-cut.

Consider the hypothetical example introduced earlier in this chapter (tables 1.1–1.3). All of the opportunities for selection in table 1.3 or eq. [1.23] will have random or environmental components. They are not unique to I_{mates} nor are they necessarily larger for mate numbers than for viability or offspring numbers. In our hypothetical example, not all individuals with large tails have genes influencing development toward larger tails. Some simply experienced better environments for tail development. As stated above, only the heritable fraction of the differences between selected and nonselected individuals can be transmitted across generations. One thing is certain: if eq. [1.23] is not greater than zero with random variations in mate numbers, it will not become greater when the random component of variation is excluded.

More to the point, since variance in mating success among females is small in most species, even random variation among males in mate numbers will result in sex differences in the opportunity for sexual selection on males (fig 1.13b). Since some males will mate more than once, other males will not mate at all. This statement is identical to Darwin's (1871, p. 332) own definition of sexual selection: "The practice of polygamy leads to the same results as would follow from an actual inequality in the number of the sexes; for if each male secures two or more females, many males will not be able to pair."

Clearly, if R remains constant and harem size H increases, an increasing number of males *must* be excluded from mating as shown by eq. [1.18] and in fig. 1.13. For this reason, I_{mates} will increase and the sex difference in the strength of selection will increase with it.

Downhower et al. (1987) raised four arguments against the use of the measure I to examine selection. Their first criticism is that I is sensitive to the units commonly used to measure fitness variance. Thus, when different units are used in different studies, they are not comparable using I. This criticism is based upon a misunderstanding of how I is used and why it is a dimensionless parameter. The choice of what to count or measure when describing selection in the currency of fitness will certainly vary from organism to organism and often from study to study of the same organism. The conceptual issue is, how do we link different studies in order to obtain an integrated picture of total selection?

Consider a specific example. Suppose that two microevolutionary research programs are investigating the body size evolution in ambystomatid salamanders. One program studies how larval body size affects the development of paedomorphic reproduction and the other investigates how adult body size influences mating success. The former study uses age at first reproduction and clutch size as measures of fitness while the latter uses mate numbers and duration of courtship. Each finds a relationship between body size and fitness and reports the direction of selection on body size, invoking *ceteris paribus* (all else being equal), as though the other study did not exist, a scenario depressingly typical of microevolutionary studies. Neither study can claim to know *the* adaptive significance of body size for several reasons including the following: (1) many of the same genes and developmental processes are involved in the determination of *both* larval and adult body size; (2) different traits and different genes are correlated phenotypically and genetically so their effects (and the effects of selection on them) are not independent of one another; and (3) natural selection operates on total fitness, yet each study has measured only a component of fitness or (as is common) a surrogate thereof. The common assumption of *ceteris paribus* not only sweeps away these concerns for any particular study but also isolates the results of different studies that should be integrated. It gives the evolutionary biologist license to carve nature up into manageable fragments and to describe selection in whatever fitness currency is expedient. Unfortunately, it also impedes the cogent and necessary integration of the separate descriptions of nature. The measure I provides a rigorous method of combining results across studies, even studies that use different fitness currencies, because it is both dimensionless and based on the single common currency of evolutionary theory, namely, the *variance in relative fitness*.

A second criticism raised by Downhower et al. (1987; see also Ruzzante et al. 1996) is that measurements of I are sensitive to differences in average fitness between populations. These authors argue that two populations may have equal fitness variance, but if they differ in average fitness, the values of I calculated for each will differ accordingly. Therefore, variation in average fitness may falsely imply that differences in fitness variance exist between populations. This criticism arises from a lack of understanding of how I is defined and what it measures, and the further confusion of I with analysis of variance. In evolutionary theory, natural selection on a trait results from the correlation between values of that trait (z) and heritable differences in relative fitness ($w[z]$, see above). The effect of natural selection can be quantified explicitly as the covariance between trait value and relative fitness whether expressed as the familiar "breeders' equation," $\Delta Z = h^2$ Cov(z,$w[z]$), or as the equally familiar Δp from population genetics When the trait is fitness itself, the *relative* change in mean fitness by natural selection, $\Delta W/W$, equals the variance in relative fitness, V_w (see eq. [1.7] above). Because the denominator of V_w is mean fitness W, differences in mean fitness do change the strength of selection even when the variance in absolute fitness is the same; *by definition*, they have to! If one wants to compare only fitness variances, which do not describe the amount of evolutionary change, then we do not recommend using I, but use of ANOVA instead. Unfortunately, as useful as ANOVA is as an experimental tool, it is not a good comparative measure of the strength of selection for evolutionary genetic studies in the field or laboratory.

A third criticism by Downhowner et al. (1987) states that measurements of I are sensitive to sampling error. Since I is based on the mean and variance in population fitness, this criticism simply restates difficulties encountered with any parameter estimation method based upon sampling. For example, the range is much more sensitive to sample size or sampling effort than is the variance, thereby conferring greater utility on the latter. This difficulty does not weaken the usefulness of our approach for studies of mating systems any more than insufficient sample sizes weaken the usefulness of other parametric statistical procedures, including ANOVA, where estimates of main effects are confounded with interactions in small experiments (Wade 1992). Indeed, Wade (1995) showed that the statistical properties of I_{mates} are the same as those of the negative binomial distribution, a distribution that has "probably been used more frequently than any other contagious distribution" (Sokal and Rohlf 1981, p. 95). This permits us to employ tests of significance when making taxonomic comparisons of different values of I, a feature we consider one of the strengths of our approach. As with most test statistics, increasing sample size increases the power of the test.

This last point addresses the fourth criticism of Downhower et al. (1987) as well. That is, no reference values exist against which a given value of I may be compared. Although several values have been published since 1979 (Wade 1979; Wade and Arnold 1980; Fincke 1986; in Clutton-Brock 1988; Yezerinac et al. 1995; Dinsmore 1985; Rajanikumari et al. 1985; Hed 1984, 1986; Clutton-Brock 1991b; Marzluff and Balda 1992; Souroukis and Cade 1993; Lande et al. 1994; Murphy 1994; Fleming and Gross 1994; Webster et al. 1995; Iribarne 1996; Morgan and Schoen 1997; Coltman et al. 1999; Stanton et al. 2000; Herrera 2000; Ferguson and Fairbairn 2001; Webster et al. 2001; Fairbairn and Wilby 2001), including those calculated by Downhower et al. (1987), we hope that our discussion of I in this book will stimulate its further use for measuring selection in natural populations and thereby provide cross-taxonomic values for the opportunity for selection. In the meantime, the similarity of the theoretical distribution of I to a negative binomial distribution with its well-known statistical properties puts the final criticism of Downhower et al. (1987) to rest.

Chapter Summary

We have seen how the two mechanisms of sexual selection, male-male combat and female choice, proposed originally by Darwin, permit the evolution of sex differences. However, understanding the Darwinian mechanisms of sexual selection does not address the apparent conflict between the macroevolutionary patterns and the microevolutionary process of sexual selection. The macroevolutionary comparison indicates that sexual selection is one of the *fastest and strongest* of the evolutionary forces, capable of producing large phenotypic differences among the males of even closely related taxa.

Our microevolutionary analysis reveals not only that the effects of sexual selection are sex-limited, but also that conflict exists in the direction of selection between the sexes as well as between different life history stages within the male sex. These are the features of a very *slow and weak* evolutionary force.

Our Quantitative Paradox of sexual selection vis-à-vis natural selection is resolved when we take into account the variance in male fitness that results from the variations among males in the numbers of mates. The success of one male in competition for mates necessarily results in the failure of one or more other males (eq. [1.18]). Thus, points are added to the distribution of mate numbers in groups such that, for every point added above the mean, one or more points are added below the mean at zero. A large variance among males in mating success ensures that selection in the male sex will be stronger than opposing viability selection in the female sex and in the male sex at earlier life stages. In subsequent chapters, we investigate how other genetic and ecological factors can enhance or diminish evolution by sexual selection. In particular, we examine the roles of spatial and temporal variation in female receptivity in their effects on I_{mates}.

2 The Ecology of Sexual Selection

We will now develop the concepts introduced in chapter 1 in greater detail, in order to apply them, quantitatively, to the evolution of animal mating systems and to a lesser degree, the mating systems of other sexual organisms. A great deal of research in animal behavior over the last three decades has focused on animal mating systems. Although this work has been very successful in identifying the *mechanisms* by which sexual selection operates within a mating system, it has not been as successful in quantifying the relative *strength* of sexual selection as it varies with mating system. Thus, we do not know whether and to what degree differences in mating system explain variations in the pattern of sex dimorphism, nor do we have a clear understanding of the relative importance of different features of a mating system. In this chapter, we develop a comparative method for measuring the variance in fitness of both males and females. We show how the *sex difference* in fitness variance (the strength of sexual selection) is strongly influenced by the variance in the distribution of male mating success. As we explained in chapter 1, identifying the mechanism of sexual selection is different from quantifying its evolutionary effect. We will demonstrate the comparative utility of our quantitative approach by applying it to examples from natural populations.

The Emlen and Oring Hypothesis

In 1977, Stephen T. Emlen and Lewis W. Oring compiled the published observations of avian and mammalian researchers through the 1970s, and presented an ecological classification of animal mating systems. The mating system classification proposed by Emlen and Oring (1977) was predictive, based on the following logic:

1. Males compete with one another for access to females:
2. Like competition for scarce resources, male reproduction is limited by the spatial and temporal availability of sexually receptive females.
3. The intensity of sexual selection depends on the rarity of receptive females in relation to the abundance of competing males.
4. Sexual selection favors male attributes that permit their bearers to find and monopolize their mates.
5. Ecological constraints on male monopolization attempts lead to a species-specific pattern of male-female associations, called a "mating system."

The underlying cause of taxonomic variation in mating systems, according to this approach, is the interaction between ecological constraints and the process of male-male competition for scarce reproductive resources. Thus, quantifying the intensity of male-male competition is equated with quantifying the strength of sexual selection and the ensuing adaptive process.

In ecology, the intensity of competition is typically quantified as resources per head, that is, as the ratio of the units of available resource divided by the numbers of competing individuals. In mating system studies, therefore, it is reasonable to expect that the intensity of male-male reproductive competition should vary with the relative rarity of reproductive resources, that is, sexually receptive females. Hence, as explained by Emlen and Oring (1977), the intensity of male-male competition for mates can be expressed as the ratio of adult males to adult females. This *operational sex ratio* or OSR (we represent this ratio as R_O; chapter 1) is widely used in studies of mating systems (Emlen and Oring 1977; Borgia 1979; Wickler and Seibt 1981; Wittenberger 1981; Thornhill and Alcock 1983; Bradbury 1985; Clutton-Brock and Vincent 1991; Clutton-Brock and Parker 1992; Arnold and Duvall 1994; Parker and Simmons 1996). Many researchers consider the operational sex ratio to be *the* measure that best quantifies the intensity of sexual selection (e.g., Reynolds 1996). Sexual selection and OSR are terms often used interchangeably (e.g., Kvarnemo and Ahnesjö 1996).

In discussing sexual selection, Darwin (1859, 1871), Bateman (1948), Williams (1966), and Trivers (1972) made observations similar to those of Emlen and Oring (1977). However, none proposed that a combination of ecological constraints on male mating behavior and the spatial and temporal availability of female receptivity could be used to classify mating systems. Emlen and Oring (1977) argued convincingly that the ecology of female reproduction determines the degree to which males are successful in reproductively monopolizing females. The ecology of female reproduction depends, in turn, upon the distribution and abundance of crucial resources, which include most aspects of nutrition and nesting. Thus, the distribution of reproductive resources fundamentally determines the intensity of sexual selection and subsequent mating system evolution under the Emlen and Oring (1977) hypothesis.

Emlen and Oring (1977) made specific predictions concerning the kinds of mating systems that would result from different features of female reproductive ecology. For example, they predicted that *polygyny*, a mating system in which individual males may reproduce with more than one female, is likely to evolve when females are spatially clumped and young are not dependent on male care. The authors further partitioned polygynous mating systems into four general forms. Under *resource defense polygyny*, males guard resources crucial to female reproduction and mate with those females aggregating around the guarded resources. In *harem polygyny*, females aggregate first around clumps of resource; then males are attracted to female aggregations and guard females directly. If males themselves are the resources, as in those cases where female mating preferences are strong, additional polygynous mating systems, such as *lekking* or *"hot spot" polygyny*, could evolve. In these systems, female mating preferences or historically successful breeding locations are hypothesized to mitigate the influence of resource ecology on mating (Bradbury and Veherenkamp 1977; Thornhill and Alcock 1983; fig. 1.1). In practice, for any of the above reasons, one or a few male "despots" may monopolize large numbers of females. In contrast, in *scramble competition polygyny*, neither females nor the resources important to their reproduction are spatially aggregated. Instead, males are temporally polygynous, encountering, guarding, and mating with individual females as they forage for resources (e.g., McCauley 1981; McCauley and Wade 1978). All systems of polygyny result in the clustering of *reproductive* resources (i.e., receptive females), either spatially or temporally, around mating males. Conversely, when females are dispersed, owing to scarce and/or widely dispersed resources, or when young are dependent on care from both parents, Emlen and Oring (1977) predicted that *monogamy* will evolve. Lastly, when resources and females are so scarce that males must form coalitions to acquire them, or when males are solely responsible for providing care of the young, *polyandry*, a mating system in which individual females may reproduce with more than one male, is expected to evolve (see review in Thornhill and Alcock 1983).

Given a spatial distribution of females, the temporal pattern of female receptivity to mating further modulates the distribution of mate numbers across males. If there is a high degree of synchrony of female receptivity, then a male's ability to monopolize many mates simultaneously may be reduced. Emlen and Oring (1977, p. 216) predicted that extreme female synchrony could limit the opportunity for multiple mating by males, whereas extreme asynchrony would enhance the "environmental potential for polygamy" or EPP. (We consider the effect of the temporal pattern of female receptivity on sexual selection in detail in chapter 3.) The spatial distribution of females is itself determined in a complex way by many factors. For example, conspecific queing may lead to the overaggregration of females or female-female competition may lead to the overdisersion of females, as per Orians' threshold model of female nesting site choice in relation to male territories. In addition, runaway sexual selection can lead to a higher level of clustering of females at resources than is "adaptive" considering only the amount of resource (see below). We do not take up here the important questions of "why" females are spatially aggregated, whether they are "adaptively" or "maladaptively" clustered, or, if overclustered, whether it is owing to conspecific queing or runaway sexual selection. These are questions deserving of further study, and are considered in more detail in later chapters. In this chapter, we will show that our method applies to *any* spatial distribution of females, whatever the proximal or ultimate causes of the particular spatial distribution.

Trivers' (1972) earlier theory of parental investment made somewhat similar predictions but was founded on the widespread occurrence of *anisogamy*, that is, the sex difference in initial parental investment in gametes that exists in most in bisexual species. Trivers (1972) argued that asymmetrical parental investment in gametes is the cause of sexual selection because such asymmetric investment manifests itself as an asymmetry in the numbers of male and female gametes. Thus, reproduction by individuals of the sex investing less (generally the males) becomes limited by the availability of individuals of the sex investing more (generally the females). Because parental investment in gametes requires resources, Trivers' hypothesis is viewed by many as the foundation of the Emlen and Oring model. Trivers' argument also bears some similarity to that of Bateman (1948), who argued that the potential variance in fitness is greater among individuals of the sex producing more numerous but smaller gametes.

Most analyses of animal mating systems since 1977 have been conducted within the Emlen-Oring-Trivers framework, and excellent reviews have been provided by Bradbury and Veherencamp (1977), Borgia (1979), Wickler and Seibt (1981), Wittenberger (1981), Thornhill and Alcock (1983), Bradbury (1985), Ostfeld (1987), Clutton-Brock and Vincent (1991), Ketterson and Nolan (1994), Arnold and Duvall (1994), Parker and Simmons (1996), and Reynolds (1996; see also Verner and Willson 1966). Several parameters have been suggested for quantifying sexual selection under the Emlen-Oring-Trivers hypothesis. However, the OSR as proposed by Emlen and Oring (1977) captures the essential concept of the dependency of sexual selection on competition for reproductive resources. The OSR can be viewed as a "*reproductive competition coefficient*" among mating males because it expresses the number of competing males, N_{males}, per unit of resource, namely, sexually receptive females, N_{females}. The greater the number of mature males relative to the number of receptive females, the greater the OSR, and, correspondingly, the stronger the intensity of male-male competition for mates (Sutherland 1985; Kvarnemo and Ahnesjö 1996).

In a concept similar to the OSR, the sex difference in gametic investment has been quantified as the ratio of maximum potential reproductive rate (PRR) of males to females (Clutton-Brock and Vincent 1991; Clutton-Brock and Parker 1992). The greater this ratio, the greater the potential difference in numbers of offspring between a successful male and a successful female. The PRR can also be interpreted as the number of females whose ova could be inseminated by a single male. This measure emphasizes the intersexual difference in maximum offspring numbers rather than the intrasexual intensity of competition or the sex difference in variance of offspring numbers. It is used often in discussions of maternal sex ratio strategy to summarize the potential fitness gain to a female of producing an offspring of one sex instead of the other (Clutton-Brock and Vincent 1991; Clutton-Brock and Parker 1992) and has been extensively used in discussions of sex-role reversed species (Vincent et al. 1994; Ahnesjö 1995; Parker and Simmons 1995; Simmons 1995; Kvarnemo 1996, 1997; Wiklund et al. 1998; Okuda 1999; Kvarnemo and Fosgren 2000; Masonjones and Lewis 2000; Ahnesjö et al. 2001). With multiple mating, the PRR can also be used to express the potential for postcopulatory reproductive competition. The PRR quantifies the maximal sex difference in reproductive fitness. When it is used to explain mating strategies, it is often assumed that every son is a "best" or "maximal" son, which is clearly impossible as we show later.

The breeding sex ratio (BSR) proposed by Arnold and Duvall (1994; Duvall et al. 1995) measures the ratio of *actually* breeding males, N^*_{males}, to breeding females, N^*_{females}, in contrast to the number of *potentially* breeding males, N_{males}, in the OSR, which is always greater than or equal to N^*_{males}. The BSR attempts to measure the *realized* male-male competition for mates, but, by ignoring nonbreeding males, it omits those males whose numbers dramatically increase the variance in male reproductive success (chapters 1 and 11). This convention is also applied when Q, the ratio of males to females qualified to mate, is first calculated, and then the OSR is calculated as Q ($PRR_{\text{males}}/PRR_{\text{females}}$) = OSR (Ahnesjö et al. 2001). As we will show, both of these approaches ignore the fact that most of the variance in male reproductive success exists between mating and nonmating males when sexual selection is strong (see below). Arnold and Duvall (1994) also suggested using a measure from phenotypic selection theory, the sexual selection gradient, as a way of isolating the statistical relationship between male trait values and mating success relative to other components of selection. In artificial selection and selective breeding experiments, the selection gradient provides a standardized measure of the strength of selection, useful for making comparisons across different experiments.

The temporal aspect of male competition for mates, which Emlen and Oring introduced as the synchronicity of female receptivity, was also recognized, but in a very different way, by Parker and Simmons (1996; see also Kvarnemo and Ahnesjö 1996; Ahnesjö et al. 2001; Wade 1979). These authors noted that, for males, display time costs energy. Thus, they added a temporal element of energy expenditure to male competition for mates. They proposed measuring the time budgets of males and females and emphasized the sex difference in the fraction of time spent on reproductive activity as critical to estimates of the intensity of sexual selection. All of these measures address aspects of the intensity of intra- or intersexual competition and its temporal modulation by the availability of females or by competing activities of males with limited time budgets. However, we will show below that, with the exception of Arnold and Duvall (1994), none of these approaches directly measures the sex difference in the strength of selection, although some, like the OSR, are often correlated with it.

In this chapter, we introduce a method for quantifying the spatial distribution of sexually receptive females and for making comparisons among species in the sex difference in the strength of selection. As mentioned above, we postpone discussion of the effect of variations in the temporal distribution of sexually receptive females until chapter 3. Our goals are to provide a statistical framework by which the predictions of the Emlen-Oring-Trivers hypothesis may be tested and extended to include all sexual species. We will focus our discussion mainly on animals, because sexual selection operates more strongly in animals than in most other sexual taxa (chapter 4). Thus, although our approach has broad application, we think animals provide the clearest examples of how our method may be used. Our framework yields many specific predictions, and, as we show in later chapters, it provides a quantitative way of classifying mating systems. Also, because we make extensive use of I_{mates}, the opportunity for sexual selection, as developed in chapter 1, our approach permits comparative analyses of mating systems that are not possible using either the measures suggested in the Emlen-Oring-Trivers hypotheses, or their recent methodological extensions.

Recall from chapter 1 that I_{mates} measures the strength of selection arising from the variance among males in mate numbers. Under assumptions explained in chapter 1, the sex difference in the opportunity for selection often arises entirely from the differences in mate numbers among males. That is, I_{mates} often equals the sex difference in the strength of selection. We now extend this concept to investigate how the spatial and temporal distribution of sexually receptive females affects I_{mates} and the other aspects of the sex difference in opportunity for selection. We illustrate the connection of our approach with concepts central to the thesis of Emlen and Oring (1977) and make explicit the role of the OSR in the sex difference in strength of selection. We also show that Lloyd's (1967) ecological measure of density-dependent competition, **mean crowding**, is a useful way to characterize the spatial distribution of receptive females because it is directly related to I_{mates}. Mean crowding quantifies the contagious clustering of females and, hence, reproduction, around males. In chapter 3, we extend this characterization to include the temporal distribution of receptive females.

Lloyd (1967) developed the concept of mean crowding, now widely used in ecological studies of natural populations. Estimates of mean crowding quantify how individuals experience the effects of crowding and competition when a population's membership is spatially distributed into clusters around patchily distributed resources. By incorporating mean crowding into the description of male-male competition for mates, our method permits us to make specific predictions about the activities males may engage in to monopolize clustered females. We can ask of any male behavior: Does it enhance or diminish the mean crowding of male reproduction? Since our method also distinguishes the mechanisms by which females may aggregate, the separate effects of male and female behaviors on sexual selection can also be quantified.

We will explicitly incorporate spatial effects of predation and viability selection, as well as female tendencies to copy the mate choices of other females, into our measurement of the spatial and temporal aggregation of female receptivity. In subsequent chapters, we will combine the spatial distribution of females not only with the temporal distribution of female sexual receptivity, but also with the tendency toward multiple mating, and examine the effects of semelparous and iteroparous female life histories. We show how our measure is affected by variations in male tendencies to guard females, to seek multiple mates, to employ alternative mating strategies, and to engage in parental care. Once we have established the effects on sexual selection of these male and female traits separately, we will then examine when and how conflict between the sexes arises and is resolved by selection. Lastly, since our method generates a dimensionless measure of the overall intensity of sexual selection, we show how it provides a basis for comparative analyses of mating system evolution.

The Mean Crowding of Sexually Receptive Females in Space

Lloyd (1967) developed a statistical framework for measuring the microspatial distribution of organisms in natural populations in relation to the ecological effects of crowding and competition. He proposed a measure m^* that identified the number of other individuals the average individual experiences as competitors. He called this measure "*mean crowding*." Lloyd noted that whenever individuals are clustered spatially and the clusters differ in size, individuals in the largest clusters experience *greater* competition than individuals in smaller clusters. Because there are more individuals in the larger clusters, by definition, the average *experience* of competition, m^*, exceeds that expected from the average density per cluster, m. Thus, Lloyd argued, average patch density is an inadequate measure of the way in which individuals experience competition and other density-dependent processes in ecology. On the other hand, if individuals are more uniformly distributed in space than random, the experienced competition may be *less* than that expected by the average density. Indeed, overdispersion is one manifestation of an ecological response to resource competition. Mean crowding captures both of these effects of the variance in size of the clusters on the intensity of competition.

We extend Lloyd's concept of mean crowding to sexual selection. We argue that, whenever females are clustered or clumped across a patchily distributed resource, OSR, EPP, PRR, BSR, Q, or other commonly used measures of reproductive competition, do not adequately reflect the variable experience of female density by males. Differently put, these measures do not accurately capture the intensity of male-male competition for mates as males experience it. Furthermore, we show that the mean crowding of females at resources (or around males, if males themselves are the resources) bears a natural relationship to the sex difference in the opportunity for selection.

In addition to mean crowding, Lloyd (1967) defined "patchiness," P, as the ratio of mean crowding to the average density per patch (m^*/m). Patchiness measures the degree to which the experience of competition is greater than or less than that indicated by the average density. It too is a useful measure for two reasons. First, whenever the distribution of individuals across patches of resource is random, patchiness equals 1. In other words, under a random distribution, mean crowding and average density are equivalent (i.e., $m^* = m$). Secondly, a parametric statistical test based on the negative binomial distribution, developed by Lloyd, can be used to test whether a measured value of P represents a statistically significant deviation from random. Wade (1995) was first to notice that Lloyd's (1967) concepts of mean crowding and patchiness apply to measurements of sexual selection like I_{mates}. Wade used Lloyd's terms to quantify the relationship between resource-defense polygyny and the opportunity for sexual selection, a theoretical framework that we use and expand throughout this book.

Resource-Defense Polygyny and the Mean Spatial Crowding of Females

In resource-defense polygyny, males defend patches of resources that are critical for female reproduction (Emlen and Oring 1977). Females tend to aggregate at resource patches, and males, by defending a patch, defend and mate with the associated cluster of females. To understand how female tendencies to aggregate influence the variance in male mating success, we must first quantify the average number of females per resource patch, that is, the average female density. We must also quantify the variance in female numbers per patch around that average. Let M be the number of resource patches containing at least one female, and let m_i be the number of females in the ith patch of resource. We could define patches based upon the distribution of males instead of females and, if we do, we obtain equivalent expressions (see below). However, we choose to define patches based on the presence of females for two reasons. First, females represent the reproductive resource being actively sought by males. Second, as a practical matter, it is often harder to find and census nonmating males since they often hide or leave breeding areas. Fortunately, the frequency of nonmating males can often be estimated from knowledge of R and H. The average density of females per patch, m, equals the sum of all females over all patches, divided by the total number of patches, or

$$m = \Sigma\, m_i/M. \qquad [2.1]$$

Note that the total number of breeding females, N_{females} or $\Sigma\, m_i$, equals the product of the average density per patch, m, and the number of patches, M. That is, N_{females} equals the product $(m)(M)$.

Following Lloyd (1967) and Wade (1995), the mean crowding of females on patches of resource is defined as

$$m^* = [\Sigma\, m_i\, (m_i - 1)]/(\Sigma\, m_i). \qquad [2.2]$$

As explained by Lloyd (1967), the expression m^* measures the degree to which individuals, in this case females, tend to be clumped on resource patches. The larger the value of m^*, the greater the degree of aggregation of females at patches of resource. When single males are able to defend and mate with such aggregations of females, then there is a direct connection between the spatial clustering of females and the strength of sexual selection.

Equation [2.2] can be rewritten in terms of two parameters that describe the spatial distribution of females across patches of resource. These are the mean density m and the variance of density among patches, V_{m}. To see this, we first recognize that $\Sigma(m_i)^2$ equals $(M)(V_{\text{m}} + [m]^2)$. This relationship is taken from the more familiar definition of the variance V_m, as equal to the mean of the squares $(\Sigma\, [m_i]^2)/(M)$, minus the square of the means $[m]^2$. We substitute this latter expression for $\Sigma(m_i)^2$ in eq. [2.2] to obtain

$$m^* = \{[M][V_m + (m)^2] - (\Sigma\, m_i)\}/(\Sigma\, m_i). \qquad [2.3]$$

Recalling eq. [2.1], we next rewrite eq. [2.3] as

$$m^* = \{V_m + (m)^2\}/(m) - 1. \qquad [2.4]$$

Rearranging terms, we have Lloyd's (1967) definition of the mean crowding, applied to the aggregation of females at resources,

$$m^* = m + [(V_m/m) - 1]. \qquad [2.5]$$

We interpret m^* in eq. [2.5] as the number of other females that the average female experiences on her resource patch. When the variance V_m is large relative to the mean, then females are clumped in space and m^* is *greater than* the arithmetic mean density m. On the other hand, if the variance in female aggregation size is small relative to the mean, then females are *overdispersed* and m^* is relatively smaller than m. When the variance equals the mean, females are randomly dispersed and m^* is equal to m.

Lloyd (1967) introduced the concept of patchiness, P, to express the degree of nonrandom aggregation on resources. When considered in the context of the spatial distribution of females, P is the ratio of mean crowding of females on patches to the average number of females per patch, or

$$P = (m^*/m). \qquad [2.6]$$

Patchiness is a measure of the relative "concentration" of females at resources. When patchiness exceeds 1, it is evidence that females are aggregating around certain clusters of resources. When patchiness is less than 1, it is evidence that females are distributing themselves more evenly than random over resources. By substituting eq. [2.5] into expression eq. [2.6] for P, we find that

$$P = 1 + [(V_m)/(m^2) - (1/m)]. \qquad [2.7]$$

Differently put, P measures the degree to which females are clumped in space.

If each patch of resource with females is guarded by a single male, and that male mates with all of the females on the patch, then the mean density of females, m, equals the average harem size, H, the mean number of mates per *successful* male. The operational sex ratio R_O equals $[N_{males} / N_{females}]$, which is the *inverse* of the average number of mates per male, R (see chapter 1). In this context, the BSR, i.e., the breeding sex ratio of Arnold and Duvall (1994), equals $[M/N_{females}]$, which can be more simply expressed as $[1/m]$ or $[1/H]$. It is also true that V_m equals the variance in mate number or harem size, V_{harem}, among successfully breeding males. These expressions help connect our formulation of the sex difference in the strength of selection as explained in chapter 1 to those measures of the intensity of reproductive competition based upon the OSR and BSR.

The Opportunity for Sexual Selection and the m^* and P of Sexually Receptive Females

In order to associate the sex difference in the opportunity for sexual selection with resource-defense polygyny, we need to calculate the variance of male mating success. The variance in mate numbers of the *successfully* mating males, i.e., those males that actually breed, is one component of m^*. However, because all breeding females are found in some patch, m^* is calculated *only* for those patches with at least one female. Thus, the calculation of m^* ignores patches without females that may be guarded by males. Hence, the variance among aggregations of females is not equal to the total variance in number of mates among all males because it omits the zero class of males, that is, the males who do not mate. This does not mean that unmated males are excluded from the analysis. It simply means that, in calculating m^*, we only need to consider the distribution of females *with successful* males. (Our approach is distinct from that applied in estimates of BSR or Q, which may include some or all of the zero-class males.) As we will see, our focus on females can be advantageous for field studies of mating systems.

We let p_0 be the frequency of potentially breeding males with no mates, and we calculate the total variance in male mating success as though we were performing an analysis of variance across two categories of males, mating and nonmating. When females are clumped on resources, some males will be capable of defending resources and some males will not. This creates two components of variance in mating success among males: (1) the variance in mating success that exists *within* the group of males that successfully defend a patch of resources and mate with females aggregating there, and (2) the variance in the average mating success that exists *between* mating and nonmating males. As in ANOVA, we take the average of the variances in mate numbers *within each of the two classes* of males, successful and unsuccessful, and add to that the variance in the average mate numbers *between* the two classes of males.

Recall from chapter 1 (eqs. [1.16] and [1.17]) that components of V_{mates} can be expressed as follows. If p_0 is the fraction of unmated males, then $(1 - p_0)$ is the fraction of mated males. Moreover, the average harem size of mated males, H, can be expressed as

$$H = R/(1 - p_0). \qquad [2.8]$$

This expression is eq. [1.16], rewritten in terms of H, the average harem size of mating males. Note that we can estimate the fraction of nonmating males, p_0, as $(1 - [R/H])$ from H and R.

The first component of the variance in male mating success is the average variance in mating success that exists *within* each of the two male categories, that is, within the class of the unsuccessful males as well as within the class of successful males; thus,

$$V_{within} = (p_0)(0) + (1 - p_0)V_{harem}. \qquad [2.9]$$

Note that because there is no variance in mating success among unsuccessful males, V_{within} equals the variance in mating success that exists *within* the group of successfully mating males, times the relative size of this group, $(1 - p_0)$. The second variance component can be understood as the variance in average mating success *between* the two classes of males, successful and unsuccessful. It is equal to

$$V_{between} = H^2 (p_0) (1 - p_0). \qquad [2.10]$$

It is clear that the variance in male mating success diminishes as males become more similar to one another in mating success. Conversely, the greater the differences in mate number among successful males, the greater will be the total variance in mate numbers. Since V_{mates} equals $(V_{within} + V_{between})$, by substitution, we have

$$V_{mates} = (1 - p_0)V_{harem} + H^2 (p_0) (1 - p_0). \qquad [2.11]$$

Note that, as H becomes large, p_0 approaches 1 and the product $(p_0)(1 - p_0)$ approaches $([H - 1]/H^2)$. Thus, for large H, V_{mates} is essentially equal to $(H - 1)$.

In order to obtain I_{mates}, we must divide V_{mates} by the square of the *mean number of mates per male*, which is the sex ratio, R^2. We find that

$$I_{mates} = ([1 - p_O][V_{harem}] + [H]^2[p_O][1 - p_O])/(R^2). \qquad [2.12]$$

Because the mean number of mates per male is $(H[1 - p_O])$, we note that (R^2) is equivalent to $(H[1 - p_O])^2$. We use this substitution to reduce eq. [2.12] to

$$I_{mates} = ([V_{harem}]/[H]^2[1 - p_O]) + ([p_O]/[1 - p_O]). \qquad [2.13]$$

Notice that I_{mates} consists of two selection opportunities. The first is measured by the ratio $(V_{harem})/(H)^2$, which is the opportunity for selection among *mating* males. Because there are varying degrees of mating success even among the successfully mating males, this term, $(V_{harem})/(H)^2$, describes the strength of selection *within* this class of mating males, owing to variations in the numbers of mates. (The opportunity for selection among mating males is also equal to the ratio $[V_m/m^2]$ because of the identity relationship between m and H. That is, the average number of females on resource patches is equivalent to the average number of mates per mating male.) The second selection opportunity is the ratio $([p_O]/[1 - p_O])$, which is the *opportunity for selection between mating and nonmating males*. The variance in mating success is $([p_O][1 - p_O])$ and the square of mean mating success is $(1 - p_O)^2$ so that the ratio $([p_O]/[1 - p_O])$ equals I_{mating}. Thus, the total opportunity for selection on male reproduction is the sum of two components. One measures selection on mating versus not mating, while the other measures the intensity of selection to mate repeatedly or to have many mates.

Since m is equal to $(R/[1 - p_O])$, we know that $(1 - p_O)$ is equivalent to (R/m). Hence, eq. [2.13] can be rewritten as

$$I_{mates} = (m/R)([V_m]/[m]^2) + ([m/R] - 1). \qquad [2.14]$$

Rearranging, we find that

$$I_{mates} = (1/R)(m^* + 1 - R) \text{ or} \qquad [2.15a]$$

$$I_{mates} = (R_O)(m^* + 1 - R). \qquad [2.15b]$$

The above equations show that when the sex ratio is male biased, so that $R < 1$ or, equivalently, $R_O > 1$, then the opportunity for sexual selection, adjusted for a biased sex ratio, is augmented. When the sex ratio is female biased, so that $R > 1$ or $R_O < 1$, then the opportunity for sexual selection is diminished. When R (or R_O) is unity, this reduces to the simpler expression of Wade (1995), namely,

$$I_{mates} = m^*. \qquad [2.16]$$

We can also express the patchiness of receptive females, P, in terms of I_{mates}. Since $P = m^*/m$, when $R = 1$,

$$I_{\text{mates}} = mP. \quad [2.17]$$

Or, since $m = H$,

$$I_{\text{mates}} = HP. \quad [2.18]$$

The Sex Difference in the Opportunity for Selection

We can now relate the opportunity for sexual selection, arising from the spatial distribution of females, to the variance in offspring numbers between males and females and thus to the sex difference in the opportunity for selection. As we showed in chapter 1 (eq. [1.20]),

$$V_{\text{males}} = RV_O + O^2 V_{\text{mates}}, \quad [2.19]$$

where O represents the average number of offspring per female and R is the sex ratio. Recall also (chapter 1, p. 29) that the opportunity for selection on males, I_{males}, equals the total variance in male reproductive success, V_{males}, divided by the squared average in male reproductive success. The average male reproductive success (assuming independence) is the product of the average number of mates per male, R or $(N_{\text{females}}/N_{\text{males}})$, times the average number of offspring per female, O, or simply (RO). Dividing eq. [2.17] by the squared average of male reproductive success, $(RO)^2$, we have

$$I_{\text{males}} = (1/R)(I_{\text{females}}) + I_{\text{mates}}. \quad [2.20]$$

We can rearrange the terms of eq. [2.18], to obtain $(I_{\text{males}} - I_{\text{females}})$, the *sex difference in the opportunity for selection*. This gives us

$$(I_{\text{males}} - I_{\text{females}}) = (R_O - 1)(I_{\text{females}}) + I_{\text{mates}}. \quad [2.21]$$

The above expression illustrates the role of the OSR, measured by R_O, in causing a difference in the strength of selection between the sexes, which can result in sexual dimorphism. It also illustrates clearly that the OSR is *only one* component of this sex difference in selection. Indeed, when R, and necessarily R_O, are equal to 1, the OSR drops out of eq. [2.19] entirely, although a sex difference in the opportunity for selection may still exist. On the other hand, if the variance in mate number among males were zero, then I_{mates} would be zero. In this case, the OSR would be the *sole* determinant of a sex difference in the strength of selection but it would not involve sexual selection owing to male variance in male numbers.

Our eqs. [2.15a, b] show that the sex difference in the opportunity for selection is a function of two factors: (1) the OSR; and (2) the opportunity for sexual selection on males, I_{mates}. The first factor accounts for differences in the number of males relative to the number of females. The second factor is equivalent to the mean spatial crowding or contagion of females around males, m^*, a quantity determined by the mean and variance in the numbers

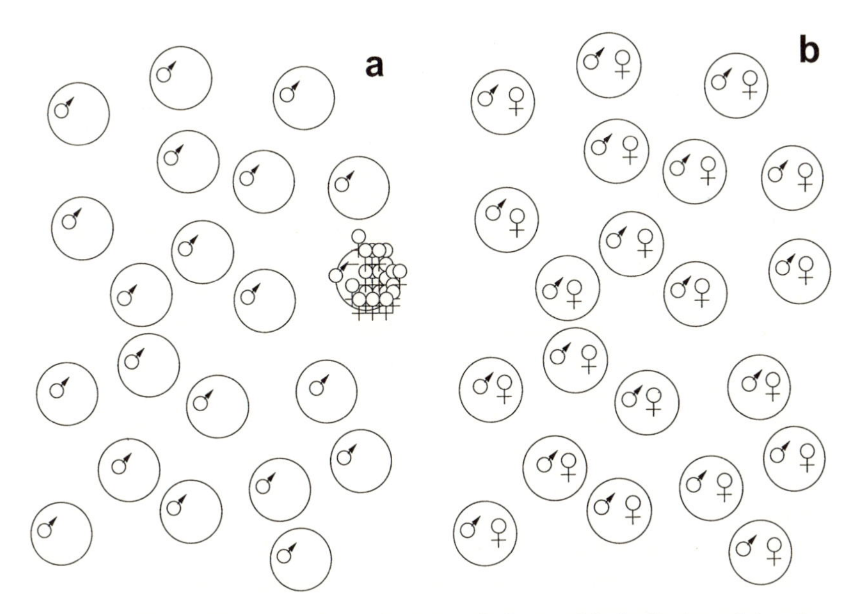

Figure 2.1. Model population for simulations of the spatial distribution of females; $N_{males} = N_{females} = 20$; $N_{patches} = 20$; (a) all females are aggregated on a single patch; (b) females are distributed evenly among patches such that each male mates.

of females on resource patches defended by males. Both factors are components of the Emlen and Oring (1977) hypothesis that the intensity of sexual selection is influenced by the spatial distribution of sexually receptive females. These components now have quantitative expression in eq. [2.15].

The Relationships among M, m, m^*, and P

We have argued that when the spatial distribution of females determines the distribution of males seeking mates, as per the Emlen and Oring (1977) hypothesis, the spatial patchiness of females is expected to influence mating system evolution. The parameters we have used to measure the intensity of selection in this context are better understood when they are used in a simple simulation. Imagine a population consisting of 20 males and 20 females ($N_{males} = N_{females} = 20$), distributed over 20 resource patches (fig. 2.1). Assume, as described above, that each male occupies a resource patch and mates with the females that are attracted to his patch. As we showed in chapter 1, we can identify the fraction of mating males as p_m, and the fraction of nonmating males as p_0. Thus, $p_m = (1 - p_0)$, and the number of mating males equals $(1 - p_0)(N_{males}) = N_{mating}$. Similarly, the number of nonmating males equals $(p_0)(N_{males}) = N_{nonmating}$. Using Eqs. [2.5]–[2.9] above, we can now identify values for M, m, m^*, and P.

First, assume that a single male attracts *all* of the females in the population to his patch. In this case, the number of mating males, $(1 - p_0)(N_{males})$, equals 1 and the number of nonmating males, $(p_0)(N_{males})$, equals 19. Because M is defined by female occupancy, we can easily see that M, the number of patches containing at least one female, equals 1, and the sum of all females in all patches (Σm_i) equals 20, the total number of females in the population $(= N_{females})$. Using eq. [2.1], we find that m, the average number of females per patch, also equals 20 (20/1), and, since the variance in the number of females per patch is zero, by eq. [2.5], m^* equals 19. The patchiness in the distribution of females on resources, P, is the ratio of m^* to m, and equals 19/20 or 0.95. We now progressively remove females from the single mating male's patch, and distribute them among the other males in the population. After each removal, we recalculate the above parameters, continuing this procedure until each male's patch contains a single female. We plot the parameter values as well as their interrelationships (figs. 2.2–2.5 below).

If we had defined patches as regions containing at least one male, then we would require information about every male, including those males without mates. We would divide all males into occupied, p_m, and unoccupied, p_0, patches based upon the presence or absence of females, and the definition of m^* as the mean crowding of the "occupied patches" would follow.

As M, the number of resource patches containing at least one female, increases, the number of mating males $([1 - p_0][N_{males}] = N_{mating})$ also increases in direct proportion (fig. 2.2a). Because the distribution of females on patches determines male mating success, the number of resource patches containing females equals the number of males guarding resource patches and mating on them. As patch number M increases, both the average number of females per patch, m, and the mean crowding of females on patches, m^*, decrease (fig. 2.2a). Note that m^* decreases less rapidly than m at intermediate values of M, because estimates of m^* include the ratio of the variance to the average number, m, of females per patch, which is decreasing. Differently put, the distribution of females is most patchy (i.e., P is greatest) at intermediate values of M (fig. 2.2b).

Now consider the relationship between the number of patches containing one or more females, M, and average harem size of mating males, H. Recall that M and $(1 - p_0)$ are directly proportional. As the number of patches containing a female(s) increases, average harem size decays (fig. 2.3a). Stated differently, as females become more evenly distributed on patches, more males are successful at mating. Conversely, as the average number of females per patch increases, average harem size increases proportionately (fig. 2.3b). This is the basis of the relationship expressed in eq. [2.8]. When male mating success depends on the distribution of females on resource patches, and when the sex ratio R equals 1, then H equals m, which equals $1/(1 - p_0)$ (fig. 2.3c).

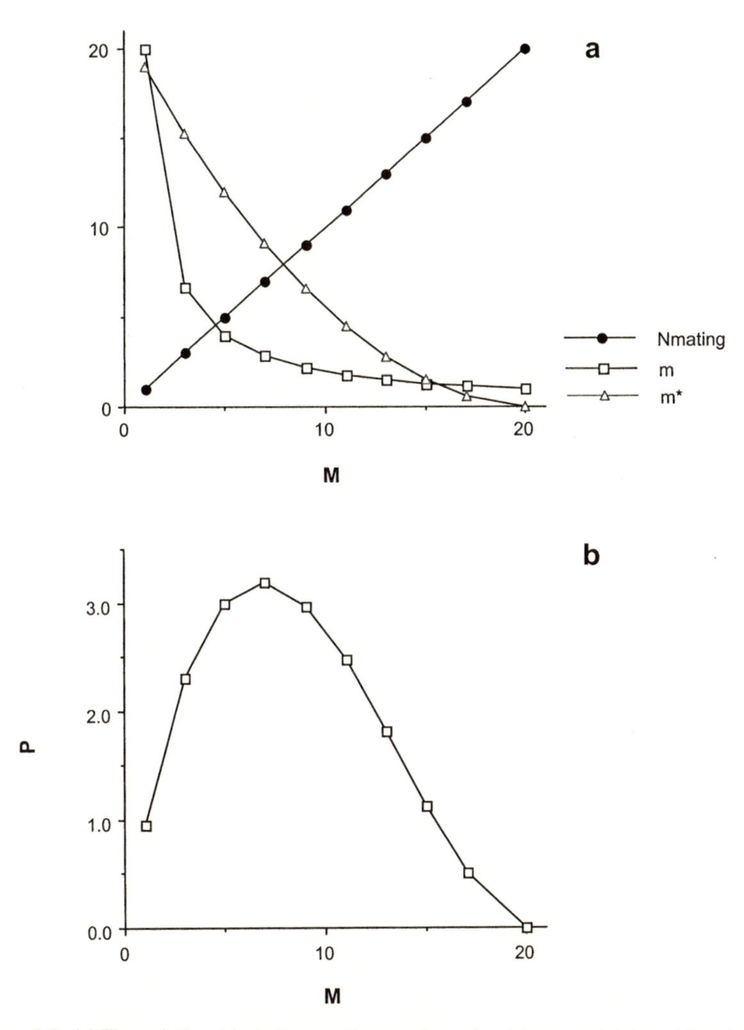

Figure 2.2. (a) The relationship between the number of patches containing at least one receptive female and the number of mating males (N_{mating}), the average number of females per patch, *m*, and the mean spatial crowding of females within intervals, *m**. (b) The relationship between the number of intervals containing at least one receptive female, *M*, and the temporal patchiness of females, *P*.

When only one male is successful in mating, the total variance in mating success among males (V_{mates}) is large because most males ($N_{males} - 1$) do not obtain any mates at all. All of the variance in mate numbers exists *between* the classes of mating and nonmating males. The variance among males decreases as the number of mating males increases because females are more evenly distributed across males (fig. 2.4a), that is, the variance in mating success between the classes of mating and nonmating males is lower.

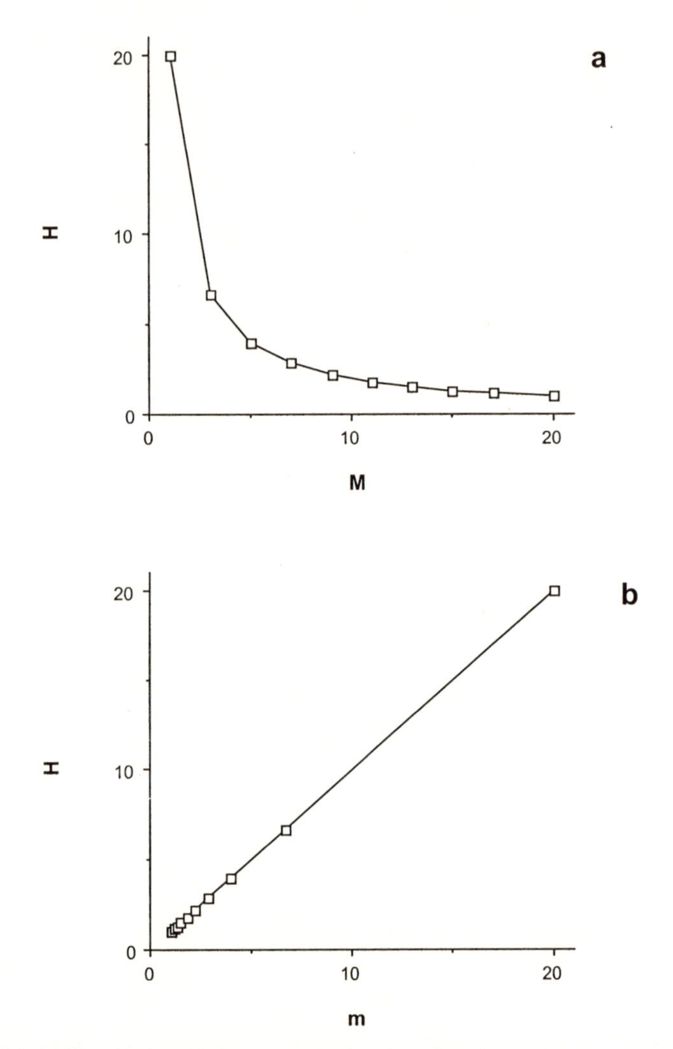

Figure 2.3. (a) The relationship between the fraction of mating males ($1 - p_0$) and average harem size (H); (b) the relationship between the average number of females per patch (m) and average harem size (H); (*continued*)

Whenever the variance in mating success among males depends on the distribution of females on resource patches, the variance in the number of resource patches containing at least one female, V_m, is equivalent to the variance in mating success among males who mate, V_{harem} (fig 2.4b). Moreover, as the mean crowding of females on patches, m^*, increases, both the variance in mate numbers among males (V_{mates}) as well as the average harem size of successful males ($H = m$) increase (fig. 2.4c). In particular, note that, when $R = 1$, m^* and V_{mates} are equivalent.

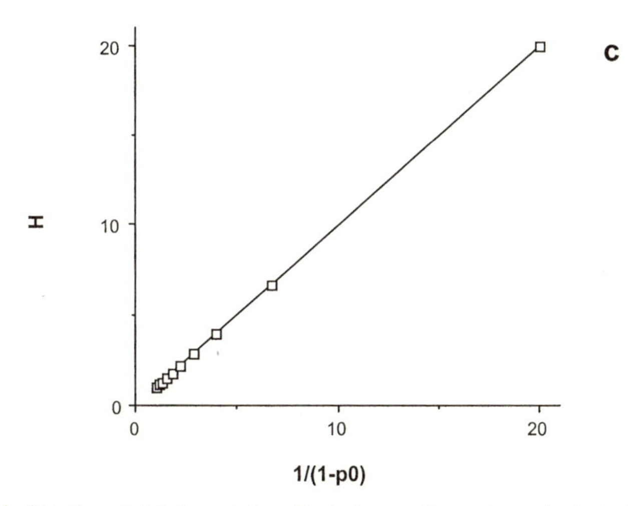

Figure 2.3. (*continued*) (c) the relationship between the reciprocal of the fraction of mating males $[1/(1-p_0)]$ and average harem size (H).

Recall that the opportunity for sexual selection, I_{mates}, is equal to the variance in mating success divided by the square of the average mating success (Crow 1958; Wade 1979, 1995). This ratio measures the relative ability of males in the population to acquire mates. The mean crowding of females, m^*, measures the degree to which females are spatially aggregated. As females become more spatially clumped, the ability of individual males to defend and mate with females within aggregations increases. Thus, an identity relationship is expected between the spatial distribution of females, m^*, and the opportunity for sexual selection, I_{mates} (fig. 2.5a). An identity relationship is also expected between the product of the average harem size of a mated male, H $(= m)$ and the spatial patchiness of females, P, that is, HP, and the opportunity for sexual selection, I_{mates} (eqs. [2.17] and [2.18]; fig. 2.5b). A similar, positive, but not quite 1:1, relationship is also expected between the spatial patchiness of females, P, and the opportunity for sexual selection on mated males, I_{harem}. This is true because the opportunity for selection on males who defend patches and mate with the females on those patches, I_{harem}, is expected to increase as the distribution of females on resources becomes increasingly patchy, as measured by P (fig. 2.5c). All three of the above expectations are met in our model population (fig. 2.5a–c).

The relationships shown in Fig. 2.5b,c (eqs. [2.17] and [2.18]) are useful for mating system analyses in situations in which the fraction of nonmating males (p_0) cannot be measured in nature. Whereas calculation of I_{mates} from field data using eq. [2.13] requires knowledge of the fractions of mating males $(1 - p_0)$ as well as of nonmating males (p_0), P, H, and I_{harem} can all be determined from the distributions of females associated with mating males alone. Thus, this approach can be applied equally well in situations in

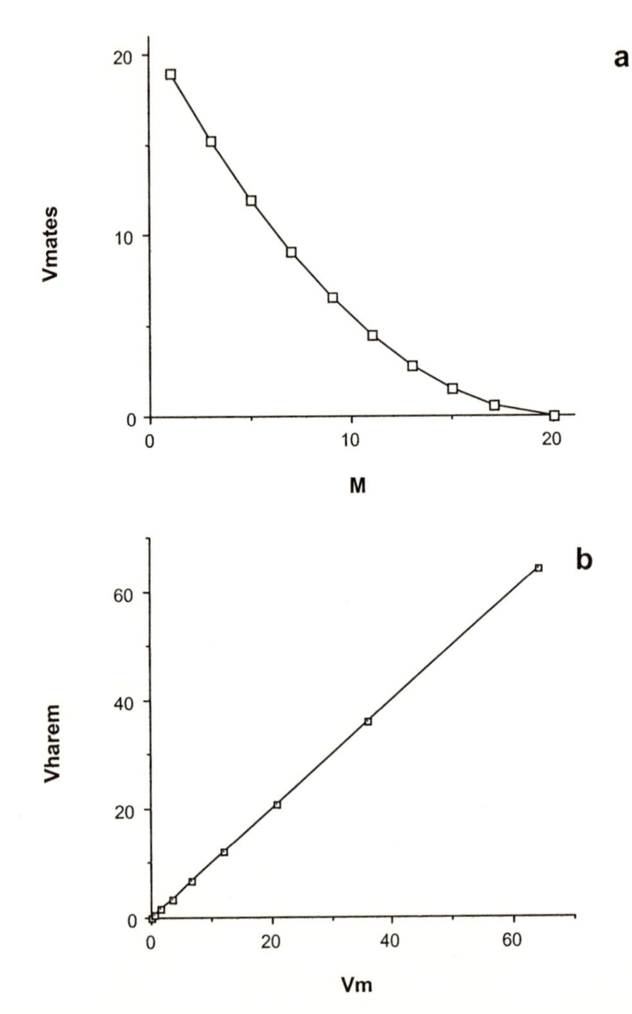

Figure 2.4. (a) The relationship between number of patches containing one or more females, *M*, and the variance in mate numbers among males (V_{mates}); (b) the relationship between the variance in the number of females per resource patch (V_m) and the variance in harem size among mating males (V_{harem}); (*continued*)

which all adult males in the population can be counted, as well as in situations in which only the mating males can be identified.

Finding the Appropriate Spatial Scale

Recognition of the appropriate spatial scale for investigating organic processes is a fundamental problem in biology, which Lloyd (1967) discussed at length in relation to sampling quadrats and mean crowding. We have consid-

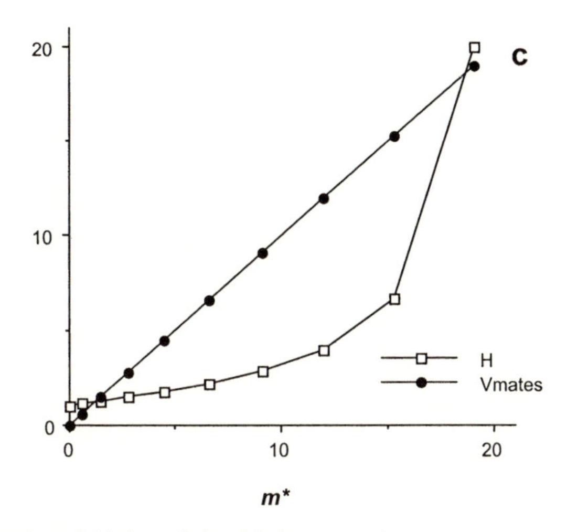

Figure 2.4. (*continued*) (c) the relationship between the mean crowding of females on resource patches (m^*), and the average and variance in mate numbers among males (H and V_{mates}, respectively).

ered situations in which females are visibly clustered on resources and when, within a cluster, one male mates with all of the aggregated females. Examples in nature include the spongocoel cavities occupied by some marine isopods (Shuster and Wade 1991a) or the territories of some lizards (Sinervo and Lively 1996), wherein clusters of receptive females are clearly separated from one another and males actively defend territory boundaries. Such aggregations can often be recognized as functional breeding units. However, because females may not mate exclusively within breeding territories, our perception of resource patchiness may differ from that of the females aggregating on it, and as well from the *realized* mating patchiness of females across males.

In such cases, the following ecological method may be useful for identifying the appropriate spatial scale for quantifying female aggregation in relation to reproduction. For ease of illustration, we will use quadrats as our unit of spatial measurement. However, as discussed by Lloyd (1967), sampling units of almost any shape lend themselves well to analyses using mean crowding, provided that these units are uniform in their dimensions. Thus, depending on the mating system examined, polygons rather than quadrats may be more appropriate sampling units. Although widely used, nearest-neighbor analyses do not translate easily into measures of selection. Moreover, regardless of how the spatial distribution of females is measured, molecular markers for maternity and paternity will still be required to establish the correspondence between the spatial ecology of receptive females and sex-specific patterns of reproduction.

Consider a scenario in which a large number of females are spatially dis-

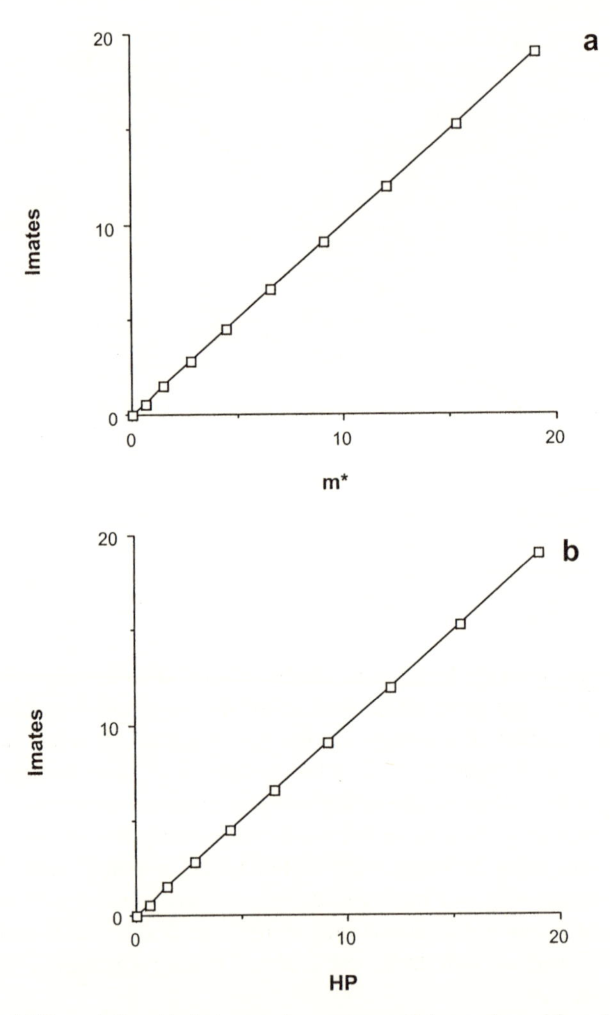

Figure 2.5. (a) The relationship between the mean spatial crowding of females (m^*) and the opportunity for sexual selection (I_{mates}) when $N_{males} = N_{females} = N_{patches} = 20$; (b) the relationship between the product of the average harem size of mated male, H ($= m$) and the spatial patchiness of females, P, and I_{mates}; (*continued*)

tributed in an area, such as a 1 km^2 quadrat (or, just as easily, a 1 m^2 quadrat). To determine the number of females present within the quadrat as well as the location of each female, we can divide the quadrat into regular sections and count the females contained in each part of the grid. If the sections within the quadrat are relatively large (say 0.50 km^2) then the average number of individuals per sector of the grid will also be large. Moreover, if individuals are not clustered within particular sections, the variance in

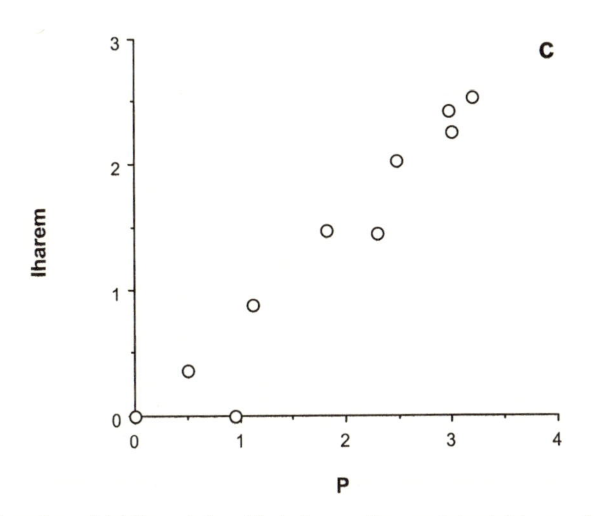

Figure 2.5. (*continued*) (c) the relationship between the spatial patchiness of females on resources (*P*) and the opportunity for sexual selection on males (I_{mates}; r^2 = 0.93, $F_{[1,9]}$ = 112.4, P = 0.001).

density among sections will be relatively small. As the area within each section is reduced (say to 0.0625 km^2), the number of sections within the grid increases. Some sections will begin to contain empty, unoccupied spaces, while others will contain individuals or aggregations of individuals. The variance in the number of individuals per section will increase compared with that determined for the first measurement at the larger grid size.

As section area is progressively decreased, the variance in the number of individuals among sections will increase up to the point at which the area within each section becomes smaller than the area that circumscribes the average aggregation. At this time a part of the grid contains only a portion of a group, and the variance in number of individuals within each section is again reduced. The variance in the number of individuals among sections will continue to decay until the area in a part of the grid is sufficient to contain only a single individual.

This procedure can be used to identify the quadrat (or polygon) size that gives the greatest spatial patchiness of females, i.e., the section area at which the variance in the number of females across parts of the grid is maximized. Is this the most appropriate scale at which to study sexual selection? The answer depends upon how the spatial patchiness of receptive females maps onto the variance in male reproductive success. In the derivation above, we assumed that (1) the presence of females defines a patch; (2) each aggregation is found and defended by a single male; and (3) all females in an aggregation mate with the male defending it. In natural populations, a single male might be able to find and mate with all females in more than one patch (i.e., the variance in male reproductive success exceeds the spatial patchiness of

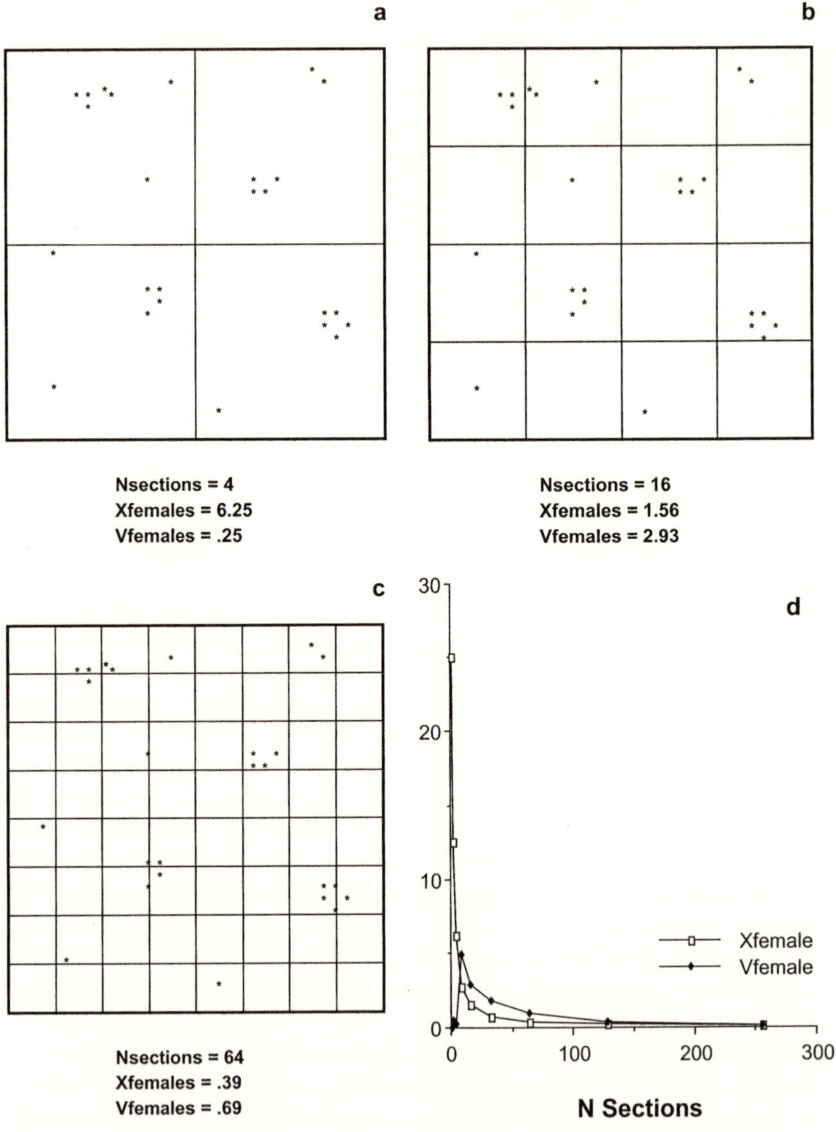

Figure 2.6. Identifying the spatial scale of female aggregation for 25 females within a quadrat; (a) the quadrat is divided into four equal sections; (b) the quadrat is divided into 16 equal sections; (c) the quadrat is divided into 64 equal sections; (d) the relationship between section number and the mean and variance in females per section; note that the variance in females per section is maximized when the section number equals 8.

females). Conversely, multiple males may fight for or sneak copulations so that the females in an aggregation share more than one sire. If so, the variance in male reproductive success will be less than that expected from the spatial patchiness of females. In a later chapter, we will argue that a runaway process may often lead females to "overaggregate," both beyond the ecological optimum determined by resources alone as well as beyond a male's ability to mate with and guard them (see Wade 1995). We recommend identifying the spatial scale that gives the greatest patchiness of the female distribution as a reasonable starting point for research. The central questions for investigation subsequently become:

1. To what extent does the variance in male reproductive success reflect the patchiness of the female spatial distribution?
2. What male traits enhance (e.g., armaments or body size) or diminish (e.g., furtive copulations or satellites) the mapping between the spatial aggregation of receptive females and the variance in reproductive success among males?

Figure 2.6 shows an example of this procedure on a small scale. In this case, the number of females in the quadrat, $N_{females}$, equals 25 (fig 2.6a–c). As the quadrat is divided into increasing numbers of sections, the average number of females per section, $X_{females}$, progressively decreases, whereas the variance in the number of females per section, $V_{females}$, first increases and then decays (fig. 2.6d). As explained above, the number of sections that maximizes the variance in the number of females per section, in this case, $N_{sections} = 8$, captures the spatial scale at which female aggregations are most recognizable and therefore are most accurately measured.

A Worked Example

To further illustrate the utility of our method, we now provide a worked example using data from a natural mating system. We will consider *Paracerceis sculpta*, a Gulf of California isopod crustacean, which breeds in the spongocoels of calcareous intertidal sponges, *Leucetta losangelensis* (fig. 2.7). Females are monomorphic in this species. Males, however, exhibit three distinct adult morphs that differ in reproductive behavior and morphology; α-males are largest and defend harems within sponges using elongated posterior appendages; β-males invade harems by mimicking female behavior and morphology; γ-males invade harems by being small and secretive (fig. 2.8; Shuster 1987, 1992). All three male morphs possess mature external genitalia and sperm-producing organs, although the relative allocation of energy toward somatic and gonadal structures differs significantly among males (Shuster 1987, 1989a). When isolated with females, the male morphs do not differ in their ability to sire young (Shuster 1989a, b), and average

Figure 2.7. A breeding aggregation of *Paracerceis sculpta*, a Gulf of California isopod crustacean in the spongocoel of a calcareous sponge, *Leucetta losangelensis*; α-males possess elongated posterior appendages (uropods) which they use to defend the entrance of their spongocoel against other α-males; females lack elongated uropods and are attracted to spongocoels containing α-males; females are particularly attracted to spongocoels containing α-males and other breeding females; spongocoels containing large aggregations of females are attractive to β- and γ-males, who invade spongocoels and steal matings from α-males; a tiny γ-male in this spongocoel is located directly below the α-male (photo by R. Lindner).

reproductive success among male types is equal over time (greater than 2 years; Shuster and Wade 1991a). Life history differences that may influence the relative contributions of α-, β-, and γ-males to the population appear to cancel because male reproductive tenure varies inversely with maturation rate (Shuster and Wade 1991a). Moreover, relative fertilization success among males varies with the density of receptive females, as well as with the frequency of other male morphs in spongocoels (Shuster 1989b; Shuster and Wade 1991a). Thus, male polymorphism appears to be maintained by frequency-dependent selection (Shuster and Wade 1991b; Shuster et al. 2001) in a manner sufficient in theory to maintain genetic polymorphism (Slatkin 1978, 1979). Indeed, the three male morphs are distinct at a single genetic locus (Shuster and Sassaman 1997).

In this example, we illustrate four things. First, we show that the spatial distribution of females with males in spongocoels can be quantified using Lloyd's (1967) measures of mean crowding and patchiness. Second, we show that when the spatial distribution of females determines male mating

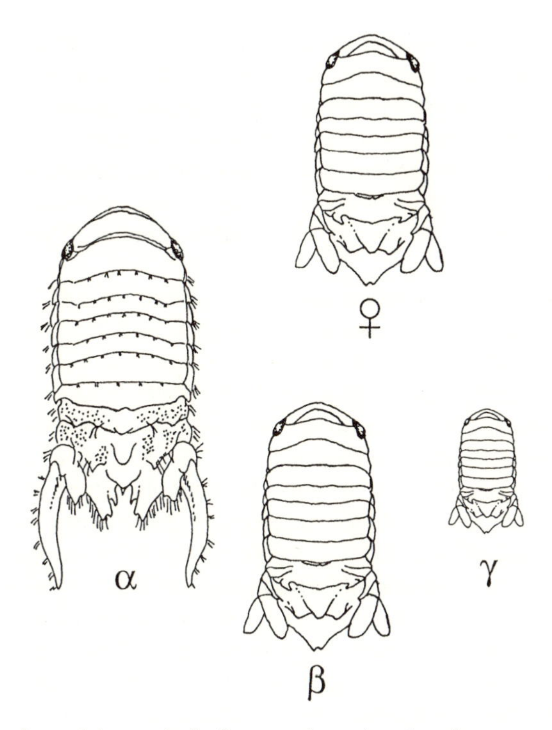

Figure 2.8. The four adult morphs in *Paracerceis sculpta*; females are monomorphic; α-males are largest, possess elongated uropods, and comprise 81% of aggregate male population samples, β-males are smaller than α-males, resemble females in their behavior and external morphology (although β-males are slightly smaller than females), invade spongocoels by mimicking female behavior, and comprise 4% of aggregate male population samples; γ-males are smallest, comprise 15% of aggregate male population and use their small size and rapid movements to invade spongocoels (from Shuster 1992).

success, the opportunity for sexual selection on mating males, I_{harem}, is correlated with the spatial patchiness of females, *P*. Third, we show that the opportunity for sexual selection, I_{mates}, is correlated with the mean spatial crowding of females, m^*. Fourth, we show that I_{mates} correlates too with the product of the average harem size, *H*, and the spatial patchiness of females, *P*. In short, the positive relationships found in our simulated population between I_{harem} and *P*, between I_{mates} and m^*, and between I_{mates} and *HP* exist in a natural population as well. Our goal in this example is to show that the formulas presented above provide a reliable method for quantifying the sex difference in the opportunity for selection arising from the spatial distribution of sexually receptive females in natural populations.

As explained by Shuster (1991b, 1992), *P. sculpta* breeding aggregations are patchily distributed. This is true in part because the sponges in which isopods breed occur beneath boulders and ledges in permanent tide pools, locations that are patchily distributed within the intertidal zone (Shuster 1986, 1991b). Receptive females in this species are attracted to spongocoels within sponges that contain established breeding aggregations (Shuster 1990; Shuster and Wade 1991b). This tendency among females to "copy" the mating decisions of other females accentuates the patchiness of breeding females within the patchily distributed sponges, and generates highly contagious distributions of females in all monthly samples (Shuster and Wade 1991b).

To document these distributions, Shuster (1986, 1989b, 1991b; Shuster and Wade 1991a,b) collected sponges from permanent tide pools in the mid-intertidal zone, 1.5 km southwest of Puerto Peñasco, Sonora, México, between October 1983 and November 1985. Each month at low tide, all sponges located within 15 randomly selected 0.25 m plots along a 100 m permanent transect were removed and their spongocoels examined for isopods. Individual isopods were identified by adult phenotype and reproductive condition. Shuster also examined the physical characteristics of sponges, of spongocoels, and of the males defending them to determine if the size, shape, or physical condition of reproductive habitats, or of the defending males, influenced the number of females in each harem. These latter questions were in vogue in the early 1980s in the form of the polygyny threshold hypothesis and the sexy son hypothesis (Verner 1964; Verner and Willson 1966; Orians 1969; Weatherhead and Robertson 1979). However, contrary to these hypotheses' predictions, neither habitat characters nor male characters had a statistically detectable influence on the number of females per male (Shuster 1986, 1990, 1991b).

Recall from the above discussion that the values of m^* and P can be calculated from the distribution of females with males, provided that females aggregate at resource patches, and that males, by defending a patch, also defend and mate with the associated cluster of females. Genetic studies have shown for *P. sculpta* that, when spongocoels contain more than one male, the paternity of the broods of individual females is mixed (Shuster 1989b; Shuster and Wade 1991a; Shuster and Sassaman, unpubl. data). However, mixed paternity has never been found in spongocoels containing a single α-male and one or more females. To illustrate our method for documenting the opportunity for sexual selection when females are patchily distributed in space, we will focus our attention, for now, on uninvaded spongocoels. These spongocoels comprise over 80% of all spongocoels in nature (ave. $\pm$ 95%CI = 0.81 $\pm$ 0.05, $N_{\text{months}} = 24$, $N_{\text{spongocoels}} = 452$; Shuster 1986; Shuster and Wade 1991a,b). Later, we will consider all of the spongocoels in the 1983–85 sample to consider the effects of multiple paternity and alternative male mating strategies on the opportunity for sexual selection (chapter 11).

If we consider only spongocoels occupied by an α-male and one or more females, then each spongocoel contains a breeding aggregation and is equivalent to a "resource patch" as described earlier. Thus, for each monthly sample, M equals the number of spongocoels containing at least one female, and m_i equals the number of females in the ith spongocoel. As shown in eq. [2.1], the average density of females per spongocoel, m, equals the sum of all females over all spongocoels, divided by the total number of spongocoels. The variance in the number of females per spongocoel, V_m, equals the statistical variance in the number of females per spongocoel, counting only those spongocoels containing females. With these values, we can now calculate the values of m^* and P for each monthly sample of isopods, using eqs. [2.5] and [2.6]. These monthly estimates are summarized in table 2.1. Also included in this table are the values of N_{females}, M, m, and V_m for each collection, as well as these parameters estimated for the aggregate sample.

There are three things to notice at this point. First, because all collections were made using a standardized sampling regime, the essential data for calculating m^* and P are the number of spongocoels as well as the number of females within each spongocoel. We need not calculate the area occupied by females in spongocoels because the spatial patchiness of females is determined by the patchiness of their breeding resource, the sponges, as well as by the tendency for females to form aggregations within spongocoels. Although not required in this instance, in situations in which female spatial distributions are less discrete, the quadrat (or polygon) approach recommended above can be used to identify the number of patches containing females. This method will often be easier than other methods for estimating spatial distributions, such as nearest-neighbor distances, and as mentioned these methods do not provide direct measures of selection. Second, in this application, only the individuals in spongocoels are breeding, which makes it much easier to account for mating males and females, as well as for unmated males (i.e., those in or guarding empty spongocoels). Third, although we can count males as well as females in spongocoels, our estimates of m^* and P are derived from the observed spatial distribution of females. Implicit in the model of Emlen and Oring (1977) is the assumption that the distribution of females will determine the breeding activities of males. Unlike mating system analyses that focus on the behavior of males, our approach is consistent with the Emlen and Oring (1977) hypothesis.

We can plot the changes in the sex ratio R and the OSR R_O, as well as m^* and V_{mates}, as a function of date of collection (fig. 2.9a,b). We see in fig. 2.9b that there is a conspicuous seasonality to sexual selection in *P. sculpta* with a peak of strong sexual selection in the fall months (October–November) and a somewhat weaker peak in the spring (March–April). There are marked sex ratio variations throughout the year (fig. 2.9a) as well as sixfold variations in m^* and we can use these variations to determine whether R or R_O is a better predictor of the strength of sexual selection. Current theory, based on the operational sex ratio ($R_O = [1/R]$), predicts that sexual selection is strongest when males are more abundant than females and weaker when females are in excess (see, for example, Emlen and Oring 1977;

Table 2.1
Calculation of m^* and P from monthly distributions of *P. sculpta* females in *L. losangelensis* spongocoels with with males, 1983–85.

Month	$N_{females}$	M	m	R	m*	P	R_O
1983							
Oct	23	11	2.09	2.08	2.09	1.00	0.48
Nov	23	9	2.56	3.14	2.78	1.09	0.39
Dec	20	11	1.82	1.06	1.40	0.77	0.55
1984							
Feb	9	6	1.50	0.58	0.89	0.59	0.67
Mar	6	5	1.20	0.16	0.33	0.28	0.83
Apr	79	33	2.39	4.66	3.34	1.40	0.42
May	20	9	2.22	1.73	2.00	0.90	0.45
Jun	41	16	2.56	1.75	2.24	0.88	0.39
Jul	48	17	2.82	4.26	3.33	1.18	0.35
Sep	25	12	2.08	1.58	1.84	0.88	0.48
Oct	57	15	3.80	3.49	3.72	0.98	0.26
Nov	2	2	1.00	0.00	0.00	0.00	1.00
Dec	2	2	1.00	0.00	0.00	0.00	1.00
1985							
Jan	12	10	1.20	0.16	0.33	0.28	0.83
Feb	18	9	2.00	1.56	1.78	0.89	0.50
Mar	28	11	2.55	1.16	2.00	0.79	0.39
Apr	22	9	2.44	2.03	2.27	0.93	0.41
May	1	1	1.00	0.00	0.00	0.00	1.00
Jun	7	4	1.75	0.69	1.14	0.65	0.57
Jul	14	7	2.00	0.86	1.43	0.71	0.50
Aug	22	10	2.20	1.56	1.91	0.87	0.46
Sep	17	7	2.43	2.25	2.35	0.97	0.41
Oct	45	13	3.46	7.48	4.62	1.34	0.29
Nov	24	10	2.40	1.84	2.17	0.90	0.42
Total	565	239					
Average	23.54	9.96	2.10	1.84	1.83	0.76	
SD	18.81	6.47	0.73	1.76	1.21	0.39	

Thornhill and Alcock 1983; Reynolds 1996; Ahnesjö et al. 2001). We find that the correlation between R and V_{mates} is high and positive ($r_{R,V_{\text{mates}}}$ = 0.82), while that between R_O and V_{mates} is high but negative (r_{R_O,m^*} = −0.71). The correlation between R and m^* is even higher (r_{R,m^*} = +0.96), while that between R_O and m^* remains highly negative (r_{R_O,m^*} = −0.90). Thus, R is a better indicator of the strength of sexual selection than is R_O. Indeed, sexual selection is weakest when R_O is highest—the opposite of

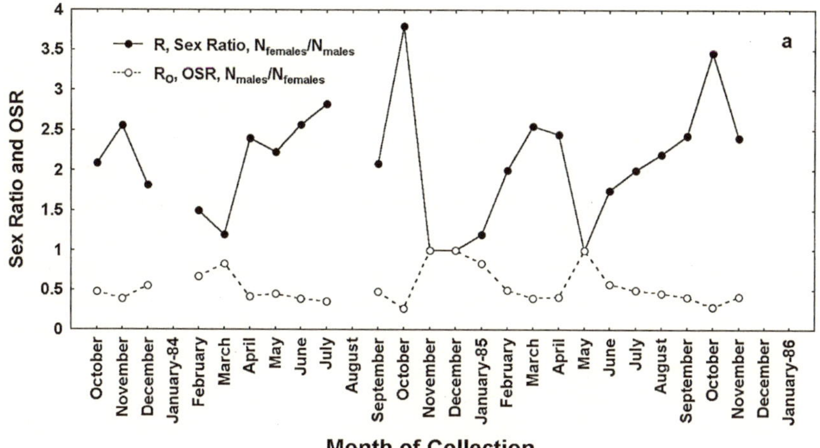

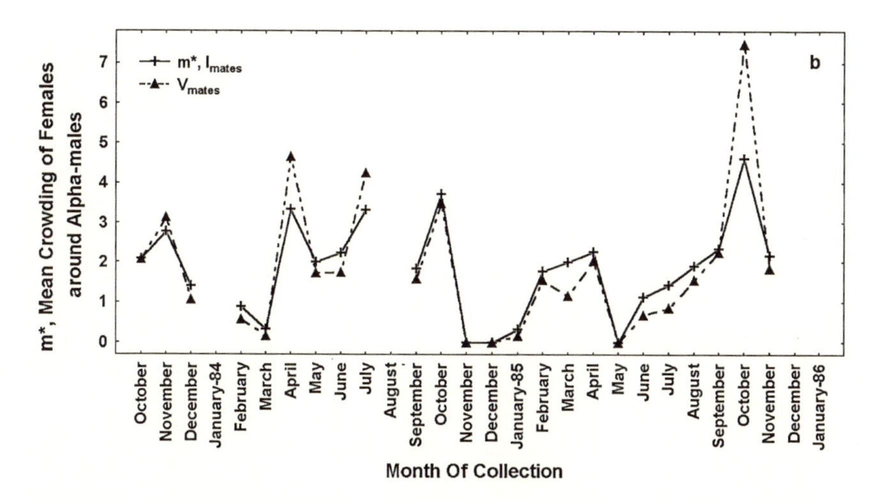

Figure 2.9. (a) Comparison of observed variation in the sex ratio R (= $N_{\text{females}}/N_{\text{males}}$) versus the operational sex ratio R_O (= OSR = $N_{\text{males}}/N_{\text{females}}$) in *Paracerceis sculpta* 1983–85. (b) Relationships between the mean spatial crowding of females m^*, the opportunity for sexual selection I_{mates}, and the variance in mate numbers among males in *P. sculpta*, 1983–85.

prediction from current OSR theory. (We will return to this point in chapter 12 in our discussion of the evolution of maternal sex-ratio biasing.)

We can now estimate the values of I_{harem} and I_{mates} directly from the distribution of males and females in spongocoels. The required populational information is summarized in table 2.2. The date of each collection is listed in column 1 of the table. We begin by identifying the fractions of mating and nonmating males. The number of α-males who defend spongocoels alone (column 3) divided by the total number of males in the collection (column 4) equals the fraction of nonmating males, p_0 (column 6, table 2.2). The number of α-males who defend spongocoels containing females (column 2) divided by the total number of males (column 4) equals the fraction of mating males, $1 - p_0$ (column 7).

The mean and variance in harem size for *all* of the α-males in each sample, mating and nonmating (H_α and $V_{H\alpha}$, respectively, columns 8 and 9, table 2.2) are equal to the statistical mean and variance in the number of females, distributed across all spongocoels containing α-males, mated and unmated. Note that H_α (column 4) is equal to the total number of females, $N_{females}$

Table 2.2
Calculation of I_{mates} from the distribution of *P. sculpta* males and females in *L. losangelensis* spongocoels, 1983–85.

Month 1	*Mated* 2	*Unmated* 3	H_α 8	$V_{H\alpha}$ 9	$H_{m\alpha}$ 10	V_{Hma} 11	I_{harem} 12	V_{mates} 13	I_{mates} 14	$I_{mates(adj)}$ 15
Oct-83	11	3	1.64	2.37	2.09	2.08	0.48	2.37	0.88	0.88
Nov-83	9	5	1.64	3.52	2.56	3.14	0.48	3.52	0.54	1.30
Dec-83	11	6	1.18	1.44	1.82	1.06	0.32	1.44	0.44	1.04
Feb-84	6	6	0.75	0.85	1.50	0.58	0.26	0.85	0.38	1.52
Mar-84	5	1	1.00	0.33	1.20	0.16	0.11	0.33	0.23	0.33
Apr-84	33	10	1.84	4.60	2.39	4.66	0.81	4.60	0.80	1.36
May-84	9	6	1.33	2.22	2.22	1.73	0.35	2.22	0.45	1.25
Jun-84	16	4	2.05	2.45	2.56	1.75	0.27	2.45	0.37	0.58
Jul-84	17	5	2.18	4.69	2.82	4.26	0.53	4.69	0.59	0.99
Sep-84	12	10	1.14	1.94	2.08	1.58	0.36	1.94	0.45	1.50
Oct-84	15	3	3.17	4.92	3.80	3.49	0.24	4.92	0.34	0.49
Nov-84	2	4	0.33	0.22	1.00	0.00	0.00	0.22	0.22	2.00
Dec-84	2	3	0.40	0.24	1.00	0.00	0.00	0.24	0.24	1.50
Jan-85	10	11	0.57	0.44	1.20	0.16	0.11	0.44	0.30	1.33
Feb-85	9	14	0.78	1.56	2.00	1.56	0.39	1.56	0.39	2.55
Mar-85	11	5	1.75	2.19	2.55	1.16	0.18	2.19	0.34	0.71
Apr-85	9	6	1.47	2.65	2.44	2.03	0.34	2.65	0.44	1.23
May-85	1	2	0.33	0.22	1.00	0.00	0.00	0.22	0.22	2.00
Jun-85	4	2	1.17	1.14	1.75	0.69	0.22	1.14	0.37	0.84
Jul-85	7	3	1.40	1.44	2.00	0.86	0.21	1.44	0.36	0.73
Aug-85	10	5	1.47	2.12	2.20	1.56	0.32	2.12	0.44	0.98
Sep-85	7	4	1.55	2.79	2.43	2.25	0.38	2.79	0.47	1.17
Oct-85	13	1	3.21	7.74	3.46	7.48	0.62	7.74	0.65	0.75
Nov-85	10	2	2.00	2.33	2.40	1.84	0.32	2.33	0.41	0.58
Overall	239	121	1.560	3.21	2.36	2.95	0.53	3.20	0.57	1.30

(table 2.1, column 1) divided by the total number of α-males, N_α (sum of columns 2 and 3). Thus, as explained above, the average mating success of *all* α-males, H_α is equal to R the population sex ratio for each sample. The average and variance in harem size among *mated* α-males in each sample ($H_{m\alpha}$ and $V_{Hm\alpha}$, respectively, columns 6 and 7, table 2.2 and colimns 3 and 4 in table 2.1) are equal to the statistical average and variance in the number of females distributed across spongocoels containing the mated α-males.

Since the opportunity for selection is equal to the variance in fitness divided by the squared mean fitness (chapter 1), we can calculate the opportunity for sexual selection on mated α-males, I_{harem}, as

$$I_{\text{harem}} = V_{Hm\alpha}/[H_{m\alpha}]^2. \qquad [2.22]$$

Substituting the values for the October 1983 sample from table 2.2, we have

$$I_{\text{harem}} = (2.09)/(2.08)^2 = 0.48. \qquad [2.23]$$

In this way we obtain estimates of I_{harem} for each collection in the 1983–85 sample (column 8, table 2.2).

Solitary males in *P. sculpta* are easy to collect and count because unmated α-males remain in spongocoels waiting for females to arrive. However, unmated males in other species are often more difficult to enumerate because they may disperse from breeding sites. When nonmating males are absent from a sample, p_0, and consequently I_{mates}, cannot be directly calculated (eq. [2.13]). However, in such situations, estimates of I_{harem} (or in this case $I_{Hm\alpha}$) can still provide a direct estimate of the intensity of sexual selection, albeit on mated males alone. As shown in the simulation above, the opportunity for sexual selection on mated males, I_{harem}, is expected to correlate positively with the spatial patchiness of females, P. This expectation is met when we plot these values for the 1983–85 samples of spongocoels containing only α-males and females (fig. 2.10). Note that the values of I_{harem} are consistently less than the values for I_{mates} (table 2.2). Thus, estimates of the opportunity for sexual selection calculated using only the mating males in the population, are conservative.

If we were unable to count the unmated α-males in this sample we might consider our analysis complete. However, since we *can* count these individuals, we can illustrate several other relationships with this data set. Recall that the variance in male mating success has two components, (1) the variance in mating success that exists within the group of males that successfully defend a spongocoel and mate with the females aggregating there, and (2) the variance in the average mating success that exists between the mating and nonmating males. As shown in eq. [2.11], these two components of variance sum to equal V_{mates}, which in this example can be expressed as

$$V_{\text{mates}} = (1 - p_0)\, V_{Hm\alpha} + H^2_{m\alpha}\, (p_0)\, (1 - p_0). \qquad [2.24]$$

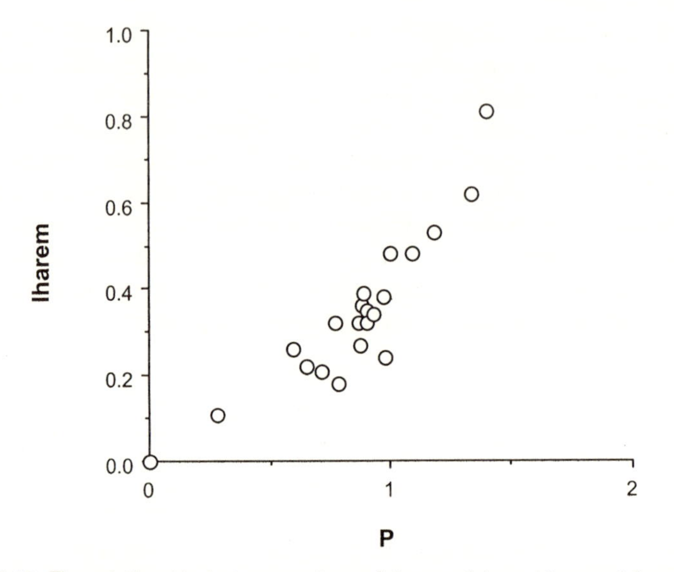

Figure 2.10. The relationship between values of the spatial patchiness of females, *P*, estimated from the distribution of females with α-males in spongocoels in *Paracerceis sculpta*, and the opportunity for sexual selection on mated α-males, I_{harem}, calculated from the mean and variance in the harem sizes of mated males ($r = 0.92$; $F_{[1,23]} = 127.13$, $P = 0.0001$).

Substituting these values for the October 1983 sample from table 2.2, we have

$$V_{mates} = (0.79)\ 2.08 + (2.09)^2\ (0.21)\ (0.79) = 2.37. \qquad [2.25]$$

Estimates of V_{mates} for each of the remaining collections in the 1983–85 sample can be obtained as well (column 9, table 2.2).

As we explained above, when the variance in mating success among males depends on the distribution of females on resource patches, increases in the mean crowding of females on patches, m^*, are expected to lead to equivalent increases in the variance in mating success among males, V_{mates}. In our simulation, the relationship between m^* and V_{mates} was 1:1 (fig. 2.4b). In our natural population we find a relationship similar to but not identical to this theoretical expectation (fig. 2.11). For a given value of m^*, the value V_{mates} is usually larger, especially for extreme values of m^*. The source of this deviation is the population sex ratio. In our simulation, *R* was fixed at 1. However, in most natural populations, *R* is more variable, and in the *P. sculpta* population, *R* was usually greater than 1 (table 2.2; Shuster 1986; Shuster and Wade 1991a,b). That is, the population sex ratio tended to be female biased. The qualitative result of a female-biased sex ratio is that more females are distributed among the mating males, which increases the variance in mate numbers relative to estimates of the mean crowding. The quan-

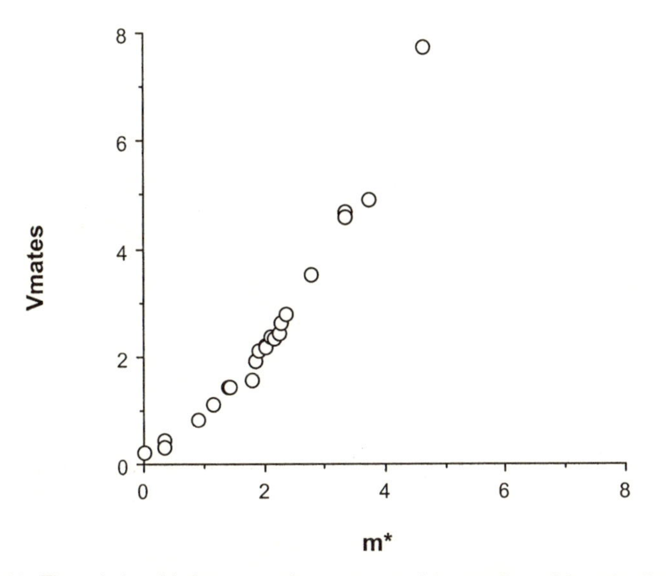

Figure 2.11. The relationship between the mean spatial crowding of females in spongocoels containing α-males and females, m^*, and the variance in mate numbers among mating and nonmating males, V_{mates}, in *Paracerceis sculpta*. The correlation is significant ($r = 0.84$; $F_{[1,23]} = 54.08$, $P = 0.0001$) but unlike in the simulation results in which $R = 1$ (fig. 2.4b) the relationship is not 1:1.

titative effects of deviations in sex ratio are most clearly seen when the relative effects of R and I_{mates} are compared.

The opportunity for sexual selection, I_{mates}, measures the intensity of sexual selection on all males, mating and nonmating. In eq. [2.13] we showed that this value can be expressed as $I_{mates} = ([V_{harem}]/[H]^2[1 - p_0]) + ([p_0]/[1 - p_0])$. Substituting the parameters calculated for mated α-males in this example, this equation becomes

$$I_{mates} = ([V_{Hm\alpha}]/[H_{m\alpha}]^2[1 - p_0]) + ([p_0]/[1 - p_0]), \qquad [2.26]$$

where $V_{harem} = V_{Hm\alpha}$, the variance in harem size among mating α-males, and $H = H_{m\alpha}$, the average harem size among mating α-males. Substituting these values from table 2.2, we find for the October 1983 sample that

$$I_{mates} = ([2.08]/[2.09]^2[0.79]) + ([0.21]/[0.79]) = 0.88, \qquad [2.27]$$

and we can now estimate I_{mates} for each collection in the 1983–85 sample (column 14, table 2.2).

There is a significant positive relationship between I_{mates} and m^* in our natural population ($r^2 = 0.84$, $F_{[1,23]} = 54.08$, $P = 0.0001$; fig. 2.12a). This relationship is identical to the relationship between I_{mates} and HP (fig. 2.12b), as it should be since $m = H$ and $P = m^*/m$. Although these relationships are highly significant, they are not 1:1, and the explanation for this

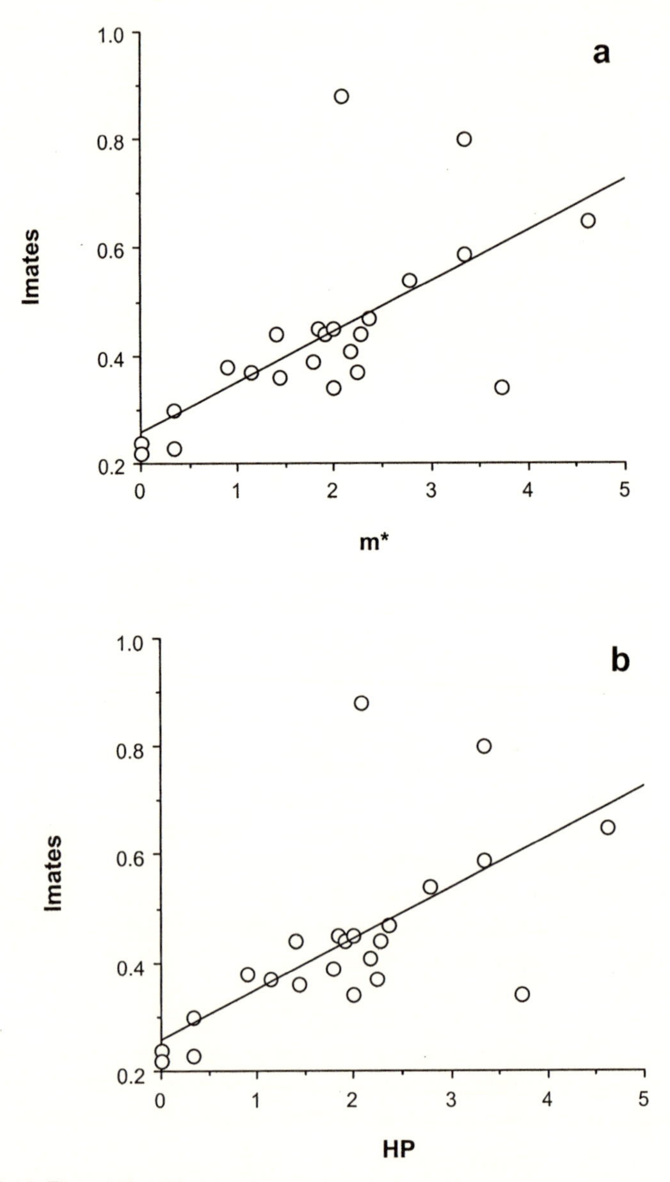

Figure 2.12. The relationship between (a) the mean spatial crowding, m^*, of females in spongocoels with α-males in *Paracerceis sculpta*, and (b) the product of harem size, *H*, and the patchiness of females in spongocoels, *P*, with estimated values for the opportunity for sexual selection, I_{mates}. These correlations are identical and significant (r^2 = 0.84; $F_{[1,23]} = 54.08$, $P = 0.0001$).

deviation is the same as that for the observed deviation between V_{mates} and m^*. We have shown (eqs. [2.15a, b]) that the opportunity for sexual selection, adjusted for the effects of a biased sex ratio, is $I_{\text{mates(adj)}} = (1/R)(m^* + 1 - R)$. Substituting the values for m^* and R estimated for the October 1983 sample, this expression becomes

$$I_{\text{mates(adj)}} = (1/1.64)(2.09 + 1 - 1.64) = 0.88. \qquad [2.28]$$

This value is identical to I_{mates} (column 14) for the October 1983 collection. However, other collections show how different I_{mates} and $I_{\text{mates(adj)}}$ can be (column 15). Following eqs. [2.16a, b], we predicted that the opportunity for sexual selection, adjusted for a biased sex ratio, $I_{\text{mates(adj)}}$, will be *diminished* when the sex ratio is female biased, that is, when R is greater than 1, or alternatively, when the operational sex ratio R_O is less than 1. When the sex ratio is male biased, so that R is less than 1, that is, R_O is greater than 1, the adjusted opportunity for sexual selection will be *augmented*. From these predictions, we expected a negative correlation between $I_{\text{mates(adj)}}$ and R, as well as a positive correlation between $I_{\text{mates(adj)}}$ and R_O. These expectations were met in our natural population (figs. 2.13a,b). Thus, in theory as well as in nature, the following relationship is clear. The magnitude of the sex difference in the opportunity for selection depends on the sex ratio, *as well as* on differences in mate numbers among males.

Chapter Summary

The Emlen and Oring hypothesis and their ecological classification of animal mating systems are founded upon the observation that the ecology of female reproduction determines the degree to which males are successful in competition for scarce reproductive resources. Although the "*operational sex ratio*" or OSR is one way to quantify the intensity of male-male reproductive competition in relation to the resource availability in the form of sexually receptive females, the OSR does not capture the sex difference in the strength of selection. In this chapter, we quantified the spatial distribution of sexually receptive females in relation to the sex difference in the strength of selection. We showed how Lloyd's (1967) ecological measure of density-dependent competition, *mean crowding*, can be used to directly relate the spatial clustering of receptive females at resources to the strength of sexual selection, I_{mates}. With our formulation, it becomes clear that the OSR captures some, but not all, of the sex-difference in the strength of selection (cf. eq. [2.21]). The OSR is the *sole* determinant of sex difference in selection only when there is no variance among males in mate numbers, i.e., V_{mates} is zero. When the success of mating males is very high (so that H is large), V_{mates} is approximately equal to $(H - 1)$. That is, it too is large.

Using a hypothetical example, we illustrated the algebraic relationships

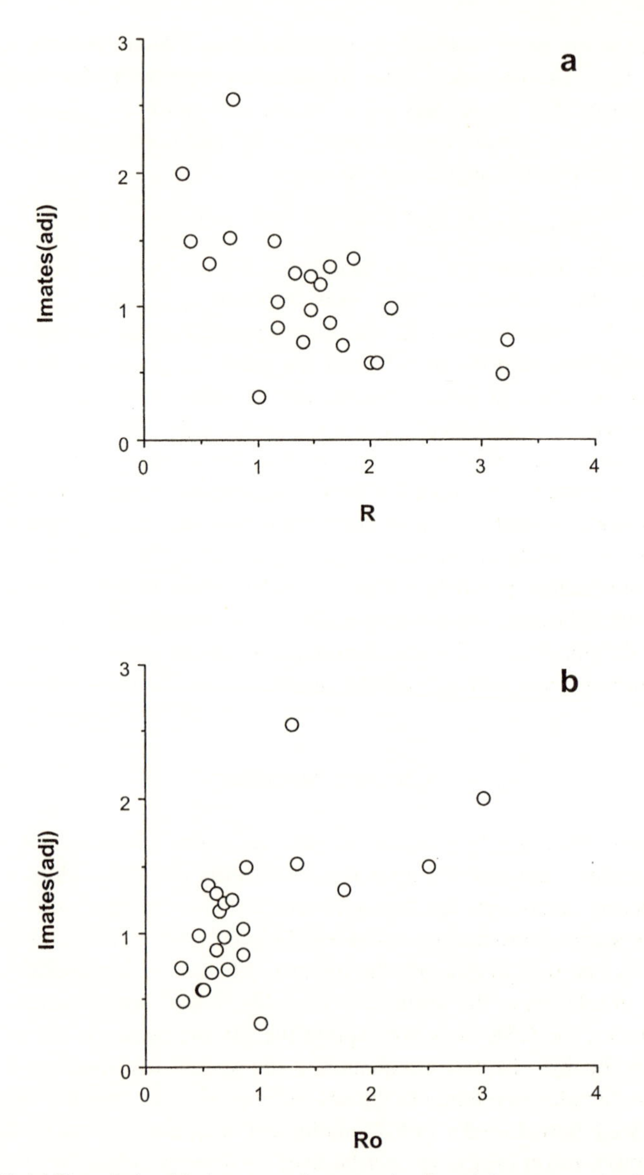

Figure 2.13. (a) The relationship between the sex ratio, R, and the opportunity for sexual selection, calculated when the sex ratio is not equal to 1, $I_{mates(adj)}$ (r^2 = 0.25, $F_{[1,23]}$ = 7.48, P = 0.012); note that the relationship is negative, indicating that small values of $I_{mates(adj)}$ are associated with large values of R (i.e., female biased sex ratios), that is, female biased sex ratios tend to diminish the opportunity for sexual selection; (b) the relationship between the operational sex ratio, R_o, and the opportunity for sexual selection, calculated when the operational sex ratio is not equal to 1, $I_{mates(adj)}$ (r^2 = 0.34, $F_{[1,23]}$ = 11.15, P = 0.003); note that large values of $I_{mates(adj)}$ tend to be associated with large values of R_o (i.e., male biased sex ratios), that is, male biased sex ratios tend to augment the opportunity for sexual selection.

among resource patch number, M, the average number of females per patch, m, the mean crowding of females on patches, m^*, and the patchiness of thc female spatial distribution. Surprisingly, the distribution of females is most patchy (i.e., P is greatest) at intermediate values of M. In addition, the opportunity for sexual selection on mated males, I_{harem}, has a positive relationship to the spatial patchiness of females, P. We moved from this hypothetical example and applied these concepts to quantify sexual selection in the marine isopod *P. sculpta*, a species whose females actively aggregate at breeding resources.

3 The Phenology of Sexual Selection

In the preceding chapter, we quantified the way in which the *spatial* distribution of females on resources influences the opportunity for sexual selection, I_{mates}. That is, we identified the degree to which sexually receptive females are *spatially* aggregated, and the *spatial scale* at which this clustering occurs. However, Emlen and Oring (1977; see also Wickler and Seibt 1981; Shuster 1991b; Duvall et al. 1995) emphasized that females may vary in the *temporal* distribution of their sexual receptivity as well. In this chapter, we show that the degree of temporal synchrony in female sexual receptivity affects both sexual selection and the sex difference in the strength of selection. These temporal effects can be understood within the same theoretical framework as the spatial effects examined in the previous chapter (Lloyd 1967; Wade 1995). We identify parameters analogous to those describing the spatial clustering and patchiness of receptive females. In particular, we define the degree to which receptive females are aggregated in *time*, or more appropriately, the *breeding synchrony*, as a critical parameter influencing the strength of sexual selection. In this way, we *quantitatively* relate the concepts of Emlen and Oring (1977) to measures of the strength of sexual selection and to the selective pressures favoring the evolution of sex dimorphism.

The Mean Temporal Crowding of Sexually Receptive Females

We begin by assuming that, on average, each reproductive female becomes sexually receptive for a species-specific length of time, as is true, for example, in many crustaceans (Wickler and Seibt 1981; Shuster 1981b, 1991a; Shuster and Caldwell 1989; Morgan and Christy 1994; Jormalainen and Shuster 1999). If we define r_i as the duration of sexual receptivity for the ith female (fig. 3.1), then the average duration of female receptivity equals

$$r = \Sigma\, r_i/N_{females}, \quad [3.1]$$

and its variance equals

$$V_{receptivity} = \{\Sigma\, (r_i^2)\}/N_{females} - (r)^2. \quad [3.2]$$

We also define the length of the breeding season, T, as the period from the beginning of receptivity of the earliest receptive female to the end of receptivity of the last receptive female. Hence, r_i must be less than or equal to T for each female.

The minimum temporal crowding of receptive females, that is, the *maximum breeding asynchrony*, occurs when females become receptive sequentially, without overlap, so that T is greater than or equal to (rN_{females}). Under maximum breeding asynchrony, we can think of the breeding season as a period of time, subdivided into T/r intervals. In this case, T/r is equal to or greater than the number of breeding females, N_{females}. This means that all males potentially compete for each female when she becomes receptive, or, in Emlen and Oring's terms, the operational sex ratio R_O is maximized at every interval and equals N_{males}.

The maximum temporal crowding of receptive females, or the *maximum breeding synchrony*, occurs when all females are receptive simultaneously and for equal durations. Under maximum breeding synchrony, we can think of the breeding season as compressed into a single interval of duration r, within which all females are sexually receptive. In this case, T equals r and $V_{\text{receptivity}}$ is zero. This means that all females become available for all males simultaneously. This situation represents less intense competition than the former situation because R_O is minimized within the single interval in which all females are receptive. In this case, R_O equals $N_{\text{males}}/N_{\text{females}}$. Remember that R_O is equal to the inverse of the sex ratio R, that is, $R_O = 1/R$ (see chapter 1).

If we assume that there is no time during the breeding season without at least one receptive female, then

$$r < T < (rN_{\text{females}}). \quad [3.3]$$

When we divide the breeding season into intervals of duration r, then the number of intervals, T/r, is bounded below by 1 and bounded above by the number of breeding females, N_{females}. This expression identifies the duration of female receptivity with respect to the duration of the breeding season. However, to understand the effects of variable synchrony in female receptivity, we must consider additional parameters.

Consider T as equivalent to the number of time intervals containing at least one receptive female. If the number of breeding females present within the same interval varies, we can express the number of females receptive in the ith interval as t_i. Thus, we can define the average number of females receptive over all i intervals, t, as

$$t = (\Sigma\, t_i)/T, \quad [3.4]$$

with variance about t equal to

$$V_t = [(\Sigma\, t_i)^2/T] - (t)^2. \quad [3.5]$$

Since T and t are defined in terms of the duration of female receptivity, r, they incorporate the species-specific patterns of selection that have shaped r (see below). The average number of receptive females, t, at any time during the breeding season equals the total duration of female receptivity ($rN_{females}$) divided by the length of the breeding season, T.

Given that the number of receptive females in the ith interval is equal to the number of females breeding at that time, t_i, or $N_{females}(t)$, the mean temporal crowding of receptive females, t^*, is defined as

$$t^* = (\Sigma N_{females}[t]\{N_{females}[t] - 1\})/(\Sigma N_{females}[t]), \qquad [3.6a]$$

or

$$t^* = [\Sigma t_i (t_i - 1)]/(\Sigma t_i) \qquad [3.6b]$$

or, more simply,

$$t^* = t + ([V_t/t] - 1). \qquad [3.6c]$$

Remember that V_t is the temporal variance in the number of receptive females. It measures the variance in the number of receptive females across intervals. The degree to which females are "aggregated" in their sexual receptivity over intervals of the breeding season is measured by t^*. It quantifies the number of other receptive females that the average female experiences when she herself is sexually receptive. When female receptivity is Poisson distributed across intervals of the breeding season, the variance in the number of receptive females per interval, V_t, equals t, and the average number of receptive females per interval, t, equals the mean temporal crowding, t^*. When female receptivity is nonrandomly distributed over the breeding season, however, the number of other receptive females the average receptive female experiences within her interval will be greater than or less than the mean.

This approach to temporal variation in female receptivity is conceptually similar to that used earlier to describe the spatial clustering of females (chapter 2). Here, our approach identifies a *slice of time* (i.e., a temporal patch) in which one or more females may be sexually receptive. It also suggests a convenient means for identifying the scale on which to measure the temporal synchrony of female receptivity, because the average duration of female receptivity describes the length of time available for a male to associate with a female, successfully transfer sperm, and assure the paternity of her brood.

An alternative means for identifying the appropriate temporal scale on which to measure female receptivity is analogous to the one based upon spatial quadrats, described at the end of the previous chapter. When the entire breeding season is divided into progressively smaller temporal intervals, the number of receptive females appearing in each interval is the fundamental unit of observation and its mean, variance, and covariance across intervals can be calculated. As described in chapter 2, this partitioning of the breeding season identifies periods when receptive females are abundant as well as when they are rare. In many circumstances, the variance in the number of receptive females among intervals will increase as the intervals become smaller. This is true because intervals longer than the average duration of female receptivity may subsume periods of time in which no females are receptive.

As interval durations become shorter, some intervals may contain more than one female, but other intervals will contain few or no females. When an interval of such duration is reached that females with overlapping periods of receptivity are separated into different intervals, the variance in the number of receptive females per interval will decrease, and will continue to do so with increasingly shorter intervals. Shorter intervals will also cause a female's receptivity to be spread across more than one interval, so that individual females may be counted more than once in our calculation of V_t. We will deal with situations in which females become receptive for more than one interval during the breeding season below.

Thus, the appropriate temporal scale at which to measure the timing of female receptivity is the interval duration at which the variance in the number of receptive females per interval is large but there are few or no empty intervals. The existence of an interval length for which the mean and the variance in number of receptive females are equal identifies a temporal scale on which the synchrony of female receptivity is effectively random. For most species, measuring the average duration of female receptivity and dividing the breeding season into intervals equal to that duration will generate large variance in the number of receptive females per interval. This approach is easier than mapping the onset and duration of receptivity for all females within a population throughout the year.

Although large variance in the duration of female receptivity could affect estimation of interval duration, we expect within-species variation in the duration of female receptivity to be low. In many species, strong stabilizing selection is likely to shape the average duration of female receptivity. Specifically, females with extremely brief periods of sexual receptivity will seldom receive sufficient sperm to successfully breed. Similarly, females with extremely long periods of sexual receptivity will be repeatedly harassed by males and will suffer decreased fitness due to physical damage, increased predation, increased exposure to sexually transmitted disease, and/or decreased ability to provision themselves and their developing ova (review in Jormalainen and Shuster 1999). Under stabilizing selection, the population variance in female receptivity durations, V_r, is likely to remain small. Thus, r alone should provide a reliable estimate of sexual receptivity durations for a particular female population (exceptions to this situation, especially those arising from interactions of r and mate mating success, are discussed in chapters 4 and 8).

The Phenology of Female Sexual Receptivity

The phenology of female sexual receptivity can be described by the relationship between the mean and variance in the number of receptive females throughout the breeding season. When the variance in the number of receptive females per interval is large relative to the mean, then receptive females are temporally clumped and t^* is large. On the other hand, when the variance in the number of receptive females per interval is small relative to the mean, then receptive females are temporally *overdispersed* and t^* is small. When the variance equals the mean, females are randomly dispersed in time so that t^* equals t. Thus, t^* measures the temporal synchrony in female breeding phenology.

The temporal "patchiness" of the female breeding phenology, S, can expressed as the ratio of the mean temporal crowding of receptive females divided by the average number of females per interval, or

$$S = t^*/t. \quad [3.7]$$

Because t^* equals $\{t + [(V_t/t) - 1]\}$ (cf. eq. [3.4c]), S equals

$$S = 1 + [V_t/(t^2) - 1/t]. \quad [3.8]$$

Differently put, S equals the degree to which the average female's experience of other receptive females exceeds the population average. Thus, when females become receptive randomly throughout the breeding season (i.e., female receptivity is Poisson distributed across the breeding season), the value of S will approach 1. However, when females are physiologically "aggregated" in their receptivity, then S exceeds 1. Like P, which provides a measure of the degree to which the average female experiences crowding by other females on resources (Lloyd 1967; Wade 1995), S measures the degree to which the average female experiences sexually receptive conspecifics in relation to the population average.

When the temporal distribution of females determines the temporal distribution of males seeking mates, and when male mating success is dependent on the distribution of females in time, as per the Emlen and Oring (1977) hypothesis, temporal patchiness is expected to influence mating system evolution. When females are synchronous in their sexual receptivity, most males will be capable of locating and defending a mate. However, when females become temporally dispersed in their receptivities, some males will be capable of defending females and some males will not. Similar to the spatial distribution of females, this creates two components of variance in male mating success: (1) the variance in mating success among mating and nonmating males, and (2) the variance in mating success among males that successfully find and defend their mates. As explained in chapters 1 and 2 (Eqs. [1.16], [1.17], [2.9]–[2.13]), these components of V_{mates} can be expressed as follows. If p_0 is the fraction of unmated males, then $(1 - p_0)$ is the fraction of mated males. Moreover, if the ratio $(N_{females}/N_{males})$ equals the sex ratio R (eq. [1.15]), then the average harem size of mated males, H, can be expressed as $H = R/(1 - p_0)$ (eq. [2.9]).

As described in chapter 2, the first component of the variance in male mating success can be understood as that which exists between mating and nonmating males. This variance is expressed as $V_{between} = H^2(p_O)(1 - p_O)$ (cf. eq. [2.10]). The second component of variance is the average variance in male mating success within each of these two male categories, which equals $V_{within} = (p_O)(0) + (1 - p_O)V_{harem}$ (cf. eq. [2.11]). Because V_{mates} equals $(V_{within} + V_{between})$, by substitution, we have

$$V_{mates} = (1 - p_O)\, V_{harem} + H^2(p_O)\,(1 - p_O). \qquad [3.9]$$

And, because the average number of mates per male, R, equals $H(1 - p_O)$, dividing V_{mates} by R^2 yields

$$I_{mates} = [1/(1 - p_O)][(V_{harem})/(H^2)] + (p_O)/(1 - p_O). \qquad [3.10]$$

Equation [3.10] describes the opportunity for sexual selection on males arising from the distribution of receptive females *in time*. This expression looks identical to eq. [2.13], which describes the opportunity for sexual selection on males arising from the distribution of receptive females *in space*. These two expressions (eqs. [2.13] and [3.10]) are equivalent in form because they each include the average mating success and the variance in mating success arising from males who mate, as well as from males who do not mate at all.

However, these expressions can differ in value because temporal and spatial patchiness have different effects on variance in male mating success. Wade (1995) noted that the algebraic relationship between the mean spatial crowding of receptive females and the average number of mates per mating male is *proportional*. That is, when the spatial clumping of females is at maximum, one or a few males could conceivably defend and mate with all of the females in the population (figs. 2.1a and 2.2a). Conversely, the relationship between the temporal crowding of receptive females and the average number of mates per mating male is *reciprocal*. That is, as females become more synchronous in their sexual receptivity, the ability of one or a few males to mate with multiple females *decreases*, particularly if the duration of female receptivity is brief (figs. 3.1a and 3.2a).

The Relationships among T, t, t^*, and S

Just as we illustrated the measurement of spatial distributions of females on resource patches defended by males, we can better understand the parameters used to measure the temporal distribution of receptive females with a simple example. Consider again a population consisting of 20 males and 20 females ($N_{males} = N_{females} = 20$), but this time imagine that receptive females are distributed, not on resource patches, but instead across a maximum of 20 time intervals, whose duration is determined by the average duration of female receptivity, r (fig. 3.1a). We will assume that all males survive the entire breeding season and can mate within each time interval, provided

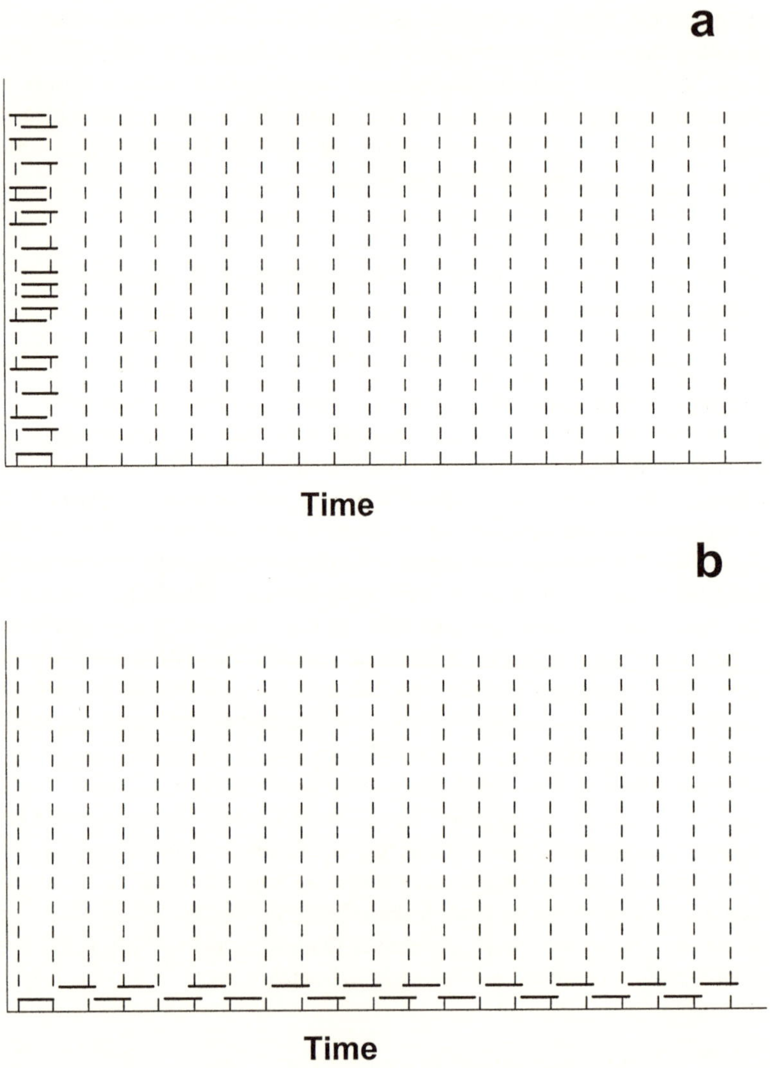

Time

b

Time

Figure 3.1. Model population for simulations of the temporal distribution of females; $N_{\text{males}} = N_{\text{females}} = 20$; $N_{\text{intervals}} = 20$; (a) all females are aggregated on a single interval; (b) females are distributed evenly among intervals such that each male mates.

receptive females are available. However, when receptive females are rare, males compete for females and only the largest males are able to successfully mate.

When mating, each male is assumed to attend a receptive female for the duration of her sexual receptivity. This prevents males from mating with other females who are receptive within the same time interval. We also assume that, while different females may become receptive simultaneously, female receptivities do not overlap adjacent intervals (we will relax these assumptions shortly). As we showed in chapters 1 and 2, we can identify the fraction of mating males as p_m, and the fraction of nonmating males as p_0. Thus, $p_m = (1 - p_0)$, and the number of mating males equals $(1 - p_0)(N_{males}) = N_{mating}$. Similarly, the number of nonmating males equals $(p_0)(N_{males}) = N_{nonmating}$. Using eqs. [3.1]–[3.4 above, we can also identify the values for T, t, t^*, and S for this population.

Let us first assume that *all* of the females in the population become receptive synchronously, that is, within the same interval t_i. In this case, the number of mating males, $(1 - p_0)(N_{males})$, equals the number of receptive females, and the number of nonmating males, $(p_0)(N_{males})$ equals 0. Because female mating behavior defines T, we can easily see that T, the number of intervals containing at least one female, equals 1, and the sum of all females in the only temporal patch (Σt_i) equals 20, which is the total number of females in the population ($= N_{females}$). Using eq. [3.4], we can see that t, the average number of females per interval, also equals 20 (20/1), and, since the variance in the number of females per interval is zero (eq. [3.5]), by eq. [3.6c], t^* equals 19. The patchiness of receptive females in time, S, is the ratio of t^* to t, and equals 19/20 or 0.95. Thus, when females are completely synchronous so that their receptivity is confined to a single interval, then S is less than but very near to 1. In the limit, as the number of females becomes large, S equals 1 for this baseline case. All other temporal distributions of female receptivity are evaluated relative to this baseline. In principle, one could argue that there are 20 intervals but only one of them has receptive females. However, intervals without receptive females are not observable unless they are flanked by intervals with receptive females. Our goal is a theory based on observable behaviors.

We can now progressively remove receptive females from this single time interval and distribute them among the other t_i intervals in the breeding season (fig. 3.1b). After each removal, we recalculate the above parameters, and continue this procedure until each interval contains a single female. We then plot the values and more clearly see the interrelationships between these parameters. In this example, we assume, as described above, that if only one female exists within a time interval the largest male will mate with her. Since, in this first hypothetical example, all males survive the entire breeding season, when females are maximally dispersed in time, a single large male could mate with each of the females in the population.

Figure 3.2a shows that as T (the number of intervals containing at least one female) *increases*, the number of mating males $[(1 - p_0)(N_{males}) = N_{mating}]$ *decreases* in direct proportion. Differently put, when the temporal distribution of receptive females within the breeding season determines male mating success, the number of intervals containing receptive females is *inversely proportional* to the number of mating males. This is the opposite of

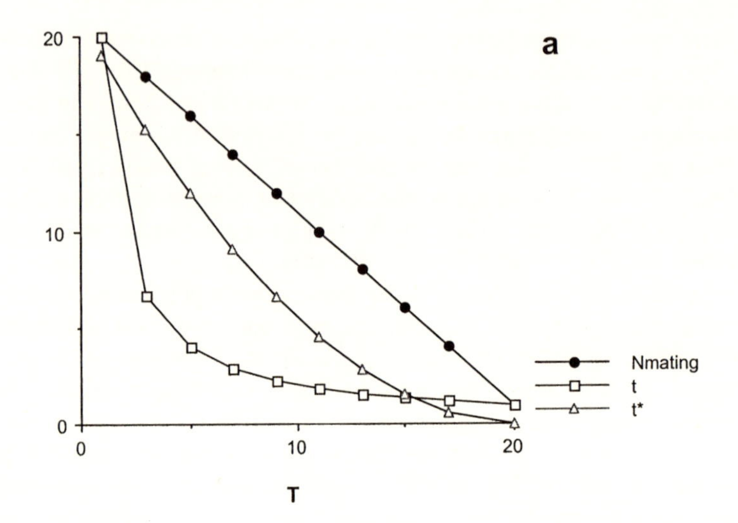

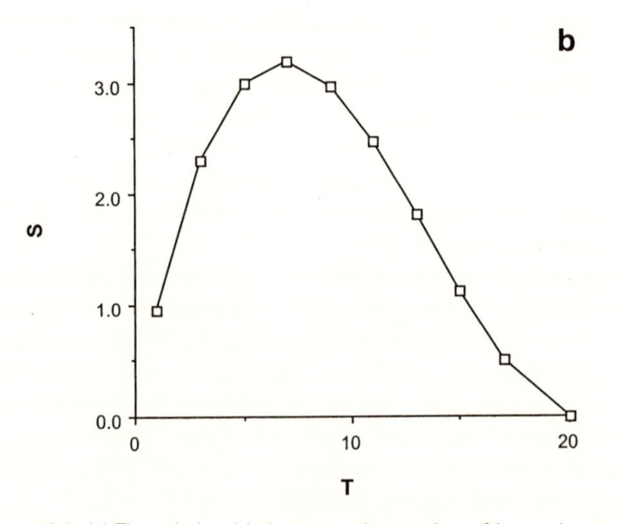

Figure 3.2. (a) The relationship between the number of intervals containing at least one receptive female and the number of mating males (N_{mating}), the average number of females per interval, *t*, and the mean temporal crowding of females wtihin intervals, t^*; (b) the relationship between the number of intervals containing at least one receptive female, *T*, and the temporal patchiness of females, *S*.

the relationship between the number of spatial patches containing receptive females and the number of mating males that is illustrated in fig. 2.2a.

Increasing the number of patches containing at least one female (T) causes both the average number of females per interval, t, as well as the mean crowding of females within intervals, t^*, to decrease (fig. 3.2a). Note that t^* decreases more rapidly than t at intermediate values of T. This occurs because estimates of t^* include the ratio of the variance to the average number of females per interval (eq. [3.6]). When some intervals contain females and some do not, this ratio is larger than when all of the females occupy the same time interval ($T = 1$) or when each interval contains one female ($T = 20$). Another way to think of this relationship is this: female receptivity is most temporally patchy at intermediate values of T (fig. 3.2b).

Consider the relationship between the number of intervals containing at least one female, T, and the average harem size, H. As the number of intervals containing receptive females increases, average harem size increases, slowly at first and then quite fast (fig. 3.3a). This function owes its shape to the relationship between male mating success and the temporal availability of females. As the temporal distribution of females becomes more even, fewer and fewer males are successful at mating because individual, competitively dominant males are able to mate with females in sequence. Conversely, as the average number of females per interval, t, increases, that is, as female receptivity becomes more synchronous, average harem size, H, drops off precipitously (fig. 3.3b). As the temporal distribution of females becomes more clumped, more males are successful at mating because individual males are less able to mate with more than one female in sequence.

When females are distributed on patches in space, V_{harem} *equals* V_m, because the variance in mating success among males depends on the distribution of females on resource patches. The more clumped females are in space, the more successful individual mating males can be (fig. 3.4a). However, when females are distributed within intervals in time, the more clumped females are, and the less successful individual males can be. When only one male is successful in mating, that is, when females become receptive asynchronously, or $T = 20$ (fig. 3.4b), the total variance in mating success among males (V_{mates}) is large, because most males obtain no mates at all ($p_0 = 0.95$). The variance among males decreases as the number of mating males increases (that is, as T decreases; fig. 3.4b), because females are more evenly distributed across males. Thus, an overall *inverse* relationship exists between the variance in the number of intervals containing at least one female, V_t, and the variance in mating success among males who mate, V_{harem} (fig. 3.4c). Consequently, as the mean temporal crowding of females, t^*, increases, the variance in mate numbers among males (V_{mates}) as well as the average harem size among successful males, H, both *decrease* (fig. 3.4d).

These relationships are the reverse of those occurring when females are distributed on patches in space (fig. 2.3a,b). In general, spatial aggregation

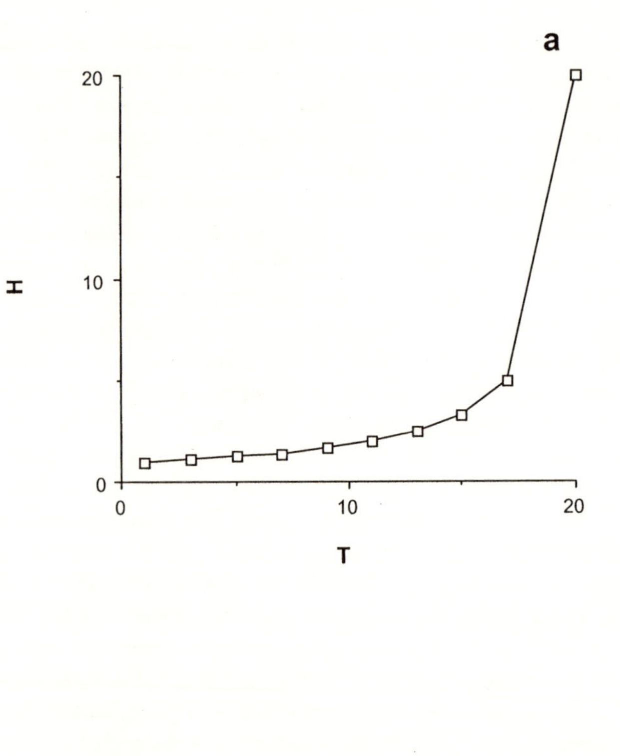

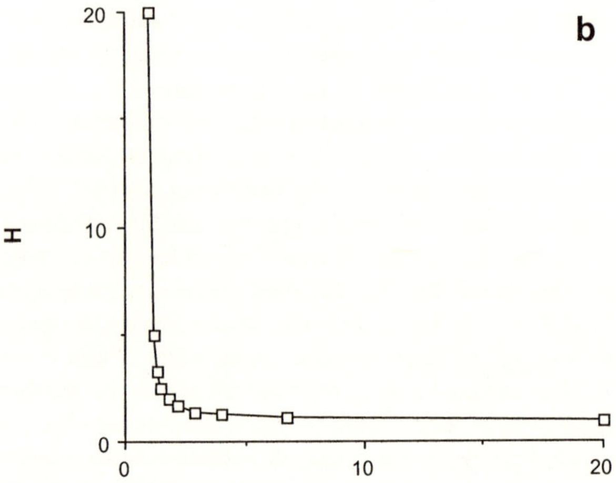

Figure 3.3. (a) The relationship between the number of intervals containing at least one female, *T*, and the average harem size of males, *H*; (b) the relationship between the average number of receptive females per interval, *t*, and the average harem size of males, *H*; (*continued*)

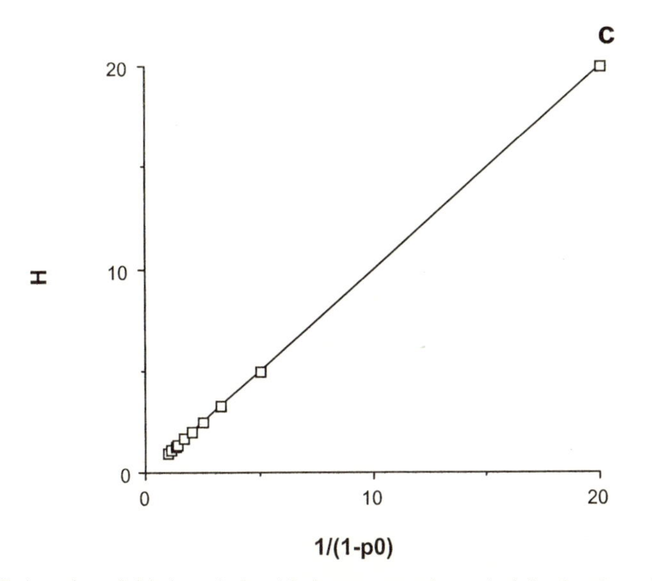

Figure 3.3. (*continued*) (c) the relationship between reciprocal of the fraction of mating males [$1/(1 - p_0)$] and the average harem size of males, *H*.

of females increases the mean and the variance in mating success among males (fig. 3.4a), whereas temporal aggregation of females diminishes the mean and the variance in mating success among males (fig. 3.4b). Note that, regardless of whether females are distributed within patches in space or distributed within intervals in time, when the sex ratio R equals 1, average harem size H equals the reciprocal of the number of mating males, $1/(1 - p_0)$; figs. 2.3c and 3.3c). Thus, as explained in chapter 1 (eq. [1.17]), the greater the number of mates secured by the average mating male, the larger must be the proportion of males with no mates at all, regardless of whether variation in male mating success is caused by female spatial or temporal distribution.

Recall that the opportunity for sexual selection, I_{mates}, is equal to the variance in mating success divided by the square of the average mating success (Crow 1958; Wade 1979; 1995). As explained in previous chapters, this ratio measures the relative ability of males in the population to acquire mates. As females become more temporally clumped, the ability of an individual male to defend and mate with more than one female within that interval decreases. Thus, an inverse relationship is expected between the temporal distribution of females, t^*, and the opportunity for sexual selection, I_{mates} (fig. 3.5a).

This relationship is more complex than the identity relationship between m^* and I_{mates} (fig. 3.5b). A similar, nonlinear relationship exists between the average number of females per interval, t, and the average harem size of mated males, H (fig. 3.3b). Again, this relationship is unlike the identity relationship that exists between H and m when females are distributed on patches in space (fig. 2.3b). Clearly, the opportunity for sexual selection on

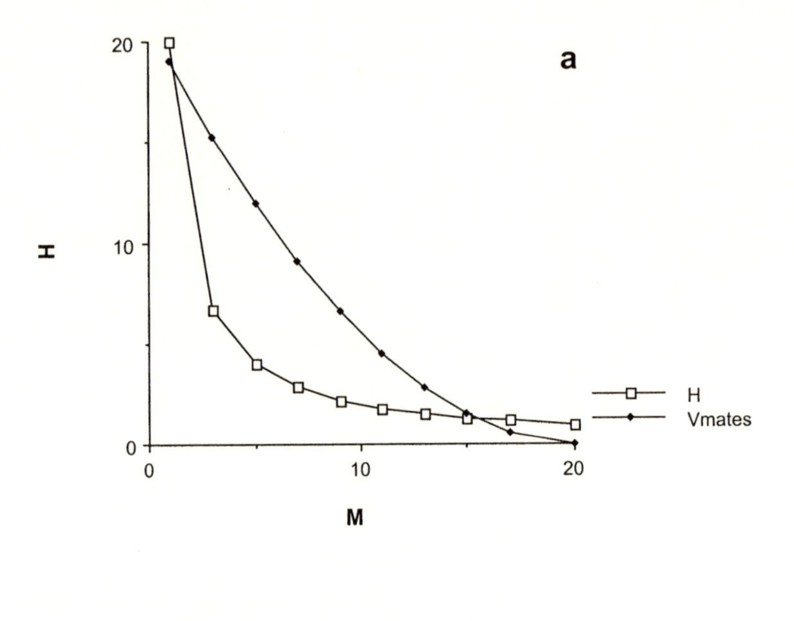

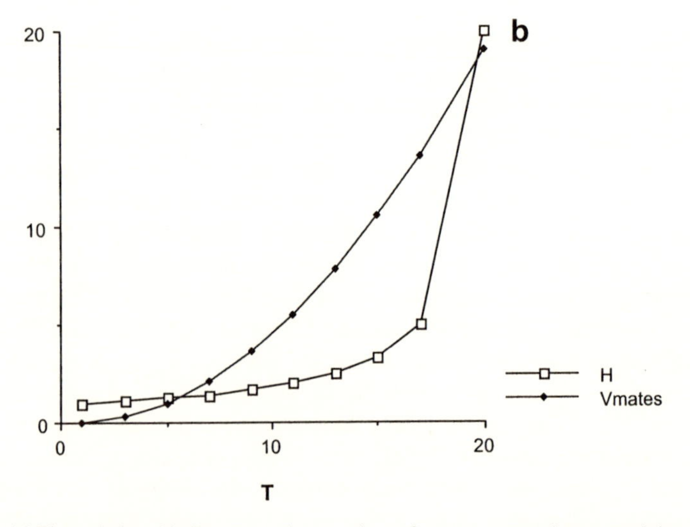

Figure 3.4. (a) The relationship between the number of resource patches containing at least one female, M, and the average and variance in mate numbers per male (H and V_{mates}, respectively. (b) The relationship between the number of time intervals containing at least one female, T, and the average and variance in mate numbers per male (H and V_{mates}, respectively). (*continued*)

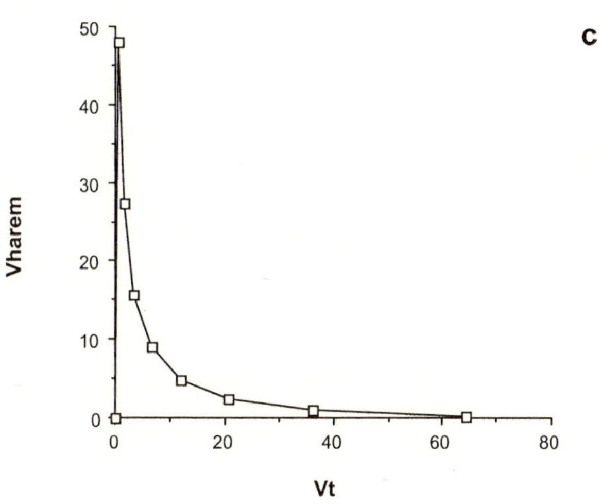

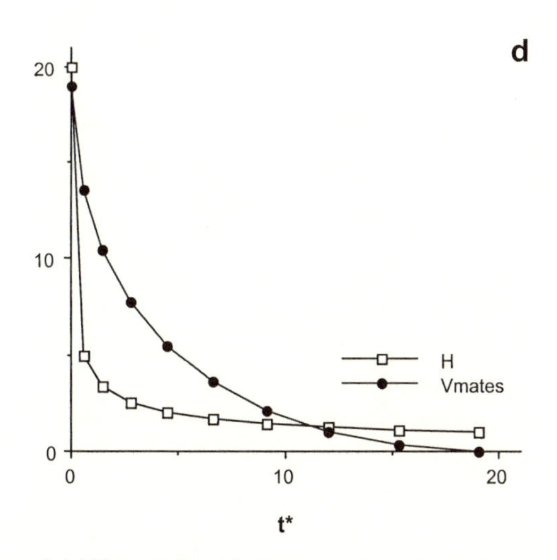

Figure 3.4. (*continued*) (c) The relationship between the variance in the number of females per interval, V_t, and the variance in harem size among mating males, I_{harem}. (d) The relationship between the mean temporal crowding of females, t^*, and the average and variance in mate numbers per male (H and V_{mates}, respectively).

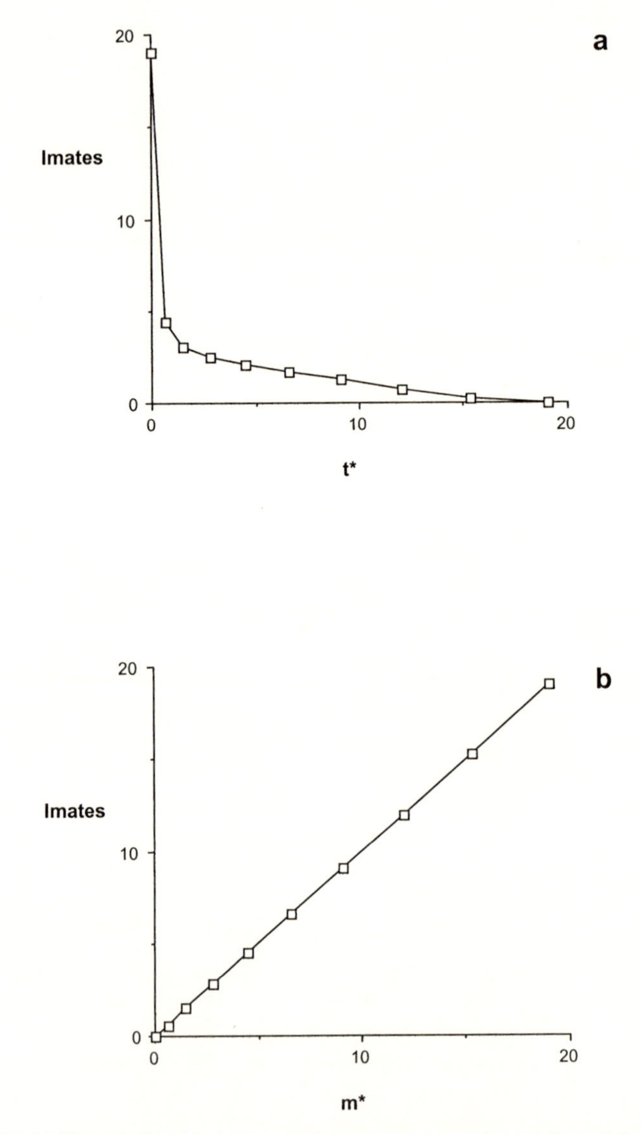

Figure 3.5. (a) The relationship between the mean temporal crowding of females (t^*) and the opportunity for sexual selection (I_{mates}) when $N_{males} = N_{females} = N_{patches} = 20$; (b) the relationship between the mean spatial crowding of females, m^* and the opportunity for sexual selection (I_{mates}) when $N_{males} = N_{females} = N_{patches} = 20$.

mated males, I_{harem}, is expected to decrease as the temporal distribution of females becomes increasingly patchy. However, the nonlinear relationships between V_t and V_{harem}, and between t and H, prevent the relationship between S and I_{harem} (fig. 3.6) from having as much predictive value as the relationship between P and I_{harem} (fig. 2.9). Thus, while we have given quan-

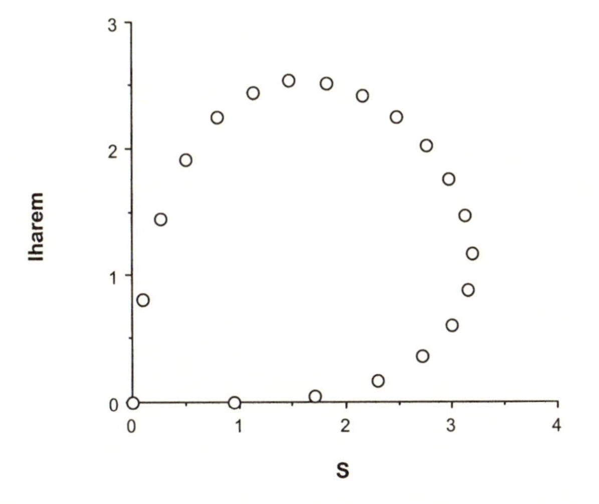

Figure 3.6. The relationship between the temporal patchiness of sexually receptive females, S, and the opportunity for sexual selection on mating males, I_{harem}; as the temporal patchiness of receptive females increases, I_{harem} increases, but the relationship is not significant ($F_{[1,19]} = 0.024$, $P = 0.88$, $r^2 = 0.036$).

titative expression to the relationship between the temporal distribution of females and the opportunity for sexual selection (fig. 3.5a), empirical estimates of I_{harem} and I_{mates} are not as easily obtained from field data of the temporal distribution of females.

The nonlinearity of the relationship between t^* and I_{mates} also complicates resolution of the specific effects of sex ratio on the sex difference in the opportunity for selection, arising from the temporal distribution of females. Consider the effect of the temporal synchrony in female receptivity on the intensity of sexual selection. This is another situation in which we believe the OSR fails to describe the actual intensity of sexual selection. Let female reproductive synchrony be so high that the denominator of R_O includes all mature females, $N_{females}$. In this case, R_O equals ($N_{males}/N_{females}$). At the other extreme of female synchrony, consider a species with the same total numbers of males and females but where only one female is receptive to mating at a time. Here, there are numerous temporally distributed bouts of reproduction, each with its own OSR. For every reproductive bout, the OSR denominator is 1 because only a single female is receptive at a time. Hence, the R_O at each bout is much higher and equals ($N_{males}/1$). The maximum intensity of sexual selection in this latter mating system, with its many bouts of male-male mating competition, each with very high OSR, will depend on which males mate successfully in each reproductive bout. If the scramble for mates results in a different male winning at each bout, then the mating system amounts to *scramble monogamy* and sexual selection will be weak (see mating system classifications in chapters 6 and 9). However, if the same few males win repeatedly, the system could be called *scramble polygyny* (or *scramble polygamy* or *scramble polygynandry* depending on female mate numbers; chapter 6) and the variance among males in mating success, as well as the strength of sexual selection, will be much higher.

Thus, *the covariance across temporal intervals in male mating success is critical.* When this covariance is low, then the intense mate competition among males within a time interval (indicated by high OSR) is diminished over the entire breeding season, T. If this covariance is high, then the intense reproductive competition within intervals becomes amplified over the breeding season. (We discuss this point and illustrate it with an example below.) The situation becomes more complex when the spatial distribution of females is also variable. For this reason, we must simultaneously consider the effects of the spatial and temporal distributions of females to identify not only how interactions between these two factors influence male mating success, but also the effects of spatially and temporally variable sex ratios.

Combining the Ecology and Phenology of Female Receptivity

Let us first consider the combined effects of the spatial and temporal distributions of receptive females on the opportunity for sexual selection, I_{mates}. This interaction is most easily understood with an example. Consider again a population consisting of 20 males and 20 females, as described above. In this situation, as in chapter 2, imagine that females settle on male territories to breed. However, also imagine that females settle on male territories during a breeding season that is divided into 20 time intervals, as described earlier in this chapter. Let us begin by assuming that females are maximally dispersed in space as well as maximally dispersed in time. That is, $m^* = 0$ and $t^* = 0$. These values are obtained using the same calculations described in the examples shown in chapter 2 and in this chapter.

Let us now consider four levels of increasing breeding synchrony such that $t^* = 0,0.5,1,3,19$ (table 3.1). These five total values for t^* are equivalent to situations in which females are distributed within $T = 20, 15, 10, 5$, and 1 intervals. Recall from fig. 3.2 that the relationship between the number of intervals containing females, T, and the number of mating males, N_{mating} ($= N_{males}[1 - p_0]$) is proportional and *negative*. That is, values of $t^* = 0$, 0.5, 1, 3, and 19 correspond to $N_{mating} = 1, 5, 10, 15$, and 20. Onto these levels of breeding synchrony, let us now map variable levels of spatial patchiness for receptive females, again over five levels. Thus, as in chapter 2, by moving females among spatial patches defended by males, we obtain $m^* = 0, 0.5, 1, 3$, and 19. These values for m^* are equivalent to situations in which females are distributed over 20, 15, 10, 5, and 1 patches. Recall from fig. 2.2 that the relationship between the number of patches containing females, M, and the number of mating males, N_{mating}, is proportional and

Table 3.1

Values for m^*_{observed} (I_{mates}) generated from the combined spatial and temporal distributions of 20 females and 20 males; each value for M, the number of females per spatial patch, generates a value for m^*, the mean spatial crowding of females, shown directly below the row of M_i values; each value of T, the number of females per temporal patch, generates a value for t^*, the mean temporal crowding of females, shown in the column to the right of the T_i values; the intersection of each value of m^* and t^* is the value for I_{mates}.

		$M \rightarrow$ 1	5	10	15	20
T	t^*	$m^* \rightarrow$ 0.0	0.5	1.0	3.0	19.0
20	0.0	0.0	0.5	1.0	3.0	19.0
15	0.5	0.0	0.5	1.0	3.0	11.5
10	1.0	0.0	0.5	1.0	3.0	9.0
5	3.0	0.0	0.5	1.0	3.0	4.0
1	19.0	0.0	0.0	0.0	0.0	0.0

positive. That is, values of $m^* = 0, 0.5, 1, 3$, and 19 correspond to $N_{\text{mating}} = 20, 15, 10, 5$, and 1.

Note that to accomplish these rearrangements in some cases it is necessary to place more than one female into the same male's patch, either within the same time interval or within different time intervals. We can address these situations by using the same rules for male mating success as described in the example above. Males are assumed to survive the entire breeding season. Thus, when females arrive on a male's patch in sequence, he mates with each of these females. However, we also assume, as is likely in nature, that a male cannot mate with two females that are present on the male's patch within the same time interval. When more than one receptive female exists in the same time interval, the females are assumed either to redistribute themselves so that there is only one female per patch, or, as is more likely, other males are assumed to invade that patch and functionally subdivide it. Paternity analyses are the best empirical method for determining the relationship between the apparent temporal and spatial distribution of mates per male and the realized distribution of mating success. It is the *realized* distribution that determines the strength of sexual selection as well as the sex difference in strength of selection.

The spatial redistributions of matings mentioned above cause a discrepancy between the value of m^* that we established initially, and the realized value of m^* that corresponds to actual matings. Thus, for each combination of values, m^* and t^*, we can identify an "observed m^*" (m^*_{obs}) that describes how matings are actually distributed in space as a result of how receptive females are distributed in time. In practice, paternity analyses are essential to estimating m^*_{obs}.

Figure 3.7a shows how m^*_{obs} changes with the temporal distribution of

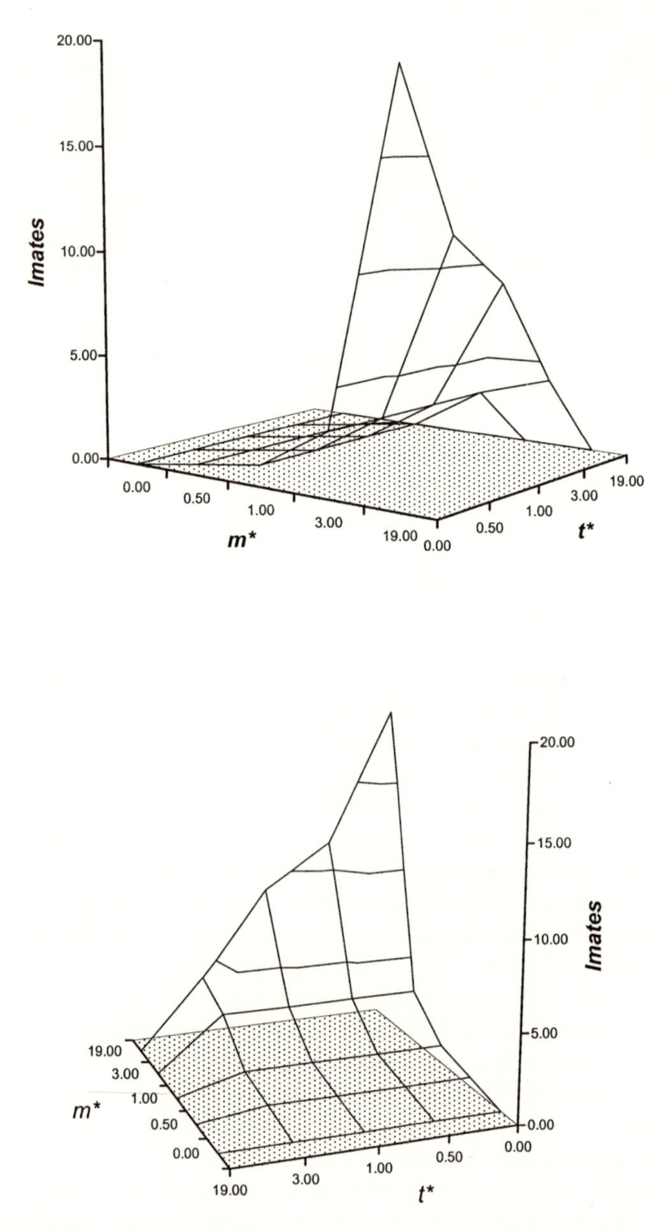

Figure 3.7. (a) The mean crowding of matings, m^*_{obs} (= I_{mates}) for a hypothetical population of 20 females, distributed over patches defended by 20 males, within a breeding season divided into 20 intervals (see text for details). The value of I_{mates} represents the effects of the spatial distribution of receptive females, as measured by the mean spatial crowding of females on resource patches, m^*, combined with the effects of the temporal distribution of receptive females, as measured by the mean temporal crowding of females within intervals of the breeding season, t^*; (b) the diagram in (a) rotated 180° to the right to show the I_{mates} surface.

females. In fig. 3.7b, this graph is rotated 180° to the right so that the surface generated by the plotted values of m^*_{obs} can be more clearly seen. When females are temporally dispersed ($t^* = 0$), m^* and m^*_{obs} are equivalent in our example, owing to the "one mate per interval" mating rule. However, as t^* increases, that is, as females become more synchronous in their sexual receptivity, the value of m^*_{obs} becomes less than m^*, to the point that when females are maximally clumped in time ($t^* = 19$), m^*_{obs} is zero regardless of the value of m^*. If, as described above, consistent rules exist for assigning male mating success, we can use either the spatial or the temporal distribution of females to calculate the values of p_0, $1 - p_0$, H, and V_{mates}. Then, using eq. [2.13] or eq. [3.10], we can calculate the value for I_{mates} as we did in the examples above (chapters 2 and 3). Either approach generates the same values for m^*_{obs} (Table 3.1). Also, and more importantly, the calculated values for I_{mates} obtained using either method show that I_{mates} equals m^*_{obs}. That is, m^*_{obs} accurately measures how I_{mates} is changed by the interaction of the spatial and temporal distributions of females. Since m^*_{obs} and I_{mates} are equivalent, henceforth we will use I_{mates} to describe the opportunity for sexual selection arising from the combined spatial and temporal distributions of sexually receptive females.

The above example confirms the conceptual model proposed by Emlen and Oring (1977), who argued that the environmental potential for polygamy (EPP) will depend on the spatial distribution of resources and the temporal availability of receptive females. Our fig. 3.7 also superficially resembles the hypothetical distribution of EPP shown in Emlen and Oring's fig. 2 (1977, p. 216), although, unlike this previous hypothesis, our example provides a *direct* estimate of the intensity of sexual selection, I_{mates}. Moreover, unlike most mating system analyses that followed Emlen and Oring (1977; see review in chapter 1), our method shows how this direct estimate can be obtained from *empirical* measurements of the spatial and temporal distributions of females.

Our approach captures and quantifies the relationship proposed by Emlen and Oring (1977) in a manner directly related to evolutionary theory. Instead of providing a subjective estimate of the potential for males to become polygynous, our method provides an empirical value for the sex difference in the opportunity for selection, I_{mates}, which can be derived directly from the observed distributions of females in space and time. Instead of focusing research on the activities that males use to monopolize females or male parental care, our approach emphasizes the process, rather than the presumed outcome, of sexual selection. Emlen and Oring proposed a hypothesis for the evolution of mating systems, but proposed no experimental approach for testing it. Our method provides an explicit methodology for testing their hypothesis that the combined spatial and temporal distributions of females determine the intensity of sexual selection on males. Furthermore, our method suggests how such data may be collected and analyzed. If a researcher can identify the temporal distribution of receptive females and can accurately assign matings to males, then I_{mates} can be estimated from these data. Alternatively, if a researcher can identify the spatial distribution of females most easily and can most accurately assign matings based on this information, then these data can provide a direct estimate of I_{mates}. We will illustrate this property of our method below. However, we will first extend our approach to consider the effects of the temporal distribution of receptive females on the operational sex ratio.

The Phenology of Sexual Selection and the Operational Sex Ratio

Clearly, the synchrony of female receptivity (measured as the mean temporal crowding of receptive females) determines whether or not the breeding season of a species consists of a series of breeding intervals. When a breeding season consists of several intervals, we need to ask the question, "What does the study of a subset of breeding intervals tell us about sexual selection and the sex difference in the *total* strength of selection?" We will show below that the answers to these questions depend upon both the temporal pattern of female receptivity and the correlation of male mating success across intervals.

In table 3.2, we illustrate quantitatively the relationship between the phenology of female sexual receptivity and the variance in male reproductive

Table 3.2

The phenology of female sexual receptivity and its relationship to the variance in mating success of males. The entries, N_{ij}, are the number of females on the territory of the ith male at the jth time interval. For this case, each female is assumed to be receptive for only a single time interval and to visit only a single male territory during her period of receptivity. (See table 3.3 for an extension to multiple intervals and table 3.4 for an extension to multiple territories.)

	Intervals (t)					
Male Territories	*1*	*2*	*3*	. . . T	*Sum*	*Mean*
1	N_{11}	N_{12}	N_{13}	. . . N_{1T}	$N_{1.}$	$N_{1.}/T$
2	N_{21}	N_{22}	N_{23}	. . . N_{2T}	$N_{2.}$	$N_{2.}/T$
3	N_{31}	N_{32}	N_{33}	. . . N_{3T}	$N_{3.}$	$N_{3.}/T$
.	.	.	.	.	.	.
.	.	.	.	.	.	.
.	.	.	.	.	.	.
K	N_{K1}	N_{K2}	N_{K3}	. . . N_{KT}	$N_{K\cdot}$	$N_{K\cdot}/T$
Sum	$N_{.1}$	$N_{.2}$	$N_{.3}$	. . . $N_{.T}$	$N_{..} = N_{females}$	$N_{..}/T = t$
Mean	$N_{.1}/K$	$N_{.2}/K$	$N_{.3}/K$	. . . $N_{.T}/K$	$N_{..}/K = R$	$(R/T) = (t/K)$

success, i.e., sexual selection. For this case, as in the previous example, each female is assumed to be receptive for only a single time interval during the breeding season, to visit only a single male territory, and to mate with its resident during her period of receptivity. With respect to the discussion above, this is equivalent to assuming that r is equal to 1. That is, our fundamental unit of time during the breeding season is equal to female average receptivity, so the breeding season is broken into T unit intervals. (Later, we will extend this simple case to permit a female to be receptive over multiple intervals and to visit multiple male territories.) First, note that the sum of all entries in table 3.2, $\Sigma\Sigma\ N_{ij}$, equals the total number of breeding females, N_{females}, and that the number of rows, K, equals the number of breeding males, N_{males}. Thus, if we assume that the breeding sex ratio, R, is unity, then K, is also equal to the sum $\Sigma\Sigma\ N_{ij}$.

Each column mean in table 3.2 is the average number of females available per male at that time interval or $R(t)$ in our notation. Remember that $R(t)$ is the inverse of the operational sex ratio at that same time interval, $1/R_O(t)$. The sum $\Sigma\ R(t)$ equals the overall sex ratio R, because the numerator of each element in the sum, $R(t)$, is the total number of males. Note, however, that the sum $\Sigma\ R_O(t)$ does *not* equal R_O, the inverse of the overall sex ratio. Instead, the sum $\Sigma\ R_O(t)$ equals $(N_{\text{males}}/H_{\text{females}})(T)$, where H_{females} is the *harmonic mean number of receptive females per time interval.* The harmonic mean number of receptive females per interval, $1/H_{\text{females}}$, is the reciprocal of the arithmetic mean of the reciprocal of the numbers of females per interval, t, or

$$1/H_{\text{females}} = (1/\text{T})\left(\sum_{\text{t}} 1/N_{\text{females}}[t]\right) \qquad [3.11]$$

Intuitively, large values of OSR occur when females are scarce, and indicate intense male-male competition. The harmonic mean of the number of receptive females weights the temporal fluctuations in female receptivity (and thus in OSR) according to the intensities of male-male competition that these fluctuations generate. The temporal average OSR (i.e., $[\Sigma\ R_O\{t\}]/[T]$) is disproportionately weighted by the lowest values of $R_O(t)$ which correspond to the intervals in which sexual selection by male-male competition for mates is strongest.

The average number of females receptive at any time during the breeding season, t, is $(\Sigma\Sigma\ N_{ij})/(T)$, which can also be written as $(N_{\text{males}})(R/T)$. The variance among the column means, $V_t\ /(K)^2$, is the temporal variance in the average number of receptive females per male. The variance among the row means, $V_{\text{mates}}/(T)^2$, is the variance among males in number of mates divided by $(T)^2$. The variance among males within a column is the variance among males in the number of mates within that time interval. Our goal is to understand the relationships between two pairs of parameters from table 3.2: (a) between V_t, the temporal variance in the number of receptive females, and

V_{mates}, the strength of reproductive selection on males; and (b) between $R(t)$, the mean number of receptive females at one time, and V_{mates}, the overall intensity of reproductive competition among males.

We know, by definition, that the total variance among cells in the number of receptive females in table 3.2 is

$$V_{N\,females} = (\Sigma\Sigma\ [N_{ij} - (N_{..}/KT)]^2)/KT. \quad [3.12]$$

Note that, as shown on table 3.2, the subscript "." refers to the sum of all values within a particular group. Thus, the total number of females in table 3.2 equals all of the females in each of the T intervals, summed over all of the K rows, or $N_{females} = N_{..}$. The variance shown in eq. [3.12] can be partitioned in two ways. First, we add and subtract $\{\Sigma R(t)\}/T$, the mean number of females per male at the jth interval, i.e., the quantities $([N_{.j}/K] - [N_{.j}/K])$ or zero inside the square brackets of eq. [3.12]. This lets us partition the variation in female numbers per cell into two components: (1) the average variance within columns; and (2) the variance among column means. The first term is the average temporal variance among males in mate numbers, $V_{mates}(t)$. At each time interval t, males differ from one another in mating success. Our term, $V_{mates}(t)$, is the average value of this variation across the breeding season. It is small when, in any given interval, all males have similar mating success. This is weak sexual selection at every interval. $V_{mates}(t)$ is large when, at any given interval, a few males have great mating success while other males have none. This is strong sexual selection at any interval. The second term is the variance among intervals in the mean number of mates per male, i.e., the temporal variance in the sex ratio, $V(R[t])$. Thus, eq. [3.12] becomes

$$V_{N\,females} = (\Sigma\{\Sigma[N_{ij} - (N_{.j}/K)]^2/K\})/T + (\Sigma[(N_{.j}/K) - (N_{..}/KT)]^2)/T, \quad [3.13a]$$

$$V_{N\,females} = (\Sigma\ V_{mates}[t])/T + V(R[t]), \quad [3.13b]$$

$$V_{N\,females} = V_{mates}(t) + V(R[t]). \quad [3.13c]$$

The term $V_{mates}(t)$ is the variance in mate numbers among males averaged over all intervals t_i. If one observes male mating success at a few time intervals selected at random from the breeding season, their average should estimate $V_{mates}(t)$. The opportunity for selection at any particular time interval t_i, $I_{mates}(t)$, equals the ratio $(V_{mates}[t]/R^2[t])$. We can now ask the question: *To what extent does this "snapshot" view of male mate competition, taken at a single interval, reflect the total intensity of competition over the entire breeding season*? We will answer this question below after we consider the relationship between $I_{mates}(t)$, the average value of $I_{mates}(t)$ or $\bar{I}_{mates}(t)$, and I_{mates}.

We now add and subtract the mean number of mates per interval for the ith male, i.e., the quantities $([N_{i.}/T] - [N_{i.}/T])$ or zero, inside the square brackets in eq. (3.12). This lets us partition the variation in female numbers among cells into two different components: (1) the average variance within rows; and (2) the variance among row means or $(V_{\text{mates}})/(T)^2$. Thus,

$$V_{N\text{females}} = (\Sigma\{\Sigma\, [N_{ij} - (N_{i.}/T)]^2/T\})/K + (\Sigma\, [(N_{i.}/T) - N_{..}]^2)/K, \quad [3.14a]$$

$$V_{N\text{females}} = (\Sigma\, [V_t(k)])/K + (V_{\text{mates}})/(T)^2, \quad [3.14b]$$

$$V_{N\text{females}} = \bar{V}_t(k) + (V_{\text{mates}})/(T)^2. \quad [3.14c]$$

The term $\bar{V}_t(k)$ measures the ups and downs of mating success experienced by the average male. It is the temporal variance in mate numbers averaged across individual males. This average variation in mating success of a single male within his territory could be measured by watching each of a few randomly chosen males throughout the breeding season. When a male's mating success in one interval is similar to or predictive of his success at other intervals, then $\bar{V}_t(k)$ will be small. It will be large if a male's success varies across intervals. It will be especially large if the identity of the most successfully mating males changes at every interval of the breeding season. In this case, some males have one interval of spectacular success preceded and followed by intervals of complete mating failure. We call this temporal variance in mate numbers experienced by an average male during the breeding season, *the local or average variance in receptivity per territory*, $\bar{V}_t(k)$.

Combining eqs. [3.13c] and [3.14c], we have

$$V_{\text{mates}} = (T)^2\{V(R[t]) + (\bar{V}_{\text{mates}}(t) - \bar{V}_t(k))\}. \quad [3.15]$$

The difference $(\bar{V}_{\text{mates}}[t] - \bar{V}_t[k])$ is of critical importance to the strength of sexual selection. If sexual selection within intervals is weak (i.e., $\bar{V}_{\text{mates}}[t]$ is small), but $\bar{V}_t[k]$ is also small, then total sexual selection can be strong. It is strong because, although the difference between winning and losing males on any given night is small, the winners are consistently better than the losers and accumulate matings throughout the breeding season or over their lifetime.

Dividing eq. [3.15] by R^2 gives us

$$I_{\text{mates}} = (T/R)^2\{V(R[t]) + [\bar{V}_{\text{mates}}(t) - \bar{V}_t(k)]\}, \quad [3.16a]$$

$$I_{\text{mates}} = I_{\text{sex ratio}} + (T/R)^2\{[\bar{V}_{\text{mates}}(t) - \bar{V}_t(k)]\}. \quad [3.16b]$$

Note first that (R/T) is the average sex ratio, t/N_{males}, so that the quantity $\{V(R[t])/(T/R)^2\}$ is the *opportunity for selection caused by the temporal variation in sex ratio* or $I_{\text{sex ratio}}$. Remember that the sex ratio at time t, $R(t)$, is the mean number of mates per male at time t_i. Thus, the variance in female receptivity among intervals contributes explicitly to the total opportunity for male mating success. $I_{\text{sex ratio}}$ is large when receptive females are abundant in some intervals and rare in others. Female reproductive synchrony (t^* high) leads to scramble monogamy and persistent pairs (see chapter 9) and, hence, large $I_{\text{sex ratio}}$. This term is small when the numbers of receptive females are evenly distributed across the breeding season (t^* is low). Furthermore, remember that S, the temporal patchiness of female receptivity, is defined as the ratio t^*/t, or $\{1 - (1/t)^2 + V(t)/(t)^2\}$. Because $\{V(R[t])/(T/R)^2\}$ is equal to $\{V(t)/(t)^2\}$, $I_{\text{sex ratio}}$ is approximately equal to $(S - 1)$.

Secondly, note that, in general, $\bar{I}_{\text{mates}}(t)$ (i.e., $V_{\text{mates}}[t]/(R[t])^2$), does *not* equal the average of the sum $\{V_{\text{mates}}(t)/(R/T)^2\}$, because the average of a ratio is not equal to the ratio of the averages. In particular, every $R(t)$ is less than R because the sum $(\Sigma\ R[t])$ equals R. An exception is the special case where the number of females available at any interval is a constant, C. When $R(t)$ equals a constant, R_C, the sex ratio over the entire breeding season, R, equals TR_C. (Note that here $[T/R]$ is $[1/R_C]$.) This means that, at each interval, $I_{\text{mates}}(t)$ equals $(V_{\text{mates}}[t]/[R_C]^2)$. Thus, in this case, $\bar{I}_{\text{mates}}(t)$ does equal the average of the sum $(V_{\text{mates}}[t]/[R_C]^2)$.

In general, we can substitute $\{[R(t)]^2\ I_{\text{mates}}(t)\}$ for $V_{\text{mates}}(t)$, and note that the proper way to average the values of $I_{\text{mates}}(t)$ is to weight each by $[R(t)/R]^2$. In this weighting, the intervals with the greater number of females (i.e., those intervals in which $R(t)$ is large) contribute more to the average opportunity for sexual selection than do the intervals with fewer numbers of females (i.e., those intervals in which R(t) is small). We see that the average "snapshot" is generally *not* representative of the intensity of male mate competition over the entire season because the factor $(R(t)/R)^2$ is always less than 1. Moreover, because each "snapshot" contains an underestimate of R, such samples will tend to overestimate the effect of sex ratio on the intensity of sexual selection on males. We will indicate this weighted average value of $I_{\text{mates}}(t)$ as $^*I_{\text{mates}}(t)$, and re-write expression [3.15] as

$$I_{\text{mates}} = I_{\text{sex ratio}} + {}^*I_{\text{mates}}(t) - (T/R)^2\{\bar{V}_t[k]\}. \qquad [3.17]$$

Finally, we note that the quantity $(T/R)^2\{\bar{V}_t[k]\}$ could be viewed as the weighted opportunity for selection *within males* or $^*I_{\text{within}}(k)$. This *within-male* source of variation in mate numbers decreases the overall variance *among males* in mate numbers. So we conclude that

$$I_{\text{mates}} = I_{\text{sex ratio}} + {}^*I_{\text{mates}}(t) - {}^*I_{\text{within}}(k). \qquad [3.18]$$

We can extend our model of the phenology of female sexual receptivity to permit each female to be receptive for multiple time intervals and to visit one or more male territories during her periods of receptivity. We let the average duration of female receptivity (the number of intervals receptive) equal r (see above). Note that this changes table 3.2 to the very similar table 3.3. Studying table 3.3, we first note that, under our current assumptions, it makes little difference to the ith male whether his mates at intervals 1 and 2, N_{i1} and N_{i2}, respectively, consist of the same or different females. In either case, they contribute in the same way to his total mating success, $N_{i.}$. We do not mean to imply that the variance in male mating success is indifferent to female searching behavior or to variations in female receptivity. The overall pattern of the entries N_{ij} results from female mate search and mate choice

Table 3.3
The phenology of female sexual receptivity and its relationship to the variance in mating success of males. The entries, N_{ij}, are the number of females on the territory of the ith male at the jth time interval. For this case, each female is assumed to be receptive for multiple time intervals and to visit one or more male territories during her period of receptivity.

	Intervals (t)					
Male Territories	*1*	*2*	*3*	. . . T	*Sum*	*Mean*
1	N_{11}	N_{12}	N_{13}	. . . N_{1T}	$N_{1.}$	$N_{1.}/T$
2	N_{21}	N_{22}	N_{23}	. . . N_{2T}	$N_{2.}$	$N_{2.}/T$
3	N_{31}	N_{32}	N_{33}	. . . N_{3T}	$N_{3.}$	$N_{3.}/T$
.	.	.	.	.	.	.
.	.	.	.	.	.	.
.	.	.	.	.	.	.
K	N_{K1}	N_{K2}	N_{K3}	. . . N_{KT}	$N_{K\cdot}$	$N_{K\cdot}/T$
Sum	$N_{.1}$	$N_{.2}$	$N_{.3}$	. . . $N_{.T}$	$N_{..} = rN_{\text{females}}$	$N_{..}/T = rt$
Mean	$N_{.1}/K$	$N_{.2}/K$	$N_{.3}/K$	. . . $N_{.T}/K$	$N_{..}/K = rR$	$(R/T) = (t/K)$

behaviors. However, given the pattern and values of N_{ij}, it matters little to a given male whether the identity of females in his territory is constant or variable from one interval to the next.

For a specific example, consider table 3.4 where we have constructed a hypothetical case of female copying behavior. We assume a sex ratio R of

Table 3.4
The phenology of female sexual receptivity with mate copying and its relationship to the variance in mating success of males. The entries, N_{ij}, are the number of females on the territory of the ith male at the jth time interval. For this case, each female is assumed to be receptive for multiple time intervals and she becomes more likely to visit the territories of males successful at the previous interval.

	Intervals (t)							
Male Territories	*1*	*2*	*3*	*4*	*5*	*Sum*	*Mean*	*Variance*
1	0	0	0	0	0	0	0	0
2	2	3	4	3	0	12	2.4	1.84
3	3	5	6	4	2	20	4.0	2.00
4	1	0	0	0	0	1	0.2	0.16
5	1	0	0	0	0	1	0.2	0.16
6	0	1	0	0	0	1	0.2	0.16
.	.	.	.	.	.	.	.	.
.	.	.	.	.	.	.	.	.
.	.	.	.	.	.	.	.	.
20	1	0	0	0	0	1	0.2	0.16
Sum	8	10	12	8	2	$40 = rN_{\text{females}}$	$8 = rt = 2 \times 4$	
Mean	0.4	0.5	0.5	0.6	0.1	$2 = rR = 2 \times 1$	$(R/T) = (t/K)$	0.944
Variance	0.64	1.55	2.34	1.14	0.19	$V_{\text{mates}} = 23.6$	1.172	

unity so that N_{females} equals N_{males} and we set this number, K, at 20. We further assume that each female is receptive twice so that r is 2. Thus, the numbers of receptive females at the five intervals are 8, 10, 12, 8, and 2, respectively, at first increasing and then decreasing sharply as might be true of a natural breeding season. The sex ratio of receptive females to males is rR, or 2, because each female enters the table twice. The average number of receptive females per interval, t, is 4 ($\{8 + 10 + 12 + 8 + 2\}/5$) and the variance in the number receptive, 11.2, is much greater than random. As a result, the sex ratio, $R(t)$, varies across time intervals, taking values of 2.5, 2.0, 1.67, 2.5, and 10, respectively. Thus, the variation in female numbers per cell, $V_{N\text{females}}$, is the sum of two components: (1) the average temporal variance among males in mate numbers, $\bar{V}_{\text{mates}}(t)$ (i.e., the average variance within columns); and (2) the temporal variance in the sex ratio, $V(R[t])$. In our example, $V_{N\text{females}}$ is 1.20, which is the sum of $\bar{V}_{\text{mates}}(t)$ or 1.172 and $V(R[t])$ or 0.028, as we expect from eq. [3.8c] above. In the second partitioning, we find that $V_{N\text{females}}$ is 1.20, which is the sum of $\bar{V}_t(k)$ or 0.256 plus $(V_{\text{mates}})/(T)^2$ or 0.944.

The interval values of $I_{\text{mates}}(t)$ are 4.0, 6.2, 6.5, 7.1, and 19.0; they are only weakly correlated with the values of $R_O(t)$. $\bar{I}_{\text{mates}}(t)$ equals 8.565 but the total I_{mates} is 5.9 ([0.944][25]/[4], remembering that the denominator, the mean number of mates per male squared, is $[rR]^2$), illustrating our earlier point that the average value of the separate intervals *overestimates* this component of selection on males. Clearly, the female copying behavior increases the consistency of male reproductive success across intervals. The most successful males at one interval (males 2 and 3 at interval 1 in table 3.3) are also the most successful at succeeding intervals, so that the average covariance in mate numbers across all pairs of intervals is 0.887.

A Worked Example

We now return to our example of *Paracerceis sculpta*, using the same samples of isopods collected near Puerto Peñasco between October 1983 and November 1985 that we described in chapter 2. Recall that in the northern Gulf of California, this species breeds in the spongocoels of an intertidal sponge, *Leucetta losangelensis*. Thus, *P. sculpta* females are contained within well-defined breeding territories defended by α-males. Females are semelparous in this species. They spend their entire preadult lives living among subtidal coralline algae, feeding, storing lipids in their digestive glands, and developing their ovaries. When ovarian development is nearly complete, females leave the algae and swim or crawl to sponges. There, females engage in a brief courtship with α-males and enter a male's spongocoel. As mentioned in chapter 2, no physical characteristics of males or of spongocoels appear to influence mating decisions by females, although the presence of gestating females within a spongocoel does appear to be attractive to unmated females (Shuster 1990, 1991b).

Within a few days, newly arrived females undergo their sexual molt. This molt is biphasic. The posterior half of the cuticle is shed first, exposing the female's genital pores. Females in this condition are sexually receptive and will mate repeatedly over the next 12–24 hours. At the end of this period, females shed the anterior portion of their cuticles, lose their functional mouthparts, and begin to shunt embryos into a ventral brood pouch, where offspring remain for the next three weeks. Juvenile isopods (mancae) leave their mothers after this gestation period as small, mobile, sexually undifferentiated individuals, who disperse from the sponge, settle on nearby coralline algae, and begin their lives as feeding, prereproductive individuals. Spent females remain within the spongocoel for several days, but eventually appear to wander away from the spongocoel or are evicted by the α-male. Spent females shelter alone or in small groups within spongocoels unguarded by α-males, or are consumed by fish. All post-parturition females die within two weeks of their single reproductive episode (Shuster 1991a, 1992).

There are eight stages that describe the physical changes females undergo during their reproductive lives (table 3.5). The number of days separating each stage is well defined. Thus, it is possible to estimate the number of days

Table 3.5
Female reproductive condition in *P. sculpta* (from Shuster 1991).

Stage	*Description*	*Days +/− S_2*
S_1	Female unmolted; cuticle well pigmented; mouthparts dark, well formed, functional; digestive glands extending to fifth segment; ovaries orange, filling body cavity dorsal to digestive glands; external brood pouch plates absent.	5.21
S_2	Female half molted; other characters as above.	0.5
S_3	Female fully molted, new cuticle dull in color, setose at segment margins; mouthparts metamorphosed, fused to head capsule; digestive glands dark, compressed into first segment; brood pouch filled with subspherical, undifferentiated embryos.	2.0
S_{4p}	Digestive glands reduced and confined to first segment; embryos elongate, orange but green at margins; eyespots absent.	6.5
S_{4e}	Digestive glands indistinct; embryos elongate, faintly green, eyespots present.	16.5
S_5	Mancae well developed, segmentation clearly visible, movements visible.	19.5
S_6	Mancae fully developed, dispersing from brood pouch.	21.5
S_7	Brood pouch empty, female spent.	24.0

before or after each collection of isopods each female in each aggregation is likely to have undergone her reproductive molt. This is accomplished for each stage by calculating a Poisson distribution of expected molting dates, using the average number of days required for each stage to proceed to or from a half-molted (S_2) condition, and the number of females in each reproductive stage (Sokal and Rohlf 1995, pp. 81–93). For example, the average number of days required before a newly arrived S_1 female initiates her reproductive molt, that is, becomes S_2, is 5.21 $\pm$ 2.91 days. If 17 S_1 females were collected in a single sample, as occurred in April 1984, then 1 of these females was likely to have molted the day after the sample was collected, 2 females were likely to have molted two days after the sample was collected, 3 females were likely to have molted three days after the sample was collected, and so on, as shown in fig. 3.8. Similarly, since 22 S_3 females were collected in the April 1984 sample, and since the average number of days required for females to proceed from stage S_2 to stage S_3 is 0.5 days, a Poisson distribution of molting dates for the S_3 females indicates that 2 of these females had undergone their molts the day the sample was collected, 5 females had molted the day before the sample was collected, 6 females had molted two days before the samples was collected, and so on (fig. 3.8). This procedure conducted for each reproductive stage generates a distribution of female receptivities for approximately 6 days before and 32 days after the sample date (fig. 3.8). From this distribution, it is possible to calculate the temporal distribution of sexually receptive females in this species for every month of the 24-month sampling period.

In this example, as in the example showing the spatial distribution of *P. sculpta* females over 1983–85, we will consider only spongocoels containing α-males and females. The duration of receptivity for each female is less than 24 hours (Shuster 1989a). Thus, for each monthly sample, T equals the number of days in which a least one female is sexually receptive (that is, in S_2 condition), and t_i equals the number of females in the ith interval. As shown in eq. [3.4], the average number of receptive females per interval, t, equals the sum of all females over all intervals, divided by the total number of intervals containing females. The variance in the number of females per spongocoel, V_t, equals the statistical variance in the number of females per interval, counting only those intervals containing females. With these values, we can now calculate the values of t^* and S for each monthly sample of isopods, using eqs. [3.6c] and [3.7]. These monthly estimates are summarized in table 3.6. Also included in this table are the values of N_{females}, T, t, and V_t for each collection, as well as these parameters estimated for the aggregate sample. For ease of viewing, we have reproduced tables 2.1 and 2.2 from chapter 2 so that the influences of the spatial and temporal distributions of females can be compared. In this chapter, these tables are tables 3.7 and 3.8, respectively.

There are several features to notice in these tables. First, the relative mag-

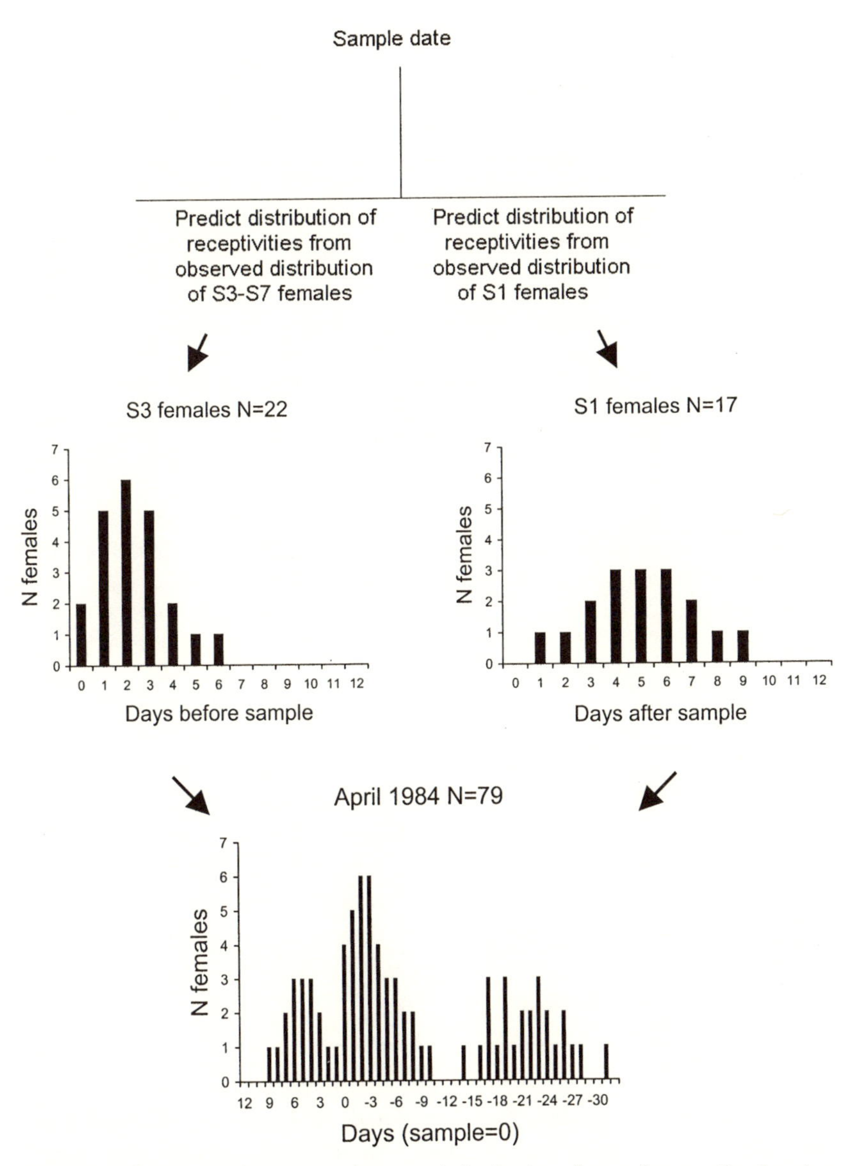

Figure 3.8. Calculating the expected temporal distribution of sexually receptive females in spongocoels in *P. sculpta* collected 1983–85.

nitudes of t^* and m^* are very different in the 1983–85 samples. In 21 of the 24 samples (86%), the value of t^* is less than 1, and in most cases t^* equals or very nearly equals zero. This is true because in most samples the average number of females per interval is approximately 1. This means that female receptivity in *P. sculpta* is highly asynchronous. Another way to visualize

Table 3.6
Calculation of t^* and S from monthly distributions of *P. sculpta* females in *L. losangelensis* spongocoels with with males, 1983–85.

Month	$N_{females}$	T	t	V_t	t*	S
1983						
Oct	23	14	1.64	0.66	1.04	0.64
Nov	23	15	1.53	0.65	0.96	0.62
Dec	20	18	1.11	0.10	0.20	0.18
1984						
Feb	9	8	1.13	0.11	0.22	0.20
Mar	6	6	1.00	0.00	0.00	0.00
Apr	79	35	2.26	1.96	2.13	0.94
May	20	19	1.05	0.05	0.10	0.10
Jun	41	27	1.52	0.47	0.83	0.55
Jul	48	32	1.50	0.50	0.83	0.56
Sep	25	21	1.19	0.15	0.32	0.27
Oct	57	30	1.90	2.02	1.96	1.03
Nov	2	2	1.00	0.00	0.00	0.00
Dec	2	2	1.00	0.00	0.00	0.00
1985						
Jan	12	10	1.20	0.16	0.33	0.28
Feb	18	17	1.06	0.06	0.11	0.10
Mar	28	22	1.27	0.20	0.43	0.34
Apr	22	20	1.10	0.09	0.18	0.17
May	1	1	1.00	0.00	0.00	0.00
Jun	7	7	1.00	0.00	0.00	0.00
Jul	14	14	1.00	0.00	0.00	0.00
Aug	22	17	1.29	0.21	0.45	0.35
Sep	17	17	1.00	0.00	0.00	0.00
Oct	45	34	1.32	0.28	0.53	0.40
Nov	24	23	1.04	0.04	0.08	0.08
Totals	565	239				

this phenomenon is to notice that the variance in the number of females per interval, V_t, is uniformly small over all of the samples, again because in most intervals only one sexually receptive female exists. When the temporal distribution of sexually receptive females is highly synchronous, that is, when t^* is less than t, the influence of the temporal distribution of receptive females on the spatial distribution of females is relatively small. Differently put, the value of I_{mates} will approximately equal the value of m^* calculated from the spatial distribution of receptive females.

Consider again the spatial distribution of *P. sculpta* females in spongocoels. Whereas in table 3.6 the values of t^* are small, in table 3.7 the

Table 3.7
Calculation of m^* and P from monthly distributions of *P. sculpta* females in *L. losangelensis* spongocoels with with males, 1983–85.

Month	$N_{females}$	M	m	V_m	m*	P
1983						
Oct	23	11	2.09	2.08	2.09	1.00
Nov	23	9	2.56	3.14	2.78	1.09
Dec	20	11	1.82	1.06	1.40	0.77
1984						
Feb	9	6	1.50	0.58	0.89	0.59
Mar	6	5	1.20	0.16	0.33	0.28
Apr	79	33	2.39	4.66	3.34	1.40
May	20	9	2.22	1.73	2.00	0.90
Jun	41	16	2.56	1.75	2.24	0.88
Jul	48	17	2.82	4.26	3.33	1.18
Sep	25	12	2.08	1.58	1.84	0.88
Oct	57	15	3.80	3.49	3.72	0.98
Nov	2	2	1.00	0.00	0.00	0.00
Dec	2	2	1.00	0.00	0.00	0.00
1985						
Jan	12	10	1.20	0.16	0.33	0.28
Feb	18	9	2.00	1.56	1.78	0.89
Mar	28	11	2.55	1.16	2.00	0.79
Apr	22	9	2.44	2.03	2.27	0.93
May	1	1	1.00	0.00	0.00	0.00
Jun	7	4	1.75	0.69	1.14	0.65
Jul	14	7	2.00	0.86	1.43	0.71
Aug	22	10	2.20	1.56	1.91	0.87
Sep	17	7	2.43	2.25	2.35	0.97
Oct	45	13	3.46	7.48	4.62	1.34
Nov	24	10	2.40	1.84	2.17	0.90
Totals	565	239				

values of m^* are much larger; on average, approximately 4.3 times larger. This is true because, in most samples, the variance among patches in the number of females, V_m, is relatively large and therefore contributes to the total value of m^*. The relative magnitudes of m^* and t^* are shown in fig. 3.9.

Another feature of this mating system contributes to the greater influence on the total opportunity for sexual selection of the spatial over the temporal distribution of receptive females. (This feature is emphasized by considering only those spongocoels that contain α-males and aggregations of females defended by α-males.) Although female receptivities overlap in this breeding

Table 3.8

Calculation of I_{mates} from the distribution of *P. sculpta* males and females in *L. losangelensis* spongocoels, 1983–85.

Month 1	*Mated* 2	*Unmated* 3	N_α 4	N_{fem} 5	p_0 6	$1-p_0$ 7	H_α 8	$V_{H\alpha}$ 9	$H_{m\alpha}$ 10	V_{Hma} 11	$I_{har\ em}$ 12	V_{mates} 13	I_{mates} 14	$I_{mat\ es(adj)}$ 15	R 16
Oct-83	11	3	14	23	0.21	0.79	1.64	2.37	2.09	2.08	0.48	2.37	0.88	0.88	1.64
Nov-83	9	5	14	23	0.36	0.64	1.64	3.52	2.56	3.14	0.48	3.52	0.54	1.30	1.64
Dec-83	11	6	17	20	0.35	0.65	1.18	1.44	1.82	1.06	0.32	1.44	0.44	1.04	1.18
Feb-84	6	6	12	9	0.50	0.50	0.75	0.85	1.50	0.58	0.26	0.85	0.38	1.52	0.75
Mar-84	5	1	6	6	0.17	0.83	1.00	0.33	1.20	0.16	0.11	0.33	0.23	0.33	1.00
Apr-84	33	10	43	79	0.23	0.77	1.84	4.60	2.39	4.66	0.81	4.60	0.80	1.36	1.84
May-84	9	6	15	20	0.40	0.60	1.33	2.22	2.22	1.73	0.35	2.22	0.45	1.25	1.33
Jun-84	16	4	20	41	0.20	0.80	2.05	2.45	2.56	1.75	0.27	2.45	0.37	0.58	2.05
Jul-84	17	5	22	48	0.23	0.77	2.18	4.69	2.82	4.26	0.53	4.69	0.59	0.99	2.18
Sep-84	12	10	22	25	0.45	0.55	1.14	1.94	2.08	1.58	0.36	1.94	0.45	1.50	1.14
Oct-84	15	3	18	57	0.17	0.83	3.17	4.92	3.80	3.49	0.24	4.92	0.34	0.49	3.17
Nov-84	2	4	6	2	0.67	0.33	0.33	0.22	1.00	0.00	0.00	0.22	0.22	2.00	0.33
Dec-84	2	3	5	2	0.60	0.40	0.40	0.24	1.00	0.00	0.00	0.24	0.24	1.50	0.40
Jan-85	10	11	21	12	0.52	0.48	0.57	0.44	1.20	0.16	0.11	0.44	0.30	1.33	0.57
Feb-85	9	14	23	18	0.61	0.39	0.78	1.56	2.00	1.56	0.39	1.56	0.39	2.55	0.78
Mar-85	11	5	16	28	0.31	0.69	1.75	2.19	2.55	1.16	0.18	2.19	0.34	0.71	1.75
Apr-85	9	6	15	22	0.40	0.60	1.47	2.65	2.44	2.03	0.34	2.65	0.44	1.23	1.47
May-85	1	2	3	1	0.67	0.33	0.33	0.22	1.00	0.00	0.00	0.22	0.22	2.00	0.33
Jun-85	4	2	6	7	0.33	0.67	1.17	1.14	1.75	0.69	0.22	1.14	0.37	0.84	1.17
Jul-85	7	3	10	14	0.30	0.70	1.40	1.44	2.00	0.86	0.21	1.44	0.36	0.73	1.40
Aug-85	10	5	15	22	0.33	0.67	1.47	2.12	2.20	1.56	0.32	2.12	0.44	0.98	1.47
Sep-85	7	4	11	17	0.36	0.64	1.55	2.79	2.43	2.25	0.38	2.79	0.47	1.17	1.55
Oct-85	13	1	14	45	0.07	0.93	3.21	7.74	3.46	7.48	0.62	7.74	0.65	0.75	3.21
Nov-85	10	2	12	24	0.17	0.83	2.00	2.33	2.40	1.84	0.32	2.33	0.41	0.58	2.00
Overall	239	121	360	565	0.34	0.66	1.560	3.21	2.36	2.95	0.53	3.20	0.57	1.30	1.57

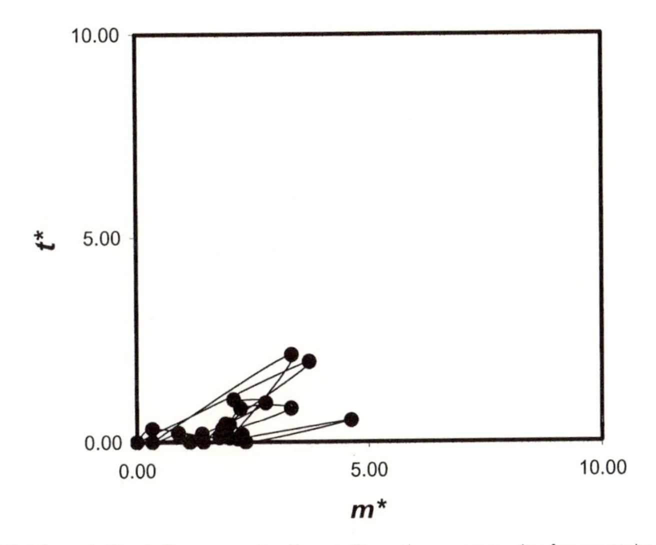

Figure 3.9. The relative influences of m^* and t^* on the opportunity for sexual selection, I_{mates}.

system (fig. 3.9), α-males are capable of excluding most other α-males from their spongocoels. Because large α-males can usurp spongocoels from smaller males (Shuster 1990), males defending female aggregations tend to be somewhat larger than the average male. Thus, the reproductive tenure of an α-male who defends his spongocoel successfully can "average over" any synchrony in female receptivity that exists within his spongocoel, and thereby mate with all of the assembled females We show in chapter 10 how this strategy of α-males is subverted by alternative mating strategies for some males in this species which have evolved the ability to invade the spongocoels without physically evicting the resident α-male.

Chapter Summary

We show in this chapter how the *temporal* distribution of female sexual receptivity affects both sexual selection and the sex difference in the strength of selection. We derive an expression describing the aggregation of female sexual receptivity over intervals of the breeding season, t^*, the mean crowding of receptivity. It quantifies the number of other receptive females that the average female experiences for the period when she herself is sexually receptive.

The mean temporal crowding t^* measures the temporal synchrony of the female breeding phenology in terms of the relationship between the mean and variance in the number of receptive females across the breeding season. When the variance in the number of receptive females per interval is large relative to the mean, then t^* is large and receptive females are temporally clumped. On the other hand, when the variance in the number of receptive females per interval is small relative to the mean, then t^* is small and receptive females are temporally *overdispersed.* Similar to the spatial patchiness discussed in chapter 2, we define S, the temporal "patchiness" of the female breeding phenology, which equals the ratio (t^*/t).

The temporal patchiness of receptive females is clearly related to fluctuations in the OSR during the breeding season. Emlen and Oring (1977) noted that male mating success and hence mating system evolution will be dependent on the temporal patchiness of receptive females. We show that the relationship between the temporal crowding of receptive females and the average number of mates per mating male is *reciprocal.* When females are synchronous in their sexual receptivity, the ability of one or a few males to mate with multiple females decreases and the variance in male reproductive success decreases. However, when female receptivity is temporally dispersed, then it is possible for some males to be serially polygamous, finding and defending a series of females distributed over the breeding season.

Because the relationship between t^* and I_{mates} is not as simple as that between m^* and I_{mates}, we show that temporal variations in the OSR fail to describe the actual intensity of sexual selection. Intuitively, large values of OSR occur when females are scarce and indicate intense male-male competition. The harmonic mean of the number of receptive females is more important, determining the effect of temporal fluctuations in female receptivity (i.e., in OSR) on the intensities of sexual selection. In addition, we show how the covariance across temporal intervals of male mating success further modulates the intensity of sexual selection. When the covariance is low, then intense reproductive competition within one time interval (indicated by high OSR) is diminished over the breeding season because the losers in one interval become the winners in later intervals. However, when the covariance is positive, then the mate numbers of early winners increase over the breeding season, inflating I_{mates}.

Lastly, we show how sexual selection measured within a single interval of the mating season can be related to the overall strength of sexual selection. In particular, the single-interval estimate is generally *not* representative of sexual selection over the entire breeding season and tends to exaggerate its actual intensity. Variations from interval to interval in the mating success of individual males decrease the overall variance *among males* in mate numbers, I_{mates}. In the chapters that follow, we will use the fundamental relationships between the spatial and temporal crowding of receptive females to categorize mating systems in terms of the sex difference in the strength of selection.

4 Multiple Matings and Postcopulatory, Prezygotic Sexual Selection

In this chapter, we investigate the consequences of multiple mating by females on sexual selection. Females of many species mate repeatedly and often store the sperm of more than one male in specialized organs, called *spermathecae*. These structures share the hallmarks of sexually selected traits: they are elaborated within species and variable across closely related taxa in ways that are not evidently adaptive outside the context of post-copulatory reproductive competition (Siva-Jothy 1987; Eberhard 1996; Pitnick et al. 1999). For example, in the featherwing beetles (family Ptiliidae), the spermatheca is considered one of the most useful diagnostic traits for identifying species (Dybas 1978). Multiple mating by females with capacities for sperm storage opens several new avenues for sexual selection: (a) postcopulatory, prezygotic reproductive competition between males (Robinson et al. 1994; Ketterson et al. 1997; Travis et al. 1997; Howard 1999); (b) "cryptic" female choice (Eberhard 1996; reviews in Choe and Crespi 1997); and (c) the interaction of postcopulatory female choice and male seminal fluid manipulations (Holland and Rice 1998). There is evidence for post-copulatory, prezygotic choice occurring in insects, mammals, and birds with multiply mated females (Birkhead and Møller 1998; Howard 1999). As a result, reproductive competition among males does not end with successful mating and sperm transfer, but instead continues after copulation. By "cryptic" female choice, we mean the selection among the gametes of different males that occurs *after mating* but *prior to* fertilization, a definition somewhat more restrictive than some (Eberhard 1996). This process is "cryptic" from the behavioral viewpoint because it is difficult to discern from observations of overt mating behaviors.

In practice, we must first begin by comparing the distribution of paternity among a female's offspring with that expected based upon the sperm contributions of her mates. Deviations from expectation can be caused by female sperm usage patterns, male ejaculatory competition, or an interaction between them. Given the variation among males in ejaculate size or duration of copulation, it must often be extremely difficult to define the "expected" pattern of sperm usage.

Clearly, the existence of elaborate sperm storage organs in females does not guarantee female "control" over fertilization. Indeed, it is often argued that the evolutionary interests of males and females are in conflict. Specifically, it is often claimed that a male is "interested" in maximizing paternity while a female is "interested" in maximizing offspring genetic quality (e.g., Holland and Rice 1998; Eberhard 1996). For example, male drosophilid flies produce over 80 proteins in their accessory glands, which are transferred at copulation. While some of these proteins act in the female to increase egg production and protect the longevity of male sperm, some decrease female receptivity and increase female mortality (Lung et al. 2002). As a result of these conflicting evolutionary goals, male-female interactions may be the norm for postcopulatory, prezygotic reproductive competition (Hosken 1999; Clark et al. 1999; Lung et al. 2002). Indeed, there is comparative evidence showing that sperm length and the length of the female's principal sperm-storage organ have diverged rapidly among species of drosophilid flies (Pitnick et al 1999) and preliminary experimental data indicate that they are coevolving traits (Pitnick, pers. comm.).

The empirical investigation of the outcome of postcopulatory, prezygotic reproductive competition between males has been greatly facilitated by the development of molecular genetic methods for paternity studies. Indeed, the number of publications on postcopulatory, prezygotic reproductive competition has grown steadily (Howard 1999) since the early reports in ladybird beetles (Nakano 1985), flour beetles (Robinson et al. 1994; Lewis and Austad 1994), and crickets (Bella et al. 1992; Howard and Gregory 1993). However, distinguishing the two component mechanisms of this gametic competition, male-male sperm competition (Clark et al. 1995) and cryptic female choice (Clark and Begun 1998), remains exceptionally difficult. In addition, there is evidence that interactions exist between male ejaculatory substances and substances in the female reproductive tract (Partridge and Hurst 1998; Clark et al. 1999; Heifetz et al. 2001; Lung et al. 2002). Whenever there are such intersexual interactions, the evolutionary landscape can permit rapid evolution (Begun et al. 2000; Swanson et al. 2001) and a wide range of equivalent combinations may be present at equilibrium, similar to the "lines of equilibria" in the classic Fisherian runaway sexual selection (Lande 1981; Kirkpatrick 1982). In this chapter, we discuss competition during fertilization within multiply mating females. In chapter 5, we will combine the mate number perspective developed in chapters 1–3 with the fertilization competition perspective explored here. There, we show how male behaviors influence the overall intensity of fertilization competition through effects on the mean and distribution of female promiscuity. We show that, because the harmonic mean number of mates per female determines the intensity of postcopulatory sexual selection (see below), male behaviors that restrict access to mates can be favored, even at the expense of fewer mates per male!

Selection among pollen in the race to fertilize ovules shares much in common with selection among sperm in the competition to fertilize ova (Delph et al. 1998; Howard 1999). Although pollen appear to play a more active role in the fertilization process, expressing 80% or more of the pollen genome during pollen tube growth, the seminal fluid of most animals contains on the order of 100 paternally transmitted gene products (Gwynne 1984; Swanson et al. 2001; Lung et al. 2002). The number of genes with male-limited expression in animals may actually exceed that in pollen.

Definitions of Promiscuity: Mate Numbers or Mating Numbers

Our approach to uniting multiple mating by females and the strength of sexual selection illustrates some of the fundamental relationships between male and female promiscuity, the sex ratio, and effects of both on the sex difference in the opportunity for selection. First, we note that there are two different ways that promiscuity is defined in the literature. One way defines promiscuity exclusively in terms of mate numbers and counts multiple matings by a given male with the same female as only a single mating. We will refer to this as *mate-number promiscuity*. Under this definition, a male or female is not promiscuous unless he or she has multiple mating partners. This definition is a behaviorally based one. It may not be appropriate when there is postcopulatory reproductive competition, because variations in sperm volume or sperm density among ejaculates, and thus the number of repeated copulations with a single female, may significantly influence male fertilization success (Shuster 1991a; Eberhard 1996; Tram and Wolfner 1998; Neubaum and Wolfner 1999).

The results of earlier chapters were based on an explicit assumption of mate-number promiscuity of males but not of females, where the average number of mates per male equals R and the average number of potential mates per female equals the reciprocal, R_O, the OSR. When the sex ratio is 1:1, R equals R_O. Thus, for a sex ratio of unity, males and females must be equally promiscuous on average. Under this definition, one sex can be more promiscuous than the other, but only if the sex ratio does not equal 1.

The second way to define promiscuity is in terms of the numbers of matings per male and per female, that is, counting each mating, whether with the same or a different partner, as a single event. We will refer to this definition as *mating-number promiscuity*. It is extremely difficult to measure in practice, especially when copulations are brief, hidden from observation, or intermittent during mate guarding. Nevertheless, it may be especially important to understanding postcopulatory, prezygotic sexual selection because paternity assurance may depend upon repeated copulations with the same female and, in many insects, paternity may depend upon the specific *order* of copulations by a female with different males (e.g., Lewis and Austad 1994; Wade et al. 1994).

With mating-number promiscuity, male and female promiscuities, P_{males}

and P_{females}, respectively, are defined as the average number of *matings* per individual. Let m_{ij} be the number of matings between the ith male and the jth female, where $i = 1, 2, \ldots, N_{\text{males}}$ and $j = 1, 2, \ldots, N_{\text{females}}$, and let the number of offspring from those same matings be o_{ij}. If we sum m_{ij} over i, $m_{\Sigma j}$, then we have the total number of matings of the jth female. Similarly, summing over j, we have $m_{i\Sigma}$, the total number of matings of the ith male. Because every mating involves one male and one female, we know that the total number of matings by females must equal the total number of matings by males:

$$\sum_{j=1}^{N_{\text{females}}} \sum_{i=1}^{N_{\text{males}}} m_{ij} = \sum_{i=1}^{N_{\text{males}}} \sum_{j=1}^{N_{\text{females}}} m_{ij}. \qquad [4.1]$$

The average numbers of matings per male, P_{males}, and per female, P_{females}, are connected through the sex ratio R, in exactly the same manner as the average numbers of offspring are (see chapter 2). Dividing the left side of eq. [4.1] by N_{males}, we obtain the average mating-number promiscuity of males as

$$P_{\text{males}} = \left(\sum_{j=1}^{N_{\text{females}}} \sum_{i=1}^{N_{\text{males}}} m_{ij} \right) \Bigg/ N_{\text{males}}. \qquad [4.2]$$

Similarly, we divide the right side of eq. [4.1] by N_{females}, and obtain the average number of matings per female, i.e., the average mating-number promiscuity of females as

$$P_{\text{females}} = \left(\sum_{i=1}^{N_{\text{males}}} \sum_{j=1}^{N_{\text{females}}} m_{ij} \right) \Bigg/ N_{\text{females}}. \qquad [4.3]$$

Substituting (N_{females}/R) for N_{males}, we have

$$P_{\text{males}} = RP_{\text{females}}. \qquad [4.4]$$

Clearly, when R is 1, under either definition of promiscuity, males and females have the same average promiscuity. Although intuition, as well as many discussions of sex differences in mating behavior, suggest that males are more promiscuous than females, the two sexes *must* engage in the same average number of matings and *must* have the same average number of mates whenever the sex ratio is unity.

The mean number of offspring of males, O_{males}, is connected to the mean number of offspring of females, O_{females}, through the ratio of mean mate numbers:

$$O_{\text{males}} = R\, O_{\text{females}} = (P_{\text{males}}/P_{\text{females}})\, O_{\text{females}} = (P_{\text{males}})(O_{\text{females}}/P_{\text{females}}). \qquad [4.5]$$

Intuitively, the average of the ratio $(O/P)_{\text{females}}$ is the number of offspring produced per female per mating. If there is no correlation between a female's offspring number and her mating number (Lewis and Austad 1994), then the mean of the ratio equals the ratio of the means $(O_{\text{females}}/P_{\text{females}})$. This ratio remains unchanged regardless of whether the offspring are randomly distributed over male mates or are all awarded to one of the mating males. When multiplied by the average number of matings per male, P_{males}, the ratio $(O/P)_{\text{females}}$ gives the average numbers of offspring per male.

We can identify the variance in promiscuity for males and females by recognizing that the variance in mating numbers equals the average of the squared mating numbers, minus the square of the average mating numbers, or

$$V_{P,\text{males}} = (\sum_i (m_i)^2/N_{\text{males}}) - (P_{\text{males}})^2, \qquad [4.6a]$$

and the variance in promiscuity for females equals

$$V_{P,\text{females}} = (\sum_j (m_j)^2/N_{\text{females}}) - (P_{\text{females}})^2. \qquad [4.6b]$$

Similarly, the variance in offspring number among males equals

$$V_{O,\text{males}} = (\sum_i (o_i)^2/N_{\text{males}}) - (O_{\text{males}})^2 \qquad [4.6c]$$

and the variance in offspring numbers for females equals

$$V_{O,\text{females}} = (\sum_j (o_j)^2/N_{\text{females}}) - (O_{\text{females}})^2. \qquad [4.6d]$$

The covariance between the number of matings and the number of offspring, $\text{Cov}(m,o)$, measures the degree to which multiple mating results in increases or decreases in offspring numbers. If this covariance is positive, multiple mating results in an increase in offspring numbers, as has been found, for example, in female flour beetles (Lewis and Austad 1994). If the covariance is negative, multiple mating causes a decrease in offspring numbers, as has been found in certain freshwater snails (van Duivenboden et al. 1985). In many plant species, the covariance between mating and offspring numbers is positive in females, owing to pollen limitation, and is positive in males as well. Although the rate of increase in offspring numbers with mate number may be faster for males than for females (e.g., Willson and Burley 1983; Willson 1990), a sexual conflict arising from a sex difference in sign of this covariance may be harder to achieve in plants than it is in animals, as we will explain below.

The covariance between mating and offspring numbers is the average of the products of individual matings and offspring numbers, minus the product of the averages of matings and offspring numbers or

$$\text{Cov}_{\text{males}} = (\sum_j [m_i o_i]/N_{\text{males}}) - (O_{\text{males}})(P_{\text{males}}). \qquad [4.7a]$$

Similarly, the covariance between the number of matings and the number of offspring for females equals

$$\text{Cov}_{\text{females}} = (\sum_i [m_j o_j]/N_{\text{females}}) - (O_{\text{females}})(P_{\text{females}}). \qquad [4.7b]$$

Lastly, the slope of the line describing the relationship between mating number promiscuity and offspring numbers is the covariance divided by the variance in promiscuity, i.e., $\beta_{\text{females}} = (\text{Cov}_{\text{females}})/V_{P,\text{females}})$. A positive regression means that selection favors increased promiscuity because it increases offspring numbers. Conversely, a negative regression means that selection favors decreased promiscuity.

The Relationship between Mate-Number Promiscuity and Mating-Number Promiscuity

We illustrate the relationship between the two definitions of promiscuity by using the hypothetical data in table 4.1a and 4.1b. In both tables, the numbers of males and females are equal ($N_{\text{males}} = N_{\text{females}} = 4$) and the total number of offspring produced is 32. In table 4.1a, the number of matings is 24; thus the average mating-number promiscuities ($P_{\text{males}} = P_{\text{females}} = [24/4] = 6.0$) and the average offspring numbers ($O_{\text{males}} = O_{\text{females}} = [32/4] = 8.0$) must also be equal across the two sexes. To investigate mate-number promiscuity, we simply replace every m_{ij} in table 4.1a with 1, as seen in table 4.1b. This converts multiple matings by a given male with the same female into a single mating, and allows us to examine mate-number promiscuity in relation to mating-number promiscuity. Note that this adjustment leaves the number of offspring per male and per female unchanged. However, the total number of matings changes from 24 to 8, so again the average number of mates per male, P^*_{males}, equals that per female, P^*_{females}, and is exactly 2.0.

Note that, in table 4.1a, the number of offspring increases as the number of matings increases *for males but not for females.* With each additional mating, a male can expect to sire two more offspring, whereas additional matings do not change offspring numbers for females. The same is true for males and females in table 4.1b. This example is identical to that in table 4.1a, except that mate-number promiscuity is the same for both sexes. Yet again, we find that offspring numbers are an increasing function of the number of mates for males but not for females. This seems counterintuitive. How can both sexes have the same average number of matings and the same average numbers of offspring, yet differ in mean number of offspring per mating? This sex difference in the number of offspring per mating suggests that males must mate more often than females to achieve the same reproductive success. *On average*, male promiscuity equals female promiscuity and male reproductive success equals that of females. How and why does the definition of promiscuity affect the sex difference in the average number of offspring per copulation? The answer lies in examining the variance in pro-

Table 4.1a

Mating-number promiscuity: Mating enhances male fitness but not female fitness. The table entries are the numbers of matings, m_{ij}, between the ith male and the jth female, where i,j = 1, 2, and 3 (i.e., $N_{males} = N_{females} = 3$) and the number of offspring from that mating, o_{ij}. The two values for each pairing are denoted as (m_{ij}, o_{ij}).

		Females Matings, Offspring				*Totals*	
		1	*2*	*3*	*4*	*Matings*	*Offspring*
	1	2,4	2,4	4,4	2,4	10	16
Males	2	0,0	3,4	3,4	0,0	6	8
	3	0,0	0,0	0,0	4,4	4	4
	4	4,4	0,0	0,0	0,0	4	4
Matings		6	5	7	6	24	
Offspring		8	8	8	8		32

$P_{males} = P_{females} = (24/4) = 6.0$
$V_{P,males} = 42.0 - 36.0 = 6.0$
$V_{O,males} = 88.0 - 64.0 = 24.0$
$Cov_{males}(m,o) = 60.0 - 48.0 = 12.0$
$(\beta)_{males} = 2.0$

$O_{males} = O_{females} = (32/4) = 8.0$
$V_{P,females} = 36.5 - 36.0 = 0.5$
$V_{O,females} = 0.0$
$Cov_{females}(m,o) = 0.0$
$(\beta)_{females} = 0.0$

Table 4.1b

Mate-number promiscuity: Mating again enhances male fitness but not female fitness. Each of the entries in Table 4.1a for m_{ij} has been replaced by 1. The number of offspring from a given mating is o_{ij}, where i,j = 1, 2, 3, and 4 (i.e., $N_{males} = N_{females} = 4$). The values for each pairing are denoted as $(1, o_{ij})$.

		Females Mates, Offspring				*Totals*	
		1	*2*	*3*	*4*	*Mates*	*Offspring*
	1	1,4	1,4	1,4	1,4	4	16
Males	2	0,0	1,4	1,4	0,0	2	8
	3	0,0	0,0	0,0	1,4	1	4
	4	1,4	0,0	0,0	0,0	1	4
Mates		2	2	2	2	8	
Offspring		8	8	8	8		32

$P^*_{males} = P^*_{females} = (8/4) = 2.0$
$V^*_{P,males} = 5.5 - 4.0 = 1.5$
$V_{O,males} = 88.0 - 64.0 = 24.0$
$V_{O,males} = (4)^2 V_{P,males} = 24.0$
$Cov_{males}(m,o) = 22.0 - 16.0 = 6.0$
$(\beta^*)_{males} = 4.0$

$O_{males} = O_{females} = (32/4) = 8.0$
$V^*_{P,females} = 0.0$
$V_{O,females} = 0.0$
$Cov_{females}(m,o) = 0.0$
$(\beta^*)_{females} = 0.0$

miscuity relative to the variance in offspring numbers and the sex difference in the covariance between promiscuity and offspring number.

The Sex Difference in the Variance of Promiscuity

We see in table 4.1a that males are 12 times more variable than females in their mating-number promiscuity ($V_{P,\text{males}} = 6.0 > V_{P,\text{females}} = 0.5$) but the sex difference in this variance is reduced considerably for mate-number promiscuity ($V^*_{P,\text{males}} = 1.5 > V^*_{P,\text{females}} = 0.0$). In both cases, however, males have a much greater variance in mating success than do females ($V_{O,\text{males}} = 24.0 > V_{O,\text{females}} = 0.0$ in both tables). For the male sex, the relationship between mating number and mate number is exactly linear with a slope of 2.0. Thus, the $V^*_{P,\text{males}}$ must be only 25% of $V_{P,\text{males}}$. For females, the relationship is not a simple linear transformation, so there is no simple equivalent transformation of $V^*_{P,\text{females}}$ into $V_{P,\text{females}}$. In general, there need be no simple transformation for males, either. What can we say about the general relationship between $V^*_{P,\text{males}}$ and $V_{P,\text{males}}$ or $V^*_{P,\text{females}}$ and $V_{P,\text{females}}$? We can say simply that, because the number of matings is always greater than or equal to the number of mates for both sexes, the variance in mate-number promiscuity, V^*_P, will always be less than or equal to the variance in mating-number promiscuity, V_P.

We are interested in how the variance in promiscuity affects the sex difference in the strength of selection. Hence, we are particularly interested in the relationship between promiscuity and offspring numbers *within* each sex. On average, when R is 1, both sexes are equally promiscuous and have equal numbers of offspring. However, a covariance between offspring numbers and promiscuity within one or both sexes can exist that causes a sex difference in the intensity of selection. Intersexual conflict in the direction of selection on promiscuity occurs when the sign of this covariance differs between males and females. It is easy to imagine that this covariance is often positive in males (more matings mean more offspring) but negative in females (more matings mean fewer offspring). Sex differences of this kind in the covariance between promiscuity and offspring numbers will determine whether one sex appears to be "promiscuous" and the other sex appears to be "coy." Clearly, both offspring numbers and promiscuity must vary among males and/or among females in order for the within-sex covariance to exist at all.

Sexual Conflict: The Sex Difference in the Covariance between Promiscuity and Offspring Numbers

In table 4.1a and b, males but not females exhibit a positive covariance between promiscuity and offspring number. The covariance between number of matings and offspring numbers for males, $\mathrm{Cov}_{\mathrm{males}}(m, o)$, is 12.0 in table 4.1a and half this, or 6.0, in table 4.1b. This positive covariance means that the more often a male mates and/or the more mates he has, the greater is the number of offspring he sires. Within males, the regression of offspring number on matings, β_{males} (i.e., the ratio of $\mathrm{Cov}_{\mathrm{males}}(m,o)/V_{P,\mathrm{males}}$), tells us how many more offspring a male can expect to gain, on average, by one additional mating. Similarly, β_{females} (i.e., $\mathrm{Cov}_{\mathrm{females}}(m,o)/V_{P,\mathrm{females}}$) is the increase in offspring numbers expected to accrue to a female with every additional mating. In table 4.1a, males gain offspring by additional matings but females do not ($\beta_{\mathrm{males}} = 2.0$; $\beta_{\mathrm{females}} = 0.0$) and, in table 4.1b, males gain offspring by having additional mates but females do not ($\beta^*_{\mathrm{males}} = 4.0$; $\beta^*_{\mathrm{females}} = 0.0$)! Thus, a male may gain more offspring by repeated matings than a female, even though *both* sexes have the same average numbers of matings, mates, and offspring.

In a natural population (as opposed to our hypothetical examples), the relationship between promiscuity and offspring number usually will be constrained by the fact that there can be no offspring without mates and matings. (Selfing hermaphrodites are a conspicuous exception.) Thus (cf. Johnson and Leone 1964, p. 394), an estimator of the slope of the regression of offspring numbers on numbers of matings is

$$\beta_{\mathrm{males}} = \left(\sum_{i=1}^{N_{\mathrm{males}}} m_{i.}o_{i.}\right) \Big/ \sum_{i=1} (o_{i.})^2. \quad [4.8]$$

In table 4.2a, we illustrate a case of intersexual conflict resulting from the fact that, for males, offspring number increases with mating-number promiscuity, but for females, offspring number decreases with promiscuity. Table 4.2b is the same example framed in terms of number of mates rather than number of matings. In both tables, the regression of offspring number on promiscuity is positive for males ($\beta_{\mathrm{males}}, \beta^*_{\mathrm{males}} > 0$) but negative for females ($\beta_{\mathrm{females}}, \beta^*_{\mathrm{females}} < 0$). Selection among males will favor increased promiscuity but selection among females will favor decreased promiscuity. Despite this difference in the direction of selection within sexes, the average promiscuity of males must remain equal to that of females. In this case, because the opportunity for selection on males for increased promiscuity ($I_{O\mathrm{males}} = V_{O\mathrm{males}}/O^2_{\mathrm{males}} = 49.5/64 = 0.77$) is stronger than that on females to decrease promiscuity ($I_{O\mathrm{females}} = V_{O\mathrm{females}}/O^2_{\mathrm{females}} = 8/64 = 0.13$), there is a net selection pressure to increase average promiscuity. Thus, while female fitness is compromised by sexual conflict in this case, the ability of females to evolve countermeasures to overcome male exploitation is limited in direct proportion to the sex difference in the opportunity for selection. (The values above were chosen explicitly to illustrate our points and, in natural populations, the frequency distributions of male and female promiscuity and their relation to offspring numbers could be even more complicated.)

Table 4.2a

Intersexual conflict: multiple mating enhances male fitness but reduces female fitness. The numbers of matings, m_{ij}, between the ith male and the jth female, where i,j = 1, 2, and 3 (i.e., $N_{\text{males}} = N_{\text{females}} = 3$) and the number of offspring from that mating, o_{ij}. The two values for each pairing are denoted as (m_{ij}, o_{ij}).

		Females Matings, Offspring				*Totals*	
		1	*2*	*3*	*4*	*Matings*	*Offspring*
	1	2,2	2,4	3,6	1,6	8	18
Males	2	2,1	1,2	1,2	1,6	5	11
	3	1,1	1,2	0,0	0,0	2	3
	4	1,0	0,0	0,0	0,0	1	0
Matings		6	4	4	2	16	
Offspring		4	8	8	12		32

$P_{\text{males}} = P_{\text{females}} = (16/4) = 4.0$
$V_{P,\text{males}} = 23.5 - 16.0 = 7.50$
$V_{P,\text{females}} = 18.0 - 16.0 = 2$
$\text{Cov}_{\text{males}}(m,o) = 51.25 - 32.0 = 19.25$
$(\beta)_{\text{males}} = +2.57$

$O_{\text{males}} = O_{\text{females}} = (32/4) = 8.0$
$V_{O,\text{males}} = 113.5 - 64.0 = 49.5$
$V_{O,\text{females}} = 72.0 - 64.0 = 8.0$
$\text{Cov}_{\text{females}}(m,o) = 28.0 - 32.0 = -4.0$
$(\beta)_{\text{females}} = -2.00$

Table 4.2b

Intersexual conflict: multiple mates enhance male fitness but reduce female fitness. The numbers of mating, m_{ij}, between the ith male and the jth female, where i,j = 1, 2, and 3 (i.e., $N_{\text{males}} = N_{\text{females}} = 3$) and the number of offspring from that mating, o_{ij}. The two values for each pairing are denoted as $(1, o_{ij})$.

		Females Mates				*Totals*	
		1	*2*	*3*	*4*	*Mates*	*Offspring*
	1	1,2	1,4	1,6	1,6	4	18
Males	2	1,1	1,2	1,2	1,6	4	11
	3	1,1	1,2	0,0	0,0	2	3
	4	1,0	0,0	0,0	0,0	1	0
Mates		4	3	2	2	11	
Offspring		4	8	8	12		32

$P^*_{\text{males}} = P^*_{\text{females}} = (11/4) = 2.75$
$V^*_{P,\text{males}} = 9.25 - 7.56 = 1.69$
$V^*_{P,\text{females}} = 8.25 - 7.56 = 0.69$
$\text{Cov}^*_{\text{males}}(m,o) = 30.5 - 22.0 = 8.5$
$(\beta)^*_{\text{males}} = +5.03$

$O_{\text{males}} = O_{\text{females}} = (32/4) = 8.0$
$V_{O,\text{males}} = 113.5 - 64.0 = 49.5$
$V_{O,\text{females}} = 72.0 - 64.0 = 8.0$
$\text{Cov}^*_{\text{females}}(m,o) = 20.0 - 22.0 = -2.0$
$(\beta)^*_{\text{females}} = -2.90$

If the most promiscuous males leave the greatest numbers of offspring and do so by mating with the most promiscuous females, the tendency toward promiscuity in males may become genetically correlated with that in females, leading to a runaway increase in the tendency to engage in multiple mating by both sexes, despite the fact that multiple mating *decreases* female fitness. We would not expect the mean promiscuity to stabilize until the gain within males (estimated as β_{males} or β^*_{males}) equals the opposing cost within females (estimated as β_{females} or β^*_{females}). On the other hand, if two males had the same numbers of mates, the male best able to prevent multiple mating by his females would sire greater offspring numbers. It is important to remember that females are a limited resource and that one male may gain offspring only at the expense of another male. Imagining that *every* male can achieve more offspring by engaging in additional copulations can lead to the contradiction that the average male has more offspring than the average female, as we show in chapter 7.

Postcopulatory Prezygotic Reproductive Competition among Males

When females mate multiple times, reproductive competition among males can continue after copulation. The mean and variance in mating-number promiscuity of females, P_{females}, and $V_{P,\text{females}}$, respectively, are critical components of the intensity of postcopulatory, prezygotic reproductive competition experienced by males. The key issue from the vantage point of a male is the number of sperm from other males with which his sperm must compete for fertilizations. We identify the mean crowding of male matings within females, m^*_P, as the key feature determining the intensity of postcopulatory, prezygotic reproductive competition experienced by males. Variation among females in the numbers of matings implies that sperm competition within females with many mates will be more intense than it is within females with fewer mates. The degree to which a particular male experiences sperm competition with other males thus depends upon the average promiscuity of his mates. The population of males experiences sperm competition to a degree dependent upon the average promiscuity of females, P_{female}, as well as the variation about this average, $V_{P,\text{female}}$.

We can quantify male-male postcopulatory, prezygotic competition using formulas from Lloyd (1967) for interspecific crowding of one species by another. If the ith male mates with the jth female a number of times m_{ij}, then his total number of matings equals $m_{i\Sigma}$ $(= \Sigma_j\, m_{ij})$ and her total number of matings equals $m_{\Sigma j}$ $(= \Sigma_i\, m_{ij})$. For this particular mate, his m_{ij} ejaculates are in competition with the $(m_{\Sigma j} - m_{ij})$ ejaculates of other males. That is, each of the m_{ij} ejaculates of this male with this female experiences $(m_{\Sigma j} - m_{ij})$ ejaculates of other males. We define the experience of sperm competition per ejaculate of the ith male as the *mean crowding of his sperm* by other mating males, or

$$m^*_{P,i} = \left(\sum_{j=1}^{N_{\text{females}}} m_{ij} \, (m_{\Sigma j} - m_{ij}) \right) \Big/ (m_{i\Sigma}). \quad [4.9]$$

The numerator of this expression is the sum, over all j females, of the product of the number of the ith male's matings with her, times the numbers of matings by her with other males. The denominator is the total number of matings by the ith male, obtained by summing across all females. Thus, $m^*_{P,i}$ is the average density of competing males per mating that the ith male experiences. For example, using the data in table 4.1a, the four males experience reproductive competition to the following degrees:

$$m^*_{P,1} = (2[6 - 2] + 2[5 - 2] + 4[7 - 4] + 2[6 - 2])/(10) = 3.4, \quad [4.10a]$$

$$m^*_{P,2} = (3[5 - 3] + 3[7 - 3])/(6) = 3.0, \quad [4.10b]$$

$$m^*_{P,3} = (4[6 - 4])/(4) = 2.0, \quad [4.10c]$$

$$m^*_{P,4} = (4[6 - 4])/(4) = 2.0. \quad [4.10d]$$

Note that *the more promiscuous a male, the greater is his experience of sperm competition.* As noted by Webster et al. (1995), the magnitude of the opportunity for sexual selection on males via sperm competition is proportional to the covariance between male mating success and sperm competitive ability. That is, males who mate most must also possess the most competitive sperm. If this relationship does not exist, multiple mating *decreases* the opportunity for sexual selection (see below). This suggests that postcopulatory, prezygotic male-male competition could itself limit male promiscuity, if the intensity of sperm competition curtails further increases in offspring numbers with additional matings.

The ratio of mean matings per female to mean mates per female, ($P_{\text{females}}/P^*_{\text{females}}$), approximates the average number of copulations per mate experienced by females. (This assumes that there is no covariance between number of matings and mate numbers, i.e., females with more mates do not also copulate more often with each male, or less often.) The experience of sperm competition by the average male, however, may be greater or less than this ratio, whenever sperm are nonrandomly distributed across females, owing to the variance in mating-number promiscuity $V_{P,\text{females}}$. When a male is able to reduce the mating-number promiscuity of his mates, say by mate guarding, depositing a sperm plug, or removing sperm of previous mates, he also experiences reduced postcopulatory sperm competition.

We can further reduce eq. [4.9] by noting, first, that $\text{Cov}_i(m_{ij}, P_j)$, that is, the covariance between male i's ejaculate number with female j, m_{ij}, and her mating-number promiscuity, P_j, equals $(\Sigma_j \, [m_{ij}][m_{\Sigma j}] - [m_i][P_{i,\text{females}}])/(\Sigma_j \, 1)$, where m_i is the average ejaculates per mate and $P_{i,\text{females}}$ is the average number of copulations of his mates. And, secondly, note that m_i^*, the mean crowding of the ith male's own ejaculates, equals $(\Sigma_j [m_{ij}][m_{ij}] - \Sigma_j [m_{ij}])/(\Sigma_j [m_{ij}])$. This last term measures the degree to which a male nonrandomly distributes (or aggregates) his ejaculates among his mates. It can exceed his average number of ejaculates per mate when he copulates much more often with some females than with others. It can be lower, if he copulates exactly the same number of times with every mate. Substituting the two quantities into eq. [4.9], we have

$$m_{P,i}^* = \{[\mathrm{Cov}_i(m,P_{\mathrm{females}}) + (m_i)(P_{i,\mathrm{females}})]/(m_i)\} - (m_i^* + 1) \quad [4.11]$$

which can be rewritten as

$$m_{P,i}^* = \{\mathrm{Cov}_i(m,P_{i,\mathrm{females}})/(m_i)\} + P_{\mathrm{females}} - (m_i^* + 1). \quad [4.12]$$

When males do not adjust their own copulation frequency to that of their mates, then the covariance term is zero, and this reduces to

$$m_{P,i}^* = (P_{i,\mathrm{females}} - 1) - (m_i^*). \quad [4.13]$$

Averaged over all males, the mean crowding of ejaculates experienced by a male, that is, *the intensity of sperm competition* experienced by the average male, equals

$$m_P^* = (P_{\mathrm{females}} - 1) - (m_E^*). \quad [4.14]$$

Clearly, sperm competition is increased by increases in the average mating-number promiscuity of females, P_{females}, and is decreased by the degree to which males cluster their own ejaculates with individual females, as measured by m_E^*.

The covariance between male matings per female and the average female mating-number promiscuity can also intensify or ameliorate sperm competition. If females vary from one another in promiscuity (i.e., $V_{P,\mathrm{females}} > 0$) and if males copulate more often with more promiscuous females, the covariance is positive and the intensity of sperm competition can exceed the average promiscuity of females. On the other hand, if males with few mates have more promiscuous mates and males with many mates have less promiscuous mates, then the covariance is negative and the intensity of sperm competition is reduced below the expected value.

The strength of this relationship determines whether sexual selection will favor increased or decreased promiscuity in males and how male matings are distributed over females. If the most promiscuous males mate with the most promiscuous females, then additional matings may not increase the variance in male reproductive success. This is true because male fertilizations per mating are diminished as a male's number of matings increases. This is a particularly important consideration in species where there is sperm mixing, such that, when a female mates with i males, each sires $(1/i)$ of her offspring. The number of offspring obtained by the ith male from his jth mate equals $(o_{ij}/P_{i,\text{females}})$. Summed over all mates and divided by the total number of mates (and assuming no covariance between o_{ij} and $P_{i,\text{females}}$), we see that O_i, the average number of offspring per mate of the ith male, is

$$O_i = (O_i)/(H_{P,i}), \qquad [4.15]$$

where $1/H_{P,i}$ is the harmonic mean promiscuity of his mates. It is equal to

$$(1/H_{P,i}) = \Sigma_j\,(1/P_{ij})/.(\Sigma_j\,1). \qquad [4.16]$$

Because the harmonic mean is influenced more heavily by small values than it is by large values, the most promiscuous females contribute very little to the average number of offspring per mate of the ith male, whereas the least promiscuous females contribute disproportionately to his fitness.

The fitness of a male is the product of the number of his mates and the average number of offspring he sires per mate. Because the latter quantity is influenced by the harmonic mean promiscuity of his mates, the postulated trade-off between remaining with the current mate and searching for additional mates is more complicated than depicted in most behavioral texts. To see this, consider the following example, which contrasts three males with different mating strategies: (1) male A, who mates with a single virgin female and, by mate guarding, sires all of her offspring; (2) male B, who mates with three females but guards none of them so that two of the females remate with two other males, while the third female mates again with three other males; and (3) male C, who mates just like male B but whose sperm have a large relative fitness advantage of 1.07 in competition with the sperm of other males. Let each female produce the same number of offspring, O, and mix the sperm of her mates when fertilizing her eggs. Male A has one mate and her mate-number promiscuity is also 1. Taking the product of mate number, offspring per female, and the harmonic mean promiscuity of his mate(s), the fitness of the male A is $(1)(O)(1) = O$. Now consider the fitness of nonguarding male B, who has three mates, but each has harmonic mean promiscuity of 0.306 (the average of 1/3, 1/3, and 1/4). Taking the product, we find that the fitness of male B is only $(3)(O)(0.306) = (0.917)O$, *less than that of male A*. Lastly, the calculation of fitness for male C is identical to that of male B except that he gains an average of (1.07)(0.306) offspring per mating. Taking the product, male C's fitness is $(3)(O)(1.07)(0.306) = (0.98)O$, also *less than that of male A*, despite male C's *superior* sperm competitive ability! Many male behaviors, such as mate guarding, depositing sperm plugs, or removing sperm of previous mates, lower the harmonic mean promiscuity of their mates. In chapter 9 we suggest that mate guarding is such a common behavior across mating systems because it has two simultaneous effects on male fitness: (1) it reduces the harmonic mean promiscuity of mates; and (2) it reduces the experience of sperm competition from other males.

Males mating with promiscuous females might have lower rather than higher fitness despite superior sperm competitive ability because sexual parasites become disproportionately clustered on the most promiscuous individuals of both sexes. In contrast, if the *most* promiscuous males mate with the *least* promiscuous females, then the reproductive success of an additional mating for a successful male can be greater than that of a single mating for a relatively unsuccessful male. The reason for this difference is twofold. The latter males seldom mate (they are not very promiscuous) and the matings they do achieve are with the most promiscuous females, so the siring payoff is small.

Our observation that a sex difference in the covariance between promiscuity and offspring numbers intensifies sexual selection is consistent with results reported by Holland and Rice (1998) in support of their "chase-away" hypothesis for mate choice. They suggest that females suffer fitness costs, i.e., reduced offspring numbers, by mating with males possessing extravagant secondary sexual characters. They propose that sexual selection favors males of outrageous extravagance but opposing natural selection on fecundity favors females who are exceptionally coy. This process is said to occur because only extreme male phenotypes are sufficient to overcome evolved female resistance to extravagant male courtship.

We see this process more simply; and thus as similar to the standard Fisherian runaway, although with a *negative* covariance between male and female promiscuity. Here, the most successful males have more offspring by virtue of having the most mates. Their sons will be more extravagant, and their daughters, because of the negative covariance between male and female promiscuity, on average, will be more resistant to mating. We have shown above that the intensity of sexual selection on males can far exceed that of viability selection on females (by nearly sixfold in our conservative example; see also chapters 1, 5, and 7), so the female character evolves despite a negative covariance between harem size and offspring number per female. At equilibrium, when the runaway is halted, a further increase in harem size of successful males is offset by the decline in fecundity within larger harems. A negative covariance between male and female promiscuity, however, could intensify sexual selection and lead to more extreme expression of male *and* female characters. Thus, a negative covariance between male and female promiscuity could explain the appearance of increased male extravagance, with its attendant high mating-number promiscuity, as well as the appearance of increased female courtship resistance, with its attendant low mating-number promiscuity without the need for significant viability selection against nonselective females. Together, these are characteristics of extreme sexual conflict that would lead females to mate only with the most extravagant, i.e., the most promiscuous, males.

Sperm Competition, Offspring Paternity, and Mate Numbers

The mean crowding of sperm within multiply mated females changes the distribution of paternity among males *within broods* but it does not necessarily affect the variance in mate numbers and the distribution of paternity *among broods*. Although variance in paternity is clearly important to the variance in male fitness, it is the variance in paternity among broods, not within broods, which is generally the largest contributor to the sex difference in the intensity of selection (see chapter 2). Thus, to understand how sperm competition contributes to sexual selection, we need to know the relationship between the intensity of sperm competition within broods, m_P^*, and the distribution of mates. Several different relationships are possible (Arnold 1994).

"Winner Fertilize All" Sperm Competition

Consider the case where m_P^* is large but only one male sires the entire brood (i.e., the outcome of sperm competition is "winner fertilize all"). If the winning male within one brood is also the winner within other broods, then male mating-number promiscuity will *increase* the variance in mate numbers among males. If the winning male is determined randomly, however, so that a male is no more likely to win in one bout of sperm competition than in another, then male mating-number promiscuity *reduces* the variance in mate numbers among males. In plants, variation in timing of pollination among flowers within an inflorescence makes it less likely that one and the same male will win fertilizations on every flower (Waser and Price 1983; Snow and Lewis 1993; Stanton 1994) even with postfertilization fruit abortion (Snow 1994). With "winner fertilize all" sperm competition, it is the *covariance in male success across bouts of sperm competition* that determines whether the variance in mate number is enhanced (covariance is positive) or diminished (covariance is zero or negative). A positive covariance means that there is effectively a sperm dominance hierarchy among males, with the highest-ranking male enjoying fertilization success whenever his ejaculate is in competition with those of lesser rivals. This increases the fitness variance among males, because the most successful males have the most competitive sperm and effectively the most mates. The least successful males may mate many times, but they lose in sperm competition and thus effectively have no mates.

A covariance of zero in the outcome of sperm competition across females means that fertilization is more like a lottery, with one lucky male fertilizing the entire brood. The most (mating-number) promiscuous males have the most lottery tickets and have more chances to sire entire broods. This too leads to an increase in the variance in fitness among males, because the most promiscuous males have effectively the greatest number of mates and the least promiscuous males have the fewest or even no mates. This perspective underlies much of current theoretical and empirical research in sperm competition (reviews in Birkhead and Møller 1998).

A negative covariance in the outcome of sperm competition across females means that the winner in one bout must be a loser in other bouts and vice versa. Here, fertilization becomes democratic with paternity equitably distributed across males. This condition diminishes the variance in fitness across males because it equalizes offspring numbers. It is difficult to imagine mating systems with this property because it appears to require that a female adjust the paternity of her brood(s) in relation to paternity of other females. However, when males invest heavily in ejaculates, like spermatophores, heavy investment by a male in one mating may necessitate his reduced investment in subsequent matings, leading to a pattern of high success with early, heavy investment but reduced later success with lowered investments. Alternatively, some kinds of frequency-dependent mating success may approximate a "winner fertilize some" system with a high variance in success across bouts.

Different values of the covariance between intensity of sperm competition and winner-fertilize-all situations can also lead to different runaway processes. When the covariance in male success across bouts is positive and males show a competitive hierarchy, then the average male will achieve greater fertilization success by limiting the mating-number promiscuity of his mates. We expect this pattern to lead to the evolution of mate guarding strategies by males (see chapter 8). Here, the most successful males are those with the most competitive sperm and the most mates, which are achieved by large numbers of matings. However, if promiscuous females are more likely than other, less frequently mating females to copulate with males with "dominant" sperm, then the most successful males will not only have sons with dominant sperm, but also daughters, which are, on average, more mate-number promiscuous.

"Winner fertilize all" bears marked similarity to the case where females become receptive asynchronously. As we saw in chapter 3, the pattern of male success across time intervals (i.e., the covariance) determines whether the sex difference in strength of selection is enhanced or diminished. If the male winning at one time interval also wins reproductive competition for receptive females at subsequent intervals (i.e., a positive covariance in mating success across intervals), then the variance in mate numbers is enhanced.

"All Fertilize Some" Sperm Competition

Alternatively, consider the case where m_P^* is large and each inseminating male sires a comparable portion of the entire brood (i.e., "all fertilize some" sperm competition). Sperm mixing in the spermatheca of female insects or

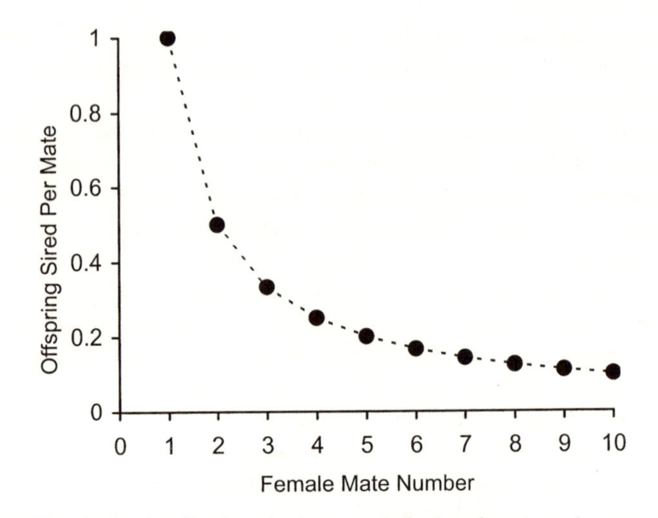

Figure 4.1. The decline in offspring sired per mating when females mix sperm from all males and apportion fertility equally.

pollen mixing on the stigma of plants might approximate this system. Although pollen competition is intrasexual selection and it may be intense (Willson 1994; Stanton 1994), it may not result in strong sexual selection owing to the indirect connection between sperm competition and the variance in male reproductive success. Moreover, pollen competition may not lead to sexual dimorphism owing to a lack of a sex difference in the strength and direction of selection, especially in hermaphroditic plants (Arnold 1994; Morgan 1994). In this instance, the opportunity for competition among ejaculates of different males, represented by the high value of m_P^*, is not realized as a high paternity variance within broods. The more promiscuous males, however, by virtue of an increased number of mates, may increase the paternity variance among broods, but, again, only under certain circumstances of covariation between male promiscuity and female promiscuity.

With random promiscuous mating and equitable fertilization, there is a steep trade-off for males (i.e., a negative correlation) between increasing mean promiscuity (i.e., mating numbers) and the expected number of offspring sired per mating (fig. 4.1). As we showed above, total male fitness is the product of mate numbers and offspring per mating. Thus, whenever an increase in promiscuity results in a proportional decrease in offspring sired per mating, the variance in male mating success is not changed. Differently put, there are many different values of mean promiscuity that are equivalent in their effects on the male variance in paternity. Thus, contrary to current logic about sperm competition, high mean crowding of sperm arising from multiple insemination of females does not automatically translate into either increased sexual selection or a sex difference in the strength of selection.

When the covariance between mating number and paternity is positive, males with higher promiscuity will be favored by selection. When the covariance is negative, males with higher promiscuity will be selected against.

The two outcomes of sperm competition discussed above, "winner fertilize all" and "all fertilize some," represent end points of a continuous spectrum of paternity with multiply inseminated females. Because high values of m_P^* are not necessarily concordant with high values of paternity, the opportunity for sperm competition maps only indirectly into the intensity of paternity selection.

Chapter Summary

There is ample evidence from a wide variety of taxa indicating that sexual selection does not end with mating and sperm transfer but rather continues as postcopulatory, prezygotic competition for fertilization. Although both sexes have the same *average* numbers of matings, mates, and offspring, males may gain more offspring by repeated matings than females, whenever there is a sex difference in the *covariance* between promiscuity and offspring number. We believe that it is these sex differences in covariance (rather than the averages) that are the source of the widely held view that "males are promiscuous" while "females are coy." Furthermore, we showed that the intensity of postcopulatory, prezygotic competition for fertilizations can be quantified by the mean crowding of male ejaculates within females, which is directly affected by the distribution of female promiscuity.

With the sperm mixing, as is typical in some insects, a male's fertilization success is determined by the *harmonic mean promiscuity* of his mates. Because the harmonic mean is more strongly influenced by small numbers than it is by large numbers, there is an inherent bias for selection to favor male behaviors that limit subsequent matings by their mates. With more active sperm competition, mate guarding also reduces the experience of sperm competition from other males. As a result, there appears to be a general tendency for selection on males to favor mate guarding even at the expense of additional matings. This finding is in contrast to many, if not most, discussions in the current sexual selection literature, which argue that sexual selection should maximize male promiscuity.

We also discussed how postcopulatory, prezygotic reproductive competition can lead to novel runaway processes, similar to those associated with Fisherian runaways of male traits and female choice based on mate numbers. Beautiful treatments of these same topics can be found in Gwynne (1984), Eberhard (1996), Birkhead and Møller (1998), and Birkhead (2000). Our contribution here is to illustrate how these aspects of male and female reproductive behavior can be quantified in relation to the sex difference in the strength of selection.

5 Female Life History and Sexual Selection

Empirical studies of sexual selection have historically focused on males (Emlen and Oring 1977; Shuster 1991b; Andersson 1994). Despite Darwin's (1871) insistence on the importance of female mate choice, early studies viewed females either as fitness prizes won in premating contests or as passive partners in mating (Parker 1970a; Smith 1984; Birkhead and Møller 1993a,b). With increased interest in female mate choice in the 1970s, coincident with the increasing prominence of female researchers in the field of sexual selection in the 1980s and 1990s (reviews in Gowaty 1997), the role of female biology in shaping the outcome of sexual selection on males has become more widely appreciated. As noted earlier, Emlen and Oring (1977) recognized that the spatial and temporal distributions of *females* influence the "environmental potential for polygamy" in most species. We have shown the quantitative effects of these female distributions on sexual selection (chapters 1–4).

Over the past two decades, studies of the role of females in mating system evolution have produced new findings. Female control of the outcome of matings with different males via "cryptic choice" is now considered common in insects and birds (Eberhard 1996; Birkhead and Møller 1993a,b; Howard 1999). "Good genes" reflected in male phenotypes are widely thought to be responsible for observed patterns of female mate choice in diverse animal taxa (Hamilton and Zuk 1982; Petrie 1994; Scheib et al. 1999). "Good genes" themselves now include those traits in males that exploit biases in the female perceptual system (e.g., Ryan 1998; Endler and Basolo 1998). Viability selection on offspring is considered the primary evolutionary force responsible for female adaptations that limit the modification of male phenotypes by sexual selection (Ryan 1998). Viability selection directly on females is also thought to produce adaptations that mitigate the negative influences of sexual conflict (Rice 1996; Holland and Rice 1998). In summary, the fitness of females is now believed to play a much larger role in sexual selection and mating system evolution.

The explosion of hypotheses involving female responses to male mating behavior has focused on explaining the phenotypic outcome of natural selection on females, when that selection is opposed by sexual selection on males. The central premise is that antagonistic pleiotropy in sex-specific fitness effects limits sexual selection and constrains the evolution of sexual dimorphism. Although viability selection is widely presumed to be as strong as, if not stronger than, sexual selection at equilibrium (Darwin 1871; Ryan 1998), there have been few attempts to quantify the *relative intensities* of sex-specific selective forces in natural populations. While males in some species possess characteristics correlated with enhanced viability (Saino et al. 1997) and males of other species appear to incur sex-specific mortality (e.g., predation on calling males in anurans and tettigoniids), the relative intensities of sexual selection versus viability selection have been poorly quantified (but see Grether 1996; Merilae et al. 1998; Hoekstra et al. 2001; Sinervo et al. 2001).

In this chapter, we will first identify the source of selection on females that arises from different patterns of female mating receptivity and reproductive life history. We quantify this selection in terms of the opportunity for selection on females and explore how it varies with context. We then quantify the sex difference in the intensity of selection and examine how it changes with mating system. This is tantamount to quantifying the degree of antagonistic pleiotropy between the sexes. Lastly, we will examine how the genetic covariance between female life history characters and male reproductive competitive ability affects the sex difference in the opportunity for selection. Positive covariances across the sexes enhance antagonistic pleiotropy and accelerate runaway processes. In contrast, negative covariances between sex-specific components of fitness not only act as constraints on runaways but should also characterize evolutionary equilibria.

The Overall Effects of Female Reproductive Life History on Sexual Selection

What do we mean by "female life history"? Although this term has been defined in several different ways (e.g., Stearns 1977; Roff 1992), our definition will focus specifically on *female reproductive life history* and on the role it plays in the evolution of mating systems. We define female reproductive life history using three characters:

1. *The number of times a female mates*: In its simplest form, we can use this trait to categorize females into two groups. Females who mate once are *monandrous*, whereas females who mate more than once are *polyandrous*.

2. *The number of reproductive events in a female's lifetime*, i.e., the distribution of fecundity over a female's reproductive lifetime. This character allows us to divide females into two additional categories. *Semelparous* females produce a single clutch of offspring, whereas *iteroparous* females produce multiple clutches of offspring.

3. *The duration of female reproductive competence*, i.e., the pattern of reproductive senescence. This trait allows us to partition iteroparous females into an additional pair of classes: *uniseasonal iteroparous* females breed multiple times in one breeding season, whereas *multiseasonal iteroparous*

females breed multiple times but over more than one breeding season (see chapter 6).

Each of these characteristics of the female reproductive life history has an influence on the sex difference in the strength of selection. The pattern of female mortality across the period of female reproductive competence can also modify the intensity of selection on females in the same way that mortality patterns of migrants modify the effective rate of gene flow (e.g., Endler 1986; Breden 1987). We discuss the effects of variations in female reproductive life history in turn here and elaborate further on variations in the duration of female reproductive competence in the next chapter.

Variation in female reproductive life history directly influences the opportunity for sexual selection on males in three ways. First, variation among females in the number of clutches *diminishes* the sex difference in the opportunity for selection, I_{mates}. This is true because $I_{mates} = I_{males} - I_{females}$ (chapter 1). As long as clutch number and egg number are independent, whenever females become variable in the number of clutches they produce, the variance in offspring numbers among females is *increased*. This causes $I_{females}$ to become larger in relation to I_{males}, and, as a result, the sex difference in the opportunity for selection, I_{mates}, is always *reduced* by I_{clutch}, as we show below. Thus, we predict that, in general, in semelparous species, the sex difference in opportunity for sexual selection, $I_{mates} - I_{clutch}$ will be greater than it is in iteroparous species. Because the change in female reproductive life history from semelparity to iteroparity increases the opportunity for selection on females, $I_{females}$, it consequently decreases the sex difference in the intensity of selection.

Variation in female mate numbers is a second way in which the female reproductive life history influences the opportunity for sexual selection on males and the sex difference in the strength of selection. Multiple mating by females, like iteroparity, *decreases* the sex difference in the strength of selection. Whenever females are polyandrous, by mating more than once, they effectively partition their progeny among different sires. Just as iteroparity partitions lifetime female reproduction into several clutches, each additional mating potentially partitions a single clutch of eggs into groups of offspring with different paternity, i.e., with each group sired by a different male. The variance in offspring numbers among males *decreases* relative to the case where a mating male sires all of the offspring of a female. This too causes the relative magnitudes of I_{males} and $I_{females}$ to change, thereby causing I_{mates} to *decrease* (recall that $I_{mates} = I_{males} - I_{females}$).

This latter relationship can be greatly complicated whenever multiple mating by females results in sperm competition or cryptic female choice. Variation among males in paternity, which is achieved postcopulation by either mechanism, is likely to change the apportionment of offspring among sires and thus affect the variance in male offspring number (Webster et al. 1995;

Eberhard 1996; Delph et al. 1996, 1998). The effect of sperm competition on the male variance in offspring numbers may not be clear *a priori*. For example, "winner-fertilize-all" sperm competition (or cryptic female choice of a single sire) reduces female mate-number promiscuity to single-male paternity of entire broods. In contrast, "winner-fertilize-some" sperm competition owing to sperm mixing (or cryptic female choice for genetic diversity) might result in the equal apportionment of paternity among all mating mates, which would greatly reduce the variance among males in offspring numbers. Thus, sperm competition and cryptic female choice can enhance or diminish the sex difference in strength of selection depending upon how the male variance in reproductive success is affected.

One way to visualize the effects on sexual selection of multiple mating by females is to consider the covariance of mates across clutches for different females. This relationship identifies the degree to which females use the same male or different males to sire each clutch. When the average covariance is 1.0, although a female may mate with many different males, the same male sires each clutch. This might occur through cryptic female choice or with male mate guarding in proximity to oviposition. In either case, single-male paternity of all broods means that the effect of female iteroparity is converted by postcopulatory reproductive competition into that of semelparity (although iteroparous, monandrous females may still be more variable in the number of offspring they produce compared to semelparous, monandrous females). Similarly, when the female average covariance is less than 1.0 but still positive, multiple mating by females does not reduce the sex difference in strength of selection as much as it would appear simply from the observation of variance of clutch numbers among females. When the covariance is zero, females essentially select a new mate at random to sire every clutch and, by the arguments above, iteroparity diminishes the evolutionary potential for sexual dimorphism.

In contrast, when the covariance is negative, females must switch mates between clutches, maximizing the variance in paternity from clutch to clutch. They might do so as a means of enhancing the genetic diversity of their progeny. In this case, the effect of iteroparity in reducing the variance in paternity is greater than would be apparent simply from observations of the variance of clutch numbers among females. Because this sort of mating system results in offspring *being more evenly distributed across males than expected by chance* (i.e., the mean crowding of offspring across males is reduced below the average), it has the strongest effect of reducing the sex difference in strength of selection. It reduces I_{males} and increases I_{females}. These are the kinds of features that might contribute to sex role reversal, as in pipefish and in phalaropes (see chapter 9; McCoy et al. 2001; Delehauty et al. 1998), where the variance in clutch number is large but the variance in egg numbers among clutches is small and, despite the potential for sperm competition, there is only one sire per clutch. These comparisons are particularly useful when considering the importance of Fisher's (1930) "runaway" process of sexual selection relative to other models of mate choice, such as those based on male genetic quality (Gowaty 1997; Olson and Owens 1998), female sensory biases (Ryan 1998), and intersexual conflict (Holland and Rice 1998). No matter how complicated the pattern of sperm usage, we can integrate its effects with those of the female reproductive life history and compare the relative intensity of viability and fecundity selection on females with the intensity of sexual selection on males. The ability to dissect maternity and paternity with molecular markers now makes this comparison possible for many species.

The third way in which female reproductive life history influences sexual selection is through the duration of female receptivity and the synchrony of receptivity among females, as we discussed earlier (see chapter 3). These two features of female reproductive biology influence sexual selection by altering the ability of males to accomplish exclusive syngamy and by changing (often decreasing) the variance in mate numbers among males. The longer a group of females remains receptive, the more opportunities males, as well as females, have to mate more than once. Like the spatial aggregation of females (see chapter 2), female temporal asynchrony can lead to a runaway process. In chapters 3 and 4, we showed how a genetic covariance can arise between the duration of female receptivity and postcopulatory male competitive ability (sperm competition). As a result, the most successful males have sons with increased sperm competitive ability, as well as daughters with a greater than average duration of receptivity. This runaway process is halted when further increases in the duration of female receptivity are opposed by natural selection (see chapter 8), although, due to the weaker intensity of natural selection relative to sexual selection, particularly during runaway processes, the location of this equilibrium may often leave females at a fitness disadvantage (chapter 4).

The Opportunity for Selection on Female Reproductive Life History

Wade (1987) investigated how multiple mating and iteroparity by females affect the opportunity for selection. He identified the component of selection that results from variance among females in their clutch numbers as I_{clutches}. It equals the variance among females in clutch number, V_{clutches}, divided by the square of the mean number of clutches per female, $(c)^2$. We will express the total variance in offspring numbers among females, V_{females}, (chapters 1–3), as the sum of three components: (1) the average variance in offspring number among clutches, $V_{\text{clutch size}}$; (2) the variance among females in the average number of clutches they produce, V_{clutch}; and (3) Cov_{Oc}, the covariance between clutch number c and clutch size O. We define O as the mean clutch size and c as the mean clutch number. Then, the total variance in female reproductive success can be partitioned as follows:

$$V_{\text{females}} = (c)^2 V_{\text{clutch size}} + (O)^2 V_{\text{clutch}} + 2(c)(O)\,\text{Cov}_{Oc}. \qquad [5.1]$$

The mean number of offspring per female is equal to the product of mean clutch size (O) and the mean number of clutches (c) as long as O and c are not correlated. When O and c are correlated, then the mean is equal to $[(c)(O) + \text{Cov}_{Oc}]$. Dividing eq. [5.1] by the square of the mean offspring per female, $(cO + \text{Cov}_{Oc})^2$, we obtain the opportunity for selection on females, I_{females}. We find

$$I_{\text{females}} = (1/a)^2\{I_{\text{clutch size}} + I_{\text{clutch}} + 2\text{Co-}I_{Oc}\} \qquad [5.2]$$

where $I_{\text{clutch size}}$ is the ratio $(V_{\text{clutch size}}/[O]^2)$, I_{clutch}, is $(V_{\text{clutch}}/[c]^2)$, $\text{Co-}I_{Oc}$ is (Cov_{Oc}/cO), and a equals the squared mean female fitness or $(1 + \text{Cov}_{Oc}/cO)^2$. When the clutch number and size do not covary, then eq. [5.2] reduces to

$$I_{\text{females}} = I_{\text{clutch size}} + I_{\text{clutch}}. \qquad [5.3]$$

The expression I_{clutch} represents the opportunity for selection on females arising from multiple reproductive events, or, alternatively, the variance in fitness among females owing to variation among females in how total fecundity is packaged into separate clutches. As explained above, this packaging of total fecundity can occur as multiple reproductive events, or by dividing individual clutches among males, by multiple mating. When females differ from one another in the number of clutches they produce, this variation becomes a major component of the female variance in fitness.

How does this latter behavior on the part of females affect selection on males and the sex difference in the opportunity for sexual selection? The answer depends upon whether or not the same male sires all of a female's clutches; that is, whether the covariance in paternity across clutches is positive, negative, or zero. If the same male sires every clutch of a given female (i.e., covariance in paternity across clutches is 1.0), the effect of producing multiple clutches on sexual selection *is less* than if a different, randomly selected, male sires each clutch of a given female (i.e., covariance in paternity across clutches is 0.0). Consider the first case. The variance in male reproductive success is equal to

$$V_{\text{males}} = V_{\text{females}} + (cO)^2 V_{\text{mates}}, \qquad [5.4]$$

and the variance in female reproductive success equals [5.1]. Dividing eq. [5.4] by the square of the mean success of males, $(mcO)^2$, where m equals the average number of mates per male, gives us

$$I_{\text{males}} = (R_0)^2 (I_{\text{females}}) + I_{\text{mates}}, \qquad [5.5]$$

where we have substituted $(1/m)^2$ for $(R_0)^2$ as in chapter 2, eq. [2.14]. When one male sires all the clutches produced by a single, iteroparous female, the sex difference in the opportunity for sexual selection, I_{mates}, is unchanged.

Now consider the second case, where a different male sires every clutch. We find that

$$V_{females} = V_{clutch\ size} + (O)^2 V_{clutch}. \qquad [5.6]$$

Dividing by the square of the mean number of offspring per female, $(cO)^2$, we obtain

$$I_{females} = (1/c)^2\, I_{clutch\ size} + I_{clutch}, \qquad [5.7]$$

where the opportunity for selection on clutch number, I_{clutch}, equals the ratio $(V_{clutch}/c)^2$. This is the component of the relative fitness variance among females that results from variation among females in the number of clutches.

When a different, randomly selected male sires every clutch, then

$$V_{males} = V_{clutch\ size} + (O)^2 V_{mates}. \qquad [5.8]$$

We substitute $(V_{females} - [O]^2 V_{clutch})$ for $V_{clutch\ size}$, and find that

$$V_{males} = (V_{females} - [O]^2 V_{clutch}) + (O)^2 V_{mates}. \qquad [5.9]$$

Finally, we divide by the square of the average male fitness, $(mO)^2$, to obtain

$$I_{males} = (R_O)^2 (I_{females} - I_{clutch}) + I_{mates}. \qquad [5.10]$$

In obtaining eq. [5.10], we used the fact that mean male fitness, mO, equals mean female fitness multiplied by the sex ratio, cOR or $[cO/R_0]$.

Thus, tendencies for females to engage in multiple mating, *and/or* to produce multiple clutches of progeny, *reduce* the sex difference in the opportunity for selection:

$$I_{males} - I_{females} = I_{mates} - I_{clutch}, \qquad [5.11]$$

where we have set R_0 equal to 1. This occurs because both tendencies simultaneously *decrease* the variance in fitness among males and *increase* the variance in fitness among females.

Wade (1987) derived the total opportunity for selection on males and females similarly as

$$I_{males} = (1/R)(I_{females} - I_{clutches}) + I_{mates}. \qquad [5.12]$$

Comparing eq. [5.10] and eq. [5.12] we see that I_{clutch} and $I_{clutches}$ are equivalent. As shown in this equation as well as above, when females produce multiple clutches and seek new mates to sire each clutch, I_{mates} no longer equals the sex difference in the opportunity for selection. We can understand this effect intuitively. When a female mates *once* and produces only a *single clutch* of offspring, her entire reproductive output is awarded to a single male. If some males mate with more than one such female and others mate with fewer, then the variance in mate numbers among mating males increases the variance in male fitness but does not affect that of females. The most successful males have a reproductive fitness equal to that of several females combined; specifically, their fitness is equal to the number mates, m, times the average clutch size, O. The least successful males have no mates and hence no fitness at all (chapter 1).

In contrast, when a female mates more than once, she does not award all of her progeny to a single male. Instead, in effect, she *partitions* her progeny into several clutches, each equal in size to the number of mates or sires. She does not change her reproductive success but she does change the variance among males in reproductive success because the number of offspring per mating is lowered for males. Even if the variance in mate numbers among mating males does not change, the variance in male reproductive success is decreased because each mating results in fewer progeny sired by each male. Thus, when females mate multiply, I_{mates} is *eroded* because the proportion of a female's clutch that is successfully fertilized by each mating male is *decreased.*

From the male viewpoint, each mating results in his siring of only a fraction of the total brood, O_{total}, of each mate he secures. This is equivalent to reducing the effective mean number of mates from R to (R/c) or, alternatively, reducing the effective variance in mate numbers among males from V_{mates} to (V_{mates}/c^2). Instead of calculating the average male's reproductive success as mO, when females mate once, the average male's reproductive success is reduced to $m(O_{total}/c)$ when females mate more than once. The multiplier, $(1/c)$ or f, equals the average of the inverse of the number of mates per female, or the *harmonic mean number of mates per female*, per mated male. The variance in the number of mates is reduced by a factor equal to f^2 or, equivalently, the number of mates per male is reduced by f. Multiple mating by females erodes I_{mates} because this activity reduces the variance in offspring numbers among the mating males or the *effective* variance in mate numbers (see chapter 4).

As an extreme example, consider the case in which a female lays *only one egg* every time she mates, and she obtains another mate at random after each laying. Recall from eq. [5.12] that $I_{males} = (I_{females} - I_{clutch}) + I_{mates}$. In this one egg–one mate example, the variance in egg numbers among females equals the variance in male mate numbers among males. When these conditions exist, $I_{mates} = I_{clutch}$, and, because I_{males} equals $I_{females}$, there is no sex difference in the opportunity for selection.

The above discussion shows that, just as there are two multiplicative components to male fitness, mate numbers, and offspring per mating, there are also two components to female fitness, clutch numbers, and offspring per clutch. We have shown how these fitness components influence the total variance in male and female fitness and how they contribute to the sex difference in fitness variance, which permits selection in one sex to be stronger than selection in the other sex. However, the effects of multiple mating and multiparity on the variance in female fitness can be somewhat more complicated when males contribute nutritionally to females during copulation. In these circumstances, the numbers of mates she has may influence the size of a female's clutch, O. We can understand these components of female fitness better by first examining what constitutes fitness for individuals in specific biological populations, and then considering which experimental designs are necessary to identify each fitness component.

Parental Effects: Assigning Fitness to Parents and Offspring

A gene expressed in the parental genome that affects the viability of all offspring, independent of the zygotic genotype, experiences only half the selection of a zygotically expressed gene with a similar, but direct, effect on viability (Wade 1998; Whitlock and Wade 1995; Wolf and Wade 2001). Whenever parents perform behaviors that affect the viability of their offspring, we need to consider carefully the different evolutionary consequences of assigning fitness to parents versus assigning fitness to offspring. This is especially important in behavioral ecology where parental effects on offspring survival can be large. Clutton-Brock (1988, p. 473) stated, "In particular, differences in offspring survival after fledging or weaning were the principal source of variation in reproductive success among breeding adults in several species." For this reason, many behavioral ecologists include components of offspring fitness, especially viability and son reproductive success, into calculations of parental fitness, despite the 50% difference in the intensity of selection for parental versus direct effects. When a single gene has pleiotropic affects on adult and offspring fitness, this difference in the strength of selection can affect our perception of the gene's evolution.

When, if ever, should components of offspring fitness be assigned to parents in evolutionary analysis? This issue has been contentiously debated for some time (Grafen 1988; Cheverud and Moore 1994) and, recently, Wolf and Wade (2001) provided an explicit theoretical treatment of the conceptual benefits and risks inherent in awarding offspring fitness to parents. They showed that, whenever there is a genetic correlation between adult and offspring fitnesses (e.g., owing to pleiotropy, linkage, or nonrandom mating), awarding offspring fitness to parents affects the evolutionary inference. In particular, it can lead to mistaking the direction as well as the strength of selection. We discuss this important issue here because in behavioral ecology it is commonplace to award offspring viability fitness to parents, when testing ESS theory with empirical data. Indeed, it has been recommended that, when mate choice of either sex is based upon "good genes" for viability, one *must* assess offspring fitness as a component of parental fitness in order to understand evolution in the context of mate choice. This recommendation, however, has not been accompanied by discussion of the different weightings necessary for direct and indirect effects in the evolutionary dynamic or of the experimental methods required for partitioning offspring viability into direct and parental effects.

Disentangling parental from offspring influences on fitness is especially problematic when parental investment in one offspring *decreases* a parent's ability to invest in other offspring, that is, when the covariance between parental and offspring fitnesses is *negative*. For most behavioral ecologists, this is the very definition of parent-offspring conflict (Trivers 1972; Clutton-Brock 1991) and there is little doubt that variations in fecundity are influenced by the quality of parental care. Moreover, it is obvious that, without parental care, offspring in many species cannot survive. Although it is reasonable to assume a causal connection between offspring fitness and parental behavior, this connection does not justify treating offspring fitness traits as components of maternal or paternal fitness in evolutionary analysis. The problem is that most attempts to assess the effects of parental care on offspring traits do so by awarding offspring fitness to parents instead of experimentally distinguishing the separate effects. This approach combines fitness effects *across more than one generation* and makes estimation of the strength of selection impossible unless one can be certain that there are only parental *or* only direct effects on offspring fitness. The magnitude of errors in the estimation of selection intensities when there are joint parental and direct offspring effects on offspring viability is discussed in Wolf and Wade (2001).

The potential confounding of maternal and direct effects can be resolved only with cross-fostering experiments or other manipulative studies that permit overall offspring viability to be partitioned into the separate causes of maternal (or indirect) and offspring (or direct) effects and their interaction. Such studies are prohibitive for many species. Thus, in our discussion below, we will assume that *offspring numbers* per mating pair reflect variations in the parental phenotypes and are not influenced by the genotypes of the offspring themselves. This assumption can be justified in some systems (Grafen 1992) and is recommended by Wolf and Wade (2001). As we have shown in previous chapters, offspring numbers translate easily into mate numbers when the variance and average number of offspring per clutch are known.

Measuring the Variance in Offspring Numbers

Our next task is to identify methods for quantifying male and female influences on the variance in offspring numbers. The most efficient scheme for accomplishing this goal is a factorial breeding design with males and females considered as separate factors, as in a two-way analysis of variance (Sokal and Rohlf 1995). This approach compares the numbers of progeny produced by a large number of male-female pairs, and thereby identifies the main effects of male and female parents, as well as the effects of parental interactions on offspring numbers. With increasing numbers of clutches produced by each pair, and with increasing numbers of mates per individual, the power of this design to distinguish the effects of each parent on offspring numbers increases. Furthermore, if groups of genetically related males and females are created, then the interaction effects of material and genetic contributions by parents to offspring numbers can be extracted as well. These experimental features are important in light of current discussions of mate choice based on the "quality" of one or both members of a mating pair (Thornhill and Alcock 1983; Petrie 1994; Gowaty 1997; Qvarnström and Forsgren 1998; Olson and Owens 1998). In these and other similar discussions, the influence of parental contributions on offspring numbers and viability, particularly those involving *interactions* among male and female factors, is assumed to be significant.

Although a two-factor, analysis of variance seems ideal for determining the degree to which genetic and material contributions of parents may influence offspring numbers, this design is problematic for most organisms and field research into mating systems. In order to execute a complete two-factor ANOVA, *all* females must be mated with *all* males in the experimental or natural population. (In more restricted chain-block designs, at least some randomly chosen females must be mated with some randomly chosen males.) This requirement is practical for only the most experimentally labile animal species (e.g., *Tribolium*, Lewis and Austad 1994; *Drosphila*, Clark et al. 1999), although it may be much easier to accomplish in plants or in external fertilizers, such as fish. Interactions may exist between the genetic and material contributions of male and female parents, and these interactions may influence offspring numbers. However, these interactions cannot be measured in most field situations, and are prohibitively difficult to measure in most laboratories. The study of such factors is therefore beyond the experimental grasp of most researchers.

A suggested solution to this difficulty is the so-called "free choice" design (Drickamer et al. 2000; Jones et al. 2000; Nilsson and Nilsson 2000), where individual males and females are allowed to select a mate from between two possible candidates, randomly chosen from the study population. The candidates are randomly chosen because the criteria for mate choice are not known *a priori*. Offspring numbers, sex ratios, and various indices of offspring viability resulting from "free choice" pairings can be compared with those obtained from random, "no choice" pairings of males and females under control of the experimenter. In peacocks (*Pavo cristatus*), such experiments revealed no fecundity differences between free choice and no choice females (Petrie 1994). However, the progeny of peacock females monogamously mated to males possessing phenotypes correlated with mating success on leks were larger, grew more elaborate trains, and survived better than random choice offspring. Other fitness components or other attributes of progeny were not compared in these experiments. In polygynandrous mice (*Peromyscus* sp., chapter 6; Drickamer et al. 2000), monogamously mated free choice females produced more litters than no choice females (X^2 = 7.602, $P < 0.01$), but did not produce significantly different total offspring numbers (X^2 = 3.839, $P > 0.05$). Females from experimental and control groups did not differ in 10 other traits related to litter production in pairwise tests. However, free choice male progeny were found to be more socially dominant than no choice male progeny (N = 16), free choice females survived better than no choice females (N = 39), and, when exposed to cold, free choice mice of both sexes built better nests (N = 202; Drickamer et al. 2000).

These results are intriguing, but the conclusions they allow are somewhat limited from a quantitative perspective, and therefore must be interpreted with caution. Consider the case where female mate choice requires that the difference between a pair of candidate males exceeds a specific value of a preference trait. For example, if the difference in body size of two males exceeds x, then the female chooses the larger male, but if it is less than x, the female does not choose or is equally likely to mate with either male. A randomly chosen pair of candidate males has, on average, only half the variance for the trait that exists within the population, i.e., the trait variance between the pair of candidates, $V_{candidates}$, equals only $(1/2)V_{population}$. Because the difference between the pair of candidates will frequently be less than x, many females will not appear to exert a choice when offered a comparison between only two males (the frequency of these outcomes is determined by the underlying distribution of the preferred trait in the male population). When "nonchoosing" females are eliminated from the analysis of data, as is often done in practice, then the parametric difference between the "free choice" and "no choice" classes is necessarily *overestimated.* That is, the fitness advantage to mate choice will appear greater in the experiment than it is in the natural population whenever nonchoosing females are eliminated from the analysis.

Excluding nonchoosing females, when estimating the effects of mate choice, is similar to excluding nonmating males from estimates of male fitness in sexual selection. Furthermore, when mate choice is determined by the interaction between a female and a male trait, offering a female a single pair of candidate males has another effect. It causes the variance in mate choice, owing to the interaction between the sexes, to appear as though it were a male effect. This too leads to an *overestimate* of the effects of mate choice on offspring fitness and, for small populations, a lack of repeatability of the results (Wade 2001). That is, interactions between the sexes will appear as single-sex effects whenever, by experimental design, one sex is allowed to vary but the other sex is held constant, as is characteristic of "free choice" studies.

There is another difficulty with free choice tests. Although free choice tests can establish a fitness advantage to mate choice, they necessarily confound parental effects with direct effects, whenever males contribute material to females while copulating, whenever females care for young, or both. Parental effects cannot be distinguished from direct offspring effects without cross-fostering. In most free choice studies, the fitness correlates differ modestly between the experimental and control groups, despite being overestimates, so that the strength of selection is unknown. The confounding of parental with offspring fitness effects also complicates interpretation of the results and no attempt has yet been made in these studies to experimentally discriminate between the two kinds of effects. If both effects were always positive or always negative, then the quantitative problem as well as our concerns would be diminished. When the effects are of opposite sign, however, they must be separated before an accurate evolutionary analysis is possible. In many circumstances, parental and offspring fitness effects of opposite sign are *expected* to be associated (Wolf 2000; Wolf and Brodie 1998; Wolf et al. 1999; Agrawal et al. 2001a). For these reasons, unambiguous evidence of male genetic quality is likely to remain elusive. As explained above, the specific effects of mate "quality" on offspring numbers can be determined only in species in which females tend to mate more than once and produce more than one clutch of offspring, in natural as well as under experimental conditions.

In summary, the "free choice" experiments may help establish the existence of female mate choice but they have several flaws that, when combined with the practice of data editing, prohibit their utility as a means of establishing a mechanism behind the choice or a parametric estimate of the strength of sexual selection resulting from it. The design confounds female choice resulting from a difference between the value of male trait(s) with an interaction between female choice and male trait(s) and, to the extent that such interactions are present, it is likely that results from one study will vary from those of others. By limiting female choice to a pair of males, the design causes the variance among test male pairs to be significantly less than that among males in the population as a whole, which reduces the average difference between males available to the female. Lastly, the design does not address the important evolutionary difference between direct genetic and indirect parental contributions to male traits. Selection experiments, in which the evolution of a random mating population is compared with that of a freely choosing population, do not suffer from these same concerns.

Nested Breeding Designs in the Laboratory and in Nature

While few species will breed according to a factorial experimental design, *nested* breeding designs provide a possible means for measuring the relative influences of male and female parents on offspring number. In such designs, the effects of males and females are presented as if they represent a *random* selection from the breeding population. Experimenters identify each of the males mating with an individual female, as well as the separate clutches of offspring produced by each female. This scheme can be used in laboratory as well as in field situations, provided that parentage can be accurately assigned, and as molecular methods for accurate paternity assignment become available for more organisms (Birkhead and Møller 1993a,b; 1998; Webster et al. 1995; McCoy et al. 2001; Jones et al. 1997, 2001a,b), these requirements will become less onerous. Indeed, if one is able to detect maternity and paternity of all offspring, then the sex difference in the variance in fitness becomes transparent.

To illustrate this experimental approach, let us imagine a species in which females normally mate more than once and have more than one clutch of offspring (table 5.1). Let us assume that, when these females become receptive, they receive sperm from a single male, oviposit shortly thereafter, and do not mate again until they have oviposited at least two more times. Alternatively, we could imagine that males guard females until they have laid an average of three clutches of eggs. After their third oviposition, females molt, shed any sperm remaining in their reproductive tract, and become receptive to another mate. We could as easily imagine females in an externally fertilizing species which spawn an average of three successive times with the same male and then locate another male, or a plant species which blooms repeatedly and whose stigmas receive pollen from a series of different male flowers.

Table 5.1 shows a diagram of three breeding episodes for three such females, in which C_{ijk} represents the *k*th clutch, sired by the *j*th male, mating with the *i*th female. This breeding scheme will look familiar to students of quantitative genetics because it has the form of a two-level, nested analysis of variance, the same type of analysis that is used to examine the resemblance of offspring to their male and female parents (Falconer 1989). The difference in our example is that we are examining offspring numbers rather than offspring phenotypes. By this approach we seek the components of the total variance in offspring numbers that are identifiable as maternal or paternal effects.

There are three sources of variance in offspring numbers that are identified using this breeding scheme. The first source of variance is the variance in clutch sizes within females, V_{within}. This quantity is equal to the variance in offspring numbers produced by each female, averaged across all females. It is expected to become increasingly large as females produce multiple clutches. Thus, using notation similar to that which we introduced earlier, we can also refer to this quantity as $V_{\text{cs,clutch}}$. The second source of variance in offspring numbers is the variance in clutch size within females due to the effects of matings with different males, $V_{\text{cs,sires}}$. This quantity equals the

Table 5.1
Breeding design for females that mate more than once and produce more than one clutch of offspring with each mate; sperm are not stored after the last oviposition with each mate.

Females (a = 3)	F_1			F_2			F_3		
Males (b = 3)	M_{11}	M_{12}	M_{13}	M_{21}	M_{22}	M_{23}	M_{31}	M_{32}	M_{33}
Clutches (n = 3)	C_{111}	C_{121}	C_{131}	C_{211}	C_{221}	C_{231}	C_{311}	C_{321}	C_{331}
	C_{112}	C_{122}	C_{132}	C_{212}	C_{222}	C_{232}	C_{312}	C_{322}	C_{332}
	C_{113}	C_{123}	C_{133}	C_{213}	C_{223}	C_{233}	C_{313}	C_{321}	C_{333}
Sire sums ΣC_{ijk}	C_{11}	C_{12}	C_{13}	C_{21}	C_{22}	C_{23}	C_{31}	C_{32}	C_{33}
Sire means C_B	C_{B11}	C_{B12}	C_{B13}	C_{B21}	C_{B22}	C_{B23}	C_{B31}	C_{B32}	C_{B33}
Dam means C_A			C_{A1}			C_{A2}			C_{A3}

Grand mean (C_{AB}) = $(1/nab)\Sigma^a\Sigma^b\Sigma^n C_{ijk}$

SS_{among} (among females) = $nb\ \Sigma^a (C_A - C_{AB})^2$

SS_{sires} (among sires within females) = $n\ \Sigma^a\Sigma^b (C_B - C_A)^2$

SS_{within} (among clutches within females) = $\Sigma^a\Sigma^b\Sigma^n (C_{ijk} - C_A)^2$

Sources of Variance in Offspring Numbers:

$V_{cs,clutch}$ = variance in clutch size within females (divided by $ab(n - 1) = MS_1$)

$V_{cs,sires}$ = variance in clutch size due to individual mating males (divided by $a(b - 1) = MS_2$)

$V_{cs,females}$ = variance in clutch size among females (divided by $(a - 1) = MS_3$)

MS_3 / MS_2 = the effect of individual females on offspring numbers

MS_2 / MS_1 = the effect of individual sires on offspring numbers

variance in the average clutch size among males across all females. Because the effects of males on female offspring numbers exist within females who mate more than once, we can partition V_{within} into two components, $V_{cs,clutch}$ and $V_{cs,sires}$. When $V_{cs,sires}$ is a large fraction of V_{within}, there will be strong selection for females to choose some males as mates and reject others. When $V_{cs,sires}$ is a small fraction of V_{within}, selection on other components of female reproductive success will be stronger than selection on female choice. A third source of variance in offspring numbers is the variance in the average clutch size among females, V_{among}. We refer to this quantity as $V_{cs,females}$. This quantity is equal to the variance in the average clutch size per female, calculated across all females, that is, it is the among-female component of variance in offspring numbers. Note that, if there are male-female interactions affecting clutch size, these will not reveal themselves in the nested design. As shown in table 5.1 and explained in more detail in Sokal and Rohlf (1995), these sources of variance can be expressed as sums of squares, which generate mean squares when divided by their degrees of freedom. Mean squares can then be compared, using F ratios, to identify the significant effects of females and males on the total variance in offspring numbers (table 5.1).

We can now identify both male and female influences on the total variance in offspring numbers as shown in table 5.1. The total variance in reproductive success among females can be partitioned into three components,

$$V_{\text{females}} = V_{\text{cs,clutch}} + V_{\text{cs,sires}} + V_{\text{cs,females}}. \qquad [5.13]$$

The experimental design above does not allow us to identify additional components of male and female contributions to offspring numbers or the interactions between the sexes or among the components. For example, we might imagine partitioning the male contribution to offspring numbers into several additional components: (1) the transfer of proteins and nutritional material, (2) the transfer of pathogens and disease, (3) the quality of defensive care, (4) the genetic contribution, and (5) the interactions among these components.

When some sires are better than others in the material transferred to females at copulation, then $V_{\text{cs,sires}}$ will contain a component of variance in offspring numbers, say $V_{\text{cs,material}}$. Similarly, males might vary from one another in their tendency to transfer pathogens, and another component of variance in offspring numbers, say, $V_{\text{cs,disease}}$, may directly influence female reproduction. It is also easy to imagine that males might vary in the incidence of pathogens or disease for genetic as well as environmental reasons. Further, it is reasonable to expect that the nutritional and disease components might covary with one another and interact positively or negatively. These male-specific causes of variation in offspring numbers are of interest in their own right and most will influence the evolutionary trajectory of the mating system and thereby the sex difference in the strength of selection. In the same manner, females could vary from one another in these same or similar components and thus influence the variance among females in offspring numbers. As is true for male parents, the contributions of female parents to the variance in offspring numbers could enhance or diminish the total variance in clutch size for genetic or material reasons.

And yet, although we can *identify* these possible sources of variance in offspring numbers among male and female parents, *measuring* these variance components is problematic. In order to measure these variance components at all, the experimenter must be able to generate a large number of clutches for males, as well as for females, and to separate the kinds of effects on offspring numbers. It is highly unlikely that we would be able to do this using a factorial experimental design with clutches randomly exposed to different kinds of male effects alone and in combination. These limitations for studies in mating system evolution exist for any species in which *all* females do not mate more than once and *all* females do not produce several clutches of offspring. A multivariate analysis using molecular markers for paternity and maternity analysis appears to be the only solution, and it too carries with it all of the limitations of correlational studies.

Other Patterns in Female Life History

If this type of female life history is problematic for understanding male and female influences on offspring numbers, what other, perhaps more experimentally tractable female life histories exist? That is, in which species is it *possible* to mate males with more than one female, mate females with more than one male, and assign offspring within clutches to individual parents? These conditions are necessary if we seek to identify the effects of material and genetic contributions by parents on offspring numbers, as well as the effects of interactions between these components on offspring numbers. The severity of these limitations for studies in mating system evolution seems overwhelming, and unfortunately these limitations are real! Remember that in the hypothetical species described in table 5.1, *all* females mated more than once and *all* females produced several clutches of offspring. As we will see, this type of female life history is atypical among animal as well as among plant species.

We can identify five other types of female life histories with respect to the relative complexity of their episodes of receptivity and fecundity (tables 5.2–5.6). We will describe them in terms similar to those shown in table 5.1 above to illustrate how well or poorly these female life histories lend themselves to the analysis of offspring numbers. For convenience, we will identify each life history as a "type." Later, we will condense these types into a simpler, more familiar terminology.

Type 1 females mate once in their lives and produce a single clutch of offspring (table 5.2). Examples of this type of life cycle may include univoltine insects such as mayflies or gnats (Sivinski and Petersson 1997), short-lived invertebrates including certain polychaetes or molluscs (Brusca and Brusca 1990), and perhaps certain annual plants with limited pollen dispersal. Type 2 females are receptive for a prolonged period, i.e., they mate more than once. They, like type 1 females, are semelparous and produce a single clutch of offspring. However, paternity within these clutches is mixed (table 5.3). Examples of this type of life cycle include semelparous, sphaeromatid isopods such as *Paracerceis sculpta* (Shuster 1990, 1992), or annual plants with moderate to long-distance pollen dispersal (Comes and Abbott 1998).

Type 3 females are receptive once in their lifetimes, but each female produces several clutches of offspring, all sired by the same male (table 5.4). Examples of this type of life cycle include monogamous, social insects such as termites (Wilson 1971, 1975), as well as long-lived monogamous or sequentially pairing birds and mammals (Jouventin et al. 1999; Winslow et al. 1993; Komers 1996). Type 4 females also mate more than once and produce more than one clutch of offspring; however, each clutch is sired by a separate male, and thus sperm are not stored between clutches (table 5.5). Exam-

Table 5.2
Breeding design for Type 1 females, which mate once and produce one clutch of offspring with each mate.

Females ($a = 3$)	F_1	F_2	F_3
Males ($b = 3$)	M_1	M_2	M_3
Clutches ($n = 1$)	C_{111}	C_{221}	C_{331}
Sire sums ΣC_{ijk}	C_{11}	C_{21}	C_{31}
Sire means C_B	C_{B11}	C_{B21}	C_{B31}
Dam means C_A	C_{A1}	C_{A2}	C_{A3}

$\Sigma C_{ijk} = C_B = C_A$
Grand mean (C_{AB}) = $(1/nab)\ \Sigma^a \Sigma^b \Sigma^n C_{ijk}$
SS_{among} (among groups) = $nb\ \Sigma^a\ (C_A - C_{AB})^2$
SS_{sires} (sires within dams) = $n\ \Sigma^a \Sigma^b\ (C_B - C_A)^2$
SS_{within} (within dams) = $\Sigma^a \Sigma^b \Sigma^n\ (C_{ijk} - C_A)^2$
Sources of Variance in Offspring Numbers:
$V_{cs,clutch}$ = does not exist because females produce only one clutch
$V_{cs,sires}$ = does not exist because females mate only once
$V_{cs,females}$ = V_{total} = population variance in clutch size among females

Table 5.3
Breeding design for Type 2 females, which mate more than once but produce one clutch of offspring; sperm are allowed to mix.

Females ($a = 3$)	F_1			F_2			F_3		
Males ($b = 3$)	M_{11}	M_{12}	M_{13}	M_{21}	M_{22}	M_{23}	M_{31}	M_{32}	M_{33}
Clutches ($n = 1$)	C_{111}			C_{221}			C_{331}		
Sire sums ΣC_{ijk}	C_{11}	C_{12}	C_{13}	C_{21}	C_{22}	C_{23}	C_{31}	C_{32}	C_{33}
Sire means C_B	C_{B11}	C_{B12}	C_{B13}	C_{B21}	C_{B22}	C_{B23}	C_{B31}	C_{B32}	C_{B33}
Dam means C_A	C_{A1}			C_{A2}			C_{A3}		

If paternity cannot be assigned,
$C_B = C_A$
Grand mean (C_{AB}) = $(1/nab)\ \Sigma^a \Sigma^b \Sigma^n C_{ijk}$
SS_{among} (among groups) = $nb\ \Sigma^a\ (C_A - C_{AB})^2$
SS_{sires} (sires within dams) = $n\ \Sigma^a \Sigma^b\ (C_B - C_A)^2$
SS_{within} (within dams) = $\Sigma^a \Sigma^b \Sigma^n\ (C_{ijk} - C_A)^2$
Sources of Variance in Offspring Numbers:
$V_{cs,clutch}$ = does not exist because females produce only one clutch; unless paternity can be assigned (see table 5.1)
$V_{cs,sires}$ = exists, but the effects of males are confounded within $V_{cs,females}$; unless paternity can be assigned (see table 5.1)
$V_{cs,females}$ = V_{total} = population variance in clutch size among females

Table 5.4
Breeding design for Type 3 females, which mate once but produce more than one clutch of offspring.

Females (a = 3)	F_1	F_2	F_3
Males	M_1	M_2	M_3
Clutches	C_{111}	C_{221}	C_{331}
	C_{112}	C_{222}	C_{332}
	C_{113}	C_{223}	C_{333}
Sire sums ΣC_{ijk}	C_{11}	C_{21}	C_{31}
Sire means C_B	C_{B11}	C_{B21}	C_{B31}
Dam means C_A	C_{A1}	C_{A2}	C_{A3}

$\Sigma C_{ijk} = C_B = C_A$
Grand mean (C_{AB}) = $(1/nab)\ \Sigma^a\Sigma^b\Sigma^n C_{ijk}$
SS$_{\text{among}}$ (among groups) = $nb\ \Sigma^a\ (C_A - C_{AB})^2$
SS$_{\text{sires}}$ (sires within dams) = $n\ \Sigma^a\Sigma^b\ (C_B - C_A)^2$
SS$_{\text{within}}$ (within dams) = $\Sigma^a\Sigma^b\Sigma^n\ (C_{ijk} - C_A)^2$
Sources of Variance in Offspring Numbers
$V_{\text{cs,clutch}}$ = variance in clutch sizes within females (divided by df = MS_1)
$V_{\text{cs,sires}}$ = does not exist because females mate only once
$V_{\text{cs,females}}$ = variance in clutch size among females (divided by df = MS_2
MS2/MS1 = the effect of individual females on offspring numbers

Table 5.5
Breeding design for Type 4 females, which mate more than once and produce more than one clutch of offspring, but only one with each mate; sperm are not stored after oviposition.

Females (a = 3)	F_1			F_2			F_3		
Males	M_{11}	M_{12}	M_{13}	M_{21}	M_{22}	M_{23}	M_{31}	M_{32}	M_{33}
Clutches	C_{111}	C_{121}	C_{131}	C_{211}	C_{221}	C_{231}	C_{311}	C_{321}	C_{331}
Sire sums ΣC_{ijk}	C_{11}	C_{12}	C_{13}	C_{21}	C_{22}	C_{23}	C_{31}	C_{32}	C_{33}
Sire means C_B	C_{B11}	C_{B12}	C_{B13}	C_{B21}	C_{B22}	C_{B23}	C_{B31}	C_{B32}	C_{B33}
Dam means C_A	C_{A1}			C_{A2}			C_{A3}		

$\Sigma C_{ijk} = C_B$
Grand mean (C_{AB}) = $(1/nab)\ \Sigma^a\Sigma^b\Sigma^n C_{ijk}$
SS$_{\text{among}}$ (among groups) = $nb\ \Sigma^a\ (C_A - C_{AB})^2$
SS$_{\text{sires}}$ (sires within dams) = $n\ \Sigma^a\Sigma^b\ (C_B - C_A)^2$
SS$_{\text{within}}$ (within dams) = $\Sigma^a\Sigma^b\Sigma^n\ (C_{ijk} - C_A)^2$
Sources of Variance in Offspring Numbers:
$V_{\text{cs,clutch}}$ = exists, but is confounded by $V_{\text{cs,sires}}$ (SS$_{\text{within}}$ divided by df = MS_1)
$V_{\text{cs,sires}}$ = may exist, but is confounded by $V_{\text{cs,clutch}}$
$V_{\text{cs,females}}$ = variance in clutch size among females (SS$_{\text{among}}$ divided by df = MS_2)
MS2/MS1 = the effect of individual females on offspring numbers

ples of this type of life cycle include mantis shrimp (Shuster and Caldwell 1989), many peracarid crustaceans (Shuster 1981a; Ridley 1983; Jormalainen and Shuster 1999), and solitary mammals (Clutton-Brock 1989).

Type 5 females mate more than once and produce multiple clutches as well, but each male sires more than one clutch, and sperm are not stored between sires. This is the female life history we have already discussed in table 5.1 above. Examples of this life cycle are common in annual plants (Delph et al. 1996), multivoltine insects (reviews in Choe and Crespi 1997), reef fish (Warner and Robertson 1978), and certain Hollywood film stars (*National Enquirer*, any issue). Type 6 females mate more than once, produce more than one clutch of offspring, and sperm mixing occurs within and between clutches (table 5.6). Examples of this type of life cycle include fruit flies (Bateman 1948), lizards (Hews 1993; Sinervo and Lively 1996), ground squirrels (Schwagmeyer and Woontner 1985), as well as group nesting birds and mammals (Dunn and Cockburn 1999; Clutton-Brock 1989). These six types of female life history are summarized in table 5.7.

The degree to which male and female influences on total offspring numbers can be identified within each of these female life histories is disappointing. In five of the six life history types, male and female effects contributing to the total variance in offspring numbers are confounded in some way. In type 1 and type 2 life histories, we can only calculate the population vari-

Table 5.6

Breeding design for Type 6 females, which mate more than once and produce more than one clutch of offspring, but sperm are allowed to mix between broods.

Females (a = 3)	F_1			F_2			F_3		
Males	M_{11}	M_{12}	M_{13}	M_{21}	M_{22}	M_{23}	M_{31}	M_{32}	M_{33}
Clutches	C_{111}			C_{221}			C_{331}		
	C_{112}			C_{222}			C_{332}		
	C_{113}			C_{223}			C_{333}		
Sire sums ΣC_{ijk}	C_{11}	C_{12}	C_{13}	C_{21}	C_{22}	C_{23}	C_{31}	C_{32}	C_{33}
Sire means C_B	C_{B11}	C_{B12}	C_{B13}	C_{B21}	C_{B22}	C_{B23}	C_{B31}	C_{B32}	C_{B33}
Dam means C_A	C_{A1}			C_{A2}			C_{A3}		

Grand mean (C_{AB}) = $(1/nab)\, \Sigma^a \Sigma^b \Sigma^n C_{ijk}$

SS_{among} (among groups) = $nb\, \Sigma^a (C_A - C_{AB})^2$

SS_{sires} (sires within dams) = $n\, \Sigma^a \Sigma^b (C_B - C_A)^2$

SS_{within} (within dams) = $\Sigma^a \Sigma^b \Sigma^n (C_{ijk} - C_A)^2$

Sources of Variance in Offspring Numbers:

$V_{cs,clutch}$ = variance in clutch sizes within females (SS_{within} divided by df = MS_1)

$V_{cs,sires}$ = exists but is confounded by $V_{cs,clutch}$ unless paternity can be assigned

$V_{cs,females}$ = variance in clutch size among females (SS_{among} divided by df = MS_2)

MS2/MS1 = the effect of individual females on offspring numbers

Table 5.7
Six types of female life history classified by reproductive episode and mate numbers.

	Mate Numbers	
Reproductive Episodes	*One*	*More than one*
One	1	2
More than one	3	4, 5, 6

ance in offspring numbers among females ($V_{cs,females}$). In both of these life history types, there is no within-dam component of variance in offspring numbers ($V_{cs,clutch}$) because females produce only one clutch. Moreover, since type 1 females mate only once, there is no sires-within-females ($V_{cs,sires}$) component to the variance in offspring numbers. Although type 2 females mate more than once, unless paternity can be precisely assigned, $V_{cs,sires}$ and the among-females component of variance in offspring numbers, $V_{cs,females}$, are confounded. If paternity can be assigned, and if clutch size is sufficiently large, then type 2 females can be considered as type 5 females. However, if clutch sizes are small or if sperm competition leads to single male paternity, type 1 and type 2 females are indistinguishable.

The within-female component of variance in offspring numbers, excluding the effects of individual sires ($V_{cs,clutch}$) can be calculated in all iteroparous females (types 3–6). However, in type 3 females, the sires-within-females component of variance in offspring numbers ($V_{cs,sires}$) does not exist because a single male sires all of the progeny in all of a female's clutches. $V_{cs,sires}$ does exist in type 4 females. However, $V_{cs,clutch}$ confounds this variance component because females produce only a single clutch of offspring with each sire. As shown above (table 5.1), $V_{cs,clutch}$, $V_{cs,sires}$, and $V_{cs,females}$ can be identified in type 5 females, and this is also possible for type 6 females, but only if paternity can be accurately assigned within and between clutches. If this is not possible, then $V_{cs,clutch}$ confounds $V_{cs,sires}$. If sperm competition occurs between broods such that, despite multiple matings by males, the sperm of a particular male is most successful in fertilizing a female's ova, then life history types 4, 5, and 6 become indistinguishable from life history type 3, in which $V_{cs,sires}$ is confounded by $V_{cs,clutch}$.

Clearly, few female life histories lend themselves easily to nested analysis of variance designs. This is why we identified type 5 life history (table 5.1) as a special case. However, we can still use the issues discussed above to understand the role that female life history characters play in shaping mating system evolution, despite the difficulty in identifying the effects of males and females on offspring numbers in many species. Although male and female effects on offspring numbers may be difficult to measure, we can still understand the relative importance of their contributions to total offspring numbers.

Adjusting I_{mates}

We discussed earlier in this chapter how I_{clutch} can erode I_{mates}. It is now clear that the three components of variance in offspring numbers among females count differently or not at all toward I_{clutch} in monogamous and polygamous species, as well as in semelparous and iteroparous species. We described six types of female reproductive life history in tables 5.1–5.7. However, these female life histories converge to four fundamental classes depending on whether paternity can be assigned or sperm competition occurs within clutches. Table 5.8 shows the four classes as well as the components of variance in female fitness that contribute to I_{clutch} for each life history class. Note that each of these components of fitness variance has been transformed to opportunities for selection by dividing by the squared grand average in offspring numbers per female.

As we have already shown, when females mate once and produce only one clutch of offspring, $I_{females}$ equals $I_{cs,females}$ and $I_{cs,clutch}$ equals zero. Thus, in species with monogamous, semelparous females, I_{mates} will remain unaffected by the characteristics of female life history. When females mate more than once but produce only one clutch of offspring, the sires-within-females component of female fitness becomes measurable ($I_{cs,sires}$). In species with polygamous, semelparous females, $I_{cs,clutch}$ is equal to $I_{cs,sires}$ and will erode I_{mates} by this amount. When females mate once but produce more than one clutch of offspring, the within-female component of female fitness ($I_{cs,clutch}$) becomes measurable and will erode I_{mates} by this amount. When females mate more than once and produce more than one clutch of offspring, the sires-within-females component of female fitness ($I_{cs,sires}$) as well as the within-female component of female fitness ($I_{cs,clutch}$) become measurable. Now, $I_{cs,clutch}$ plus $I_{cs,sires}$ will erode I_{mates}. If the single clutch produced by each polygamous, semelparous female is sufficiently large that a within-females component of female fitness can be identified, then these females are indistinguishable from species with polygamous, iteroparous females, i.e., females who mate more than once and produce more than one clutch of offspring.

Table 5.8
Variance components contributing to I_{clutch} by female life histories when classified by reproductive episodes and mate numbers.

	Mate Numbers	
Reproductive Episodes	*Monogamous (= 1)*	*Polygamous (>1)*
Semelparous (=1)	0	$I_{cs,sires}$
Iteroparous (>1)	$I_{cs,clutch}$	$I_{cs,clutch} + I_{cs,sires}$

With accurate paternity assignment and large clutch sizes within-females, sires-within female, and among-females components of variance in female fitness can be identified in each of these female life histories. Both $V_{cs,sires}$ and $V_{cs,clutch}$ contribute to I_{clutch}, and the larger these components become, the more the opportunity for sexual selection, I_{mates}, will be eroded. The magnitude of $I_{cs,sires}$ relative to I_{clutch} is likely to vary among species. However, in general, greater variance in offspring numbers is likely to exist among females who mate once and produce multiple clutches of offspring (monogamous, iteroparous females) than among females who mate more than once and produce only one clutch of offspring (polygamous, semelparous females), unless male contributions to offspring numbers are extreme.

Three Conclusions

Three conclusions are now possible from the above discussion. First, in most species, the contributions of male genetic "quality," or, for that matter, any material or genetic contributions by males to the total variance in offspring numbers among females, are *extremely difficult to quantify*. This is an important point because the contribution of male quality to female fitness is assumed to be significant in the majority of recent publications discussing this issue (reviews in Gowaty 1997; Ryan 1998; Olson and Owens 1998; Jennions and Petrie 2000). We have shown that, when males do contribute genetic or material benefits to females, the contributions will be empirically detectable only in species in which females mate more than once and produce more than one clutch of offspring.

In many cases, the effects of sires and dams on offspring numbers are confounded. In laboratory situations, it may be possible to separate these confounded effects by experimentally forcing females into life histories that involve multiple mating, cross-fostering, and multiparity. However, conclusions drawn from these experiments must be tempered by how severely experimental protocols diverge from the life histories of individuals in natural populations. To be meaningful, experimental breeding designs should replicate the breeding systems of the focal species in nature.

The second conclusion is that, regardless of whether male contributions to female fitness can be identified, these effects of male "quality" lie *within* $I_{cs,sires}$, which lies *within* I_{clutch}, which lies *within* $I_{females}$ (see eq. [5.12]). This means that the opportunity for selection on male quality will generally be *smaller* in magnitude than the opportunity for selection on females, which will generally be *smaller* than the sex difference in the opportunity for selection, I_{mates} (see eq. [5.12]). This is true, as explained in chapter 1, because the fitness of the average mating male is equal to the average number of mates per mating male, m, times the average number of offspring per female, O. Thus, when m is larger than 1, the variance in offspring numbers among mating males, V_{mO}, will be larger than the variance in offspring numbers among females, V_O, by a factor of m. As fewer males mate and more males do not, I_{mates} becomes extremely large compared to the opportunity for selection on females, I_{females}. In such cases, I_{clutch}, and particularly $I_{\text{cs,sires}}$, will become extremely small.

Thus, any female tendency that reduces the number of mating males, such as female mate choice, is expected to increase the fitness variance among males *disproportionately* compared to the fitness variance among females. Small effects on fitness over long periods of time do produce evolutionary effects. However, when there are several simultaneous effects on male fitness variance, the larger factors will predominate and interfere with the effects of selection on the traits of smaller effect. If the small effect is uniform, it has a better chance of affecting the evolutionary process than if it is highly contingent, i.e., a pure male effect versus a male-female interaction effect. Interactions, however, are more effective at causing differences among subdivided populations (Wade 2002).

This brings us to our third observation. The only circumstances in which the above conditions will *not hold* will be when m equals 1, that is, when *all* males are able to mate. However, even in this unusual situation, the component of female fitness arising from matings with different sires ($I_{\text{cs,sires}}$) will remain elusive unless females mate with multiple males or we can use markers to identify paternity.

These observations reinforce our conclusion that the effects of male parental care and male genetic quality on offspring numbers can be clearly identified by experiment only when females mate more than once and produce multiple clutches of offspring, that is, only in species in which females are *polygamous and iteroparous*. It is in such species that the value of I_{clutch} is largest and therefore most effective in reducing the sex difference in the opportunity for selection. The greater the reduction in the opportunity for sexual selection, the greater the importance of male genetic and material contributions to offspring numbers. This is why experiments designed to examine the effects of mate quality on offspring numbers should replicate the natural breeding system of the species studied as closely as possible. The importance of male contributions to offspring numbers depends on the proportion of the total variance in offspring numbers such contributions represent. The greater the variance in fitness among males, the smaller the contribution "male quality" can make to the total fitness variance.

Our observations also show that in most species, the effects of variance in mating success among males will often far exceed the effects of viability selection on females and their progeny. This means that in species in which sexual selection on males is strong and $V_{\text{cs,sires}}$ is a small component of V_{females}, female mate choice, if it exists, is likely to be based on arbitrary male characters (chapters 1 and 4). When the number of mates per mating male equals the number of mates per mating female ($m = 1$), there is no opportunity for sexual selection. However, this situation, as mentioned, is unstable. If ecological circumstances arise in which females begin to have more mates than males, I_{clutch} can become very large, and the sexes may begin to reverse, both in the degree to which each sex provides parental care, as well as in their external phenotypes (chapter 8).

Chapter Summary

Hypotheses involving female responses to male mating behavior currently focus on the phenotypic outcomes of natural selection on females, when that selection is opposed by sexual selection on males. The intensity of such viability selection is widely assumed to equal or exceed that of sexual selection, and although speculation on these issues abounds, few attempts have been made to quantify the relative intensities of these sex-specific selective forces in nature. We show that selection on females arises from different patterns of female receptivity and reproductive life history. The two multiplicative components of male fitness are mate numbers and offspring per mating. The corresponding two components of female fitness are clutch numbers and offspring per clutch. Indeed, we show that these components of female fitness variance *can* erode the sex difference in the opportunity for selection. However, our investigation of the possible methods for measuring these fitness components leads to the following conclusions.

First, when offspring fitness is assigned to parents, as is the case when female mate choice or parental behavior is presumed to influence offspring "quality," serious quantitative errors can arise when predicting the strength and direction of selection. Second, mate choice experiments that compare the effects of "free choice" with "no choice" matings not only confound parental and offspring contributions to fitness, but also tend to exaggerate male contributions to observed patterns of mate choice. Third, whereas factorial breeding designs can identify male and female contributions to offspring fitness, as well as the interacting, "complementary" parental effects presumed to drive mate preferences, such designs are impractical for all but the most tractable laboratory species. Fourth, although nested breeding designs can identify sire and dam effects on offspring numbers, this approach too is complicated by the fact that male and female influences on offspring numbers are often confounded.

In species in which male influences on female fitness can be extracted, we show that the degree to which viability selection can oppose sexual selection depends on the relative magnitude of the variance in male and female fitness. Our analyses indicate that viability selection constrains sexual selection *only* when sexual selection is weak. Thus, male "quality" is expected to influence female fitness most when sexual selection is negligible. Consequently, when sexual selection is strong relative to viability selection, female mate preferences are likely to be arbitrary. In general, male quality will be most important when variances in male and female fitness are approximately equal in magnitude. However, such mating systems may be unstable, and when fitness variance arises in one or the other sex, distinct sexual phenotypes will emerge, or, in some cases, reverse in character.

6 The ΔI Surface

In the first three chapters, we developed a statistical approach for describing the opportunity for sexual selection on males arising from the spatial and temporal distributions of receptive females. We have used the same approach to translate the differences in mate numbers and reproductive episodes in female life histories into opportunities for selection on females. Sexual dimorphism results from the *sex difference* in the opportunity for selection, which we call ΔI. We will now begin to combine the above methods to develop a more testable theory for mating system evolution.

We saw in chapters 4 and 5 how multiple mating and multiparity by females each reduce the magnitude of I_{mates} resulting from female spatiotemporal distributions. We will combine the sex-specific selection intensities into ΔI, as a means for visualizing the sex difference in the opportunity for selection. We illustrate how ΔI can be used to understand the taxonomic variation in mating systems arising from diverse sources of selection. Our approach extends the Emlen and Oring hypothesis by identifying ΔI, as the key to understanding the evolution of sexual dimorphism. Our approach generates more specific categories of mating systems than are currently in use (see also chapter 9), and thus explains more of the possible variation in mating system diversity than nonquantitative, inductive approaches based on parental investment theory. Our approach also indicates when apparently different mating systems should be expected to give rise to comparable levels of sexual dimorphism.

Multiple Regression and Female Spatiotemporal Distributions

Recall from chapter 3 that, under many conditions, the opportunity for selection on males, I_{mates}, arising from the spatial and temporal distributions of females equals m^*_{obs}, the observed mean crowding of *matings*. As we have shown, the intensity of sexual selection arising from each aspect of the female distribution of receptivity can be measured independently (Wade 1995; chapter 3). However, to identify within-species variation in the values of I_{mates} as well as to characterize seasonal patterns in I_{mates} between species, we need a method that (1) summarizes values empirically determined from specific paired values of m^* and t^* into a single estimate of m^*_{obs} ($= I_{\text{mates}}$) and (2) permits us to examine quantitatively the effects of varying (m^*, t^*) on the sex difference in strength of selection.

We have already visualized the interaction between m^* and t^* by plotting their values on two horizontal axes and then plotting values of m^*_{obs} on a

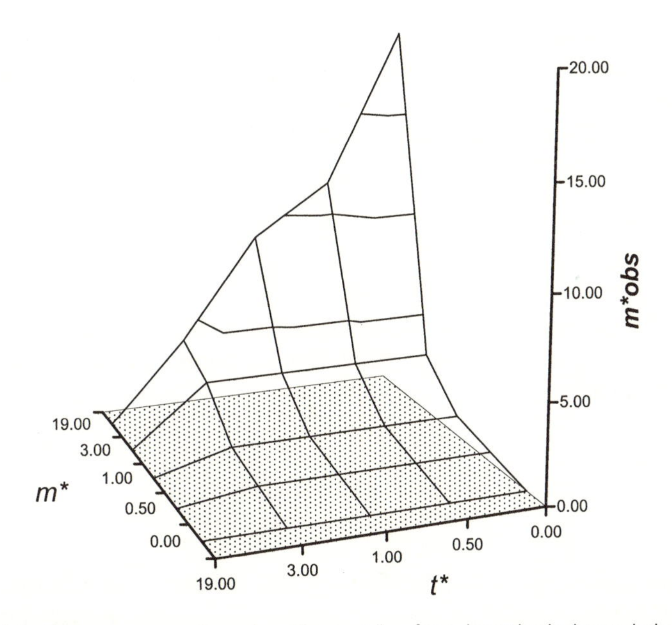

Figure 6.1. The mean crowding of matings, m^*_{obs}, for a hypothetical population of 20 females, distributed over patches defended by 20 males, within a breeding season divided into 20 intervals (see chapter 3 for details). The value of m^*_{obs} represents the effects of the spatial distribution of receptive females, as measured by the mean spatial crowding of females on resource patches, m^*, combined with the effects of the temporal distribution of receptive females, as measured by the mean temporal crowding of females within intervals of the breeding season, t^*.

third, vertical axis (fig. 6.1). Other authors have used similar schemes to identify the "environmental potential for polygamy" (Emlen and Oring 1977; Duvall et al. 1995), although the scale on which the EPP is measured is not consistently defined. Our method differs from earlier analyses because it uses internally consistent measures of the spatial and temporal patchiness of females, to provide an empirical measure of the opportunity for sexual selection, I_{mates}. As we will see, measurements of the sources as well as the magnitude of I_{mates} are useful tools for predicting patterns in mating system evolution within and among species (chapter 9).

We are interested in examining variation in I_{mates} as a function of seasonal and taxonomic variation in m^* and t^*. In a given species, the spatial and temporal distributions of females are likely to vary within and between breeding seasons. We have shown that such snapshot measurements can overestimate the value of I_{mates} (chapter 3). However, seasonal and taxonomic variations in the relative magnitude of estimates of I_{mates} are of interest in their own right, especially in comparative investigations within and among species. The appropriate statistical framework for describing such variation within species, as well as between species, is multiple regression.

We use the conventional partial regression equation from Sokal and Rohlf (1995), which estimates the value of the dependent variable $\hat{Y}$ as a function of k independent variables $X_1, \ldots, X_k$.

First, each Y and X_j variable is standardized as y' or x',

$$y' = (Y - \overline{Y})/s_Y \qquad [6.1]$$

and

$$x_j' = (X_j - \overline{X}_j)/s_{Xj}, \qquad [6.2]$$

where $\overline{Y}$ and $\overline{X}$ represent sample averages, and s_Y and s_X represent the standard deviations of each sample. Each standard partial regression coefficient is expressed as

$$b'_{Yj\cdot} = b_{Yj\cdot}\,(s_{Xj} \,/\, s_Y). \qquad [6.3]$$

Thus, the standardized multiple regression equation is

$$\hat{y}' = b'_{Y1\cdot}x'_1 + b'_{Y2\cdot}x'_2 + \cdots + b'_{Yk\cdot}x'_k. \qquad [6.4]$$

This expression gives the rate of change of the dependent variable, Y, in standard deviation units, per one standard deviation unit of independent variable X_j, with all other independent variables held constant.

Our goal is to apply this partial regression approach to identify the relative effects of female spatial and temporal distributions on the opportunity for sexual selection. Graphically, we can plot the simultaneous effects of m^* and t^* on the elevation of I_{mates} from the two-dimensional m^*, t^* surface. The conventional equation (eq. [6.4]) thus becomes

$$\hat{y}' = b'_{Y1\cdot}x'_1 + b'_{Y2\cdot}x'_2 \qquad [6.5]$$

where x'_1 equals the standardized spatial patchiness of receptive females, and x'_2 equals the standardized temporal patchiness of receptive females. Each $b'_{Yi.}$ is the standardized partial regression coefficient of the relationship between the spatial or temporal distribution of females and the opportunity for sexual selection, I_{mates}. That is, each b' gives the expected rate of change of the dependent variable I_{mates} in standard deviation units per one standard deviation unit of each independent variable while holding the other independent variable constant.

The I_{mates} Surface

Note that measuring the values of m^* and/or t^* generates an empirical estimate of I_{mates} for a particular species. The multiple regression approach identifies the statistical relationship, in terms of a, b_{Y1}, and b_{Y2}, that exists for multiple measurements of m^* and t^* and thus for estimates of I_{mates} across species. It addresses the issue of how much of the variation in I_{mates}

across species is explained by variation in m^* and t^*. The regression is useful in identifying or explaining *potential causes* of species differences in sex dimorphism. The empirical approach asks, "Does I_{mates} vary among species?" whereas the regression approach asks, "Why and to what extent does I_{mates} vary among species?"

Every pair of m^* and t^* values generates a point on a third, vertical axis for I_{mates}. The elevation of this point provides a population-specific estimate of the intensity of sexual selection on males in terms of I_{mates} at the time of sampling. Note that extreme spatial dispersal of receptive females and extreme temporal synchrony in female receptivity each contribute little to I_{mates}, whereas extreme spatial clumping and extreme temporal asynchrony can make I_{mates} very large (fig. 6.1). Intermediate values of m^* and t^* each contribute moderately to the elevation of the I_{mates} surface. For a given species, it is important to emphasize that I_{mates} will not be a single point. Just as McCauley et al. (1988) showed that the key parameter of kin selection theory, r, the genetic correlation within groups, could not be treated as a species-specific constant, so too I_{mates} will vary within and among breeding seasons or breeding aggregations.

Measurements of within-season variation in the spatial and temporal distributions of females will describe an ellipsoid or perhaps a more complicated shape on the I_{mates} surface for each species. The slope of the ellipsoid will depend on how m^* and t^* interact within a single breeding season (fig. 6.2). Between breeding seasons, the shape of the ellipse as well as its slope on the I_{mates} surface may change. However, the degree to which the values of m^* and t^* change for individual species, as well as the degree to which the relationships between these variables change between seasons, will be constrained by life history characters and viability selection common to all females in each population. Thus, each species will generate a unique, yet coherent cloud of points whose variation may be described along the m^*, t^*, and I_{mates} axes (fig. 6.3). Confidence limits on the values of I_{mates} for each species are obtained by estimating the average and variance in the value of I_{mates} throughout the breeding season, or by examining the variation in I_{mates} over several breeding seasons. In some cases, variation within species will exceed the variation among species. However, we expect that in many cases sufficient variation in I_{mates} among species will allow interspecific if not interpopulation classification of mating systems.

The Sex Difference in the Opportunity for Selection, ΔI

Recall from chapters 4 and 5 that the total opportunity for selection on polygynous, semelparous species (that is, species in which females mate once, males mate more than once, and females produce a single clutch of offspring; see table 5.3) can be expressed as

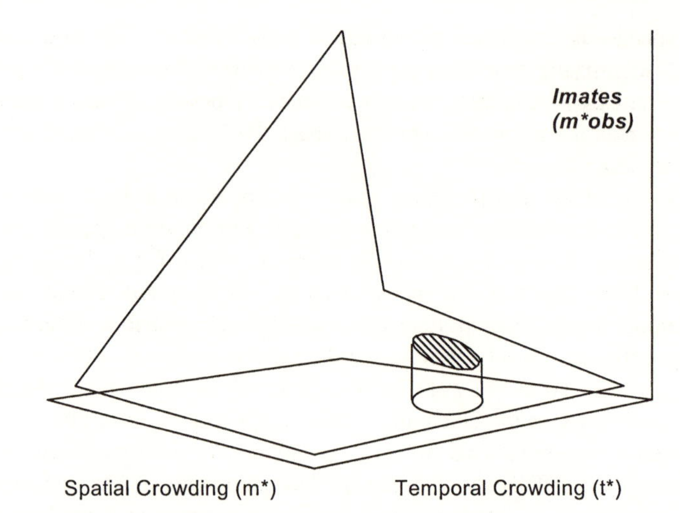

Figure 6.2. The I_{mates} surface; this surface arises from the intersection of points on the m^* axis (spatial patchiness) with points on the t^* axis (the reciprocal of temporal patchiness). In a given species, the spatial and temporal distributions of females are likely to vary within and between breeding seasons. Within-season variation in the spatial and temporal distributions of females will describe an ellipsoid or perhaps a more complicated shape on the I_{mates} surface (hatched area). The slope of the ellipsoid relative to the horizontal will depend on how m^* and t^* interact within a single breeding season.

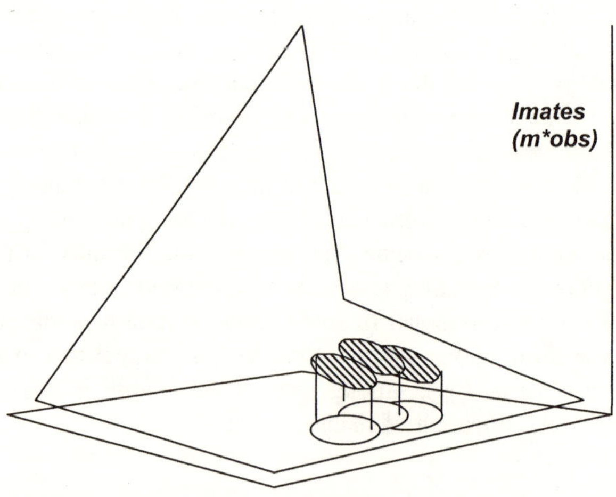

Figure 6.3. The degree to which the values of m^* and t^* change for individual species, as well as the degree to which the relationships between these variables change between seasons, will be constrained by life history characters and viability selection common to all females in each population. Thus, each species is expected to generate a unique yet coherent a cloud of points whose variation may be described along the m^*, t^*, and I_{mates} axes (hatched areas).

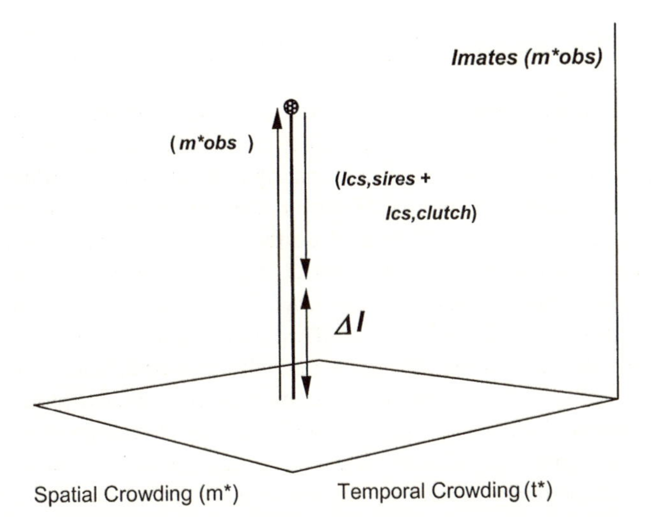

Figure 6.4. The sex difference in the opportunity for selection arises from the spatial and temporal distributions of females, adjusted by multiple mating and multiple breeding episodes by females. The adjusted sex difference in the opportunity for sexual selection is ΔI.

$$I_{males} = 1/R\ (I_{females}) + I_{mates} \qquad [6.6]$$

where $I_{females}$ equals the opportunity for selection among females. For polygamous, iteroparous species (that is, species in which females as well as males mate more than once and females produce more than one clutch of offspring), the total opportunity for selection can be expressed as

$$I_{males} = 1/R\ (I_{females} - I_{clutch}) + I_{mates} \qquad [6.7]$$

where I_{clutch} equals the opportunity for selection on females arising from multiple matings ($I_{cs,sires}$) as well as from multiple reproductive events among females. If the sex ratio, R, is assumed to equal 1, eq. [6.7] can be rearranged so that

$$\Delta I = I_{males} - I_{females} = I_{mates} - I_{clutch}. \qquad [6.8]$$

The right side of this equation shows the sex difference in the opportunity for selection, ($I_{males} - I_{females}$) which equals I_{mates} in the absence of among-female variation in clutch number.

This term, ΔI is the sex difference in the opportunity for sexual selection adjusted for the fact that, in many species, females mate more than once and/or produce multiple clutches of offspring (fig. 6.4). For monandrous, semelparous females (that is, females who mate once and produce a single clutch of offspring; table 5.3), ΔI and I_{mates} are equivalent.

The Relative Magnitudes of I_{mates} and I_{clutch}

In many cases, the opportunity for selection on males arising from the variance in mate numbers, I_{mates}, is expected to exceed the opportunity for selection on females arising from multiple mating and multiple reproductive episodes, I_{clutch} (chapters 1 and 4). However, the magnitude of the difference between I_{mates} and I_{clutch} will depend on the spatial and temporal distributions of females and the degree to which I_{clutch} enhances variance in offspring numbers among females (see below). The sex difference in the opportunity for selection, ΔI, will be smallest when polyandrous, iteroparous females are spatially dispersed and temporally clumped, that is, when m^* and t^* are both small and females mate more than once and produce more than one clutch of offspring (I_{clutch} is large). The sex difference in the opportunity for selection, ΔI, will be largest (i.e., I_{mates} will be eroded least) in species with spatially clustered, monandrous, semelparous females, i.e., when m^* is large, t^* is small, and females mate only once and produce a single clutch of offspring. In semelparous as well as in iteroparous females, the magnitude of $I_{cs,sires}$ (i.e., the opportunity for selection on female fecundity due to the effects of different sires on offspring numbers, chapter 5) will be affected by the spatial and temporal distribution of males, since these factors determine the rate at which females are likely to encounter males.

A General Classification of Mating Systems Based on Mate Numbers

The degree to which females mate with more than one male, as measured by $V_{cs,sires}$, can affect the variance in offspring numbers among females as described in chapters 4 and 5. As with the number of times females breed, the magnitude of $V_{cs,sires}$ depends on the degree to which females mate more than once. The number of times individuals mate already provides a general framework for the classification of mating systems (Emlen and Oring 1977; Bradbury and Vehrencamp 1977; Thornhill and Alcock 1983). However, these classifications can be refined and quantified using $V_{cs,sires}$ and V_{mates}, both of which affect the sex difference in the strength of selection, ΔI.

For example, *monogamy* is now used to describe circumstances in which females (or males) (1) mate only once in their lives (Hughes et al. 2000; Luis et al. 2000), (2) mate more than once in their lives, but ostensibly with only one partner (Burda et al. 2000; Taylor et al. 2000; Kokko 1999; Cooper et al. 1997), (3) mate with only one partner per season, but may change mates between seasons (Lubjuhn et al. 2000; Goldizen et al. 2000), and (4) mate with several partners within seasons, raise offspring with only one mate per season, and change primary mates and secondary copulatory partners between seasons (Soukup and Thompson 1998; Gowaty and Buschhaus 1998).

Polygyny, *polygamy*, and *promiscuity* are all used to describe mating systems in which males as well as females mate more than once (Thornhill and Alcock 1983; Choe and Crespi 1997; Amat et al. 1999; Fitzsimmons 1998). Yet polygyny and polygamy can also refer to mating systems in which females mate few or many times within their lives (Darwin 1871; Thornhill and Alcock 1983). *Polyandry* and *polygynandry* are used to describe mating systems in which females have multiple mates (Davies and Hartley 1996; Rose et al. 1997; Jones and Avise 1997; Berglund et al. 1989; Butchart et al. 1999; Vehrencamp 2000; Nievergelt et al. 2000; Marks et al. 2002), but, again, the relative number of mates per sex, and thus the degree to which differences in the distribution of mate numbers can contribute to the sex difference in the opportunity for selection, often remain undefined. To further complicate matters, when discrepancies exist between the true meaning of the mating system descriptor and what actually happens when males and females mate, the term "social" has been added, allowing *social monogamy*, *social polygyny*, *social polygynandry*, and other oxymorons to populate the current literature on mating systems (see above references).

To standardize this terminology, and to make it more useful for identifying the degree to which differences in mate numbers among females enhance $V_{cs,sires}$, we propose the following scheme (table 6.1). We will later identify specific mating systems that are subsets of this general framework (chapter 9). In our lexicon, the term *monogamy* (*monos* = one; *gammos* = marriage; Greek), can refer *only* to mating systems in which each sex has a single mate for life. This term makes sense from the standpoint of measuring $V_{cs,sires}$ because it represents circumstances in which $V_{cs,sires}$ as well as V_{mates} equal zero. We consider the term *polygyny* (*polus* = many; *gunē* = female; Greek) to refer *only* to mating systems in which females mate with a single male in their lives, but males may mate with more than one female. As explained in chapter 1, when some males mate with more than one female, other males must be excluded from mating. Thus, with polygyny, while $V_{cs,sires}$ equals zero, V_{mates} is large.

We consider the term *polygynandry* (*polus* = many; *gune* = female; *andro* = male; Greek) to refer *only* to mating systems in which both sexes are variable in their mate numbers, but males are more variable than females. In such cases, V_{mates} will be larger than $V_{cs,sires}$. We consider *polygamy* to refer to *only* those mating systems in which both sexes have multiple partners and approximately equal variance in mate numbers; thus $V_{cs,sires}$ and V_{mates} are approximately equal. We consider *polyandrogyny* to refer to *only* those mating systems in which both sexes are variable in their mate numbers, but in which females are more variable than males. In such cases, $V_{cs,sires}$ will be larger than V_{mates}.

With polygyny, polygynandry, polygamy, and polyandrogyny so defined,

Table 6.1
A general classification of mating systems based on male and female mate numbers.

Category	Definition	*Variance in Mate Number* Females $V_{cs,sires}$	Males V_{mates}
Monogamy	Each sex mates once	0	0
Polygyny	Females mate once; males are variable in mate numbers	0	+ +
Polygynandry	Both sexes have variable mate numbers; male mating is more variable than female mating success	+	+ +
Polygamy	Both sexes have variable mate numbers; male mating success is approximately equal to female mating success	+	+
Polyandrogyny	Both sexes have variable mate numbers; female mating success is more variable than male mating success	+ +	+
Polyandry	Males mate once; females are variable in mate numbers	+ +	0

polyandry (*polus* = many; *andro* = male; Greek) can refer *only* to mating systems in which females are variable in their mate numbers, but males mate for life with a single female. In such cases, V_{mates} equals zero, but $V_{cs,sires}$ is large because some females mate more than once and other females fail to mate at all. *Promiscuity* may be used to describe any mating system in which individuals mate more than once. However, this term is a general one and therefore is not considered as part of the specific definitions described above. We defined mating-number promiscuity and mate-number promiscuity in chapter 4 to show how these definitions of promiscuity are related and how each affects the sex difference in the strength of selection.

When females mate with only one male, as in monogamy or polygyny, $V_{cs,sires}$ is equal to zero. $V_{cs,sires}$ is nonzero, however, for all other general categories of mating systems. Similarly, V_{mates} is zero in monogamy or in polyandry, but is nonzero for all other general categories of mating systems (table 6.1). This terminology is in accord with that introduced in chapters 4 and 5, in which females alone are described as *monandrous* and *polyandrous* based on whether they have one or more than one mate. Similar descriptions of males (*monogynous* [= one female] and *polygynous* [= many females]) are also possible when males have one or more than one mate. These last four terms *do not* describe mating systems themselves. They only describe the mate numbers of individual females or males. Thus, these terms are also

Table 6.2
Terms describing differences in mate numbers and breeding episodes among females.

	Breeding Seasons / Breeding Episodes Per Season			
	1		*>1*	
Mate Numbers	*1*	*>1*	*1*	*>1*
1	monandrous semelparous	monandrous uniseasonal iteroparous	monandrous multiseasonal uniparous	monandrous multiseasonal iteroparous
>1	polyandrous semelparous	polyandrous uniseasonal iteroparous	polyandrous multiseasonal uniparous	polyandrous multiseasonal iteroparous

appropriate for use in describing how females must secure mates depending on the number of reproductive events in their lives (tables 6.2 and 6.3).

Male and Female Longevity

Within populations, the duration of the interval between breeding episodes can also affect the sex difference in the opportunity for selection. If circumstances favoring reproduction occur rarely or are seasonal, the duration of the interval between seasons favorable to breeding may exceed the life span of most individuals (fig. 6.5a). When the life span of most individuals is longer than the interval between breeding episodes, individuals are iteroparous, breeding more than once in their lives (fig. 6.5b).

The circumstances in which semelparity and iteroparity may evolve are presumed to result from a trade-off between the relative abundance of resources available for egg production, and the duration of the female's repro-

Table 6.3
Components of variance in offspring numbers among females attributable to differences in mate numbers and breeding season numbers, where *i* refers to multiple breeding episodes that occur within breeding seasons, and *j* multiple breeding episodes that occur between breeding seasons.

	Breeding Seasons / Breeding Episodes Per Season			
	1		*>1*	
Mate Numbers	*1*	*>1*	*1*	*>1*
1	0	$V_{(clutch)}$	$\Sigma^j V_{(clutch)j}$	$\Sigma^j\Sigma^i(V_{clutch})_{ij}$
>1	$V_{(cs,sires)}$	$\Sigma^i V_{(cs,sires)i} + V_{(clutch)i}$	$\Sigma^j V_{(cs,sires)j} + V_{(clutch)j}$	$\Sigma^j\Sigma^i(V_{cs,sires} + V_{clutch})_{ij}$

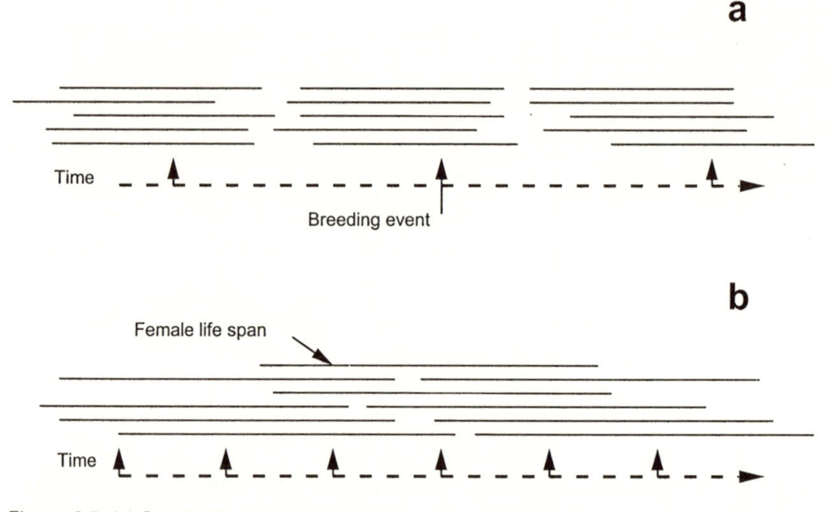

Figure 6.5. (a) Species in which the duration of the interval between breeding episodes exceeds individual female life spans; (b) species in which the duration between breeding episodes is less than individual female life spans; dashed horizontal line represents time, solid horizontal lines represent female life spans, and arrows pointing up represent breeding episodes.

ductive life (Stearns 1977; Roff 1992). Within this framework, it appears that when female lifetimes are short and resources are scarce, reproduction is rarely possible. However, when female lifetimes are short and resources are abundant, semelparity is likely to evolve (fig. 6.6). When female lifetimes are long and resources are scarce, females may store resources and breed in a single, suicidal burst. This breeding strategy resembles semelparity. However, since females in such species usually produce several subclutches of offspring, often with more than one male parent (Kirkendall and Stenseth 1985), such *uniseasonal iteroparity* is functionally similar to iteroparity proper. Iteroparity occurs when resources are abundant and females live long enough to turn environmental surpluses into multiple clutches of eggs (table 6.2; Fig. 6.7).

In chapter 5 we identified opportunities for selection arising from four categories of female life history (table 5.8). The scheme shown in table 6.3 generates four additional categories of variation in female offspring numbers, although these categories are, in fact, nested within the variance in offspring numbers that exists for polyandrous, iteroparous females ($V_{cs,sires} + V_{cs,clutch}$). This expanded terminology provides a useful means for partitioning the total variance in offspring numbers into each seasonal component, and, clearly, considerable within-female variance in offspring numbers can arise among individual females who breed different numbers of times in different numbers of seasons. As described in chapter 4, these variance com-

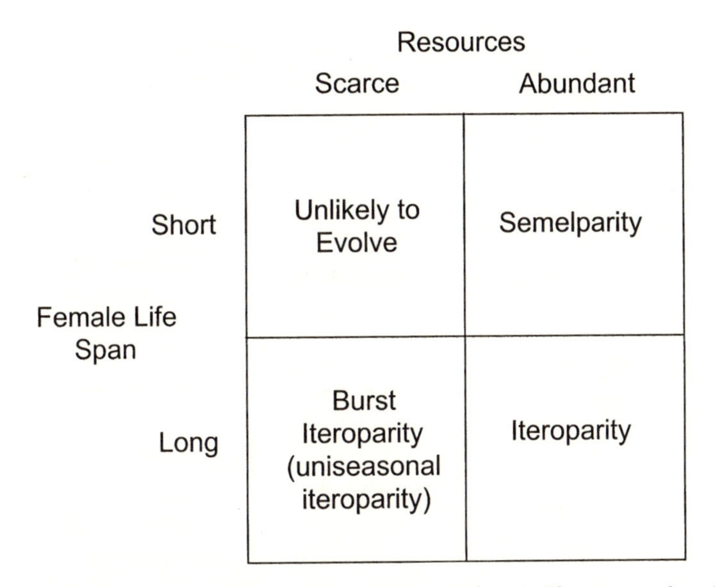

Figure 6.6. The influences of resource availability and female life span on female life history strategy.

ponents contribute to the variance in offspring numbers within females (V_{clutch}). Thus, as shown in chapter 5, the variance components identified in table 6.2 are easily converted to opportunities for selection by dividing each component by the squared average in offspring numbers per female. Thus, the contribution of seasonal variation to total selection can be understood separately from that caused by variance in mate numbers, allowing the relative influences of sexual selection and viability selection on females to be more clearly identified.

With our standardized terminology, the overall effects of female life history on the opportunity for sexual selection become clearer. As explained in chapter 4, our use of the term *semelparity* refers specifically to females who produce only one clutch of offspring in their lives. Our use of the term *iteroparity* refers to females that reproduce more than once and can include the range of variation described above (Kirkendall and Stenseth 1985; tables 6.2 and 6.3). Researchers often have more detailed information on female life span and the duration of the interval between breeding episodes than they do on resource abundance or on the relative amounts of energy males and females devote toward mating effort or parental effort. These latter factors may be important for predicting life history variation itself. However, for predicting the types of mating systems likely to evolve under different spatial and temporal distributions of sexually receptive females, information on female life span and the duration of the interval between breeding episodes is more useful (chapter 9).

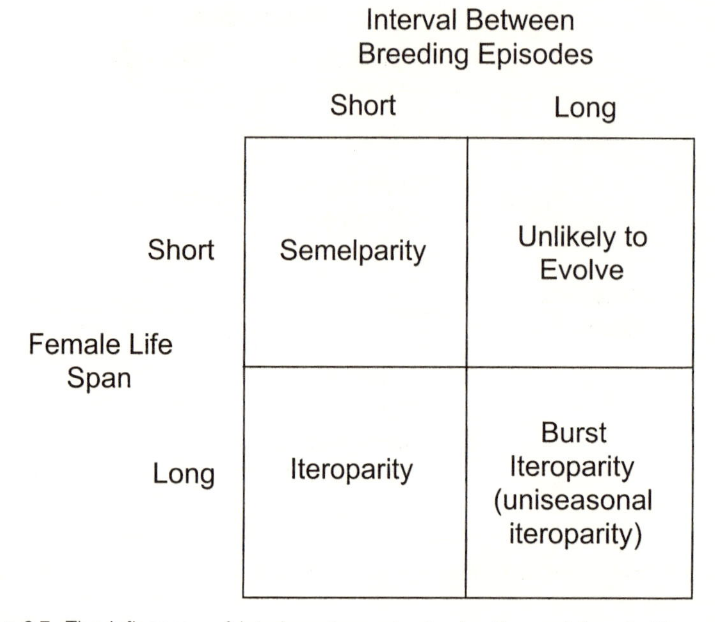

Figure 6.7. The influences of interbreeding episode duration and female life span on female life history strategy.

When female life span and the interval between breeding episodes are both short, females are expected to breed once and die (i.e., become semelparous; fig. 6.7). Iteroparity is unlikely to exist under such circumstances because females will not survive between breeding episodes (fig. 6.5). When female life span is short and the interval between breeding episodes is long, neither semelparous nor iteroparous females are expected to exist (fig. 6.7). Short-lived females with either reproductive strategy are unlikely to track breeding opportunities that are widely separated in time. When female life span is long and the interval between breeding episodes is short, females are expected to become iteroparous because on average they will produce more offspring in their life spans than long-lived semelparous females (Roff 1992). Lastly, when female life span and the interval between breeding episodes are both long, females are expected to exhibit burst iteroparity, that is, to produce their several clutches of offspring over a short period of time (fig. 6.7). This last category can include multiseasonal uniparity or multiseasonal iteroparity (table 6.2).

Long-lived males are likely to encounter somewhat different spatial and temporal distributions of females from one breeding season to the next. Among-season variation in I_{mates} as experienced by long-lived males can enhance the among-male variance in mate numbers and thereby enhance the sex difference in the opportunity for selection. Thus, in any specific mating system analysis, the total opportunity for sexual selection must be summed over all breeding seasons in which the same males appear (chapter 3). The value of I_{clutch} may also be influenced by variation in female life span. The contribution of $I_{cs,sires}$ to I_{clutch} will be measurable if long-lived females mate with one male per breeding season, as well as if females mate with multiple males within each breeding season. The corrosive effects of $I_{cs,sires}$ on I_{mates} in each case will depend on the degree to which sire numbers contribute to the variance in offspring numbers among females, as well as on the degree to which sperm competition occurs within and between breeding seasons. In such species, it may be useful to consider instantaneous rather than discrete measures of ΔI.

However, in general, if increased life span among males increases the number of males who successfully mate, I_{mates} will *decrease*. This is true because increased numbers of mating males decreases the variance in mate numbers among males (chapters 1 and 5). If increased life span has no effect on the number of mating males, but instead increases the variance in mate numbers among these individuals, I_{mates} will *increase*. Similarly, if increased life span decreases the number of mating males, because long-lived males exclude other males from mating across generations, decreased numbers of mating males will increase the variance in mate numbers and so increase I_{mates}. If increased female life span increases the number of females who successfully acquire mates, I_{clutch} will decrease and I_{mates} will be eroded less. On the other hand, if increased life span has no effect on the number of breeding females, but instead increases the variance in clutch numbers among these individuals, I_{clutch} will increase and I_{mates} will be eroded more.

Clutton-Brock and Parker (1995) have discussed the effects of "times in" versus "times out" for breeding adults on the operational sex ratio, and have made similar inferences about the effects of increased life span on the intensity of sexual selection. Our approach gives quantitative expression to these concepts in terms of opportunities for selection, allowing researchers to go beyond qualitative comparisons of the environmental potential for polygamy among species, as well as to specifically address semantic arguments over the relative importance of natural and sexual selection (e.g., Koenig and Albano 1986).

Our method identifies the sources of variance in offspring numbers and thus partitions the opportunity for selection among males and females due to differences in life history as well as differences in longevity. This approach identifies the specific influences these factors have on offspring numbers for both sexes. Our method uses a dimensionless estimator, ΔI, to measure the relative intensity of selection in these contexts. As explained above, this feature allows comparisons of selection intensities within and between species, as well as within and between contexts. We will consider the specific effects of variation in population sex ratio on the opportunity for sexual selection in chapter 8.

Chapter Summary

Previous approaches for cross-taxomonic comparisons of mating systems are qualitative or imprecise. We present methods, consistent with the approach presented in previous chapters, that revise and improve on the Emlen and Oring scheme for mating system classification. Our approach generates more specific categories of mating systems than are currently in use, and thus explains more of the possible variation in mating system diversity than nonquantitative, inductive approaches based on parental investment theory. Our approach also indicates the circumstances under which apparently different mating systems should be expected to give rise to comparable levels of sexual dimorphism.

We use multiple regression to describe within- and between-season variation in the values of m^* and t^* and show how this partial regression approach can identify the relative effects of female spatial and temporal distributions on the opportunity for sexual selection. We next show how the relative magnitudes of I_{mates} and I_{clutch} describe a species-specific characteristic for each mating system. We call this surface, ΔI, the sex difference in the opportunity for selection arising from female spatiotemporal distributions and adjusted by variance in female fitness arising from differences in the numbers of mates and reproductive episodes among females. Its dimensionless quality allows within- and among-species classification of mating systems.

We next provide a general classification scheme for mating systems based on mate numbers and their effects on ΔI. According to this system, all mating systems can be classified by the degree to which variance in mate numbers among females, measured by $V_{cs,sires}$, as well as variance in mate numbers among males, measured by V_{mates}, may affect the sex difference in the strength of selection, ΔI. We consider how male and female longevity contribute to or erode the value of ΔI. We specify methods for identifying female semelparity, degrees of female iteroparity, and methods for bringing such information to bear on the total variance in offspring numbers among females. Researchers often have more detailed information on female life span and the duration of the interval between breeding episodes than they do on the relative amounts of energy males and females devote toward mating effort, parental effort, or resource abundance. Thus, this approach provides a practical as well as testable method for investigating mating systems in nature.

7 Conceptual Difficulties in Mating System Research

Mating systems are classified by the activities males use to monopolize mates (Emlen and Oring 1977; Bradbury and Vehrencamp 1977; Borgia 1979; Thornhill and Alcock 1983; Clutton-Brock 1989; Arnold and Duvall 1994; Reynolds 1996). As advocated by Emlen and Oring (1977), this scheme assumes that male mate-monopolization behaviors evolve in response to the spatial and temporal distributions of receptive females. As we have shown, variation in the spatial and temporal distributions of receptive females is the most direct way to create the sex difference in fitness variance, which drives the evolution of sexually dimorphic traits. Nevertheless, recent authors have begun to question the utility of the Emlen-Oring hypothesis, because it appears to require ad hoc qualification when applied to specific individual species (Reynolds 1996). For example, mechanisms by which members of one sex interfere with "free" mate choice by members of the other sex appear to characterize many animal mating systems, but were not specifically predicted by Emlen and Oring (1977; chapter 4).

Recent research has focused on how males influence or subvert mate choice by females, by means of forced copulations, coercive male guarding, the exchange of copulations for resources, or male parental care. The inability of the Emlen-Oring hypothesis to specifically account for these and similar observations has increasingly led mating system research toward descriptions of interactions between *individual* males and females. This is often combined with speculation about the possible fitness advantages to the particular interacting pair of individuals of the behaviors they express. To us, it appears that mating system research has moved away from describing why certain behaviors are adaptive in a population and become increasingly centered on the individual and why, within the special context of each individual, its own behaviors are adaptive. We find this change in perspective difficult from an empirical viewpoint for several reasons. First, it diverges from the traditional ecological genetic approach in a way that hinders testing of theory. In ecological genetics, phenotypic variation is categorized and, subsequently, the researcher determines whether or not there is an association between fitness and phenotypic category and, if so, how strong is the association. In the new approach, each individual is presumed a priori to act adaptively at every moment of its life, and the research challenge is to explain why the behavior of each individual is adaptive *even though it is different from that of other individuals*. Individual variation in behavior is justified case by case as adaptive in terms of special contexts, specific to each individual. Adaptation is no longer the "onerous concept" of Williams (1966), but rather a property of the individual and its behavior at all times. Whenever two individuals are observed to behave differently, it is because it is adaptive for each to do so. The standard evolutionary constraints on adaptive perfection, such as genotype × environment interaction, pleiotropy, gene flow, or mutation-selection balance, are not considered relevant.

Second, the current approach appears to discard the concept of "relative fitness," in that individuals expressing less than optimal behaviors are, nonetheless, described as "making the best of a bad job" given the unfortunate context in which they find themselves. In traditional ecological genetics, variance in relative fitness associated with different behavioral categories is the definition of selection. Measuring the variation in fitness and associating it with phenotypic categories is equivalent to measuring how selection is acting on the phenotype (Wade and Kalisz 1990). Since mean relative fitness always equals 1, whenever there is variance in relative fitness, some individuals must have fitness less than 1. In phenotypic selection theory, if these same individuals also disproportionately express a particular behavior, that behavior, by definition, is "selected against." The ecological genetics approach seeks to measure the causes of natural selection as they are currently operating.

In contrast, the new focus in behavioral studies appears to assume, not only that all traits are adaptive, but also that all individuals achieve maximal inclusive fitness at all times. Models in which inclusive fitness is optimized are only a small part of general evolutionary theory, but they have become a major part of behavioral evolutionary theory. This can be problematic, especially in cases of sexual selection where interactions between males and females affect offspring numbers or viability. In general evolutionary theory, fitness would be formally assigned to mating pairs rather than to individuals. Notably, whenever fertility selection is stronger than viability selection, fitness at the level of the individual is not necessarily maximized. Assuming that it is will lead one to the wrong conclusions (see below for examples). Indeed, the greater the degree of context dependence of individual fitness, the *less likely* it is that a model which considers only the individual in its local context will give the correct evolutionary dynamic (see, for example, Agrawal et al. 2001b).

Third, much of the behavioral variation observed among individuals is considered environmentally induced but adaptive in context. That is, behavior is accorded a much greater degree of "adaptive plasticity" than any other phenotype. Adaptive phenotypic plasticity is a controversial topic in evolutionary biology, where there is debate over whether it is an indirect by-product of local selection and genotype × environment interaction ($G \times E$) or whether it is favored directly by natural selection acting on genotypic norms of reaction per se (cf., Schlichting and Pigliucci 1998). In either case, adaptive plasticity is considered a derived or an evolved property of the phenotype, and there are specific circumstances (involving $G \times E$, environmental frequencies, mating patterns, and gene flow among environments) that make it more or less likely to evolve. In many behavioral studies, it seems to us, these considerations are circumvented by the assumption that adaptive plasticity is a fundamental property of behaviors. The assumption is justified by an appeal to ESS models, that is, game theory models that identify the *evolutionary stable strategy* or ESS (Maynard Smith 1982), where "flexible" strategies often beat inflexible ones in variable environments (although this is not true in game theory in general).

In evolutionary theory, however, genetic phenomena, like dominance, pleiotropy, $G \times E$, and epistasis, are essential to understanding the evolution of phenotypic plasticity (e.g., Schlichting and Pigliucci 1998, p. 39). These are all genetic phenomena poorly handled by ESS models. Furthermore, the assumption that adaptive behavioral plasticity extends even to very rare social contexts ignores the well-established roles that environmental frequency and predictability play in the evolution of adaptive reaction norms. Consider a gene that is advantageous only in a very rare social environment, experienced by only a tiny fraction, say 1%, of all individuals. Consider, further, that the same gene is disadvantageous, by an amount t, in a common social environment, experienced by the majority of individuals, say the remaining 99%. The fitness advantage, s, enjoyed by those few individuals who have the gene and who experience the rare social environment does not by itself determine whether the gene will become fixed or lost. Ignoring drift and other evolutionary forces, natural selection favors the gene only if its average effect on fitness $\{0.01s + 0.99t\}$ is positive. For this hypothetical gene, the positive effect in the rare environment would have to be 100-fold greater than its deleterious effect in the common environment. If the advantage accrued only to males while the disadvantage accrued to both sexes, the positive effect on males in the rare environment would have to be 200-fold greater than its deleterious effect on both sexes in the common environment. If it were an X-linked gene and females were homogametic, then the advantage to males in the rare environment would have to be even greater, because two of every three copies of the gene would occur in females. In a harem polygynous system, where only one in ten males mated, the advantageous effect would have to be an order of magnitude greater still (i.e., 2000-fold higher).

Limiting the expression of the gene to the special, advantageous social environment would clearly change how the average effect of the gene is calculated and would mitigate, perhaps entirely, the opposing effects of t. However, context-limited gene expression is itself an evolved property. A modifier gene, M, that caused the context-limited expression of the behavior gene is itself only favored in the rare social context, and then only in those individuals with the behavior gene. The average fitness effect of the modifier M depends *multiplicatively* upon *both* the rare social context and the rare genetic context. Thus, the modifier experiences even weaker selection than the behavior gene, whose evolution seems problematic. Like the famous arguments between Fisher and Wright on the evolution of dominance modifiers, pleiotropy, linkage, mutation, and population size all play important roles in the evolution of rare genes with very small average fitness effects. None of these effects are handled well in ESS models.

When the fitness advantages of particular reaction norms are not compared empirically to alternative patterns of behavioral expression, it can be difficult to avoid the "circular reasoning typical of the 'adaptationist programme' (Gould and Lewontin 1979)" (Schlichting and Pigliucci 1998, p. 54). We agree with Lande (2000, p. 344) that "Many students of sexual selection still adhere to erroneous definitions of fitness and the panselectionist belief that all phenotypic evolution is adaptive." For these reasons, we are reluctant to assume that each individual is expressing either a uniquely adaptive reaction norm or one aspect of an all-purpose adaptive reaction norm, suitable for any environmental contingency. In this chapter, we will discuss the current, individual-focused approach to the study of mating system evolution as it applies in different aspects of sexual selection ranging from mate choice to parental care. Our intent is to identify some of the current hypotheses that seem at variance with standard evolutionary genetic theory.

Males First?

Tests of the Emlen-Oring hypothesis, relating patterns of female receptivity to mating systems and sexual selection, can be contingent on the sequence in which data are gathered. As mentioned above, Emlen and Oring (1977) classified mating systems by the activities males use to monopolize mates. This scheme is consistent with parental investment theory (Trivers 1972), which views male activities as the signature of species-specific differences in initial parental investment, with males typically investing much less than females as a direct correlate of anisogamy. Male competitive breeding activities are often easier to observe and measure than are the distributions of receptive females or the details of female life history, because males, to a greater extent than females, vary in conspicuous coloration, morphology, or behavior among closely related species (see chapter 1; Andersson 1994). It is not surprising then, that most analyses of animal mating systems *begin* with descriptions of male mating behavior and reproductive biology, rather than with investigations of female receptivity or life history. After all, it is the among-male taxonomic pattern of divergent phenotypes that suggests an explanation based on sexual selection (Dunn et al. 2001).

However, when mating systems are classified in terms of male behaviors, subsequent investigations of the evolutionary processes governing the differences among breeding systems tend also to assume a male-biased causal context. Studies of female distributions, life histories, and behaviors, which come *after* a species has been assigned a mating system, tend to gather and interpret female life history data in a manner consistent with and supportive of the prior classification based on male behaviors (see Ostfeld 1987). This *confirmatory* approach makes it difficult to falsify the hypothesis that male activities evolve *in response* to female spatiotemporal distributions (Shuster 1991b).

The Emlen-Oring hypothesis suggests that female spatiotemporal distributions drive the evolution of male mating behaviors, although mating systems are more often classified in terms of the outcome of this evolution, i.e., male mating behavior. We suggest that this encourages a tendency toward circular reasoning about cause and effect, which may impede hypothesis testing, especially when almost any observation can be given an ad hoc adaptive interpretation a posteriori. The methods of the preceding chapters provide a means for quantifying these causal processes. We suggest that a mating system classification based on *process* may ultimately prove more useful and more insightful than one based on *outcome*.

Adaptive Phenotypic Plasticity

As we discussed above, the individual-focused approach to the study of mating system evolution presumes a high level of adaptive behavioral plasticity and a close fit between observed behavior(s) and its adaptive benefit to the performer. Although this approach provides a plausible explanation for some aspects of behavioral diversity, it provides little guidance in regard to the evolutionary origins of behavioral traits or on their initial spread through populations. In theory, adaptive plasticity in behavior involves a pattern of phenotypic expression that is contingent on particular environmental conditions (Schlichting and Pigliucci 1998). The evolution of such flexibility requires either the evolutionary modification of behavioral traits *after* their origin and spread, or the assumption that behaviors with adaptively contingent expression arise *de novo*. We consider the latter unlikely because studies in genetics and artificial selection have shown that sex-limited gene expression is most commonly a derived property of a phenotype.

Because the genetic correlation between the sexes tends to be high for most phenotypic traits studied in most organisms (cf. Boake 1991; Partridge 1994; Lynch and Walsh 1998, pp. 718–726), selection within one sex results in a correlated response to selection in the other sex (e.g., Delph, in prep.). Darwin (1871) referred to this as "the laws of equal transmission of characters." In the context of human sexual selection, Darwin (1871) stated that "*It is fortunate that the law of equal transmission of characters to both sexes prevails with mammals; otherwise it is probable that man would have become as superior in mental endowment to woman, as the peacock is in orna-*

mental plumage to the peahen." Fisher's explanation for the mild expression in females of male sexually selected traits involved the rapid spread in both sexes of a character advantageous to male mating success and the subsequent evolution of sex-limited expression of the trait owing to opposing selection in females (see also Lande 2000). If sex-limited expression is not a *de novo* property of most mutations, then it seems less likely that other modifications of gene expression, such as those restricted to one sex and limited to certain social environments but not others, arise *de novo*. Below, we discuss the evolution of adaptive plasticity in standard evolutionary genetic theory (Levins 1968; Slatkin 1978, 1979; Denno 1995; Fairbairn 1994; Roff 1994, 1995; Schlichting and Pigliucci 1998) and which elements of the theory apply well and which apply poorly to behavioral plasticity.

The evolution of behavioral plasticity requires three steps. First, modifier alleles capable of changing the expression of a behavior in response to different environments must exist. Second, individuals differing at these modifier loci must experience the environments under which the modified expression confers an adaptive advantage. Third, selection on a modifier allele must be sufficiently strong and consistent in relation to other evolutionary forces acting on it that the allele spreads through the population. Under these conditions, behavioral plasticity will evolve. Strong or consistent selection means that the environment in which the modifier is adaptive must be common or highly predictable. If the environment is rare, then most individuals with the modifier will not experience the environment, will not express the modifier, and will not reap the selective advantage. That is, the strength of selection on the modifier is correlated with the abundance or rarity of the special environment that provides individuals with the opportunity to express the plastic phenotype. In addition, the magnitude of the fitness advantage, given expression, influences the rate of evolution. Thus, adaptive behavioral plasticity is more likely to evolve in response to some kinds of environmental variation than others, namely, those environments that are *common* and *highly predictable*.

There are many examples in nature of dispersal polymorphisms, such as the winged and wingless morphs in insects, which clearly represent adaptively plastic phenotypes (reviews in Fairbairn 1994; Roff 1996; Zera and Denno 1997; Baskin and Baskin 1998; Schlichting and Pigliucci 1998; Greene 1999). In most of these cases, a decline in local resource abundance, experienced by all individuals (i.e., a common environment), precipitates the adaptive expression of the winged, dispersal morph and emigratory behavior, often at a considerable fecundity cost. In such circumstances, modifiers of development are under strong selection, they are expressed by most individuals, and the evolution of adaptive plasticity is not problematic. Similarly, the evolution of diapause mechanisms, permitting temporal escape from a deteriorating environment, is not problematic because it involves a deleterious change in the environment (often drought or temperature) experienced by all individuals in the population. In both examples, all individuals in the population are under intense selection to "seek a better opportunity." There are additional examples of adaptive plasticity in coloration, sometimes involving cryptic and aposematic coloration (e.g., geometrid moths [Greene 1999] and desert locusts [Sword 1999]).

On the other hand, when environmental contingencies are rare or are experienced only by a small fraction of individuals in a population, then most copies of modifier alleles never encounter the unique circumstances in which they might be adaptive. When these individuals are further limited to one sex, selection is proportionately halved again. In such circumstances, selection on the modifier, owing to the special environment, is likely to be weak. Evolution of the modifier is then likely to be governed by its pleiotropic effects on other traits. Reasoning along these lines arose in the early history of evolutionary genetic theory in the debate between Fisher and Wright over the evolution of dominance (Provine 1971). Wright (1929a,b) argued that, because new mutations are rare, dominance modifiers do not often occur in genetic backgrounds that are heterozygous for new mutations, because the frequency of these backgrounds is on the order of the mutation rate itself.

In these circumstances, the evolution of alleles capable of modifying dominance of new mutations at other loci is more likely to be influenced by their pleiotropic effects on other traits. This would be especially true in small populations where random genetic drift would interfere with only a modest average selective advantage. The difficulty posed by rarity of the selective environment is ignored in many adaptive scenarios for behavioral plasticity, where the fitness advantage of a sex-limited behavior is contingent upon a myriad of special social circumstances. Overall, we find it difficult to accept the predicated high level of adaptive behavioral flexibility without also adopting a postmodern, Panglossian worldview in which the actions of every individual, in every circumstance, are considered adaptive, unless proven otherwise.

There are some circumstances, however, in which even a sex-limited trait might commonly experience a predictable environment. Remember that the frequency of unsuccessfully mating males, p_0, is an increasing function of the harem size, H, the average mating fitness of successfully mating males: $p_0 = (1 - [R/H])$. When H is large, *most* males (i.e., a large fraction of the total population; chapter 1) experience poor mating success. Adaptive behavioral plasticity could evolve in these circumstances as long as the fitness advantage of the plasticity is sufficiently strong. We will address this issue in more detail in chapter 11.

Mate Choice and Good Genes

The concept of adaptive mate choice for the purpose of obtaining "good genes" which will increase offspring viability is another example of a topic enjoying an ad hoc proliferation of special theories. The original "good genes" theory of female mate choice posited the existence of variation among males for genes affecting offspring viability (Trivers 1972). Females selecting mates with "high viability" genes are expected to bear more viable offspring than randomly mating females. The earlier, alternative theory of arbitrary mate choice, based on Fisher's runaway process (Fisher 1930; see below), was proposed, at least in part, to explain the apparent decline in male viability resulting from exaggerated traits, conspicuous mating behaviors, and male-male combat. It states that the existence of a female mating preference creates a stronger selective pressure on males than opposing viability selection (see chapter 1). Female mating preferences coevolve with male reproductive phenotypes owing to the genetic association established by nonrandom mating. Lande (1981) and Kirkpatrick (1982) further showed that Fisher's runaway process leads to a line of stable equilibria, determined by the opposing and balancing forces of male viability and male reproductive success due to female mate choice (Lande 2000). The variety of equivalent equilibria under Fisher's process offers an explanation for the rapid diversification of male traits among small populations and among closely related species (Lande 1981, 2000). Ryan (1998), Basolo and Endler (1995), Endler and Théry (1996), Endler and Basolo (1998), Burley (1988), Burley and Symanski (1998), and others have further shown that, while the male traits preferred by females may be arbitrary with respect to fitness, they may not be at all arbitrary from the perspective of the female sensory system.

Nevertheless, considerable literature has accumulated in recent years arguing that females select male phenotypes that demonstrate the bearer's "good genes" (Siva-Jothy and Skarstein 1998; Westreat and Birkhead 1998; Lesna and Sabelis 1999; Scheib et al. 1999; Roulin et al. 2000; Alonzo and Sinervo 2001; Brooks and Kemp 2001; Doty and Welch 2001; Kokko 2001) and, further, that males carry "handicap traits" or "costly signals," which serve as indicators of high-viability genes (Zahavi 1977; Evans and Thomas 1997; Keyser and Hill 2000; Oestlund-Nilsson 2001; Dunn and Cockburn 1999; Zahavi and Zahavi 1997). In many of these arguments, mate choice is seen as a type of communication between males and females. Furthermore, in communication, every exchange between signaler and receiver is viewed as a game-theoretic process, in which the receiver gains or loses fitness in proportion to the "honesty" of the signal (Kodric-Brown and Brown 1984; Bradbury and Vehrencamp 2000; Badyaev and Hill 2000; Ditchkoff et al. 2001; chapter 8). At equilibrium, neither the signaling nor the receiving strategy is presumed to have a fitness advantage and, on average, the receiver must gain from responding to the signal or nonreceivers will invade and the game ends without communication.

Game-theoretic approaches to mate choice, parent-offspring conflict, and parental care of offspring tend to overlook four essential features of evolutionary theory:

1. The average fitness of males (signaling strategy) and females (receiving strategy) *must always be equal* (when R is 1) and are not equal just at equilibrium.

2. With fertility selection, fitness gains do not accrue separately to males or to females, as though they were opposing strategies, but instead accrue to the *mating pair*, especially when interactions between male and female genotypes affect pair fertility as posited explicitly in many mate choice models.

3. Selection on males can be much stronger than selection on females because the variance among males in mate numbers can be so large. Thus, the "balance" between selection in males and in females cannot be discerned solely from observing pairwise interactions in the laboratory or field.

4. Mating fitness concepts or data analyses that confound parental genotypic effects on offspring viability with direct offspring genotypic effects can so obscure the evolutionary dynamic that *the direction of evolution may be incorrectly inferred.* The game-theoretic analysis of parent-offspring conflict is especially prone to this mistake (Wolf and Wade 2001).

Quantitative Genetics as an Alternative to Game Theory

Quantitative genetics offers an alternative approach to game theory for the study of behavioral evolution. For example, in quantitative genetics, the evolution of communication is modeled as the coevolution of two traits, a signaling trait and a receptor trait. The coevolutionary process depends not only upon the phenotypic and genetic correlation between the signal and receptor traits. It also depends upon the distribution of mating types that results from intersex communication, as well as upon the fertility of the resulting pairs. Whenever the interaction of signal and receiver creates a nonrandom pattern of mating and, furthermore, when mating results in variations among pairs in fitness, then whether for "good genes" per se, or good gene combinations, a genetic correlation between signal and receptor is established, even in the absence of pleiotropy.

In regard to tests of "good genes" theories of mate choice, empirical analyses do exist which have found at least weak correlations between male phenotypes and some (often indirect) assessments of offspring fitness (Ditchkoff et al. 2001; Doty and Welch 2001; Jones et al. 2000; Nowicki et al. 2000). However, the reported data tend to come from artificial breeding designs not representative of those that exist in nature (Petrie 1994; chapter 5). In at least one apparent case of good genes based on the outcome of pollen competition in *Silene alba*, Delph et al. (1998) has shown that, regardless of paternal genotype and thus of presumed male quality, females provision early-fertilized seeds to a greater degree than later-fertilized seeds. The lack of unambiguous evidence for adaptive mate choice based on "good genes" for offspring viability is consistent with the predictions of Fisher's theory (Lande 1981, 2000).

Another explanation for the observed patterns of female preferences, which we consider plausible, is the "immediate gain hypothesis," wherein females exert a choice of mates in order to avoid sexually transmitted parasites or diseases (Kirkpatrick 1985; Ryan 1998). This hypothesis has received much less empirical attention than the gene-based parasite hypothesis, wherein the gain for choosy females lies in their disease-resistant offspring (Hamilton and Zuk 1982). Whereas the immediate gain hypothesis confers a consistent, direct fitness benefit to female choice, the "good, disease-resistant genes" hypothesis requires a rather delicate frequency-dependent alternation of host and parasite genotypes as well as male traits that signal, not only health, but health resulting from the ability to resist disease. Although direct benefit is a simpler hypothesis than "good genes", this explanation, like Fisher's, is seldom considered in articles claiming to have demonstrated adaptive mate choice.

Rather than accept Fisher's theory, or address results that show little evidence of genetic quality (Boake 1986; Houde 1992; Jia and Greenfield 1997; Jones et al. 1998; Barber and Arnott 2000; Jia et al. 2000), researchers, convinced of the merits of the hypothesis, have (1) expanded and changed the definition of "good genes" or (2) proposed that female choice of good genes exists, but is somehow thwarted or hidden by male machinations or forced copulations. The expanded version of female mate choice now includes females selecting mates to obtain genes good for offspring genetic diversity (different genes are good; Jones 1987; Murie 1995; Brown 1997; but see Weatherhead et al. 1999), for offspring local environment (different genes are good for different offspring; Zuk et al. 1990; Johnson et al. 1993; Møller 1997, Møller and Alatalo 1999; Kempenaers et al. 1997; Lesna and Sabelis 1999; Doty and Welch 2001; Kokko 2001), and for complementarity to the female's own genotype (different genes are good for different females; Amos et al. 2001; Kempenaers et al. 1999; Penn and Potts 1999; Johnsen et al. 2000; Møller 2000; but see Tregenza and Wedell 2000).

Any deviation from random mating, in any direction, can be and has been construed as evidence of female choice of "good genes" with little regard for quantitative issues such as the strength of selection, the strength of indirect selection on exaggerated male traits, or alternative hypotheses. Some hypotheses simply do not explain either of the two major patterns in nature: the evolution of exaggerated male traits or the divergence of male (but not female) traits between closely related species. For example, the hypothesis of female choice for offspring genetic diversity postulates that exaggerated male traits are indicators of a healthy immune system or of immune system genes as reflected in variation (or sometimes the lack of it) at major histocompatibility (MHC) loci, the genetic factors associated with self-recognition in species with cell-mediated immune systems (Saino et al. 1997; Siva-

Jothy and Skarstein 1998; Westneat and Birkhead 1998; Penn and Potts 1999; Roulin et al. 2000; Johnsen et al. 2000). This now widely accepted hypothesis fails to address the fact that it is quite difficult to convert *balancing selection* on genes in the MHC into strong *directional selection* on other genes controlling male breeding phenotype. It is even more difficult to have *balancing selection in both sexes* at one or a few genetic loci associated with *opposing directional selection between the sexes*, which is necessary to explain sex differences in phenotype. It is difficult for us to see how adaptive mate choice favoring genetic diversity at some loci can explain the evolution of extreme sexual differences caused by alleles at other loci. "Trivial" characters, not those evolving by natural selection, are what led Darwin to postulate the existence of female mate choice.

Hypotheses of "mate choice for offspring genetic diversity" differ little in practice from rare male mating advantage. Like balancing selection, this phenomenon does not lead to the kind of strong directional selection responsible for the rapid evolution within and diversification among taxa of exaggerated male traits. When mate choice for genetic diversity reaches the point where each female has her own "best" mate, then strong directional selection on any trait as a correlated response is extremely unlikely. We have suggested that female mate preferences may be arbitrary. Let us now consider the good genes and runaway hypotheses in more specific detail to determine which explains more of the available evidence on female mate choice and thus the evolution of elaborate male characters.

Fisher's Runaway Process of Sexual Selection

In 1930, R. A. Fisher added a new dimension to Darwin's proposal of sexual selection by female choice. Fisher pointed out that evolutionary changes could occur in *both* sexes during sexual selection. He proposed that this coevolution of the sexes would be self-accelerating and turn sexual selection into a "runaway" process. He postulated that individual females differ from one another in mating preference for genetic reasons. That is, some females are *choosier* when selecting mates. The differences among females in mate preference would lead some less choosy females to accept male mates with certain characters, say short or long tails, but would lead other more choosy females to accept as mates only males with extreme characters, say, in this case, extremely long tails.

In males, Fisher postulated (as Darwin had) that there was variation in some character on which females based mate choice. In our example, males might vary from one another in the length of their tails, but no matter what the preference of the individual female, she tends to accept a male with a long tail as a mate. Thus, males with the longest tails are the most successful at obtaining mates. Short-tailed males are restricted in their mating to the smaller number of less choosy females. The large-tailed males have more mates and thus leave more offspring (cf. chapters 1 and 2). Fisher recognized that this mating pattern had two effects on the numerous offspring of males with long tails. The first effect is that the sons of these males will have longer tails than the average male because they have inherited genes for long tails from their fathers. The second effect is that the daughters of these males will be of greater than average choosiness compared to daughters of less successful males with less choosy mates because they have inherited genes from their mothers that predispose them to prefer long-tailed males as mates.

Differently put, females with the most extreme preferences will mate with the most phenotypically extreme males. These same males will be the more successful at reproduction because they are accepted as mates by a greater number of females. Mating is *not random* in regard to male traits and female choosiness; rather, it is *assortative*. The genes for extreme values of the male and the female traits are brought together by the act of mating. Thus, not only will male tail length increase over evolutionary time as a result of the mating success accruing to preferred males, but the mating preferences of females will also become more extreme.

We can visualize this process using the same terminology we have used since chapter 1, albeit with slightly abbreviated notation. Consider a population consisting of N_{males} and $N_{females}$, *where the sex ratio* R is 1. As in shown in chapter 1, the total number of females divided by the total number of males yields the average number of mates per male. If each female chooses her mate randomly with replacement the probability that she will choose a given male equals $1/N_{\text{males}}$. The variance in mate numbers among males will be

$$V_{\text{mates}} = N_{\text{females}}(1/N_{\text{males}})(1 - [1/N_{\text{males}}]). \qquad [7.1]$$

Or, rearranging,

$$V_{\text{mates}} = N_{\text{females}}\,(N_{\text{males}} - 1)/(N_{\text{males}})^2. \qquad [7.2]$$

Since the opportunity for sexual selection, I_{mates}, equals the variance in mate numbers, divided by the squared average of mate numbers, we find for our randomly mating population that

$$I_{\text{mates,random}} = [N_{\text{females}}(N_{\text{males}} - 1)/(N_{\text{males}})^2]/(R)^2. \qquad [7.3]$$

When the sex ratio equals 1, this reduces to $(N_{\text{males}} - 1)/(N_{\text{females}})$, which is essentially R_O or $(1/R)$ when the number of females is large.

Suppose now that females tend to mate only with certain males in the population, that is, mating becomes nonrandom. Let p_m represent the fraction of males bearing the preferred character and p_0 $(= 1 - p_m)$ represent the fraction of males that lack it. Let s represent the degree to which females prefer particular males. Now, when a female selects a mate, her probability of mating with a preferred male is increased above $(1/N_{\text{males}})$ by s and her probability of mating with a nonpreferred male is decreased below $(1/N_{\text{males}})$ by s. This creates two classes of males, p_m males who are more likely to mate successfully, and p_0 males who are not. As shown in chapter 1, we can calculate the value of I_{mates} explicitly if we know the distribution of mates among successfully mating males as well as the proportion of males belonging to this class. Following Wade and Pruett-Jones (1990, eq. [6], p. 5750), we can identify the opportunity for sexual selection arising under this form of nonrandom mating as

$$(I_{\text{mates,preferred}}) = (I_{\text{mates,random}}) \frac{(N_{\text{males}} + N_{\text{females}}s)}{(N_{\text{males}} + s)} \qquad [7.4]$$

In this expression, the magnitude of the female preference, s, determines the proportions of the male population that will consist of mating and non-mating males (Shuster and Wade 1991b). It is clear that the value of $I_{\text{mates,preferred}}$ is always greater than $I_{\text{mates,random}}$ with positive values of s. If the tendency to be a preferred male is heritable, the opportunity for sexual selection under nonrandom mating, multiplied by the frequency of such males in the present generation, generates the frequency of preferred males in the next generation. With positive values of s, the frequency of the preferred character will increase. The greater the value of s, the more rapid this increase will be.

If female tendencies to prefer particular males are heritable, genetic covariance between the female preference and the male character will arise among the progeny of females who mate with particular males. As V_{mates} becomes greater due to nonrandom mating by females, these males will increase in frequency, and, due to the genetic correlation between the female preference and the male character, the average female preference will increase as well, leading to further increases in V_{mates}.

Theoretical models show that the evolution of extreme male characters can be very rapid under this kind of process (Lande 1981; Kirkpatrick 1982). As female mating preferences become more extreme, males with more exaggerated characters have more mates, males with less extreme characters have fewer mates, and selection on males becomes increasingly strong. However, the choosiest females mate with the most successful males, and the genes for extreme female preference become associated with the genes for the male trait. In this sense, the female choice genes appear to "hitchhike" on the reproductive success of the genes affecting the extreme male trait. However, these female choice genes do more than simply hitchhike. In fact, they *intensify* the process of male competition for mates because they are the determinants of male mating success in the next generation. Thus, female choice is more than just a mechanism that permits sexual selection to take place; it can *accelerate* the rate of evolution by sexual selection.

Should Females Choose Arbitrary Male Traits or Good Genes?

The evidence that females choose mates and that there is heritable variation among females in mating preferences has itself generated new controversies in evolutionary biology. In particular, to postulate adaptive female choice of mates, we must answer the question: If females choose, what male characters *should* they use in deciding to accept or reject a male as a mate? Two very different answers (see above) have been put forward and are currently the focus of great debate over which, if either, is true (Halliday and Arnold 1987; Thornhill and Alcock 1983; Petrie 1994; Gowaty 1997; Qvarnstroem and Forsgren 1998; Olson and Owens 1998; Møller and Alatalo 1999).

The first hypothesis is that female choice is adaptive because the offspring of females inherit "good genes," from the chosen fathers, which enhance viability. Under this hypothesis, the sometimes bizarre and elaborate courtships of different species are considered mechanisms whereby a female screens candidate males for mating. Here, strenuous and prolonged courtship is like a behavioral polymerase chain reaction (PCR), amplifying the female's view of the viability genes her offspring might inherit from among the courting, candidate males (Grafen 1990; Zahavi and Zahavi 1997).

The second hypothesis is called "arbitrary" female choice. The phrase "arbitrary female choice" is meant to suggest that the male character on which the mating decisions of females are based is arbitrary with respect to heritable differences in male viability fitness. That is, the variations in the male characters used to make mating decisions are not associated in any predictive way with genes affecting viability (at least not initially, before the character is exaggerated by evolution through sexual selection, and the association between degree of expression and viability becomes negative). Under this hypothesis, the *existence* of female choice itself defines which males are fit and which are not with regard to mating success. It does not matter at all which male characters the females are using to make mating choices; it matters only that a preference exists. When a mating preference exists, even if only a weak bias in females toward accepting certain kinds of males as mates, then males will differ from one another in mating success and runaway sexual selection will ensue. The "good genes" hypothesis argues differently. Females choose mates whose ornamentation predicts the heritable viability fitness of their sons, which is maximally, due to the existence of two sexes, half the total additive variance in viability.

The logic of both scenarios is plausible, so how are *we* to choose between them? One way is to assume that the "good genes" theory is true and examine its consequences. Imagine a population with heritable variation in body size (fig. 7.1). Further, imagine that there *is* an "optimal" intermediate body size with regard to fitness. Very small individuals, male or female, *and* very large individuals have lower fitness than individuals of intermediate size.

Lastly, imagine that the average size in the entire population lies below the optimum body size. In this circumstance, a female *could* improve the fitness of her progeny by choosing her mate wisely. We can quantify the consequences of different mating decisions on the part of females and show that this is true.

What is the wise choice, with respect to fitness of the offspring? Is it the "good genes" choice of the optimal male? Given that the average female is below the optimum body size, if she chooses to mate with a male *above* the average body size, then her offspring will inherit a higher fitness than they would if she mates at random and does not exert a choice. If she chooses a larger than average male as a mate, she will leave more offspring owing to his large size. From the perspective of a mating female, there are definitely some males with good genes and some with bad genes in this population! The genes manifest themselves in the size differences among males. And, most importantly, they are heritable or transmissible to the offspring of choosy females. It is good for a female to exert a choice of mates in this hypothetical situation. But, will this lead to adaptive or arbitrary female choice? How choosy *should* a female be? How much above the average should her mate be to give her offspring with the greatest gain in fitness? The mating preferences of the females with the greatest gain in offspring fitness will be favored under sexual selection, by definition. What should the average female do?

Suppose the average female mates with a male with the best genes, who *has exactly the optimum body size* as the good genes hypothesis predicts she should do (fig. 7.1a). Because her offspring will inherit a body size *intermediate* between her own, which is below the optimum, and that of her mate, which is at the optimum, her offspring will still have an average body size *below the optimum* (fig. 7.1a). True, they will not be as far below optimum as their mother, but she could have made a better choice with respect to the viability of her offspring and achieved a greater fitness gain for them.

Suppose she chooses a male with a body size *above the optimum* and beyond the average for the population (fig. 7.1c). Her offspring will have an even higher fitness, because the average size of her offspring will lie closer to the optimum than those of the female selecting the optimal male. Why so? Because offspring body size is the average of the size of the mother, below the optimum, and the size of the father, above the optimum. In this case, the average size of her offspring lies closer to the optimum and they are fitter than the offspring of a female choosing a male at the optimum. By simply resisting most mating attempts, females effectively exert a mating preference for large, vigorous males (McCauley 1981). As a result, females that prefer extreme males, that is, males with sizes above the optimum, will give birth to fitter offspring. Their daughters will inherit not only a good body size but also the extreme mating preference of the mother.

This, in turn, will trigger Fisher's "runaway process" because the exis-

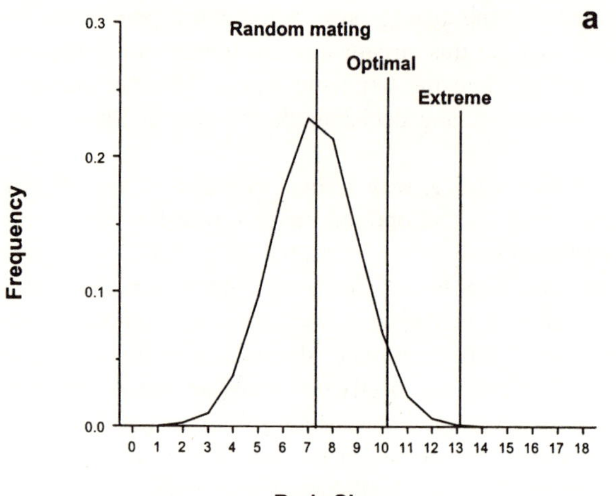

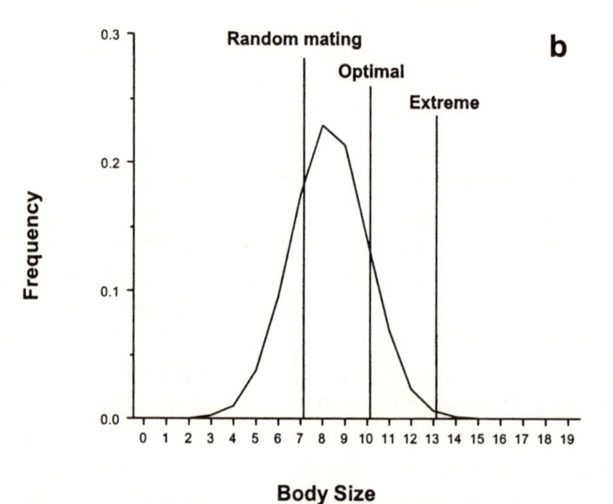

Figure 7.1. The distribution of body sizes among offspring in a hypothetical population in which females have the option of selecting among males who vary in body size; (a) average sized females selecting males of average size are likely to produce offspring of average, but suboptimal size; (b) average sized females selecting males of optimal size will produce larger but still suboptimal sized offspring; (*continued*)

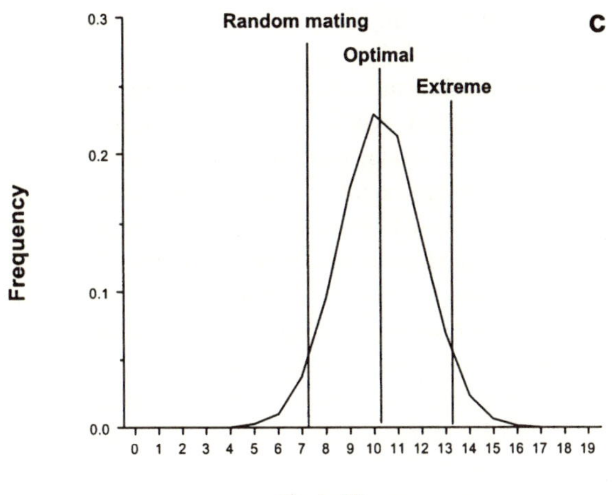

Figure 7.1. (*continued*) (c) only females choosing extremely large males as mates will produce offspring of optimal size; this condition can lead to a runaway between male body size and female preference for males exhibiting extremely large size.

tence of a female mating preference itself confers high fitness in the form of enhanced mating success on extreme males. When we set out to construct a "good genes" scenario for evolution by sexual selection, it immediately becomes a "runaway" process! The male character favored by females does not reach the optimum for viability fitness and settle down there. The female mating preference favored by sexual selection is for extreme rather than optimal traits in males. This will occur despite the cost in viability fitness to females in future generations as mating preferences become ever more extreme. We conclude that male sexually selected traits may have no direct positive relationship with viability fitness. There is good theoretical reason to consider the exaggerated phenotypes of males as "arbitrary" or "trivial" characters as Darwin did (chapter 1).

Mate Choice and Sensory Biases

Nonrandom mate choice caused by preexisting sensory biases in females, and the subsequent production of signals by males that exploit such biases, appear to be excellent examples of arbitrary male characters evolving by runaway sexual selection. However, there is now debate over whether sensory biases in females, and the signals males use to exploit them, will generate genetic correlations between senders and receivers at all (Ryan 1998). This argument has two parts. The first part of the argument is based on the observation that individuals *in different species* can send and receive signals, e.g., insect pollinator systems. Ryan (1998) argues that a genetic covariance cannot occur between signal senders and signal recipients because "correlations between genes determining the plant's signal and the insect's response to that signal are not possible," evidently because no genetic exchange is possible between the participants (but see Goodnight 1991 and Wolf et al. 2001). Thus, according to this hypothesis, genetic correlations are not necessary for the coevolution of sender and receiver in communication systems.

The second part of the argument is based on the observation that female preferences for extreme visual or auditory signals by male fish and frogs, which facilitate signal transmission, appear to *predate* the evolution of that signal in males. From this, Ryan (1998) argues "it is now clear that traits and preferences often do not co-evolve via genetic correlations, that female mating preferences for a given male trait are influenced by adaptations and constraints outside of the context of female responses to that particular trait, and that receiver biases can explain much of the diversity in male signaling phenotypes."

Contrary to the first argument, a genetic covariance *can* exist across species in metapopulations and *will* affect the coevolutionary trajectory of the interacting species (Goodnight and Craig 1996; Goodnight 1991, 2000; Wolf et al. 2001; Wade 2003). Interspecific interactions are a special case of indirect effects and genotype × environment interactions. Evolution "when the environment contains genes" can be fundamentally different from adaptation to abiotic local environments (Wolf et al. 2001), especially in subdivided populations. It is true that genetic correlations across species are not necessary for coevolution to occur, but it is also true that, when they do occur, the rate of coevolution is faster and the diversity of its outcomes greater, very much like sexual selection (Thompson 1999; Gomulkiewicz et al. 2000). Indeed, precisely the kind of genetic associations between host and parasite genotypes that Ryan (1998) argues against play a large role in other theories of sexual selection (e.g., Hamilton and Zuk 1982).

Ryan (1998, p. 1999) suggests that

> contrary to coevolution through genetic correlation, a trait and a preference in sexual selection—or, more generally, a signal and a receiver in animal communication—can evolve out of concert, with the evolution of one component lagging behind that of the other. . . . If a receiver has a bias toward responding to certain signal parameters, such as louder sounds or brighter colors because they are easier to detect, we would expect the evolution of louder or brighter signals without assuming the need for genetic correlations between trait and preference, as required by indirect selection. . . . Sensory exploitation is a theory that males evolve traits to exploit preexisting receiver biases, rather than preferences and traits coevolving via a genetic correlation.

However, even if there is no genetic correlation a priori, when there is genetic variation in the degree of sensory bias among females, the nonrandom mating that results from the male exploitation of the bias, and the presumed fitness payoffs that arise from it, will establish a genetic correlation. Given the evolutionary speed with which visual systems deteriorate by mutation and indirect selection in cave environments (e.g., Jernigan et al. 1994), it is likely that there is heritable variation in the degree of sensory bias among females. In addition, in many models of sexual selection, it is postulated that signal perception is influenced by the interaction of signal and environment. For example, water turbidity is believed to influence the intensity of coloration in fish (Stoner and Breden 1988; Reznick et al. 1996; Houde and Endler 1990) and, similarly, ambient light levels are assumed to influence the evolution of plumage coloration in birds (Marchetti 1993; Endler and Théry 1996). Whenever an interaction between signal and environment affects receiver stimulation, then there will be $G \times E$ for the fitness consequences of signal perception. Because $G \times E$ for the fitness is a common way to maintain genetic variation (e.g., Frank and Slatkin 1990), we expect that this situation would act to maintain genetic variation among females in sensory bias too. Analysis of quantitative variation for these traits could provide a clearer picture of how this mechanism works (Boake 1986, 1991, 1994; Barber and Arnott 2000; Emerson 2000; Jia and Greenfield 1997; Jia et al. 2000). However, few research efforts in sexual selection have been based on the foundation of quantitative genetics.

The observation that sensory biases may *predate* genetic correlations between signal and receiver, because they predate the existence of the signal, does not eliminate the role of genetic correlations in runaway evolution once the signals exist. Correlations between female mate preferences and extreme male characters are among the most well documented aspects of the sexual selection literature (reviews in Andersson 1994; Ryan 1998). If female signal preference and male signal production co-occur in the same species and are both heritable, nonrandom mating will establish a genetic covariance between these characters. It is clear that sensory biases exist in females for certain colors, sounds, temperatures, smells, or aggregations (reviews in Ryan 1998; Basolo 1995). If males vary in traits evocative of the existing sensory bias, and if particular signals enhance a female's probability of mating with the males that produce them, then this pattern of nonrandom mating will associate the stimulatory, signal-producing alleles in males with the alleles for the greatest bias in females, and the runaway process will ensue. Kirkpatrick and Barton (1997) reached the same conclusions, although their estimates suggest that these genetic correlations will not be very strong.

Ryan (1998) argues that sensory biases in females evolve by natural selection because they are functional in signal reception, possibly in the context of maternal-offspring interaction. That is, female mating preferences for a given male trait are adaptations influenced by constraints lying outside the context of sexual selection. The female adaptations and constraints Ryan describes must therefore arise by single-sex, among-family selection. As we have noted (chapter 1), selection in these contexts is weak, and consequently, the evolution of sensory biases under natural selection will necessarily be slow. And, even with weak genetic correlations, the indirect effects of sexual selection may not be trivial (Agrawal et al. 2001b).

We have shown how sexual selection can often be stronger than opposing natural selection on female fecundity and viability (chapter 5). The runaway process involves selection on both sexes at once, and some of the stronger selection on males, through nonrandom mating, leads to indirect selection on the female mating preference. When this occurs, the process becomes self-accelerating, intensifying selection on both sexes. We expect that as this process intensifies it may readily overcome weaker, maternal, or among-family viability selection. There is little reason to dispute the hypothesis that sensory biases may arise under natural selection (Ryan 1998). However, if genetic correlations between female signal preferences and male signal production form as soon as nonrandom mating occurs, if they persist because they are self-reinforcing, and if they exceed in intensity the source of selection that originally produced them (Lande 1981; Kirkpatrick 1982), then it is difficult to dismiss the effect of runaway sexual selection on these traits.

Runaway Sexual Selection and Sexual Conflict

What stops runaway sexual selection? Fisher (1930) argued that the runaway process ceases when males become so extreme that they are crippled by their adornments and are removed by natural selection. That is, the evolutionary exaggeration of male traits stops when opposing viability selection balances sexual selection. Fisher (1930) made no specific predictions about how selection would act on female preferences at the end of the runaway process. However, it is reasonable to infer that, since the female preference evolves by indirect selection, when the evolution of the male trait is halted, so is the evolution of the female preference. Many researchers also assume that viability selection acting on females, through increased difficulty in finding acceptable mates or antagonistic pleiotropy, may also oppose sexual selection on males (Andersson 1994; Whiteman 1997; Ryan 1998; Holland and Rice 1998; Knowles and Markow 2001). However, because natural selection acting on each sex separately is a slower and less efficient process than accelerating runaway sexual selection (chapters 1, 4, and 5), we believe that Fisherian runaway sexual selection is most often limited by opposing viability selection on males after male phenotypes become exaggerated, or by other factors which diminish the variance in mate numbers among males. The ability of sexual selection to drive phenotypes well beyond an optimum favored by viability selection also explains why extreme male characters and extreme female preferences for them often persist at high frequency in natural populations.

As we mentioned in chapter 4, Holland and Rice (1998) suggest that cessation of the runaway process occurs when females choosing extreme males incur a fitness penalty, and the direction of selection on mating preferences reverses. "Chase-away" selection, they argue, leads to a *reduction* of female preferences for extreme males, coinciding with the evolution of *more extreme* characters in males to overcome increased female resistance. Hyperextreme males thereby stimulate females to "ignore" their evolved reluctance, and, in conflict with the females' best interests, cause them to mate with hyperextreme males. As evidence in support of their hypothesis, Holland and Rice (1998) report a reduction in the frequency of females with preferences for extreme males, coincident with the persistence of extreme males in certain populations. Thus, the chase-away hypothesis provides an explanation for the persistence of extreme males in natural populations, as well as a means by which the runaway process may slow down and stop.

What other possible mechanisms could explain this observation? We suggest that the coevolution of female mating resistance and male vigor is itself the result of a Fisherian runaway process. Whenever females resist male mating attempts, females effectively exert a mating preference for large, vigorous males that can overcome the resistance (McCauley 1981). If males vary from one another in their ability to overcome female resistance, and females vary from one another in ability to resist, the elements of a classic Fisherian runaway process are in place. The most persuasive males will have more mates than other males and will thus leave more offspring. The sons of these persuasive males will inherit their father's tendency to persuasion and the daughters of these males, on average, will be more resistant to male mating attempts. At equilibrium, most males will be able to persuade most females and the variance among males in mate numbers will stabilize. The observed pattern will be coy females, highly resistant to male mating attempts, and persistent and persuasive males, with no apparent correlation between the two traits. Despite the genetic association of male persistence and female resistance, no chase-away process is necessary to produce the observed pattern.

In another example from chapter 5, we showed how a negative covariance between male mate number and female fecundity can counterbalance opposing sexual selection. There (see also Wade 1995), we suggested that female fecundity might diminish in larger harems owing to a runaway process between male harem-holding abilities and female tendencies to aggregate. Such a genetic correlation would lead females to overaggregate on resources and thereby suffer the fitness-lowering consequences of resource competition. In this situation, females suffer decreased fitness when mating with the most successful males, but no chase-away process has occurred.

The two examples above require a single evolutionary step, the establishment of a genetic correlation between a quantitative trait in each sex. The chase-away hypothesis requires at least four evolutionary steps. A Fisherian runaway process must first arise between the sexes that causes males to develop extreme phenotypes. Matings with extreme males must next become so harmful to females that opposing viability selection causes the runaway to decelerate. Female mating preferences for less extreme male phenotypes must next be favored by viability selection, and lastly male phenotypes must become less extreme. Even with these additional steps, the chase-away hypothesis fails to explain how the relationship that initiates and drives the runaway process, i.e., the positive genetic correlation between extreme male characters and extreme female mating preferences, is reversed and made to proceed in the opposite direction. When the genetic correlation is zero, indirect selection on the female mating preference ceases and the runaway is halted, but the variance in male mating success does not vanish. It is not clear to us how or why single-sex, among-family viability selection on females, which was overwhelmed by the variance in male mating success at the start of the runaway process, will come to predominate and initiate a reversal at the end of the process.

The chase-away process requires that alleles for female *resistance* become associated with alleles for *less extreme* male characters. As we showed in chapter 4, a negative covariance between male and female mating-number promiscuity can arise if females mate infrequently and, when they do mate, they do so only with the most promiscuous males. This leads to the association of the most choosy and reluctant females with the most promiscuous males, that is, males with extraordinary mating success. This association is the opposite of the one proposed by Holland and Rice (1998) for chase-away selection, which requires that some females relent and mate with less extreme males, and also that these males leave more offspring than males with larger numbers of more resistant females. That is, the viability fitness of females preferring less extreme males, w_{less}, must not only exceed that of females mating with more extreme males, w_{more}, but also the ratio of ($w_{\text{less}}/w_{\text{more}}$) must exceed the ratio of the mate numbers of the most extreme males, m_{more}, to that of the less extreme males, m_{less}. Only when the product of $(w_{\text{less}})(m_{\text{less}})$ exceeds $(w_{\text{more}})(m_{\text{more}})$ can the genetic correlation be reversed.

Initial Parental Investment

Parental investment theory (Trivers 1972) lies at the foundation of most current mating system research (Emlen and Oring 1977; Bradbury and Vehrencamp 1977; Borgia 1979; Wickler and Seibt 1981; Wittenberger 1981; Thornhill and Alcock 1983; Bradbury 1985; Clutton-Brock and Vincent 1991; Clutton-Brock and Parker 1992; Arnold and Duvall 1994; Parker and Simmons 1996; Reynolds 1996). According to this view, differences in gamete size between the sexes represent differences in initial parental investment, and these inequities are the ultimate source of sexual selection. The term *anisogamy* refers to the tendency among males to produce many small sperm and for females to produce fewer, larger ova (Trivers 1972; Parker et al. 1972).

The origin and maintenance of sexual reproduction is a central problem in current evolutionary theory because, all else being equal, asexual populations have a twofold fitness advantage over their sexual counterparts. Models for the evolution of sexual reproduction are generally preceded by the evolution of anisogamy, the process by which unequal-sized gametes evolve from isogamous or phenotypically indistinguishable gametes. The twofold cost of sex exists, at least in part, because of anisogamy, since it is the production of large ova by the female sex that limits population growth rates. At the same time, anisogamy generates sexual selection whenever the number of small gametes produced by males exceeds the number necessary to fertilize the ova of a single female. Thus, sexual selection resulting from the variance in mate numbers of the sex producing small gametes does not exist in asexual populations. Sexual selection, born of anisogamy, can itself be a sufficient explanation for the evolution of sex as long as deleterious mutations are more harmful to males than to females (Agrawal 2001a).

Gamete dimorphism is thought to have arisen from ancestral populations of isogametic producers, whose population-wide gamete size was normally distributed. Parker et al. (1972) showed that when total reproductive energy among individuals is fixed and syngamy occurs at random, disruptive selection favors two phenotypes: (a) producers of gametes that maximize fertilizations by producing many small gametes (i.e., "males"), and (b) producers of large gametes that provide nutritional investment to developing zygotes (i.e., "females"). The relatively few ova produced by females constitute a limited resource for which all males must compete. Male tendencies to be more competitive in their attempts to mate, to be less discriminating in their mate choices, as well as to be less parental toward offspring than females, are all attributed to cytoplasmic differences in gametic investment and architecture under this theory (Alexander and Borgia 1979; Thornhill and Alcock 1983; Kvarnemo and Ahnesjö 1996). These generalizations have been applied, not only to animal mating systems, but also to circumstances involving pollen or gametophyte competition in plants (Willson 1994; Grant 1995; Delph et al. 1998), and to mating types in protists (Parker et al. 1972; Alexander and Borgia 1979).

From the parental investment perspective, the intensity of sexual selection should correlate with the intensity of male-male competition for mates. Because differences in initial parental investment in gametes are considered the ultimate source of among-male competition, attempts to measure the intensity of sexual selection have mainly focused on the ratio of male parental investment to that of females. This historical emphasis can be traced to Trivers' statement: "The single most important difference between the sexes is the difference in their investment in offspring. The general rule is this:

females do all of the investing; males do none of it" (Trivers 1972, p. 207). Although exceptions exist (see below), this statement provides the conceptual foundation for most research investigating conflict between the sexes.

It is evident in most sexual organisms, including humans, that, on average, females invest more time and energy in offspring than do males. For many, this logic extends an additional step: male reproductive success is limited by the number of matings, whereas female reproductive success is limited by time and energy (Waage 1997). The essence of this idea can be found in Bateman (1948, p. 365), who stated:

> in most animals the fertility of the female is limited by egg production which causes a severe strain on their nutrition. In mammals the corresponding limiting factors are uterine nutrition and milk production, which together may be termed the capacity for rearing young. In the male, however, fertility is seldom likely to be limited by sperm production but rather by the number of inseminations or the number of females available to him. . . . In general, then, the fertility of an individual female will be much more limited than the fertility of a male. . . . This would explain why in unisexual organisms there is nearly always a combination of an undiscriminating eagerness in the males and a discriminating passivity in the females.

Trivers (1972), attempting an empirical test of these ideas compared the maximal number of male gametes to the maximal number of female gametes across several taxa. Others have estimated the ratio of the maximum potential reproductive rate (PRR) of males to females (Clutton-Brock and Vincent 1991; Reynolds 1996).

In some studies, parental investment in gametes is confounded with parental investment in sex ratio. Trivers and Willard (1973) predicted that, because males must be in good condition to compete successfully for mates, family sex ratio varies with the ability of females to invest in male versus female offspring. Under the Trivers-Willard hypothesis, females in good physical condition should produce a preponderance of sons, while females in poor condition should tend to produce more daughters (Kruuk et al. 1999). A change in offspring sex ratio with maternal condition has been reported for several vertebrates, although, in most studies, population density, food availability, or female rank have served as indirect indicators of maternal condition. For example, female-biased sex ratios have been reported to occur at high population density in red deer (Kruuk et al. 1999) and with decreased food supply in snow geese (Cooch et al. 1997), but also with high food availability in Seychelles warblers (Komdeur 1996). Social conditions affecting maternal condition are also associated with reports of female-biased sex ratios, such as decreased male attractiveness (collared flycatcher, Ellegren et al. 1996), low male mate quality in the blue tit (Svensson and Nilsson 1996), solitary breeding by female root voles (Aars et al. 1995), low maternal rank (Saharan arrui, Cassinello and Gomendio 1996), or high maternal rank (yellow baboons, Altmann et al. 1977).

In evolutionary genetic models, it is very difficult for selection to favor genes that systematically skew sex ratio, because the underrepresented sex at the time of mating has a fitness advantage (Fisher 1930; Futuyma 1998). Models of environmental sex determination (Bull and Charnov 1989) predict that skewed primary sex allocation occurs when there are sex differences in the fitness gains associated with incubation environment. More importantly, the predicted direction of the sex ratio bias is *toward the sex produced in the poorer environment*. In this case, it would be females produced in the low-fitness environment of mothers in poor condition. Thus, if poor maternal condition can be viewed as a special environment unfavorable to males, then these same species should exhibit mild female-biased sex ratios (see also Freedberg and Wade 2002). This type of bias in the primary sex ratio requires a positive covariance between sex ratio and an environmental aspect of fitness. A covariance between a population trait, like sex ratio, and local fitness is the definition of group selection (Wade 1995). Thus, the findings of Bull and Charnov (1989) are consistent with the predictions of Williams (1966) that sex ratio is a trait that should respond to group selection (Colwell 1981). We will return to the topic of sexual selection and sex ratio evolution in chapter 12. There we show that sexual selection always creates evolutionary pressure to bias the sex ratio toward females and the strength of this pressure is proportional to the strength of sexual selection.

Despite their popularity in recent years, these approaches to the analysis of sexual selection are inadequate for understanding the evolution of mating systems. The confounding of multiple investments, in gametes, in offspring sex ratio, and in parental care, clouds rather than clarifies theoretical issues of the sex difference in the strength of selection, the origins of anisogamy, and the evolution of parental care. As explained earlier (chapter 1), ratios of gamete investment do not measure the existing sex difference in the strength of selection or the intensity of sexual selection (Wade 1979; Wade and Arnold 1980). Anisogamy may permit sexual selection and competition for mates, but the degree of anisogamy, that is, the sex difference in gamete size, does not equal the intensity of sexual selection, because it is not necessarily proportional to the variance in relative fitness of either sex. Indeed, anisogamy is likely to be the result rather than the cause of sexual selection (see below). Measuring sex-specific selection requires estimates of relative fitness before and after each episode of selection for each sex (Crow 1958, 1962). Neither ratios of parental investment, nor correlations between female dominance status and offspring sex ratio, nor estimates of operational sex ratio (OSR) equal the variance in relative fitness for either sex within populations. In some cases, as we showed in chapter 3, a measure like the OSR can influence the strength of sexual selection, but the effect of sex ratio is not equal to, and is usually less than, the sex difference in the variance in relative fitness.

If these measures do not equal the intensity of selection within species, then they may not be the best independent variables for gauging variation in the strength of sexual selection in comparisons among species. For example, species differences in gamete production are more likely to be the consequence of variation in mating system across species rather than the cause of it. The large size of male testicles in chimpanzees appears to be the *result* of sexual selection in a polygamous mating system (Smith 1984). The polygamous mating system of chimpanzees is unlikely to have evolved in order to allow male chimps to capitalize on their testicle size.

There is also the concern of how best to standardize sperm production rates versus egg production rates across species (Clutton-Brock and Vincent 1991; Reynolds 1996; Kvarnemo and Ahnesejö 1996). Given sex differences in resource utilization and behavior within species, it is not clear how sperm produced per unit of time or energy, should be equated with ova produced per unit of time or energy. Whatever the relevant units for each sex, when expressed as ratios of gametic investment (e.g., Clutton-Brock and Vincent 1991), they do not necessarily "cancel out" to give a dimensionless measure suitable for interspecific comparisons. Not only do ratios in general have poor statistical properties, owing to the possibility of correlations between numerator and denominator, but the suggestion to use ratios of *maximum* rates of gamete production introduces another statistically undesirable feature. Extreme values, like maxima or minima, are notoriously sensitive to sample size and sampling effort, heightening the issue of confidence limits on the estimates for any particular species. For these reasons, we find that ratios of gametic investment are conceptually inadequate for estimating the sex difference in strength of selection, and that such measures are statistically dubious for multiple reasons.

Prezygotic Versus Postzygotic Investment

An additional difficulty with the parental investment approach to the study of mating systems is that it focuses on postgametic or postzygotic investment of reproductive effort. There is little doubt that males in many species invest smaller amounts of energy than females in terms of gametic mass or parental care. However, these same males may invest considerable energy toward structures or behaviors that are useful in gaining access to mates. For example, the male courtship calling of anurans, which converts metabolic energy into acoustic energy, is both expensive and inefficient. Ryan (1988; Bucher et al. 1982) has shown in the Tungara frog, *Physalaemus pustulosus*, that less than 1.5% of the metabolic energy used in calling is converted to acoustic energy; 98.5% of the energy expended is wasted. Mechanisms of attraction have other costs to male viability as well (Krebs and Davies 1987; Davies and Halliday 1979; Backwell et al. 1995; Balmford et al 1994). Arak (1983) has collated the expenditure of energy by males of different species of anurans. The common toad (*Bufo bufo*) is typical of species that breed explosively. Males of *B. bufo* experience a daily weight loss (1.07%) while males of the natterjack toad (*B. calamita*), a lek breeding species, lose 0.32% of their weight each day. Territory maintenance, i.e., calling and fighting with other males, costs a male green frog (*Rana clamitans*) approximately 0.18% of its body weight per day throughout a breeding season that may extend several weeks.

Thus, the mode of male competition for females may determine the variance in male mating success and thus influence the opportunity for sexual selection. Moreover, the variance in male mating success changes as soon as male competition begins, often before any gametes are produced. This means that the level of male-male competition for mates determines the amount of energy remaining for male investment in gametes or offspring. Males in most species partition their limited reproductive energies into some *combination* of intrasexual competition and male parental care. If the level of prezygotic investment toward mate competition determines or affects the level of postzygotic investment by males, circumstances could exist in which male competition is reduced. In such cases, postzygotic investment by males could equal or exceed that of females. The parental investment approach to the study of reproductive behavior considers only the latter portion of male investment in its measures of the intensity of sexual selection (Trivers 1972; Clutton-Brock and Vincent 1991).

Parker et al. (1972) interpreted the results of their theoretical investigation on the evolution of anisogamy as consistent with parental investment theory. Yet the evolutionary dynamics of gamete dimorphism are specifically excluded from Trivers' (1972) hypothesis. In parental investment theory, relative gamete sizes represent a fixed, ancestral state. With this assumption in mind, it is difficult *not* to conclude that male competitive activities associated with mate acquisition must arise from low energetic investment in gametes. However, when the relative amounts of prezygotic and postzygotic investment in offspring are considered simultaneously in each sex, it is clear that a high level of male-male competition could itself favor low parental investment in offspring (Wade 1979). It is less clear how low levels of parental investment by males could be the cause of high levels of male-male competition (cf. Trivers 1972).

Bateman (1948) showed with *Drosophila* that female fitness saturates with increasing mate numbers, whereas male fitness increases with increasing mate numbers. These well-known results are consistent with the ultimate predictions of both Parker et al. (1972) and Trivers (1972). However, reanalysis of Bateman's (1948) observations using the opportunity for selection approach (Wade 1979; Wade and Arnold 1980) shows that the intensity of sexual selection in Bateman's experiment is due to a sex difference in the variance in mate numbers, not to differences in initial parental investment, despite Bateman's speculation on relative investment in gametes and its possible effects on mating behavior by males and females (see above). Bateman's fundamental result was that some males mated more than once and many males did not mate at all, whereas most females mated about the same number of times. In short, there was greater variance in mate numbers among males than among females. This result *alone* is both necessary and sufficient to explain the sex difference in offspring numbers.

Wade (1979) suggested that differences in initial parental investment among the sexes might best be explained in this light as well. That is, males may be less parental than females because males *increase* their access to mates by *decreasing* their initial parental investment. If this hypothesis is true, initial parental investment between the sexes is a *secondary* rather than a *primary* cause for sex differences in mating and parental behavior. Differently put, differences in initial parental investment among the sexes are likely to *arise from* rather than *lead to* sexual selection favoring increased mate numbers in males and emphasis on parental care by females.

The results of Parker et al (1972) are usually interpreted as confirmation of parental investment theory. However, we believe that they are more consistent with Wade's (1979) hypothesis than they are with that of Trivers (1972). Individuals with smaller than average gamete size had greater fertilization success. Individuals with larger than average gamete size suffered fewer zygote failures. Both types were favored in the Parker et al. (1972) model. Such increases in gamete size were possible, according to Parker et al., because the probability of large gametes colliding with the more numerous small gametes was high. Clearly, this result could not be obtained unless selection favored individuals of *both* mating types, whose levels of postzygotic investment negatively correlated with prezygotic levels of mate competition. Thus, an explicit prediction of Parker et al. (1972), one that is *inconsistent* with parental investment theory, is that the level of postzygotic investment among mating types will depend on the level of prezygotic investment in mate competition within mating types.

We suggest that some of the difficulties now attributed to the Emlen and Oring (1977) hypothesis can be avoided if researchers focus their efforts on obtaining direct measurements of selection, rather than on obtaining indirect estimates of relative parental investment. The approach we have proposed above and will elaborate below permits researchers to *predict*, rather than merely *describe*, how males respond to female distributions and life history characteristics. By quantifying the source and intensity of selection, researchers can move straight to the conceptual core of the Emlen-Oring hypothesis, without invoking parental investment theory. As we have shown in the first six chapters of this book, the sex difference in the opportunity for selection arises from female spatiotemporal distributions and female life history characters, *not* from differences in relative gametic investment. This may appear to be a matter of proximate versus ultimate causation (Williams 1966). However, as argued by Wade (1979, 1987; Wade and Arnold 1980), as demonstrated by Parker et al. (1972), and as we will describe below, sex differences in initial parental investment most likely result from rather than cause sex differences in the variance in mate numbers.

Parental Care and Investment in Offspring: The Evolution of Male Parental Care in the Context of Sexual Selection

Clutton-Brock (1991) asserted that parental care enhances offspring survival during development (fig. 7.2), and that female parental care is evolutionarily derivable from a "non-care-giving" ancestral state, conditional on the trade-off between current offspring survival versus future reproduction. Male parental care is conspicuously less common in most taxa than is female parental care (Clutton-Brock 1991). The standard explanation for the taxonomic bias toward female parental care is based on the investment arguments of Trivers (1972) that were formalized in the ESS theory of Maynard Smith (1977). The fundamental argument is this: Because males invest relatively little in offspring, owing to anisogamy, the gain to a male from additional matings vastly exceeds the potential gain to him in fitness from the increment in viability added by caring for existing young.

Variation among taxa in the incidence of paternal care (e.g., Clutton-Brock 1991) is often explained by certainty of parentage (Trivers 1972; Alexander and Borgia 1979), which is greater for females than males in species with internal fertilization. More generally, parental care with external fertilization is explained by gametic proximity, wherein the sex closest to the eggs at fertilization cares for them (Trivers 1972; Williams 1975; Dawkins and Carlisle 1976). Under the gametic proximity hypothesis, offspring become fitness hot potatoes, wherein the nearest parent, often the male, is stuck in the "cruel bind" of either caring for the young or squandering its reproductive investment and opportunity. The latter hypothesis is the current explanation for the increased incidence of male parental care in fishes and amphibians (but see Beck [1998] for taxonomic evidence contradicting this view).

There are many additional arguments connecting offspring survival, parental care, and sexual selection (e.g., Gross and Shine 1981; Tallamy 2001). It has been argued not only that male parental care is one of the direct benefits of female mate choice, but also that it is an indirect "good genes" benefit owing to the enhancing effect of caring sons on future grandchildren (Møller and Thornhill 1998). In the section below, we examine the logic of these arguments under the constraints that necessarily connect the mean and variance of male and female fitness as shown in the earlier chapters. These

Figure 7.2. Male giant water bug (Belostomatidae) defending his brood (photo by E. D. Brodie III).

constraints are not explicit in the ESS theory of alternative strategies including parental care and, to the extent that they are violated, the theoretical conclusions must be considered suspect.

Theory of Parental Care, or, Lake Woebegon, where the Fitness of Every Deserting Male is Above Average

In any ESS model, as in population genetics theory, in order for a population to become polymorphic for two (or more) strategies, at equilibrium, each of the strategies must have equal average fitness. For instance, in hawk-dove or prisoner's dilemma games, polymorphism requires equal fitnesses at equilibrium (Maynard Smith 1982; Dugatkin 1998). In using ESS models to investigate male and female reproductive strategies, stable within-sex polymorphisms also require the equilibration of fitnesses. However, there is another constraint necessary to game-theoretic models of reproductive strategies that must obtain *at all times*, regardless of whether a population is or is not at equilibrium. This is the constraint that the average fitness of males must equal that of females multiplied by the sex ratio, R. This constraint makes the application of ESS theory to the evolution of reproductive strategies *different* from its application to other kinds of interaction where the constraint of equal fitness is necessary only at the end of the analysis, to achieve polymorphic equilibrium. Differently put, ESS models of male-female reproductive strategies must include constraints in the fitness payoff matrix such that the average fitness for males equals that for females *at every iteration.*

We now consider the payoff matrix most frequently used to investigate the evolution of parental care (Maynard Smith 1977) and also to "explain" the trade-off in male behavior between caring for offspring versus finding additional mates. The fitness matrix in table 7.1 makes a series of reasonable hypotheses about the consequences to male and female fitness of caring for the young or deserting the young. And, in the typical analysis of this payoff matrix, there are four possible ESS solutions, depending upon the ecological conditions under which the game is played: (1) Males and females desert; (2) males desert, females care; (3) males care, females desert; and (4) males and females care.

The fitness entries for females in the matrix (table 7.1) assume that there are two kinds of females: (1) caring females, who lay W_c eggs, and (2) deserting females, who lay W_d eggs. Let the frequency of caring females be f_c and that of deserting females be $(1 - f_c)$ or f_d. Because deserting females can lay another clutch of eggs with the energy they save by not providing parental care to the young, it is further assumed that $W_c < W_d$. It is also

Table 7.1
The pay-off matrix to males and females in the ESS model of the evolution of parental care (Maynard Smith 1977; Clutton-Brock 1991, table 7.1, p. 105).

		Females	
Males		*Cares For Young*	*Deserts*
Cares for Young	Female Fitness	W_cV_2	W_dV_1
	Male Fitness	W_cV_2	W_dV_1
Deserts	Female Fitness	W_cV_1	W_dV_0
	Male Fitness	$W_cV_1\ (1 + p)$	$W_dV_0\ (1 + p)$

assumed that the survivorship of offspring with maternal care, S_c, exceeds that of the offspring of deserting females, S_d, i.e., that $S_d < S_c$, because caring enhances offspring survival. Considering only the female sex, it is clear that the strategy with the highest average fitness, W_cS_c or W_dS_d, will predominate in the population. Females should desert when W_dS_d exceeds W_cS_c and should remain and care for the young when W_cS_c exceeds W_dS_d. This is often described as a trade-off for females between investment in present versus future reproduction and some have said that caring for present offspring imposes an "opportunity cost" in the form of sacrificing future reproduction. A stable polymorphic population with caring and deserting females will exist only if the average fitnesses of both strategies are equal (i.e., $W_cS_c = W_dS_d$).

The analysis is somewhat complicated because offspring survivorship depends upon the behavior of both parents, not just upon the behavior of females. This dependence upon both parents can be addressed by setting offspring survival to one of three values, V_2, V_1, or V_0, depending upon the *number* of parents providing care, two, one, or none, respectively. This means that care given by a male has the same effect on offspring survivorship as care given by a female. Even with this refinement, S_d still tends to be less than S_c because the viability of caring females is some average of the larger values, V_2 and V_1, whereas offspring viability of the deserting, noncaring females is the average of the two lower survivorship values, V_1 and V_0. Nonrandom mating by females, contingent upon a judgment made prior to mating on the likelihood of a male caring or deserting, can greatly complicate calculating the fitness average. The hypothesis that male deception confounds female mate judgment adds yet another layer of complexity.

The analysis of table 7.1 also assumes that there are two kinds of males: (1) caring males, who have one mate but, by caring, increase offspring survivorship; and (2) deserting males, who do not increase offspring survivorship but, instead, get the fitness benefit of extra matings with probability p. In the literature considering such circumstances, no explicit model for pairing deserting males with females is given, nor is it discussed where the "extra" females might come from in order for deserting males to acquire additional mates. Like the case for females, whether a male should stay and provide care for his young, or desert and find another mate, depends upon the relative increment to male fitness. When the survival gain to present offspring, S_c, exceeds the future gain of additional matings, p times average female fecundity fitness, then males are expected to provide care for the young. When the inequality is reversed, they should not. Similar to the case for females, the calculation of the average benefit is generally confounded by the presence of multiple female strategies and nonrandom mating resulting in mate choice based upon a promise or indication of future care.

In many teaching presentations, it is argued (somewhat circularly) that additional matings must confer a greater increment to male fitness than paternal care because males of so many species do not provide care. In addition, fitness gain through additional matings is a strategy available to males but not to females, unless, like many plants, they are sperm limited. It is often concluded that this asymmetry "explains" conspicuous sex difference in behavior of promiscuous males and caring females. At this point in many discussions, the existence of all four possible outcomes in nature is taken as evidence in support of the analysis, and taxonomic variation in the frequency of different outcomes is explained by sex differences in gamete investment. We believe that this ESS approach has fundamental problems and, when the problems are corrected, the results are congruent with standard evolutionary theory.

To analyze the flaws in the ESS approach, we take table 7.1 as our starting point. Two strategies for each sex allow a minimum of four kinds of matings. (We say "minimum" because some males could desert and remate in a manner not specified in the ESS model.) Let the frequency of the four mating types be given by $G_{c,c}$, $G_{c,d}$, $G_{d,c}$, and $G_{d,d}$ where the subscripts denote whether an individual cares (c) or deserts (d), with females listed first and males listed second. Thus, $G_{d,c}$ is the frequency with which caring males are paired with deserting females. The sum of the four frequencies must equal 1; if it does not, then the frequency distribution of mating pairs is incomplete. We will not make any assumptions about the frequency of the four mating types; what follows applies to random mating as well as to any pattern of nonrandom mating.

Using the sex-specific fitnesses from the payoff matrix (table 7.1) and the four mating type frequencies, we can calculate the mean fitness of males, W_{males}, and females, W_{females}. Starting with the females, we find that their mean fitness equals

$$W_{\text{females}} = G_{c,c}(W_cV_2) + G_{c,d}(W_cV_1) + G_{d,c}(W_dV_1) + G_{d,d}(W_dV_0). \quad [7.5]$$

It is not specified in the pay off matrix what happens to the initial mating type frequencies $G_{i,j}$, or to female fitnesses when some males desert and find new mates. Thus, at some very fundamental level, table 7.1 is incomplete in regard to the biology being modeled. Implicit in the payoff matrix is the assumption that additional matings by deserting males do not change female fitness in any way. This implication is evident in two features of table 7.1: (1) no entries are posted for remating females; and, (2) the payoffs for remating males are given as p times the fitness of caring or deserting females, W_cV_1 or W_dV_0, respectively.

Mean male fitness is somewhat more problematic than that of females. If we use the payoff matrix of table 7.1 and the mating type frequencies, we find that

$$\begin{aligned} W_{\text{males}} = {} & G_{c,c}(W_cV_2) + G_{c,d}(W_cV_1[1 + p]) \\ & + G_{d,c}(W_dV_1) + G_{d,d}(W_dV_0[1 + p]). \end{aligned} \quad [7.6]$$

Table 7.1 requires a sex difference in the mean fitness of the two sexes equal to

$$\Delta W = W_{\text{males}} - W_{\text{females}} = p\{G_{c,d}(W_cV_1) + G_{d,d}(W_dV_0)\}. \quad [7.7]$$

Because all of the terms on the right-hand side of eq. [7.7] are positive, mean male fitness must exceed mean female fitness, $W_{\text{males}} > W_{\text{females}}$. This is what we mean by our reference to the average fitness of deserting males in the heading of this section. *When the sex ratio is unity, this condition is not possible.* Hence, when R is 1, the fitness matrix of table 7.1 can be reconciled with standard evolutionary theory *only if* either p equals zero or the frequency of deserting males $(G_{c,d} + G_{d,d})$ is zero. That is, either deserting males gain no additional matings or there are no deserting males. In either case, the ESS analysis is undermined. If the sex ratio is not 1, then eq. [7.7] implies that females are more abundant than males and that some females wait to mate with deserting males. From chapter 1, we know that

$$\Delta W = W_{\text{males}} - W_{\text{females}} = (W_{\text{females}})(1 - R), \quad [7.8]$$

so that a sex difference in mean fitness requires a sex ratio R greater than 1.

The reason behind our conclusion and the flaw in the ESS analysis is straightforward. The payoff matrix for males in table 7.1 awards a higher fitness to males than it does to females. No sex ratio restrictions like those we have explained above have been placed on the ESS analysis, either in discussions in the literature, or in its application to various taxa. We are left with the question: Are the inferences about the evolution of sex-differences in parental care true if the theory on which they are based contains internal contradictions? The taxonomic patterns reviewed in Clutton-Brock (1991) exist, but the classic explanation for them is founded on flawed (or, at least, incomplete) theory.

We can change table 7.1 to make the entries consistent with standard evolutionary theory and the biology being modeled. Some other considerations must be added to the standard payoff matrix to make it internally consistent, but what they are is not entirely clear because the options are many. Here, we address the simplest case, wherein deserting males, whose frequency is m_d $(= G_{c,d} + G_{d,d})$, take a fraction of matings away from both kinds of care-giving males (table 7.2, row 2). This is consistent with the fundamental idea of chapter 1; that is, When one male has more than the average number of mates, some other male or males must have no mates at all. Differently put, if desertion is a male strategy whose fitness payoff derives from additional matings, these additional matings must come at the expense of other males. Here we postulate that they come at the expense of males practicing the alternative strategy of providing offspring care. Note that we must also keep track of two kinds of deserting males, those successful and those unsuccessful at remating.

It is consistent with the intent of table 7.1 to assume that a fraction p of

Table 7.2
The adjusted pay-off matrix to males and females in the ESS model of the evolution of parental care (Maynard Smith 1977; Clutton-Brock 1991, table 7.1, p. 105).

Males		Females: Cares	Females: Deserts
Cares	Female Fitness	W_cV_2	W_dV_1
	Male Fitness	$W_cV_2(1 - pm_d)$	$W_dV_1(1 - pm_d)$
Deserts	Female Fitness	W_cV_1	W_dV_0
	Male Fitness Fails to Remate	W_cV_1	W_dV_0
	Male Fitness Remates	$W_cV_1 + (G_{c,c}W_cV_2 + G_{dc}W_dV_1)$	$W_dV_0 + (G_{c,c}W_cV_2 + G_{dc}W_dV_1)$

deserting males is successful at remating. In table 7.2, we have taken these rematings away from the care-giving males, i.e., from the $G_{c,c}$ and $G_{d,c}$ family types, in proportion to their occurrence. However, it might be more reasonable that the extra matings involve deserting females than caring females. If so, then we can take the extra matings of the deserting males only from that fraction of care-giving males whose mates have deserted them (table 7.3). Neither change (table 7.2 or 7.3) involves changes in female fecundity fitness or offspring survivorship dependent on female care. That is, additional matings with deserting males do not change either the number of eggs laid by deserting females or the viability of their offspring. These matings do change the distribution of offspring among male types, removing some of the offspring of deserted, care-giving males and awarding them to deserting males. However, unlike the classic case, this equal exchange does not cause a difference in mean fitness between the sexes.

What is the outcome for the adjusted payoff matrix? The easiest case to analyze is that in table 7.3, although a similar analysis of table 7.2 leads to

Table 7.3
The adjusted pay-off matrix to males and females in the ESS model of the evolution of parental care (Maynard Smith 1977; Clutton-Brock 1991, table 7.1, p. 105).

Males		Females: Cares	Females: Deserts
Cares	Female Fitness	W_cV_2	W_dV_1
	Male Fitness	W_cV_2	$W_dV_1(1 - p)$
Deserts	Female Fitness	W_cV_1	W_dV_0
	Male Fitness Fails to Remate	W_cV_1	W_dV_0
	Male Fitness Remates	$W_cV_1 + (G_{dc}W_dV_1)$	$W_dV_0 + (G_{dc}W_dV_1)$

the same conclusions. We use exactly the same fitness logic as the ESS theory: deserting males will be able to invade a care-giving male population only when the mean fitness of care-giving males, $W_{c,\text{males}}$, is less than that of deserting males, $W_{d,\text{males}}$. Substituting from table 7.3, we find that

$$W_{c,\text{males}} < W_{d,\text{males}}, \tag{7.9a}$$

$$\{W_d V_1(1 - p)\} < \{W_d V_0 + pG_c W_d V_1\}, \tag{7.9b}$$

$$\{V_1(1 - p)\} < \{V_0 + pG_c V_1\}. \tag{7.9c}$$

If we standardize viability, so that V_1 equals 1 and V_0 equals $(1 - s)$, then this reduces to

$$\{(1 - p)\} < \{1 - s + pG_c\}. \tag{7.10}$$

If we assume that we are modeling invasion, then we can also set G_c equal to 1, and so obtain

$$\{(1 - p)\} < \{1 - s + p\}. \tag{7.11}$$

This reduces further to the condition

$$(s/2) < p. \tag{7.12}$$

In contrast, under these same assumptions about viability, the classic model where females desert requires that

$$(s/[1 - s]) < p. \tag{7.13}$$

where we have translated the classic condition (see Clutton-Brock 1991 and table 7.1) $P_0[1 + p] < P_1$ into our terms by substituting V_0 for P_0 and V_1 for P_1). The correct condition, eq. [7.12], is *always* easier to satisfy than the incorrect classic one, eq. [7.13], because, for all values of s, $(s/2) < (s/[1 - s])$. In standard evolutionary theory (Wade 1998, 2001), the evolution of an allele, with a maternal or a paternal effect on fitness of s, depends upon the value of $(s/2)$, because the allele is expressed in only half the individuals in the population. Alternatively, its evolutionary fate depends upon whether $(s/2)$ is greater than or less than 0 because it is a type of among-family selection (Cheverud and Moore 1994; Wade 1998, 2001). When a gene has two pleiotropic effects, one a direct effect of s_d and the other a sex-limited parental effect of the same magnitude, s_p, then the condition for the evolution of this allele is that the sum of the fitness effects, $(s_d + [s_p/2])$, exceed 0 (cf. Wade 1998, 2001). Thus, the form of eq. [7.12] is exactly the same condition as the one governing whether or not a direct effect exceeds a paternal effect in standard evolutionary genetic theory.

When the extra matings involve care-giving females, then we need more information than is given in the classic model. The central question is whether or not caring females lay more eggs and increase W_c to W_d if, although caring for a first clutch, they mate again with a deserting male. If not, then female fecundity fitness does not change. If the female has already been deserted, then these matings also do not affect the distribution of offspring among male types; offspring are simply being exchanged between deserting males. Does offspring viability change when a caring female mates again with a deserting male? It could *if* she was mated to a care-giving male and he subsequently withdrew or reduced his level of care. If so, then mean female fitness is reduced below that of eq. [7.5]. If not, then no changes are necessary. In summary, if male mating behavior simply exchanges offspring among males (within or between male types) without changing either female fecundity or offspring survival, then eq. [7.5] holds. When cuckolded care-giving males reduce or withdraw care, then mean female fitness is lowered.

Chapter Summary

Mating system research has moved increasingly away from the analysis of adaptations within populations, and increasingly toward descriptions of interactions between individual males and females. According to this recent approach, not only is each individual presumed to behave adaptively, but the research challenge is to explain why the behavior of each individual is adaptive, despite its differences from other individuals. Behavior is accorded a much greater degree of "adaptive plasticity" than any other phenotype, and the concept of "relative fitness" has been discarded in favor of the hypothesis that individuals expressing less than optimal behaviors "make the best of a bad job" given their unfortunate circumstances. In current mating system research, adaptation appears to have become a property of all individuals at all times. In this chapter, we examine some of these current hypotheses.

We note that many tests of the Emlen-Oring hypothesis begin with descriptions of male mating behavior and reproductive biology, rather than with investigations of female receptivity or life history. Studies of female biology, which come after a species has been assigned a mating system, tend to confirm rather than falsify the hypothesis that male activities evolve in response to female spatiotemporal distributions. We suggest that a mating system classification based on *process* may ultimately prove more useful and more insightful than one based on *outcome*.

Behavioral flexibility is assumed to be ubiquitous within and among taxa. However, conditions favoring the evolution of phenotypic plasticity are, in fact, quite specific and often onerous. Thus, adaptive behavioral plasticity is more likely to evolve in response to some kinds of environmental variation than others, namely, those environments that are common and highly predictable. When environmental contingencies are rare or are experienced only by a small fraction of individuals in a population, then most copies of modifier alleles never encounter the unique circumstances in which they might be adaptive.

The "good genes" theory of female mate choice posits the existence of variation among males for genes affecting offspring viability. Females selecting mates with "high viability" genes are expected to bear more viable offspring than randomly mating females. According to this hypothesis, mate choice is viewed as communication between males and females in which every exchange between signaler and receiver is viewed as a game-theoretic process, in which the receiver gains or loses fitness in proportion to the "honesty" of the interaction.

We identify four essential features of evolutionary theory that are often overlooked with such game theory approaches. These include (1) the average fitness of the sexes must be equal at all times, not just at equilibrium; (2) with fertility selection, fitness accrues to mating pairs, not separately to interacting males and females; (3) selection can act differentially on each sex depending on its fitness variance; consideration of pairwise interactions may not reveal fitness variance; (4) analyses that award offspring fitness to parents are confounded and can incorrectly estimate the magnitude and direction of evolutionary change.

We explain the features of Fisher's runaway hypothesis, discuss why it is such a powerful evolutionary force (it involves both sexes), and how it accelerates the rate of evolutionary change (in each generation, selection becomes more intense). We compare the predictions of "good genes" and "arbitrary" mate choice models and show that good genes models quickly lead to runaway processes. There is good theoretical reason to consider the exaggerated phenotypes of males as "arbitrary" or "trivial" characters as Darwin did. Nonrandom mate choice caused by preexisting sensory biases in females, and the subsequent production of signals by males that exploit such biases, appear to be excellent examples of arbitrary male characters evolving by runaway sexual selection. In contrast to current attempts to place mate choice involving sensory biases into a separate category of sexual selection, we suggest that this process too involves Fisherian runaway processes, and that genetic correlations in sender-receiver relationships must exist, whether they involve separate sexes within species, or interacting individuals between species.

Sexual conflict and "chase-away" selection have been proposed as mechanisms by which Fisherian runaway processes slow down and stop. We show how these hypotheses are unnecessarily complex, involve viability selection on females, and therefore do not adequately explain this phenomenon. Again, we show that runaway processes, in this case involving negative genetic covariances between male and female traits, provide simpler, more easily testable, and therefore more satisfactory explanations for the persistence of extreme male characters as well as female preferences for them in natural populations.

Parental investment theory lies at the foundation of most current mating system research. The current explanation for male-female differences in mating behavior and parental care is gamete dimorphism, the fact that, by definition, males produce many, motile sperm while females produce fewer, nonmotile ova. This difference has led many researchers to attempt to quantify sexual selection in terms of the ratio of the maximal number of male gametes to the maximal number of female gametes across several taxa. Others have estimated the ratio of the maximum potential reproductive rate (PRR) of males to females, and still others have suggested that females may adjust their sex ratios based on their perception of their own physical condition. We argue not only that the results of such studies suffer from statistical difficulties, but also that this approach may confuse cause and effect. The large size of male testicles in chimpanzees, for example, appears to be the *result* of sexual selection in a polygamous mating system; the polygamous mating system of chimpanzees is unlikely to have evolved in order to allow male chimps to capitalize on their testicle size.

Anisogamy is thought to have arisen from ancestral populations of isogametic producers, whose population-wide gamete size was normally distributed. We argue that this explanation accords well with our hypothesis that initial parental investment between the sexes is a *secondary* rather than a *primary* cause for sex differences in mating and parental behavior.

Male parental care is conspicuously rarer in most taxa than is female parental care (Clutton-Brock 1991). The standard explanation for the taxonomic bias toward female parental care is based on the investment arguments of Trivers (1972) that were formalized in the ESS theory of Maynard Smith (1977). The fundamental argument is this: Because males invest relatively little in offspring, owing to anisogamy, the gain to a male from additional matings vastly exceeds the potential gain to him in fitness from the increment in viability added by caring for existing young. We examine the logic of these arguments under the constraints that necessarily connect the mean and variance of male and female fitness as shown in the earlier chapters. These constraints are not explicit in ESS theory or alternative strategies and, to the extent that they are violated, the theoretical conclusions must be considered suspect.

8 Behavioral Influences on ΔI

In this chapter we describe factors that may influence the opportunity for sexual selection, ΔI, other than those specifically arising from the spatial and temporal distributions of females, or from the details of female life history. In the previous chapter, we were critical of adaptive explanations for variants of reproductive behavior expressed by only a few individuals and only in rare contexts. We also drew attention to what we consider inconsistencies in the logic of some arguments regarding the selective forces presumed to shape mating systems and therefore mating behavior. Our goal in this chapter is to provide a theoretical framework, consistent with that presented in the early chapters, which describes how behavioral variations within each sex can further modify the sex difference in the opportunity for selection and thus influence the evolution of sex dimorphism and mating system evolution. After completing our framework in this chapter, we will provide specific examples of mating systems as tests of our hypotheses in chapter 9.

Prolonged Female Receptivity

In chapter 3, we argued that, in most species, strong stabilizing selection will shape the average duration of female receptivity, r (eq. [3.1]). We reasoned that females with extremely brief periods of sexual receptivity may seldom receive sufficient sperm to breed successfully. Conversely, females with extremely long periods of sexual receptivity may be repeatedly harassed by males and suffer decreased fitness due to physical damage, increased predation, increased exposure to sexually transmitted disease, and/or decreased ability to provision themselves and their developing ova (Rowe et al. 1994; Jormalainen and Shuster 1999). Under stabilizing selection, the population variance in the duration of female receptivity, V_r (eq. [3.2]), is likely to be constrained, although small values of V_r do not necessarily imply synchrony of receptivity across females. Given stabilizing selection, average female receptivity r should be a reliable stereotype of sexual receptivity duration for the female population, except in two circumstances.

In the first situation, we expect selection to prolong receptivity to arise when relatively few males successfully acquire mates, so that H and V_{mates} are large, and sexual selection is intense (i.e., I_{mates} is large). Under such conditions females are usually assumed to produce male progeny possessing the successful, competition-associated traits of their fathers (Weatherhead and Robertson 1979; Thornhill and Alcock 1983). However, other outcomes of extreme polygyny are that individual females receive few sperm, have low genetic diversity among their progeny, or may not be mated at all (Knowlton and Greenwell 1984; Warner et al. 1995a; Westneat and Sargent 1996; Mesnick 1997; Gowaty 1997). All of these factors may contribute to a *negative covariance* between harem size and average female fecundity. It is this negative association between harem size and female fecundity that can lead to directional, rather than stabilizing, selection on the duration of female receptivity.

If females defended by dominant males suffer decreased insemination success, within-harem selection may favor females with prolonged receptivity, because prolonging r increases a female's probability of successful mating when sperm are locally scarce. Unlike situations discussed in the previous chapter, in which females improve their own viability fitness by changing mates, our argument here is based upon within-harem selection among females, where the strongest selection favoring prolonged receptivity occurs within the largest harems. Under these conditions, prolonged receptivity also offers a greater window of reproductive opportunity for the invasion of harems by males able to avoid exclusion by dominant males (e.g. sneakers and/or female mimics; Shuster 1992). Just as synchrony of female receptivity diminishes I_{mates}, so too does prolonged female receptivity. Although the duration of female receptivity may primarily be influenced by resource-dependent environmental factors, the situation above may permit females, as a group, to avoid male monopolization and hence diminish I_{mates}. Not only can a negative covariance between harem size and average female fecundity slow or halt the runaway process, but the ensuing within-harem selection favoring an increased duration of female receptivity can result in further erosion of I_{mates}.

Our proposal that prolonged female receptivity and multiple insemination decrease I_{mates} is contrary to simulation results obtained by Webster et al (1995). They showed that extra-pair matings may *increase*, rather than decrease, the opportunity for sexual selection. This assumption is typically made whenever multiple mating occurs (Parker 1998). However, the increased opportunity for sexual selection shown in Webster et al. (1995) occurs *only* when there is a positive covariance between the number of progeny a male sires within his harem and the number of successful extra-pair matings he obtains. Differently put, extra-pair copulations (EPCs) increase the variance in reproductive success among males only when already successful males become even more successful through EPCs, while unsuccessful or marginally successful males become even less so.

In our scenario, males otherwise excluded from reproduction are able to increase their reproductive success by taking some fertilizations *away from* successful males. This diminishes the variance in male reproductive success. Thus, we differ from Webster et al. (1995) in the sign of the covariance in male success across harems associated with extra-pair copulations. As we pointed out in chapter 4, this covariance is critical to determining the strength of sexual selection because it can enhance or erode the variance in male reproductive success. Our intent is not to criticize the model of Webster et al. (1995) but rather to reemphasize the importance of the covariance in male success across harems or across intervals. It is this covariance that determines whether or not a given scenario increases or decreases the sex difference in the variance of fitness.

The predictions of Webster et al. (1995) identify a second set of circumstances in which selection may increase r, the average duration of female receptivity. Prolonged female receptivity permits mate-number promiscuity by females, which is a necessary prerequisite of sperm competition. With multiple mating by females, the possibility of another runaway process is clear. Those males with the most competitive sperm will sire more offspring than those with less competitive sperm and so have higher fitness. On average, the mates of these males will be more promiscuous. Thus, not only will these males leave more offspring, but also their sons will have more competitive sperm and their daughters will be more promiscuous. Recall that simply prolonging the duration of female receptivity can be equivalent to an increase in mate-number (as opposed to *mating*-number) promiscuity (chapter 4). In this way, a change in the mean of the temporal distribution of female receptivity initiates a change in the effective variance of the spatial distribution of receptivity. Thus, selection to prolong the period of female receptivity permits multiple mating, which in turn may favor the most highly competitive male gametes, as we have suggested in chapter 4.

Curtsinger (1991) argued that "sexy" male sperm and multiple matings by females are unlikely to coevolve because only strong linkage disequilibrium between prolonged female receptivity and competitive male gametes would be sufficient to maintain these characters in the same population. The strength of indirect selection on female traits, like mate-number promiscuity, however, depends upon the product of the genetic correlation of traits across the sexes and the strength of selection on males. Although the direction of indirect selection is determined by the sign of the genetic correlation resulting from linkage disequilibrium or pleiotropy, the strength of indirect selection is not. Theoretical and empirical analyses (Lande 1981; Kirkpatrick 1982; Wade and Arnold 1980; Houde 1992; Hall et al. 2000; Dugatkin and Kirkpatrick 1994; Breden and Wade 1991; Jernigan et al. 1994; Wade 1995; Ciuetta and Clark 2000) indicate that genetic correlations between male traits and female preferences (in this case sperm competitive ability and the tendency for females to mate repeatedly) are initiated and persist because of the combination of nonrandom mating and the overwhelming advantage gained by males with preferred characters (e.g., sperm that are highly successful in the female reproductive tract). If females vary in mate numbers and males vary in sperm competitive ability, paternity will be nonrandom even with random mating. This runaway process is less contingent on nonrandom mating to initiate a genetic correlation than some of the others discussed earlier. It is the consequence of nonrandom mating, namely, nonrandom paternity, that is critical in combination with the sex difference in strength of selection.

This example of runaway sexual selection is analogous to one proposed by Wade (1995) in which female tendencies to aggregate and male tendencies to defend aggregations may coevolve in intensity until males are no longer able to successfully defend large female aggregations. Male viability selection is not required to stop this runaway process. Rather, it ceases owing to limits on a male's ability to defend and sequester large numbers of females away from other males. In the case of multiple mating and sperm competition, female receptivity may become so prolonged that competitive sperm experience only a temporary advantage, as when females continue to mate but do not provide a means for sperm to remain viable.

Alternatively, the runaway process between sperm competitive ability and female promiscuity may stop when multiple matings by females favor a diversity of sperm competition strategies, which persist by frequency dependent selection. These may include spermatophores, which protect and nourish sperm within the female reproductive tract (Gwynne 1984), or differences in sperm number, if large numbers of sperm per ejaculate are likely to outcompete ejaculates with smaller numbers of sperm (Parker 1978a,b; 1998). In some female reproductive tracts or aquatic media, different sperm shapes may be more likely than others to navigate or be more successful in blocking other sperm (Sivinski 1980; Gomendio and Roldan 1993; Pitnick 1996; Snook 1997). Males may also produce secretions that enhance the success of an individual's own sperm, but decrease the success of other sperm, or modify the chemical environment of the female reproductive tract (Rice 1996). When there are a number of equivalent, frequency-dependent, postcopulatory male strategies, a runaway process becomes much less likely, because it is difficult to maintain a covariance between mean female promiscuity and a "mean" successful male strategy. Moreover, with increased numbers of mating males, the sex difference in the opportunity for selection is reduced.

Male Monopolization of Females

We discussed above how mate-number promiscuity can reduce the opportunity for sexual selection, I_{mates} (chapter 4). This is true, for example, when the sperm of each mating male fertilizes some ova. As noted by Webster et al. (1995), if only certain males win in sperm competition, I_{mates} can become very large. Yet sperm competition of any kind is possible only if $I_{cs,sires}$ is nonzero, that is, if females mate more than once (chapter 5). Thus, one way in which males may prevent sperm competition altogether, even if females are inclined to mate more than once, is by guarding their mates throughout the period of female receptivity. Stated differently, mate guarding can reduce

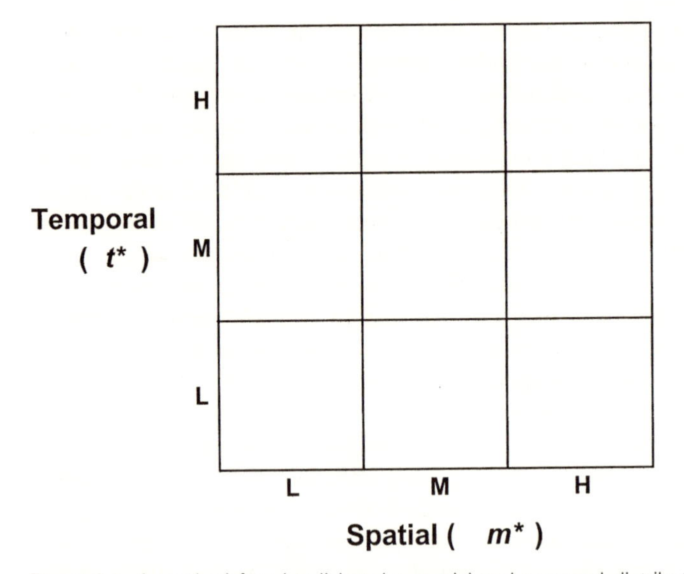

Figure 8.1. A method for visualizing the spatial and temporal distributions of sexually receptive females in terms of the mean crowding of receptive females in space (m^*) and the mean crowding of receptive females in time (t^*). L, M, and H refer to low moderate and high values for m^* and t^*; thus the nine sections in the diagram refer to unique combinations of the spatial and temporal distribution of sexually receptive females.

or eliminate mate-number promiscuity while still allowing continued mating-number promiscuity.

Mate guarding appears to offer costs and benefits for males as well as for females (Sparkes et al. 1996; Jormalainen and Merilaita 1993, 1995; Jormalainen et al. 1999). Futhermore, many behaviors performed by males and by females in response to mate guareding can change the apparent spatial and temporal distribution of receptive females and its relationship to the sex difference in the opportunity for selection (fig. 8.1). We can visualize the possible distributions of females in space and time using a modification of the diagram we introduced in chapter 6 to describe the ΔI surface. We are interested in examining how the spatial and temporal distributions of females influence male mate guarding, and thus how these changes can further influence the contour of the ΔI surface. Thus, we will illustrate combinations of m^* and t^*, with the diagram in fig. 8.1, where m^* varies along the horizontal axis and t^* along the vertical axis. Each axis is divided into sections, representing low, intermediate, and high values for the parameters m^* and t^*, dividing the diagram into nine squares. These values are used for convenience to show the *extremes* of female spatiotemporal distributions. Since the effects of m^* and t^* are considered simultaneously in our model, each square provides a unique combination of these values.

We will use also this diagram throughout this chapter and the next to discuss how variation in m^* and t^* affect selection on males and females. Note that the conditions we describe for each square are by necessity, quite general. We acknowledge that, in nature, many species will lie at the margins of the regions we describe, and will therefore possess a composite of the features we associate with each zone. We hope that the limitations of our framework will stimulate researchers, not to reject it out of hand, but instead, to reexamine their data and describe the particular situations in greater detail.

Simply put, mate guarding reduces the ability of females to mate more than once. If a male guards a female successfully, he fertilizes all of her ova. If a male unsuccessfully guards his mate, or if he leaves her in search of other mates before her receptivity is complete, the male's fertilization success with that female will be eroded due to matings by other males. Distinct sperm utilization patterns by females appear to exist among animal taxa. Although sperm mixing is widespread, last male sperm precedence is considered common in insects and first male sperm precedence is most often documented in crustaceans (Parker 1970a,b; Walker 1980; Ridley 1983; Diesel 1989; reviews in Smith 1984; Koga et al. 1993; Birkhead and Møller 1998; but see Shuster 1989a; Jormalainen et al. 1999). Yet even in these species, incomplete male mate guarding can result in mixed paternity within broods. Futhermore, in a surprising number of experiments designed to examine patterns of sperm precedence, each female is allowed to mate with only two males. Commonly held notions of sperm utilization by females appear to break down when females are allowed to mate with more than two males, and at least in darkling beetles (*Tenebris molitor*, Drnevich in press; Lewis and Jutkiewicz 1998; Radwan 1997; Zeh and Zeh 1994) sperm precedence gives way to sperm mixing with multiple mating. Thus, the approximate fertilization success of a male mating with a promiscuous female can realistically be expressed as the reciprocal of the number mating males, $1/M$ (fig. 8.2).

Stated differently, the fertilization success of each male equals the reciprocal of that female's mate number promiscuity, $1/P_j$, where P_j represents the mate-number promiscuity of the jth female (chapter 4). The overall fertilization success of a male equals the harmonic mean of the mate number promiscuity of his mates, or

$$1/H_M = 1/M\ [\sum_j 1/P_j]. \qquad [8.1]$$

Consider again the example we described in chapter 4 of the male with three mates. The first female has two mates (one plus him), the second has three mates (two plus him), and the third female has four mates (three others plus him). The reproductive success of our focal male is the harmonic mean of the promiscuity of his mates, when each female allocates her offspring to her M mates as $1/M$. Differently put, this male sires one Mth of the offspring of each of his mates. Let each female have O offspring. How many offspring

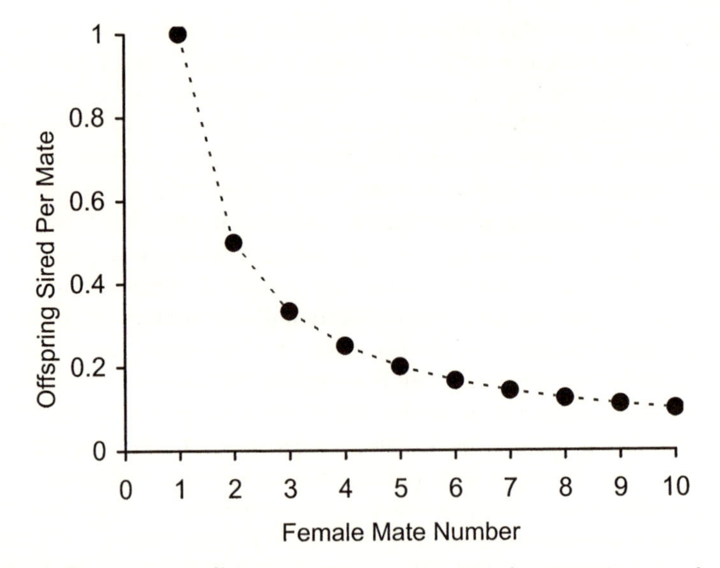

Figure 8.2. The decline in offspring sired per mating when females mix sperm from all males and apportion fertility equally.

per mate does this male sire on average? The answer is an average of $([(1/2) \times O] + [(1/3) \times O] + [(1/4) \times O])/3$ offspring per mate. This reduces to $\{O/3\} \times \{(1/2 + (1/3) + (1/4)\}$, which equals $(0.36)(O)$. Another way to view 0.36 or $(1/3)([1/2] + [1/3] + [1/4])$ is as $(1/H_M)$, the harmonic mean promiscuity of the three females, who have one, two, and three mates, respectively. The harmonic mean, $1/H$, is defined as the "reciprocal of the arithmetic mean of the reciprocals" and the harmonic mean promiscuity of these three females is $1/H_M = (1/3)([1/2] + [1/3] + [1/4])$. Thus, each male obtains (O/H_M) offspring per mating. If a male can reduce the average number of mates per female in his harem toward one, then H_M is smaller and the gain from each mating, O/H_M, is larger.

The cost of mate guarding to males is typically assumed to be lost mating opportunities as though the fitness of *any* male can be increased without constraint. Consider our male with his harem of three females, whose harmonic mean promiscuity is now 0.36, i.e., 2.78 mates per female. If this male abandons his harem to seek an additional copulation outside of his harem, then two changes in his fitness must be considered: (1) his loss of offspring due to the increased promiscuity of the abandoned females in his harem; and (2) his gain in fitness through the additional mating he secures. If each female in the original harem mates with one additional male, then $1/H_M$ is reduced to 0.26 or 3.83 mates per female $= (1/3)([1/3] + [1/4] + [1/5])$. That is, the loss per female is *greater than* converting $1/H_M$ $(=0.36)$ to $1/[H_M + 1]$ $(=0.27)$. If his additional mate is herself as promiscuous as his average mate, $H_M + 1$, then the gain from the additional mate is $O/[H_M + 1]$ while the loss is $(HO)/[H_M + 1]$, where H is the average harem size of other mating males. Clearly, attempts at additional matings can be costly when sexual selection is already strong.

Any male behaviors that enhance the promiscuity of females diminish male fitness in a *nonlinear* manner through $1/H_M$. The most obvious way a male can increase the promiscuity of his mates is to abandon them immediately after mating, with the goal of seeking other, possibly promiscuous, females. When viewed quantitatively, the trade-off between mate guarding and male promiscuity does not so obviously favor male promiscuity. Indeed, the potential advantages of male promiscuity cannot be evaluated without considering the harmonic mean number of mates for each of the male's mating partners. With sperm mixing, whenever H, the harem size, exceeds H_M, the mean female promiscuity, then increasing H by 1 at the expense of increasing H_M by 1 *always reduces male fitness*. Under these conditions, male mate guarding is always favored over male promiscuity.

Compared to predictions since Bateman (1948) regarding the inherently promiscuous mating behavior of males in most animal species, the above relationship places unexpected limits on the mating behavior of males. The fitness of roving males must exceed that of guarding males if they are to invade such a population. Thus, the rate at which roving males encounter and mate with females must exceed the reciprocal of the harmonic mean fertilization rate. The higher the fertilization rate roving males achieve with each female, the fewer females they must encounter. The lower the fertilization rate of roving males, the more females they must encounter to invade a population of guarding males. Conversely, guarding males can invade a population of roving males when the enhanced fertilization success attendant to mate guarding exceeds the product of mate numbers and average fertilization success for roving males. We will see below that for most spatial and temporal distributions of females, guarding males are favored over roving males. However, considerable variation exists in the form of male mate guarding, as well as circumstances in which roving is favored.

Small m^* and Male Monopolization of Females

Spatial dispersion of females favors mate guarding independent of the timing of female receptivity (m^* = low, t^* = low to high; fig. 8.3). This is true because male encounter rates are low when females are dispersed in space. Increasing reproductive synchrony among females further reduces the number of potentially receptive females searching males can encounter. Reproductive asynchrony among females could allow males to move from one receptive female to the next (m^* = low; t^* = low; fig. 8.3). However, if females are spatially dispersed and breed asynchronously, roving males must

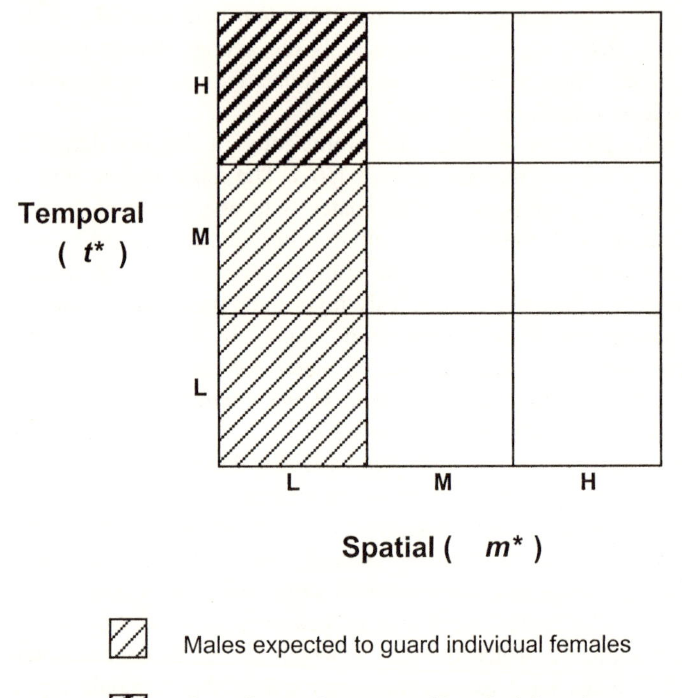

Figure 8.3. Spatial dispersion of females favors male mate guarding regardless of the timing of female receptivity. The shaded area represents the spatial distribution of females in which males are expected to guard individual females before mating.

encounter receptive females at a rate equal to the reciprocal of their fertilization success, if they are to invade a population of males who guard each female they encounter.

A male who guards a female for the duration of her sexual receptivity is expected to fertilize all of the ova produced by the female he guards. Roving males must match this fertilization success, either by locating a large number of mates, or by increasing their fertilization rate. The latter requirement can be accomplished by the use of sperm plugs (see chapter 4), or more simply by guarding each female encountered until she becomes unreceptive, that is, by becoming a guarding male. Whether a searching male can locate a sufficiently large number of receptive mates to compensate for the loss of fertilizations to sperm competition is the crux of the matter. Sperm plugs enhance a male's fertilization success with mated but abandoned females, but so too does guarding a female until she becomes unreceptive. Whether mate guarding or leaving a sperm plug is the more effective strategy depends upon the

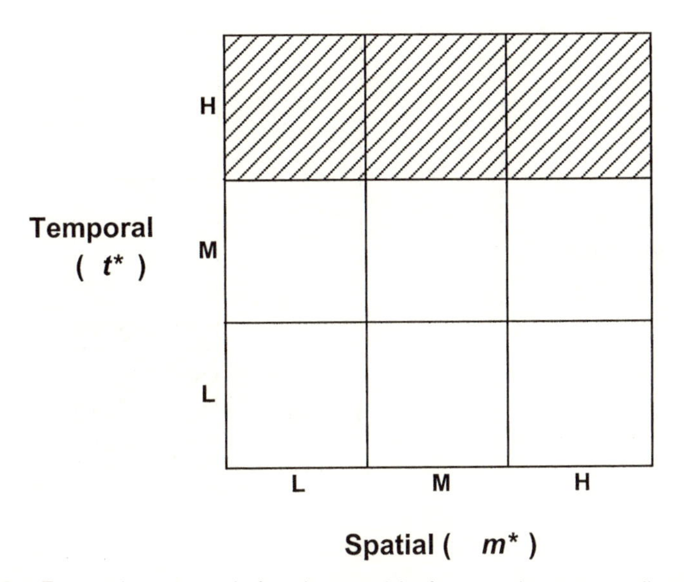

Figure 8.4. Temporal synchrony in female receptivity favors male mate guarding regardless of the spatial distribution of receptive females. The shaded area represents the temporal distribution of females in which males are expected to guard individual females before mating.

relative costs, encounter rates, and searching risks that attend each male mating strategy.

Large t^* and Male Monopolization of Females

High temporal synchrony of female receptivity (t^* large) also favors mate guarding, independent of the value of m^* (fig. 8.4). When m^* is near zero, females are spatially dispersed. Male encounter rates are small when females are dispersed in space, and high reproductive synchrony among females further reduces the number of potentially receptive females that a searching male can encounter. As we have explained, searching males must encounter new mates at a rate greater than the reciprocal of their fertilization success because male fitness is the product of mate numbers and fertilizations per mate. This requirement again restricts roving by males.

With high values of t^*, most females are likely to be guarded by males, making encounter rates with available females by searching males necessarily low, even when females are in close proximity. In many species, mate guarding behavior is less of an energy investment than creating a sperm plug. However, as mentioned above, mate guarding appears to offer costs and benefits for both sexes (Sparkes et al. 1996; Jormalainen and Merilaita

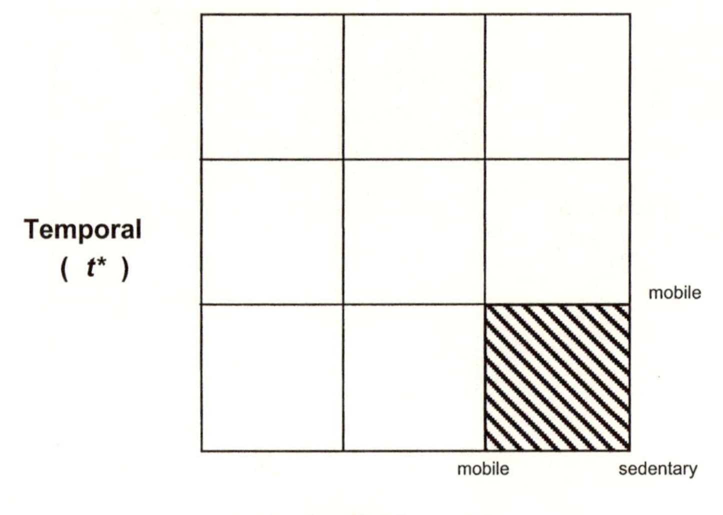

Figure 8.5. Spatial aggregation of females combined with temporal synchrony in female receptivity favors male guarding of female groups. Males will tend to be sedentary at extreme levels of female spatial crowding and temporal asynchrony. Males will tend to be more mobile as females become less crowded in space or more synchronous in their receptivity.

1993, 1995; Jormalainen et al. 1999) and, as we argue below, these costs play a significant role in allowing the invasion and spread of alternative male mating strategies.

Large m^*, Small t^* and Male Monopolization of Females

A highly aggregated spatial distribution of females combined with temporal asynchrony in female receptivity (m^* = high; t^* = low; fig. 8.5) will also favor mate guarding, although, as described in chapter 2, in such circumstances, males are likely to defend groups of females rather than individual mates. When females are spatially clustered, males who find and guard groups of receptive females not only enjoy unusual reproductive success, but also lower their search costs and limit sperm competition (Jormalainen 1998). Mate guarding may prevent sperm competition altogether, even when females are inclined to mate more than once, because repeated copulation can reduce or eliminate female mate-number promiscuity despite continued mating-number promiscuity. An interaction also exists between the frequency of male mate guarding and the availability of females to mate with other, roving males. When a male guards a female, he reduces her availability to other males, guarding or not guarding. Males able to guard clusters of females more severely reduce the availability of receptive females to other males. Since the adaptive value of roving strategies depends upon the encounter rate with available females, males guarding harems can limit the degree to which this alternative strategy may evolve.

If t^* determines the encounter rate of males with receptive females, why then is mate guarding nearly ubiquitous? Let us consider the extremes of the range of t^*. When spatially clustered females are asynchronous in receptivity (m^* = high; t^* = low; fig. 8.3), then males guarding harems have minimal search costs as well as minimal transit times between receptive mates. Within harems, asynchrony provides a succession of receptive females in close proximity favoring sedentary guarding males. The greater the mating success of the harem-guarding males (i.e., the larger the harem size, H), the greater the fraction of males excluded from mating, p_0 (since $p_0 = 1 - [R/H]$, see chapter 4), and the larger the value of I_{mates}.

In contrast, when receptive females appear within a narrow interval of time (t^* = high; fig. 8.3), males are expected to guard and mate with a single female. Males attempting to guard an entire cluster of females will lose part of the reproductive tenure of every female in the harem, because the narrow window of receptivity causes a trade-off between guarding and copulating. Males that abandon their first mate to search for others before r, the average tenure of receptivity, expires permit remating by the female and risk losing fertilizations to sperm competition.

Roving males must encounter new mates at a rate greater than the reciprocal of their fertilization success, or their fitness will be less than that of guarding males (fig. 8.2). Increased spatial clustering of females enhances encounter rates and may allow roving males to persist with guarding males. However, both breeding synchrony of females across the cluster, and guarding by males, limits the number of available females, making the effective encounter rates for searching males negligible. Sexual selection, measured by I_{mates}, is weak in this instance. The reproductive synchrony of females affects the intensity of sexual selection, which is higher when t^* is low than when t^* is high. Nevertheless, male mate guarding in some form is favored over the range of t^* values. Although the evolutionary outcome of these female distributions, male mate guarding, is the same in each case the selective process that produces it is different at each extreme (I_{mates} high versus I_{mates} low). Clearly, in mating system evolution, outcome is not a unique signal indicating process.

In circumstances in which males defend groups of females (fig. 8.5), the greater the spatial clumping and temporal asynchrony of receptive females, the more sedentary guarding males can afford to be. Transit times between females are reduced and males can anticipate the receptivity of *individual* females. Depending on the mating success of males defending female groups, males who guard individual females for the duration of their sexual receptivities may also appear in such populations. The greater the mating success of individual, harem-guarding males, the more other males attempting to hold harems will be excluded from mating. Under these conditions, the mating success of males who guard individual females will exceed that of males that are unsuccessful as harem defenders. Thus, as harem holders become more successful, the frequency of male-female pairs is expected to increase (chapter 9).

Intermediate Values of m^* and t^* and Male Monopolization of Females

As groups of receptive females become less clumped in space or in time, male guarding will increasingly involve speed and maneuverability. Decreased spatial clumping or increased female breeding synchrony are expected to erode the ability of dominant males to defend female groups. Yet dominant males are expected to continue to defend harems, despite losses in fertilization success, as long as their mate numbers exceed the reciprocal of their fertilization success (fig. 8.2; eq. [8.1]). The proximity of receptive females keeps mate numbers high. However, in these groups as well, males who defend individual females for the duration of their receptivity are expected to appear with greater frequency as female receptivity becomes more synchronous, or as receptive females become more dispersed in space.

Thus, roving behavior is likely to occur at high frequency among males when females are moderately clumped in space as well as moderately asynchronous in the timing of their receptivity (m^* moderate, t^* moderate; fig. 8.6). Under these circumstances, high encounter rates with receptive females can compensate roving males for low average fertilization success (fig. 8.2). Moreover, when females are moderately clumped in space and moderately synchronous in the timing of their receptivity, males who transfer fewer sperm per female, and thereby increase their mate numbers, are expected to experience higher average fitness than males that attempt to provide all the sperm each female needs, but succeed in transferring it to fewer females.

The ability of roving males to tolerate relatively low fertilization rates when the density of receptive females is high provides an alternative to the hypothesis that intense mate competition among males *always* favors individuals who produce sufficient sperm in each ejaculate to fertilize all of a female's ova (Trivers 1972; Thornhill and Alcock 1983; Andersson 1994). Sperm competition models extend this hypothesis by assuming that males compete most successfully by increasing their sperm numbers (Parker 1998). Although males in some species appear to produce a superabundance of sperm (Birkhead and Fletcher 1995; Westneat et al. 1998), variation in the number of sperm per ejaculate is known in a wide range of taxa (Shuster 1989a; Levitan and Peterson 1995; Shapiro et al. 1994; Baker and Belis 1995; Heinz et al. 1998; Wedell and Cook 1999; Kiflawi 2000). These re-

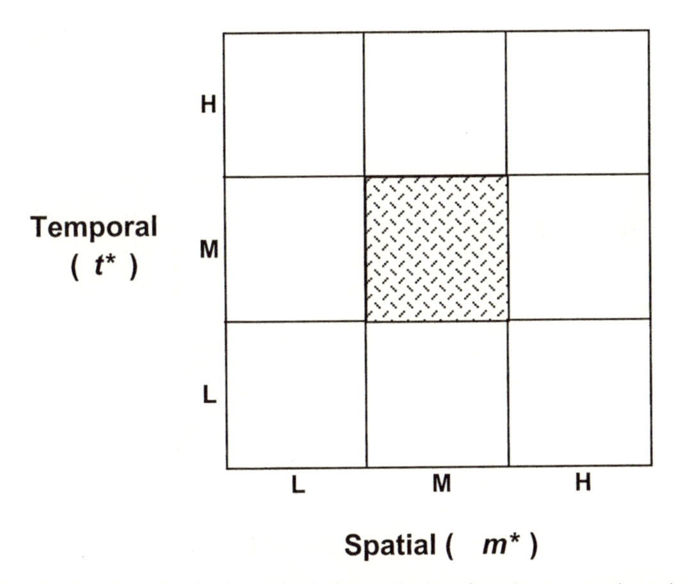

Figure 8.6. Searching behavior by males is favored when females are moderately aggregated in space as well as moderately asynchronous in the timing of their sexual receptivity.

sults suggest that males can and do allocate energy toward sperm production in ways other than predicted by the lottery hypothesis (Parker 1998).

Roving behavior may also arise to lesser degrees when females are moderately clumped in space and asynchronous in the timing of their receptivity (m^* = moderate, t^* = low; fig 8.7), or when females are moderately synchronous in the timing of their sexual receptivity and highly clumped in space (m^* = high, t^* = moderate; fig. 8.7). In the first situation (m^* = moderate, t^* = low; fig 8.7), males are expected to tend small groups of females and consort with individual mates as they become receptive. Since females are only moderately aggregated in space, transit times are increased and males must move further between the females in their cluster to locate receptive individuals. The size of the female group will depend on the maximum number of receptive females a male can mate in sequence, and sperm depletion in guarding males is likely as the number of females visited increases. If insemination of females at the boundaries of male ranges is incomplete, males who float between groups or remain near female groups as subordinate males may persist at frequencies commensurate with the fraction of fertilizations they receive (see chapter 9).

In the second situation (m^* = high, t^* = moderate; fig. 8.7), males are likely to tend individual females since females are aggregated in space and female receptivities moderately overlap. However, such moderate asynchrony in female receptivity among females in close proximity could favor

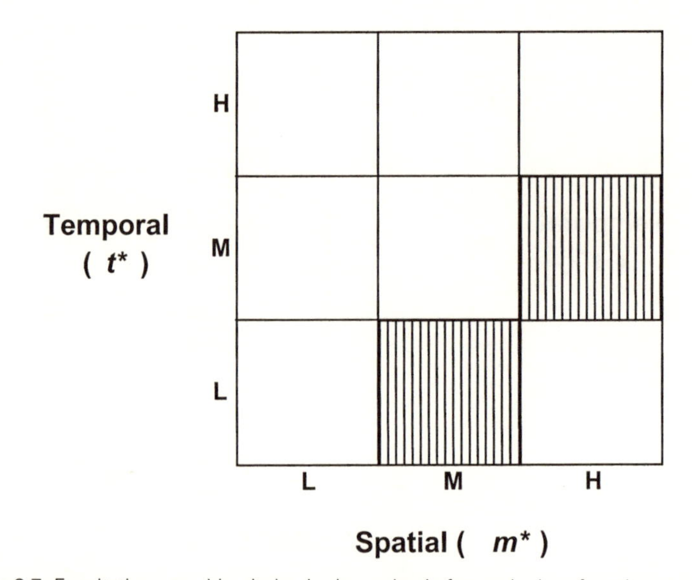

Figure 8.7. Facultative searching behavior by males is favored when females are moderately clumped in space and asynchronous in the timing of their receptivity, as well as when females are highly aggregated in space and moderately asynchronous in the timing of their receptivity.

occasional roving by paired males if such behavior results in even small numbers of fertilizations (chapter 9). When H is small, paired males who opportunistically rove could sire more offspring than paired males who do not. In such circumstances, roving behavior among ostensibly paired males could persist at high frequency, a situation that is often called "social monogamy" (Gowaty 1997).

From these considerations, it is now clear that mate guarding by males in some form is favored under most spatial and temporal distributions of females (fig. 8.8). However, male roving behavior may also be favored at the boundaries of all circumstances in which mate guarding persists, if spatial and temporal distributions of females intermediate to those we have defined leaves some females unguarded. If rovers become increasingly successful in mate acquisition, the variance in mate numbers among males will decrease because the fraction of nonmating males is decreased. And yet increased numbers of rovers can also increase opportunities for multiple insemination and sperm competition. As explained above, sperm competition can cause I_{mates} to increase beyond the values predicted from the spatial and temporal distributions of females alone (chapters 4 and 7). Mate guarding abrogates female tendencies to mate more than once and thereby eliminates the contribution of $I_{cs,sires}$ to I_{clutch}. Mate guarding and sperm competition are thus

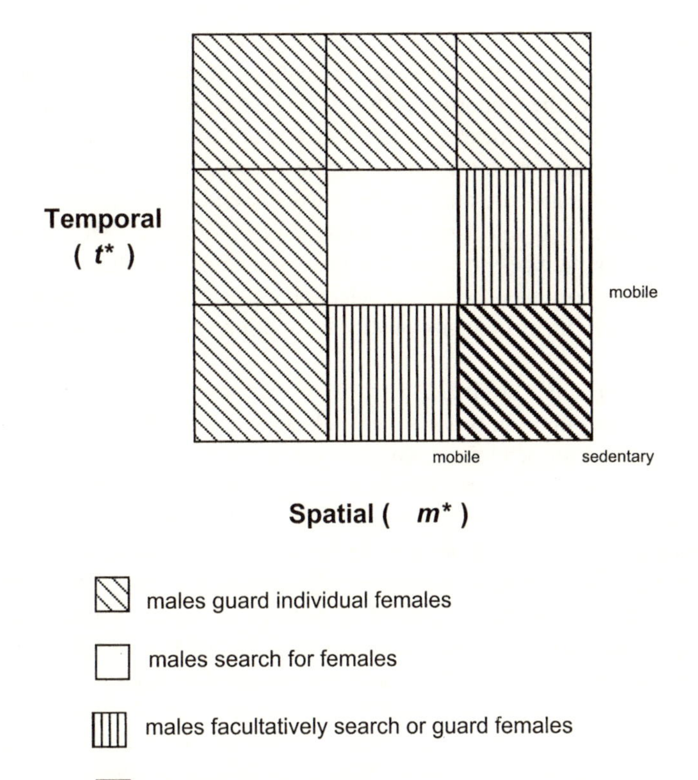

Figure 8.8. Summary of conditions in which mate guarding is expected to occur.

similar in their ability to mitigate the effects of multiple mating on the sex difference in the opportunity for selection.

Female Copying

Female tendencies to copy the behavior of other females can enhance the opportunity for sexual selection (Shuster and Wade 1991b). Copying behavior is favored when information sharing decreases the time spent or the risk incurred by individuals searching for resources (Kiester 1979; Wade and Pruett-Jones 1990). The advantages of information sharing are likely to increase as resources required by females become increasingly patchy. Thus, copying behavior, when used by females as a means for locating patchy resources, is expected to increase the patchiness of female distributions as well. If resource patches are defendable by males, female tendencies to copy the resource utilization behavior of other females and thus aggregate at resources, will increase the variance in mating success among males and thereby increase the opportunity for sexual selection (Shuster and Wade 1991b; Wade 1995). However, when females are patchily distributed, more than one male may be attracted to female aggregations (chapter 3). When aggregations of males are attracted to female assemblages, females have further opportunities to select mates nonrandomly (Altmann 1990). Thus, tendencies among females to copy the mate selections of other females can enhance the opportunity for sexual selection *above* that which can be identified by periodic measurements of the distributions of females in space and time.

If resource patches are fixed in space, an estimate of the effects of female copying on the opportunity for sexual selection can be obtained by calculating the distribution of male mating success at each patch, and comparing it with the distribution of mate numbers per male if females mated at random (Wade and Pruett-Jones 1990; Shuster and Wade 1991b). Alternatively, if patches move or if females are attracted to males themselves, the distribution of mate numbers per male (or offspring numbers per male; see chapters 1 and 4), compared to that expected if mating occurs at random, can also provide an estimate of female mate copying. In light of our discussion of female mate choice, nonrandom mating is expected to lead to runaway sexual selection (see chapter 4). Female mate copying can also lead to a runaway process between female tendencies to copy and male abilities to defend female aggregations (Wade 1995). Which mechanism is responsible for the nonrandom distribution of mates per male can be determined by detailed experiments similar to those conducted by Dugatkin (1998; Dugatkin and Godin 1993; Dugatkin and Kirkpatrick 1994)).

Female copying associated with resource acquisition could involve different cognitive processes than female copying associated with mate choice. For this reason, Pruett-Jones (1992) advocated the description of copying in the context of resource use as "conspecific cueing," and reserved the term "female copying" for circumstances involving mate choice. Further subdivisions of intraspecific influences on mating behavior are discussed by Westneat et al. (2000). The cognitive processes responsible for different mechanisms of mate choice are difficult to distinguish, and similar population genetic consequences arising from conceptually distinct copying mechanisms may render their identification moot. However, it remains worthwhile to consider the circumstances in which female copying, in either context, may or may not occur.

Whether females locate resources using shared information can be inferred from the distributions of females in space and time (fig. 8.9). Spatial dispersion of resources is expected to cause females to distribute themselves widely in space. When females are spatially dispersed, not only will females have few opportunities to copy the resource locating behavior of their con-

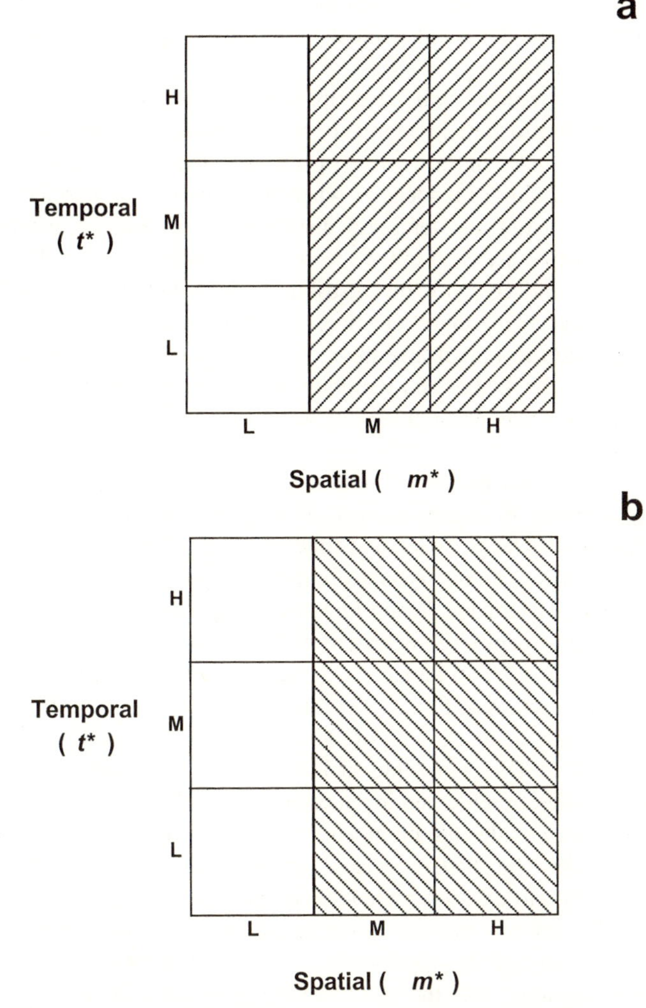

Figure 8.9 (a) The spatial distribution of females in which females are likely to copy the behavior of other females in locating patchily distributed resources; (b) the spatial distribution of females in which males are likely to be attracted to female aggregations, thereby allowing female mate copying behavior.

specifics, but if resources are truly dispersed, females will benefit little by aggregation. However, when resources are patchily distributed, more than one female may successfully exploit a resource patch. In such circumstances, females will arrive at resource patches more rapidly if they are attracted to other resource-seeking females, at least until the aggregation becomes large enough to exhaust the patch. Thus, it is reasonable to suggest that when females are patchily distributed in space they are likely to be associated with patchily distributed resources (Emlen and Oring 1977). When resources *are* patchy, copying behavior in the context of resource acquisition is likely to be favored among females.

As mentioned, females will have opportunities to mate nonrandomly if males are attracted to female aggregations. If females tend to copy the behavior of other females in the context of resource acquisition, it is likely that females will show similar tendencies when selecting mating partners. Figure 8.9a shows the spatial and temporal distributions of females in which females are likely to copy each other's resource-location behavior. This distribution overlaps entirely with the distribution in which females will have opportunities to copy the mate selections of other females, if spatially aggregated females tend to attract assemblages of males (fig. 8.9b). However, as described above, the spatial and temporal distributions of females can influence the mating behavior of males in ways that may ultimately constrain female behavior.

We have shown that males will tend to guard females and thereby restrict female opportunities to engage in mate copying in a majority of the possible distributions of females in space and time (fig. 8.8). When females are spatially clumped as well as synchronous in their receptivity, males are expected to guard individual females. When females are spatially clumped and asynchronous in their receptivity, males are expected to guard female groups or the resources around which females aggregate. When males guard individual females for the duration of female receptivity, females will have little opportunity to copy the mate choices of other females. Similarly, when males guard groups of females, guarding males will restrict female mate choices. Only when females are clumped in space and males do not attempt to defend females, or when females are clumped in space and males defend resources instead of females (fig. 8.10; see also section below), will females remain sufficiently mobile to exercise mate choice. At these times, nonrandom mating by females, caused by female tendencies to copy one another's mate selections, can increase the opportunity for sexual selection on males above that which is apparent from the spatial and temporal distributions of females alone. Methods for calculating this increase are explained elsewhere (Wade and Pruett-Jones 1990; Shuster and Wade 1991b).

As discussed in chapter 4, female tendencies to copy the behavior of other females may covary with male mate guarding abilities. Such genetic correlations are expected to lead to runaway sexual selection for large harems of females that are defended by large and pugnacious males, or perhaps toward explosive breeding aggregations (Wade 1995; chapter 9). Alternatively, genetic covariance between female mate copying behavior and male tendencies to display in groups can lead to runaway sexual selection for leks or breeding hot spots (see below). In each of these cases, when female copying behavior leads to increased variance in mate numbers among mating males, I_{mates} will increase beyond the value measurable from the spatial and tempo-

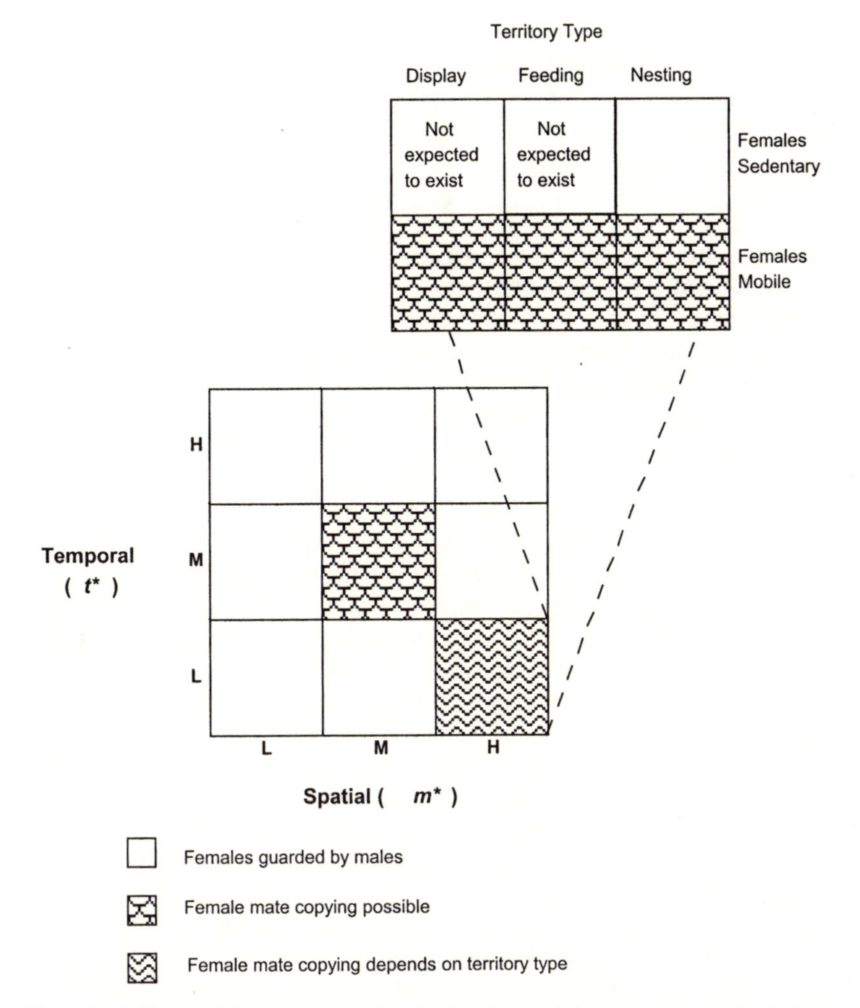

Figure 8.10. The spatial and temporal distributions in which female mate copying behavior will be constrained or unconstrained by male guarding behavior.

ral distributions of receptive females. However, if female aggregations arising from female copying behavior also attract aggregations of males, the number of mating males may increase. As the number of mating males increases, the value of I_{mates} will decrease.

Male Monopolization of Resources

Males are capable of defending many types of resources in the context of mating. These resources include feeding sites, nesting sites, and display sites on leks. The degree to which these resources are patchy in their distribution will determine the degree to which females aggregate in space and time, the degree to which mates may obtain multiple mates, and the degree to which I_{mates} can increase.

Feeding Site Defense

If feeding sites are dispersed in space, or if food sources are transient in time, the spatial and temporal distributions of receptive females are likely to remain dispersed. In such cases, males are expected to defend individual receptive females rather than the sites to which females are attracted (fig. 8.3). Similarly, if females aggregate at food sources, but are synchronous in their receptivity, males are expected to defend individual females, despite the spatial proximity of numerous receptive females (fig. 8.4). Moderate spatial and temporal clumping of food sources may encourage females to form loose assemblages. However, such circumstances are likely to favor roving males rather than resource-tending males for reasons explained above (fig. 8.6).

Roving by some portion of the male population is also favored (a) when females are spatially aggregated and moderately asynchronous in their receptivities (m^* = large; t^* = moderate; fig. 8.7), as well as (b) when females are asynchronous in their receptivites and moderately aggregated in space (m^* = moderate; t^* = low; fig 8.7). In each of these circumstances, males are likely to defend females rather than food sources, because in the presence of rovers direct guarding of females is likely to provide greater fertilization success than resource defense. Thus, male defense of feeding sites is expected to evolve *only* when feeding sites (1) are highly clumped in space, (2) remain available for prolonged periods, and (3) are visited asynchronously by females (fig. 8.11).

Although all females attracted to a particular feeding site may not be simultaneously present at that site, with respect to male mating success, these females are functionally aggregated in space, and become available for mating asynchronously in time. As explained above, females could become sedentary and remain at feeding sites. However, the food source may become depleted and male mate numbers subsequently limited (fig. 8.11). Males guarding resources are therefore not expected to allow females to remain at feeding sites for prolonged periods. When males defend food sources required by females, males are expected to remain attached to sites and females are expected to remain mobile, often with forceful encouragement by males.

Nesting Site Defense

As with feeding site defense, the spatial and temporal distributions of females on nests will determine whether males defend nests or females. Fe-

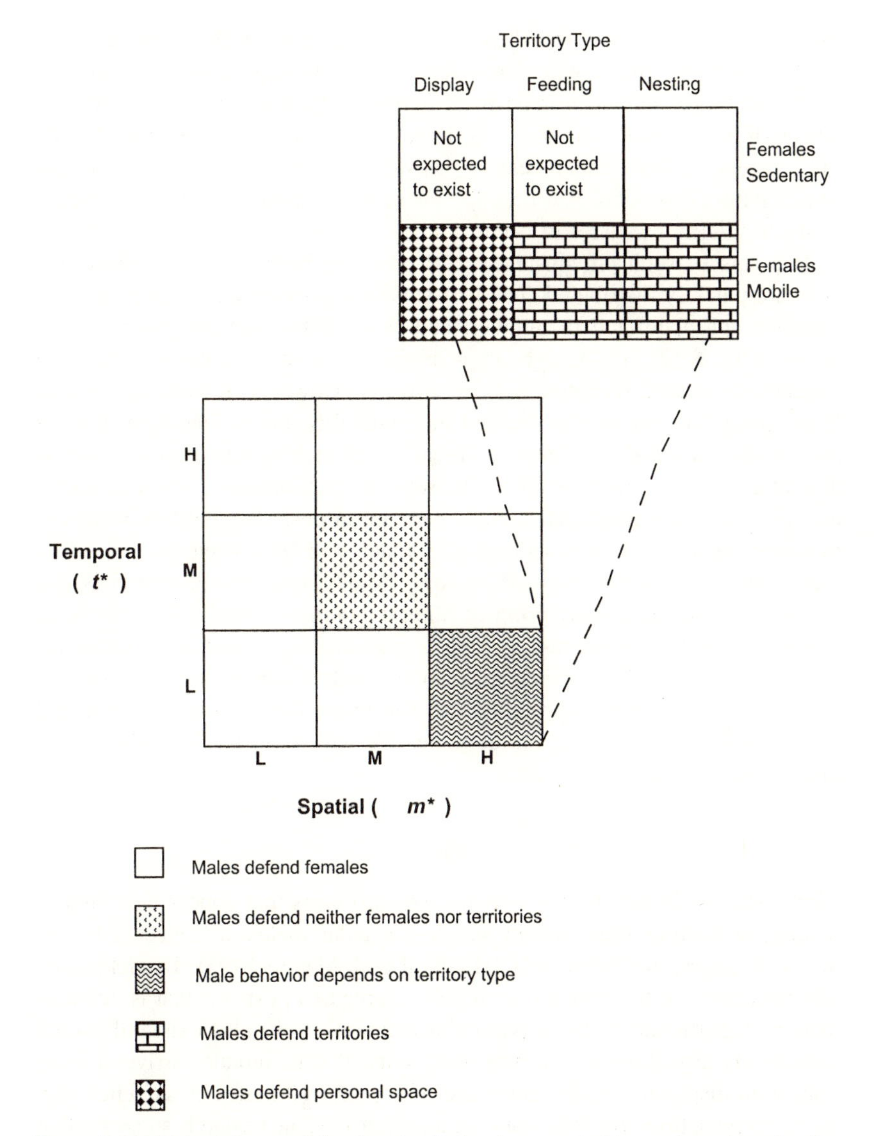

Figure 8.11. Male defense of feeding territories is expected to evolve when feeding sites are patchy in space and not transient in time. Females could be either sedentary or mobile with respect to feeding sites. However, since male mating success relies on the persistence of the food source, males are likely to drive off sedentary females.

males are expected to protect their nests and young under most spatial distributions. Whether males will also defend nests will depend on the degree to which nest sites influence male mate numbers. In most cases, as described above (figs. 8.3, 8.4, 8.6, and 8.7), males will defend females directly and only appear to show attachment to nests. However, if male mating success requires his possession of a nesting site before courtship begins, we expect nest site fidelity by males to occur.

Two sets of such circumstances are possible, both occurring when nest sites are patchily distributed. The first situation exists when nesting sites are clumped space, limited in number, and females become receptive in synchrony (fig. 8.12, m^* = high; t^* = high). In most such cases, males are expected to defend receptive females directly. However, if nests are crucial to offspring success and are limited in availability, males who control nests may be the only males to breed, whereas males lacking nest sites will not be able to attract and retain females. The second situation exists when nest sites are fixed in space and females either remain on nests and become receptive in sequence, or visit sites sequentially, mate, and then leave to seek other mates (fig. 8.12; m^* = high, t^* = low). Males who defend nest sites can accumulate the eggs or offspring of numerous females, and will continue to increase their mating success as long as the nest site remains favorable for offspring survival. Thus, when male mating effort and male parental effort coincide in enhancing offspring number for males, the opportunity for sexual selection on males can become extreme. We will return to this issue in our discussion of male parental care.

Leks

Circumstances in which males display on territories that appear unrelated to feeding or nesting sites, and in which particular males are selected by females as mates, are called *leks* (Höglund and Alatalo 1995). By definition, leks occur only when females are highly aggregated in space, that is, females mate with particular males at particular locations. Leks also exist only when females are asynchronous in their receptivity, that is, females arrive at locations with displaying males, and mate with preferred males in sequence (fig. 8.13). Explanations for leks vary widely, but current research suggests that leks form in "hot spots," that is, locations in which the ranges of numerous females overlap due to female patterns of movement or resource use. By definition, resources are not clumped at lek locations (reviews in Bradbury et al. 1985; Höglund and Alatalo 1995).

Male mating success is extremely variable in lekking species (Höglund and Alatalo 1995). Explanations for extreme variance in mate numbers have recently focused on male displays as indicators of male quality or parasite resistance (Hamilton and Zuk 1982; Petrie 1994). Although the opportunities for natural and sexual selection on offspring numbers have been calculated in few lekking species (Wade 1979; Wade and Arnold 1980; Arnold and

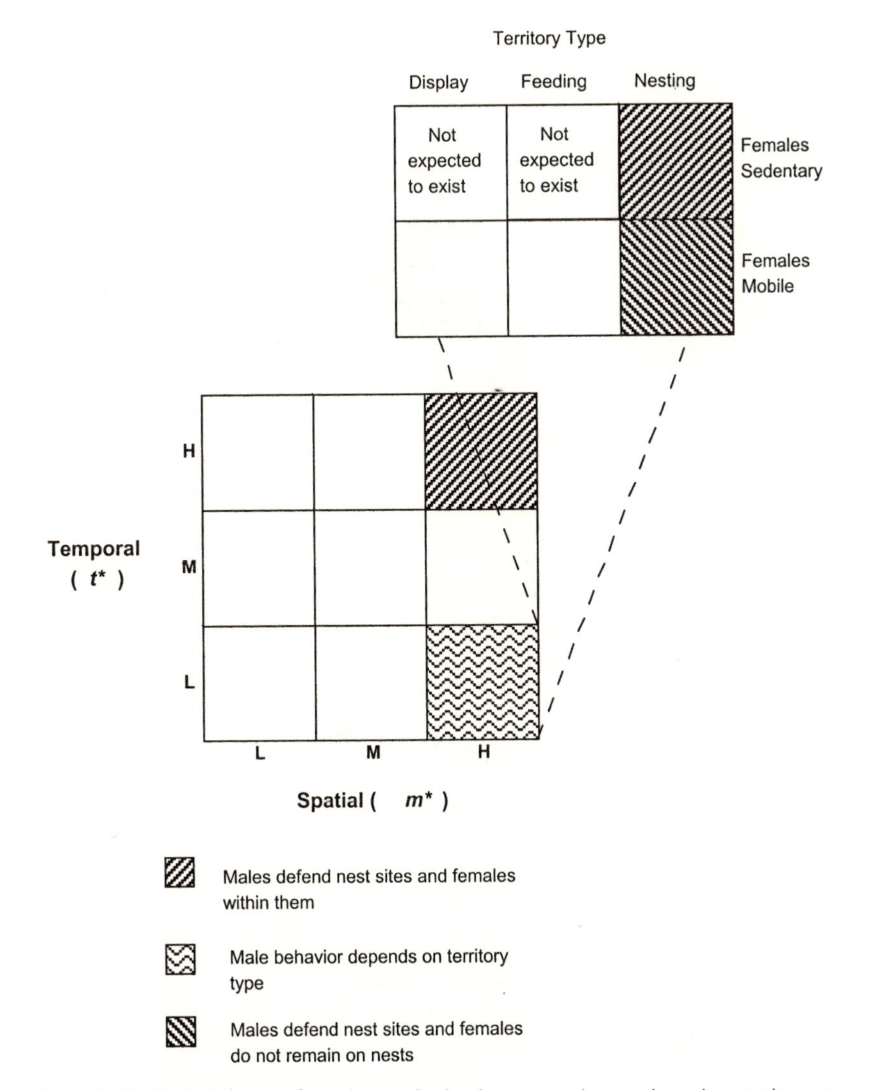

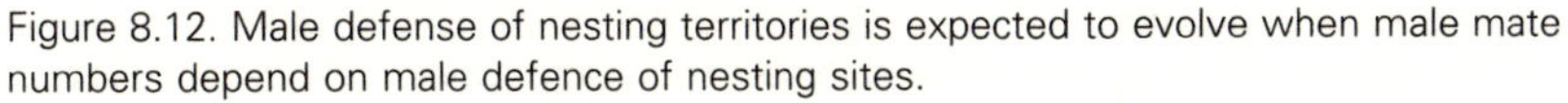

Figure 8.12. Male defense of nesting territories is expected to evolve when male mate numbers depend on male defence of nesting sites.

Wade 1984a,b; see Kruijt and deVos 1988), when variance in male mating success is extreme, as it is on leks, the proportion of total selection that is due to viability selection on females is likely to be small. As we have argued in chapters 1, 5, and 7, the larger the sex difference in the opportunity selection becomes, particularly if variance in mate numbers is caused by nonrandom mate choice, the more likely mate choice criteria will become arbitrary. It is interesting therefore, with respect to the evolution of leks, that more species with extreme spatial and temporal clumping of females do not con-

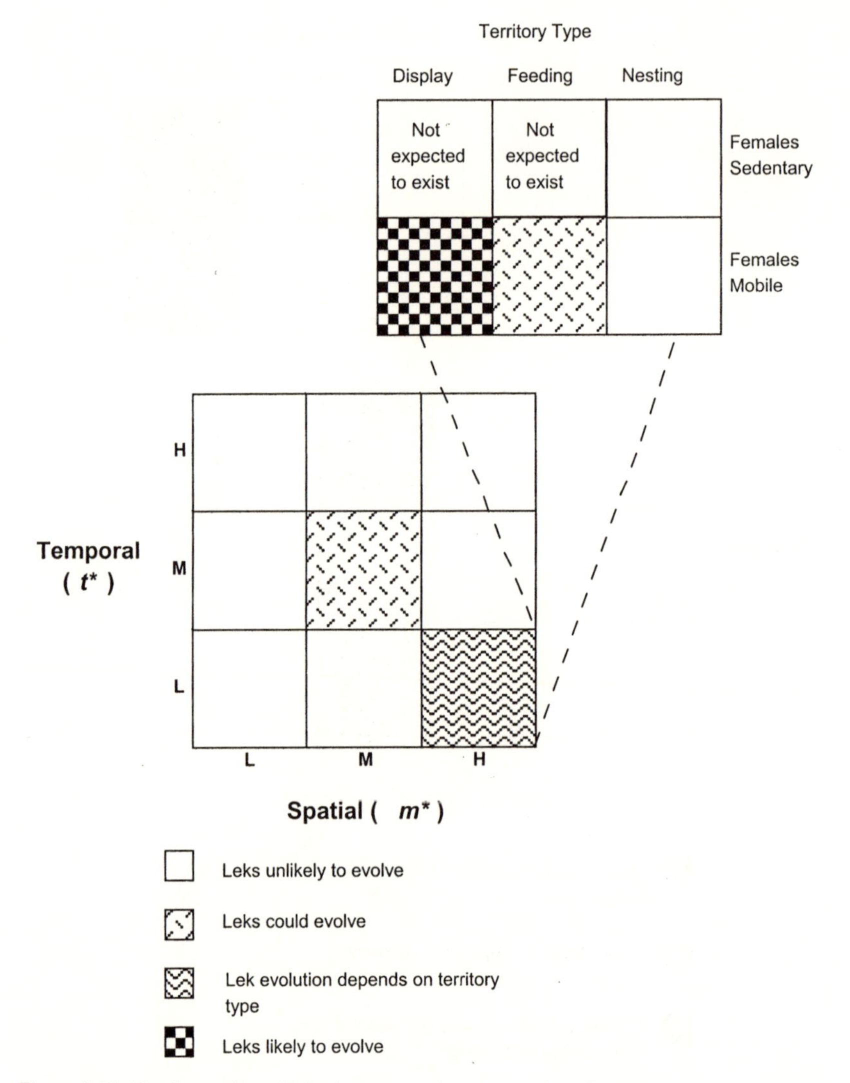

Figure 8.13. The formation of leks is expected to occur when female spatial and temporal distributions are highly clumped and female aggregation does not involve feeding or nesting sites.

verge toward this form of mating system. We will return to our discussion of leks in chapter 9.

Persistent Monandry

As explained above, the term *monandry* describes a mating system in which females mate with only one male in their lives. In light of recent molecular studies of paternity in birds, the concept of monandry may seem outdated. Yet monandry *does* exist in short-lived invertebrates (Shuster 1989a,b; Robinson and Allgeyer 1996; Fischer and Fiedler 2000; Jones 2001), in certain birds (Mauck et al. 1995) and mammals (Sever and Mendelssohn 1985; Brotherton et al. 1997), as well as in protistan species in which sexuality is achieved by the transfer of genetic material from one cell to another (Raikov 1995; Dacks and Roger 1999). In fact, monogamy may therefore represent the ancestral state of all mating systems (Dacks and Roger 1999). We have shown that if females mate more than once with the same male, this condition is equivalent to monandry (chapter 4). However, with the production of multiple gametes, multiple mating becomes possible, and among multicellular organisms, circumstances in which a female may fertilize all of her ova with the sperm of a single male may seem remote. The widespread occurrence of monandry among multicellular organisms makes the question important. Indeed, when can monandry persist?

As shown in figs 8.3–8.7, and 8.10, the spatial and temporal distributions of receptive females determines when and for how long males will attempt to guard their potential mates. Differently put, males in some circumstances will attempt to retain females for the duration of their receptivity, while in others, males will sacrifice fertilizations to enhance their mate numbers. Under these constraints, *monandry* can persist only when individual males guard individual females for the *duration* of female receptivity.

This requirement is likely to be met in four sets of circumstances (fig. 8.14). First, when females are spatially dispersed and synchronous in their receptivity (m^* = low, t^* = high; fig. 8.14), males are expected to locate females *before* they become sexually receptive and to remain with their mates until female receptivity is *complete* (Shuster 1981b; Ridley 1983; Jormalainen 1998). Synchronous breeding by females, combined with spatial dispersion of receptive females (m^* = low, t^* = high; fig. 8.14) prevents males from locating multiple mates, and, as explained above, mate guarding minimizes multiple insemination. Under these conditions, in species in which females breed only once and mate guarding is effective, monandry will commonly occur. Also under these conditions, monandry may be routine in species with prolonged parental care, or with short interbreeding intervals. In such species, persistent pairs may save time and energy spent by individuals who attempt to locate new mates after every breeding episode. These mating systems are most likely to evolve when females are dispersed in space and synchronous in their sexual receptivity for reasons explained below.

Monandry is also possible when females are spatially aggregated and synchronous in their receptivity (m^* = moderate to high, t^* = high; fig 8.14). Here, too, each female is likely to be located and guarded by a male before she becomes receptive, and males are likely to remain with their mates for the duration of female receptivity. Spatial aggregation of females is likely to cause males to assemble at female aggregations, and mate guarding will

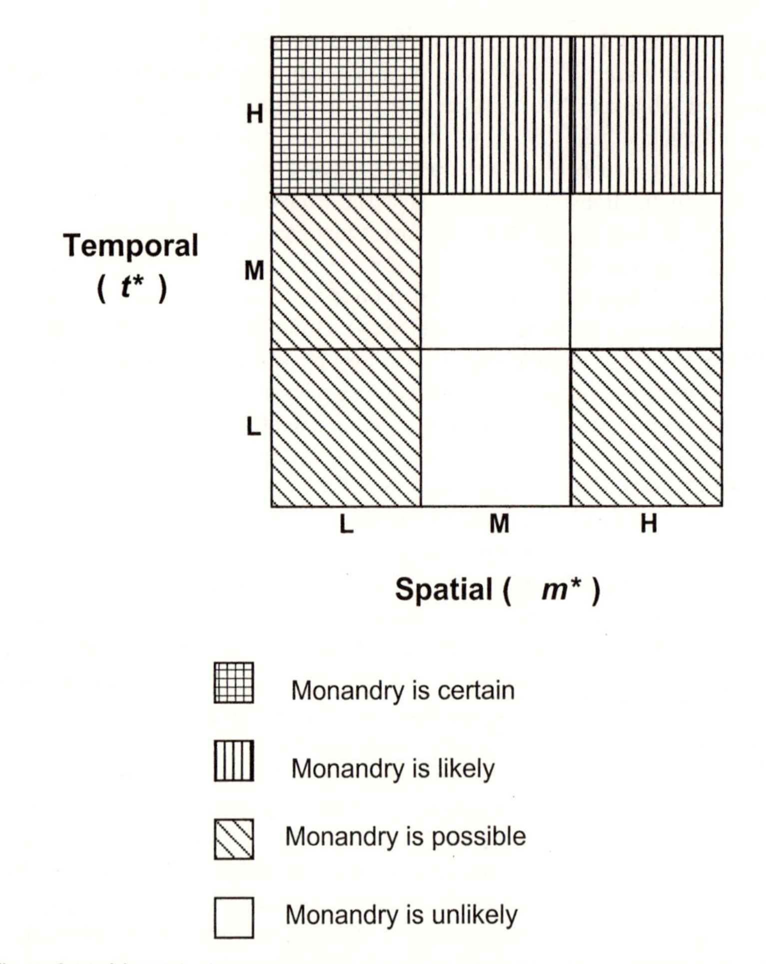

Figure 8.14. Monandry is expected to persist when females are guarded by males for the duration of their sexual receptivity.

minimize a male's chances of being displaced by another male, although if the number of males and females at such locations is large, direct competition among males may be minimal (see below). Despite the close proximity of pairs, synchrony in female receptivity prevents males from multiple mating. Again, in species in which females breed only once, monandry can persist. Persistent pairs may occur less frequently in this situation than in the previous one because circumstances favoring female aggregation and synchronous reproduction may not persist for long periods of time.

Monandry is possible when females are spatially dispersed and asynchronous in their receptivity since each female is likely be closely guarded by a male (m^* = low, t^* = low to moderate; fig. 8.14). However, since multiple males are likely to be attracted to individual receptive females, when female receptivity occurs, the assembled males are likely to attempt to mate simultaneously, with multiple paternity as the usual result (Brockmann et al. 1994, 2000). Although selection favoring effective mate guarding could minimize sperm competition, whereby males remain with females for the duration of their receptivity (Jormalainen et al. 1999), males are unlikely to remain with individual females over several breeding episodes since other females may become available. Thus, under these conditions, monandry is most likely when females mate once and are semelparous. Many species which breed asynchronously are also iteroparous (Morin 1987; Clutton-Brock 1989; Møller and Ninni 1998). Thus, monandrous, asynchronously breeding females may be rare in nature.

In general, monandry is unlikely to persist when males can rove between females (m^* = moderate to high, t^* = low to moderate; fig. 8.14). When males rove, they abandon females early in order to locate other mates. Abandoning females before their receptivity is complete may cost roving males fertilizations, but, as explained above, this loss is acceptable if roving allows males to achieve mate numbers that exceed the reciprocal of their average fertilization success (fig. 8.2). Incomplete insemination of females, as well as opportunities for multiple mating among males, can lead to multiple insemination of individual females. Abandoned, incompletely inseminated females who seek additional mates may have higher fecundity than females who remain monandrous. Roving males may also, via their multiple copulatory attempts, successfully inseminate previously mated females who already have adequate supplies of sperm. When multiple mating occurs, a runaway process between sperm competitiveness and female tendencies to mate more than once can ensue (chapters 4 and 7). Once multiple mating arises, monandry will disappear.

Monandry will usually be eliminated when females are guarded in groups. When females are spatially clumped and asynchronous in the timing of their receptivity, dominant males can exclude subordinate males from harems. However, harem-holding males may deplete their sperm stores by guarding and mating with multiple females. Thus, females who prolong the duration of their receptivity (i.e., mate more than once) may have higher fecundity than females whose receptivity durations remain short. Such females will be mated more than once by the dominant male or will be mated by subordinate males behaving as satellites (Clutton-Brock 1989). Guarded females with prolonged receptivity will avoid losses in fecundity experienced by guarded females with shorter periods of receptivity, and monandry will soon be eliminated. As is true with all female adaptations, the larger the value of I_{clutch}, the faster the female population will respond to viability selection associated with male guarding behavior. The smaller the value of I_{clutch}, the slower the populational response.

However, these same circumstances may favor males who tenaciously defend a single female rather than attempt unsuccessfully to defend female groups or seek matings as satellite males. As explained in chapter 9, males who secure a single mate can experience higher fitness than males unsuccessfully attempting either of the other two strategies. Thus, the fourth situation favoring monandry may be unexpectedly widespread. In all situations in which variance in male mating success prevents some males from mating, monandry is expected to persist.

Mate Guarding and Sexual Conflict

We have defined sexual conflict as a sex difference in the covariance between offspring numbers and mate-number promiscuity (chapter 4). This situation can result from the distributions of receptive females in space and time, as well as from the details of female life history which influence female mate and mating numbers (chapters 1–5). We expect sexual conflict to be greatest when the sex difference in the opportunity for sexual selection, ΔI is large; that is, in species in which females are spatially clumped, temporally asynchronous in their receptivity, mate once, produce only one clutch of offspring, and tend to copy the mate choices of other females. We expect sexual conflict to be nonexistent when there is no sex difference in the opportunity for sexual selection; that is, in species in which females are spatially dispersed, temporally synchronous in their receptivity, mate more than once, and produce multiple clutches of offspring. In each of these cases, aggressive interactions between males and females may not be obvious. In the first case, despite the potential for male exploitation of females, the magnitude of the sex difference in the opportunity for selection may limit the ability of females to respond to sexual exploitation by males. In the second case, no aggressive interactions are expected because no context for sexual conflict may exist. Again, as with mate guarding, distinct selective processes ($I_{mates(high)}$ versus $I_{mates(low)}$) produce a similar behavioral result.

Even when sexual conflict is obvious, as it is when males attempt to guard females to ensure paternity and females resist male guarding attempts, conflict resolution in favor of one sex or another (Fairbairn 1993; cf. Rice 1996; Jormalainen and Merliata 1998; Jormalainen 1998) will depend on the sign and magnitude of ΔI. If I_{mates} exceeds I_{clutch}, males will tend to exploit female distributions at the expense of female fecundity. If I_{clutch} exceeds I_{mates}, females will tend to exploit male distributions at the expense of male fecundity. Exploitation is expected be resolved in favor of the sex whose fitness variance is greatest. Moreover, the degree to which each sex can respond to exploitation by the other will depend on the relative magnitudes of I_{clutch} and I_{mates}. It is interesting to note that because sexual exploitation, by definition, imposes a fitness cost on members of the exploited sex, members of the exploiting sex also suffer a fitness cost via their exploited mates (Chapman and Partridge 1996; Jormalainen et al. 2001). Thus, rather than leading routinely to an accelerating evolutionary arms race, we expect sexual exploitation to be self-limiting.

Male Combat and Female Choice

Sexual dimorphism is expected to evolve when selection acts differently on members of one sex than it does on members of the other sex (chapter 1; Slatkin 1984). As we have shown, differential reproduction among members of the same sex *causes* sexual selection. Darwin (1874) recognized that sexual selection could operate in two contexts: (1) Combat, usually among males, in which individuals compete for access to mates, and (2) mate choice, usually exercised by females, in which individuals of one sex prefer certain members of the opposite sex as mates. Numerous authors have attempted to tease these selective contexts apart to provide explanations for patterns of sexual dimorphism among species. Although most studies conclude that combat and choice as sources of selection are hopelessly confounded (Andersson 1994; see Murphy 1998), we suggest that certain basic patterns of sexual dimorphism can be identified within the framework we have described.

That male combat can generate a large sex difference in the opportunity for selection is well established (Clutton-Brock 1991b; Andersson 1994). However, it can be less powerful than selection in the context of female mate choice. Female mate choice *accelerates* the process of sexual selection because both male characters and female preferences coevolve (chapters 1, 4, and 7). The degree of aggregation of (possibly reluctant) reproductive resources achieved by male-male combat may be less than that occurring when resources actively accumulate around males (i.e., female choice).

Darwin (1874) recognized that male characters evolving by female mate preferences tend to be more elaborate (or in Darwin's words, more "trivial") than those evolving in the context of male-male competition. As we have explained, the degree to which male characters are favored in the context of male combat or female choice is expected to depend on the combined spatial and temporal distributions of females, as well as on female tendencies toward multiple mating and multiple breeding episodes. The larger the sex difference in the opportunity for selection, ΔI, that is, the larger I_{mates} is relative to I_{clutch}, the more likely male characters will be exaggerated by female mate preferences. The smaller the sex difference in the opportunity for selection, the less likely male exaggeration will be.

Consider male body size as a character under selection. Large male body size may be required for the defense of mates over a range of female spatial and temporal distributions. However, when females are spatially dispersed and breed synchronously (I_{mates} is small), female preferences for large males are unlikely to increase I_{mates} because differences in mate numbers among males are restricted by the spatial and temporal distributions of females. That is, when females are spatially dispersed and breed synchronously, most males are able to mate. Here, the strength of selection, even if females prefer large males, will be small. However, when females are spatially clumped and breed asynchronously (I_{mates} is large), any tendency for females to prefer larger males as mates can augment I_{mates} beyond its value due to the spatial and temporal distributions of females alone. This is true not only because many, relatively smaller males are already excluded from breeding in such mating systems, but also because female preferences can further skew mating success within the class of mating males as described above. This asymmetrical potential for exaggeration of male combat structures is possible not only in circumstances favoring large male body size, but also for any other male character advantageous in male-male interactions (e.g., coloration, displays, weapons, pheromones, vocalizations), provided I_{mates} is already large (see below).

Female mate preferences and runaway selection may modify preferred male characters so that they are no longer useful for their original purpose. Suppose that large antler size is advantageous to males in battle. If females begin to prefer large antler size because of the advantage it may provide their sons in battle, such a preference will increase, not only the frequency of males with unusually large antlers, but also the frequency of females that prefer extreme males as mates. The intensity of selection on antler size is increased beyond the context of male combat owing to female preferences. In this way, the strength of runaway selection on males using antlers to attract females could exceed the strength of intrasexual selection on males using antlers to fight (chapter 7). This runaway may cause displaying males to mate more successfully than fighting males. Males with extreme antlers may become less effective fighters, but such losses will be overcome by increased mating success due to female preferences for males simply possessing extreme antlers. When this occurs, antlers, although still bearing awesome resemblance to functional weaponry, will become increasingly less functional in active combat and increasingly arbitrary relative to selective contexts other than extreme female preferences. This is a somewhat different explanation for the preponderance of male display over combat from the game theoretic hypothesis (Maynard Smith 1982; Dugatkin and Reeve 1998).

Sexual Dimorphism

If degrees of sexual dimorphism can be predicted from the combined effects of female spatiotemporal distributions and the details of female life history (chapter 6), it should also be possible to use these same variables to identify the forms that sexual dimorphism may take. As first reviewed by Darwin (1874), there are numerous categories of sexual dimorphism, whose treatment is beyond the scope of this book. However, we can consider four major categories of sexual dimorphism because they include many observable forms of sexual differences. The four categories include (1) male attachment structures, (2) differences in body size, (3) weaponry, and (4) conspicuous displays.

Male Attachment Structures

When the spatial and temporal distribution of females favors male guarding of individual mates, structures that enhance males' ability to retain their grasp on females are likely to be favored by sexual selection. Males are expected to defend individual mates in a wide variety of female spatial and temporal distributions (fig. 8.15). However, the degree to which attachment structures are crucial to male mating success may vary within this range of selection opportunities.

When females are spatially dispersed as well as synchronous in the timing of their receptivity (m^* = low, t^* = high; fig. 8.15), males are expected to begin guarding their mates well before females become sexually receptive (Shuster and Caldwell 1989; Jormalainen 1998). Males are also expected to remain with their mates throughout the female receptive period (Jormalainen and Shuster 1999). When spatially dispersed females become receptive in synchrony (m^* = low, t^* = high; fig. 8.15), mating males are unlikely to locate additional mates before all females become unreceptive, as described above. Thus, when females are spatially dispersed and temporally clumped, most males will mate successfully. Similar circumstances exist when females are aggregated in space as well as in the timing of their sexual receptivity.

Aggregated distributions of receptive females are likely to attract males and could so increase the intensity of attempts by single males to displace guarding males. Yet, if female receptivities are synchronous, males are more likely to associate with an individual female and not attempt to displace other males. Circumstances favoring long-term mate guarding could favor the evolution of structures that allow males to cling to females. However, since most males will successfully acquire mates when males are attracted to aggregations of simultaneously receptive females, under these conditions, clasping organs are not expected to evolve by sexual selection (fig. 8.15).

When females are spatially dispersed, male guarding of individual females will be favored regardless of whether females are moderately to highly asynchronous in their receptivities, as described above. However, as female receptivities become increasingly asynchronous, males are likely to gain additional mates by searching for other females after their first mate becomes unreceptive. As explained above, spatial dispersion of females combined with temporal synchrony in female receptivity are expected to make male-

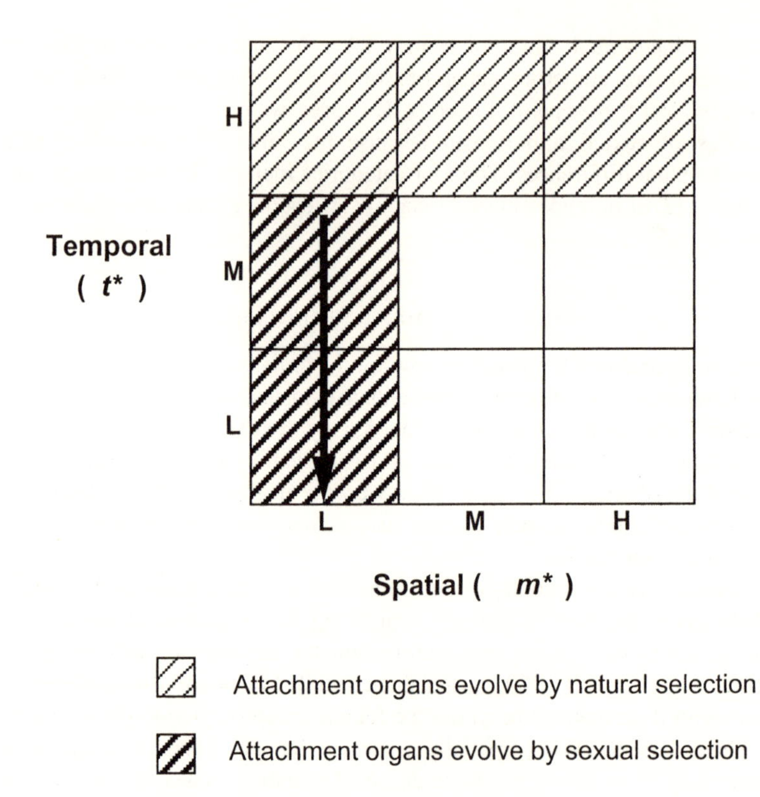

Figure 8.15. The spatial and temporal distributions of females in which attachment organs are expected to evolve in males. When I_{mates} is small, natural selection is expected to be primarily responsible for the evolution of male attachment organs; as I_{mates} increases in magnitude, sexual selection is more important in these structures' evolution. In the latter case, male attachment organs are most likely to be observed when females are temporally synchronous in their receptivity and spatially clumped, or when females are temporally asynchronous in their receptivity and spatially dispersed.

male interactions infrequent and of low intensity. However, as females become increasingly *asynchronous* in their sexual receptivity, more males may be attracted to isolated, receptive females. The development of male attachment structures is expected to become increasingly pronounced as the receptivities of spatially dispersed females become increasingly asynchronous (m^* = low, t^* = moderate to low; fig. 8.15). Thus, the evolution of male attachment structures in response to male-male competition is most likely when females are highly dispersed in space and breed asynchronously in time (m^* = low, t^* = low; fig. 8.15).

Darwin (1874) puzzled over whether male attachment organs could arise by natural or sexual selection. He reasoned that if successful mating required attachment structures when wave action or other environmental influences acted to separate mating pairs, then clasping organs in males must evolve by natural selection. On the other hand, if claspers assisted males in retaining mates during male combat, sexual selection must be involved in their modification. Consideration of the sex difference in the opportunity for selection, arising from the spatial and temporal distributions of females, provides a solution to this dilemma. When females are spatially dispersed and female receptivities are asynchronous (m^* = low, t^* = low; fig. 8.15), male interactions will frequently occur and sexual selection will favor clasper modification to a degree proportional to I_{mates}. However, when females with synchronous receptivities are spatially dispersed or spatially clumped (m^* = low to high, t^* = high; fig. 8.15), the sex difference in the opportunity for selection will be negligible because all mature males are likely to locate and guard individual females. In such circumstances, there is no sex difference in the opportunity for selection and natural selection alone can be considered responsible for the evolution of structures by which males retain their grasp on females.

Differences in Body Size

When males guard females to prevent mating attempts by other males, male combat is likely to occur. Large male body size as well as territoriality and aggressiveness are expected to be favored among males in such circumstances (review in Andersson 1994). As described above, we expect males to be larger, more territorial, and more aggressive than females when the spatial and temporal distributions of females favor mate guarding (but see below). This is true when receptive females appear asynchronously in time, regardless of their spatial distribution (m^* = low to high, t^* = high; fig. 8.16), and is particularly true when receptive females appear asynchronously in time and are spatially clumped (m^* = high, t^* = low; fig 8.16a).

If searching behavior is favored among males, large body size may be less important than it is when males guard individual females or groups of mates. This is true under two conditions: (1) When receptive females appear with moderate asynchrony in time and are moderately to highly clumped in space (m^* = moderate to high, t^* = moderate; fig. 8.16b), and (2) when receptive females are moderately clumped in space and appear asynchronously in time (m^* = moderate, t^* = moderate to low; fig. 8.16b). Male searching behavior could also be favored if females are highly clumped in space and appear asynchronously in time, but this male strategy is likely to be employed only by satellite males. However, such activity by males is expected to favor extremely small and mobile individuals. Thus, circumstances that simultaneously favor large body size and small body size in males, particularly when I_{mates} is large, may produce male populations that are dimorphic

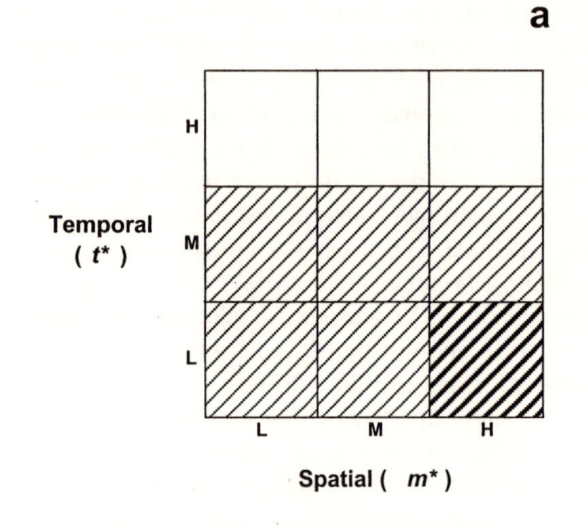

 Large male body size favored

 Extremely large male body size favored

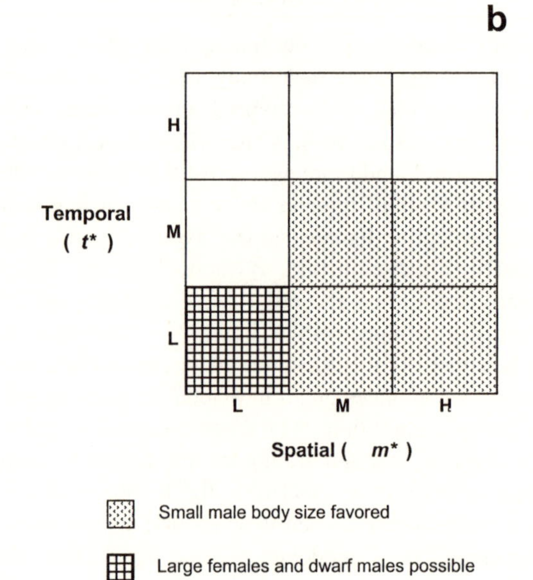

Figure 8.16. The spatial and temporal distributions of females favoring large male body size, small male body size, and extreme sexual dimorphism in size involving large females and dwarf males.

or polymorphic in size (see chapters 10–12). Male searching behavior may also be favored when females are spatially dispersed and asynchronous in their receptivity (m^* = low, t^* = low; fig. 8.16b). However, such conditions also favor male mate guarding as well, as explained below.

These patterns will be most easily visualized when I_{mates} is only weakly eroded by I_{clutch}, that is, when females tend to be monandrous and semelparous (chapters 5 and 6). However, if females tend to mate more than once and produce multiple clutches of offspring, the effects of sexual selection on male body size may become less conspicuous. Extreme female fecundity can be favored when juvenile survivorship is low as in many marine invertebrates (Brusca and Brusca 1990) or in terrestrial invertebrates such as spiders (Vollrath 1998). If relatively few females breed successfully due to preadult mortality, the opportunity for selection on large female body size can be intense because relatively few females contribute to the next generation.

If I_{mates} as well as I_{clutch} favor increased body size, sexual dimorphism will not be obvious because large size will be favored in both sexes. However, if fecundity selection favoring large size among females and sexual selection favoring small size among searching males are simultaneously favored, sexual size dimorphism can become extreme. Such cases are likely to lead to the evolution of extremely large females and dwarf males (Andrade 1996; Vollrath 1998). We expect this pattern of sexual dimorphism to exist when receptive females are dispersed in space and appear asynchronously in time (m^* = low; t^* = low, fig. 8.16b). Males in such species must move between receptive females to successfully mate, but may also need to defend receptive females when they find them. Despite this competition, I_{clutch} may still equal or exceed I_{mates} if preadult mortality among females is high. In some species, numerous tiny males may congregate and compete for matings by various means while attached to a single, huge female (Funch and Kristensen 1995; Hoeg 1995; Legrand and Morse 2000; Pascual 2000; Luetzen et al. 2001). If several males mate successfully, I_{mates} may be further eroded since multiple mating by females decreases the opportunity for sexual selection. Stated differently, tendencies toward multiple mating by females prevent mate monopolization by males, and can cause I_{clutch} to exceed I_{mates}.

Weaponry

Horns, antlers, or other extensions of male anatomy may also evolve in the context of male mate guarding, although in many species exhibiting weapons both males and females possess them (Geist 1966; Geist and Bayer 1988). When weapons occur in both sexes, they appear to be used in contests with conspecifics as well in interactions against predators. Thus, the context in which such weaponry has evolved is likely to predate the evolution of weapons in the context of sexual selection, through similar viability selection on both sexes. When weapons are slightly more exaggerated in males, it is reasonable to infer that a sex difference in the opportunity for selection on this character exists (chapter 1; Darwin 1874; Slatkin 1984).

Four general classes of male weapons appear to exist. (1) *Horns* consist of unbranched, pointed extensions, including elongated teeth, fangs, spines, or spurs. These appendages appear to be useful for piercing or fencing with rivals. Pain or risk associated with impalement are likely to induce intruders to move away from potential impalers. Thus, with time, the threat of impalement alone may be sufficient to keep competitors away from mates (Shuster and Caldwell 1989). (2) *Antlers* consist of branching or curving extensions, including enlarged mandibles, legs, or chelae. These structures appear to be useful for grappling with and displacing rivals, often with a quick toss of the ensnared opponent (Eberhard 1979). (3) *Bosses* consist of thickened and rounded extensions of hardened material. These structures appear to be useful for ramming, pushing, and thus displacing rivals (Geist and Bayer 1988). (4) *Blades* consist of flattened extensions, often with narrowed exterior margins. These structures appear to be useful for prying rivals off balance, since greater surface area may be applied to the object to be displaced. Narrowed exterior margins on flattened extensions may be useful for displacing rivals attached to hard surfaces (Shuster, pers. obs).

In general, weapons are likely to be elaborated in males when physical size alone is not sufficient to provide mate-guarding individuals with sole access to defended females. The primary advantage gained by the possession of body extensions appears to be enhanced leverage or concentration of force (Geist 1966; Clutton-Brock 1982). Whereas body size can be useful in physically covering guarded mates or in pushing competitors aside (Shuster 1981b), horns, antlers, or other extensions appear to allow their users to push or drive rivals away more efficiently (Shuster 1991b; Emlen 2000). Increased leverage in pushing appears to be useful in contests on open ground, as in ungulates (Geist 1966; Clutton-Brock 1982; Geist and Bayer 1988), stag beetles (Eberhard 1979), and fiddler crabs (Backwell et al. 2000). However, elongated appendages also appear to provide leverage advantages during contests within cavities, as in cane beetles (Eberhard 1979), dung beetles (Emlen 1996; 2000), and sphaeromatid isopods (Shuster 1992). Certain sphaeromatids inhabiting hard substrates possess bladelike anterior or posterior extensions that may be useful in prying rivals off of their mates (Bruce 1994; Shuster, pers. obs). Bladelike extensions may also be useful for applying lateral force and thus for lateral displacement of rivals. Extensions useful for impaling rivals are likely when escalated contests occur, since opponents may not simply be displaced, but are also injured or killed (Darwin 1874).

The four categories of weapons described above are not mutually exclusive, and combinations of some or all four types may be recognizable in the appendages of particular species. However, in general, sexually dimorphic weaponry is expected to evolve in species in which females become spatially clumped and whose receptivities are asynchronous in time (m^* = moderate

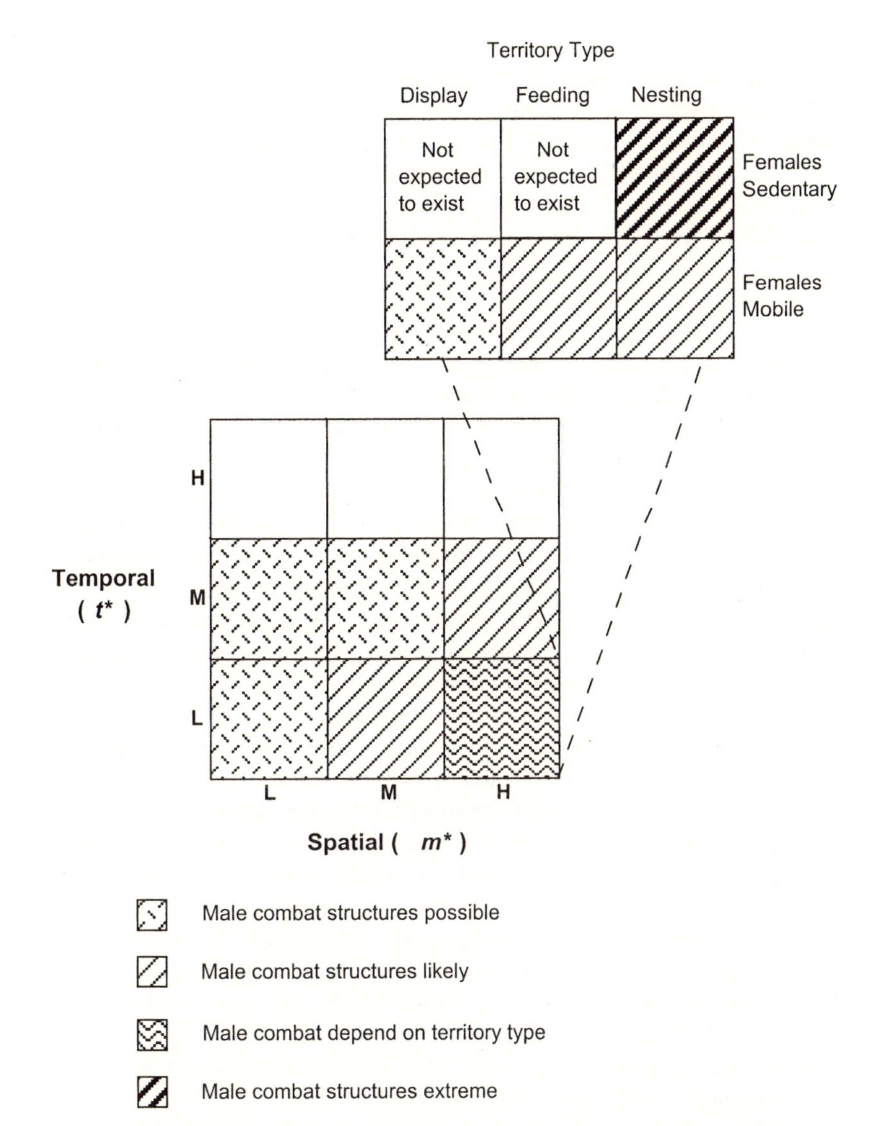

Figure 8.17. The spatial and temporal distribution of females favoring the evolution of male combat structures; horns or other sharp unbranched appendages are most likely useful in escalated contests, and thus these structures are most likely to be observed when female temporal asynchrony and spatial aggregation are extreme.

to high, t^* = moderate to low; fig. 8.17). Combat at such times may be of sufficient intensity among males that, even if females tend to mate more than once, relatively few, well-armed, and aggressive males will mate successfully. The more escalated contests over mates can become, the more dangerous weapons can be. Thus, horns are expected to appear when I_{mates} is extreme (m^* = high, t^* = moderate; fig. 8.17), whereas antlers, bosses, and blades are more likely to be observed at less extreme values of I_{mates} (m^* = moderate, t^* = low; fig. 8.17). Weapons are expected to arise or become elaborated in males in the same contexts as increases in male body size. Thus, we expect to observe weapons in highest frequency among cursorial species, in species which breed in cavities, and in species in which males defend fixed sites where receptive females aggregate (fig. 8.17).

Displays

Symbolic communication is thought to evolve when inexpensive information transmitted at a distance (e.g., movements, colors, or sounds) serves to prevent more expensive interactions between individuals in the signal's absence (e.g., searching, fighting, or mating attempts; Rohwer and Rohwer 1978; Moynihan 1982; Krebs and Dawkins 1984). Signals can be modified by natural or by sexual selection (Ryan 1998). Thus, individuals in any monogamous or polygamous species may exchange displays that communicate the locations of food, nests, offspring, or enemies. However, the greater the sex difference in the opportunity for selection, the more exaggerated the signals broadcast in the context of mate acquisition signals can become, and the more exaggerated the response of receivers can be (Ryan 1998).

Signals evolving under natural selection appear to convey information honestly. That is, the sender of the signal accurately communicates its motivation to fight, cooperate, or mate. In turn, the receiver of the signal either avoids or engages the sender appropriately. Such "truth in advertising" (Kodric-Brown and Brown 1984) is maintained when sender and receiver both benefit from interacting in the least expensive and most mutually beneficial way. Thus, "bluffing" is prevented by selection favoring signal receivers who routinely probe signal senders (Steger and Caldwell 1983). If threats do not correspond with physical strength, or if appeasement displays do not correlate with cooperation, these signals will soon be ignored by receivers and eliminated from sender repertoires (see chapter 7).

Such reasoning can be misleading when applied to sexual signals (chapter 7). A commonly held notion, attributable to Zahavi (1975, 1977; Zahavi and Zahavi 1997) is that extreme male signals *must* reliably communicate exceptional male quality. Production of extreme signals is presumably easier for males in good physical condition; thus, according to Zahavi's (1975) hypothesis, extreme signals must communicate reliable information about male quality to females. Parental investment theory generates similar predictions. According to this hypothesis, females are expected to be more choosy than males when selecting mates due to their greater initial investment in offspring (Trivers 1972). Since females show measurable preferences for particular male signals, parental investment theory requires that these signals re-

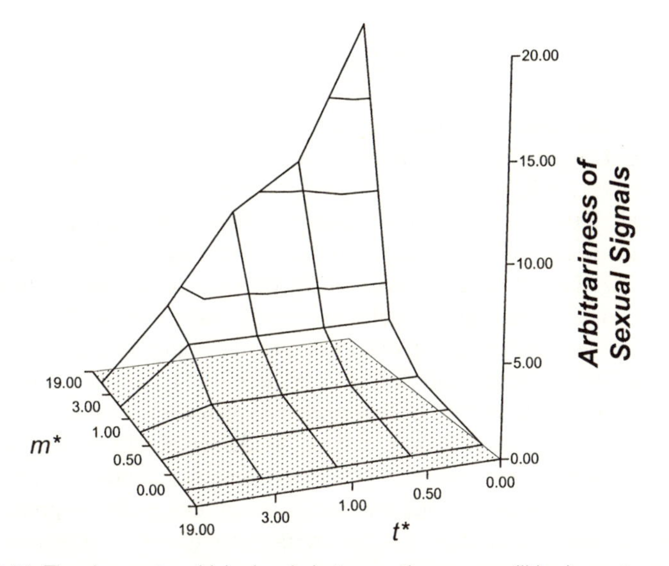

Figure 8.18. The degree to which signals between the sexes will be honest or arbitrary is expected to follow the ΔI surface; the smaller the sex difference in the opportunity for selection, the more honest signals are expected to be; the greater the sex difference in the opportunity for selection, the more arbitrary signals are expected to be. Note: to enhance visualization of this surface, this diagram is rotated 90° counterclockwise relative to the other diagrams in this chapter.

flect genuine fitness gains that females acquire by mate selection. Both of these hypotheses assume that viability selection is the primary source of selection on male signals as well as female responses. However, as explained above (chapters 1, 5, and 7), if the effects of viability selection on signals are minute compared to the effects of sexual selection acting on senders as well as on receivers, the link between signal honesty and receiver behavior becomes unnecessary.

We have shown that, compared to the intensity of runaway sexual selection, viability selection on females associated with mate choice is often weaker (chapters 1, 5, and 7). We therefore expect the intensity of selection on visual, auditory, and chemical displays to follow the ΔI surface (fig. 8.18), being less exaggerated and communicating more reliable information when the sex difference in the opportunity for selection is small, and becoming increasingly exaggerated and arbitrary when ΔI is large. Unlike signals under viability selection, we also expect extreme male displays to be arbitrary with respect to male genetic quality. Such signals will persist in male populations only because they attract female attention and thereby enhance male mating success (Christy 1995).

Male Parental Care

In most animal species, females provide exclusive care for developing eggs and young. Less often, both parents share in these duties (reviews in Ridley 1978; Trivers 1985; Clutton-Brock and Parker 1992; Alcock 1995; Drickamer et al. 2002). Taxa in which males provide primary or exclusive parental care are rare, including only seven families (14%) of amphibians, 40 families (7%) of fish, three families (4%) of hemipteran insects, three species (< 1%) of nereid polychaete worms, a single species (< 0.1%) of opilionid spider, and in all of the Pycnogonida (King 1973; Ridley 1978; Gross and Shine 1981; Thornhill and Alcock 1983; Arnaud and Bamber 1987; Sella 1990; Bain 1992). The current explanation for the rarity of exclusive male care is anisogamy; that is, fundamental differences in initial parental investment prevent males from becoming parental (Trivers 1972; Clutton-Brock and Parker 1992; chapter 7). According to this hypothesis, not only are females, with their greater initial investment in offspring, more inclined to provide care, but the small per gamete investment in offspring by males is expected to predispose males to abandon parental responsibilities more readily than females (Trivers 1972; Thornhill and Alcock 1983). Patterns of male parental care have been linked to resource distributions (Emlen and Oring 1977; Clutton-Brock and Parker 1992), modes of fertilization, and levels of paternity confidence (Gross and Shine 1981; Scott 1990; Whittingham et al. 1992; Schwagmeyer et al. 1999).

Contrary to hypotheses invoking differences in initial parental investment, we expect male parental care to evolve when the indirect fitness to males who care for young exceeds the direct fitness of males who rove and provide no parental care (chapter 7; Wade and Shuster 2002). Within the constraints on I_{mates} generated by species-specific values of m^* and t^*, such circumstances will be much more common than predicted by parental investment theory (see below). We predict that male parental care will occur most commonly in two types of situations: (1) When receptive females are dispersed in space and appear synchronously in time (m^* = low, t^* = high; fig. 8.19), or (2) when receptive females are clumped in space and dispersed in time (m^* = high, t^* = low; fig. 8.19).

In situation 1 (m^* = low, t^* = high; fig. 8.19) I_{mates} is near zero. Males are likely to have few opportunities to mate more than once, and thus are likely to have sufficient residual reproductive energy to enhance their fitness by providing parental care to their young. This hypothesis differs from the model proposed by Emlen and Oring (1977) which emphasizes resource availability. These authors argued that males must assist females in acquiring resources to successfully rear young; thus monogamy evolves when resources are highly dispersed. This may be true. However, an alternative ex-

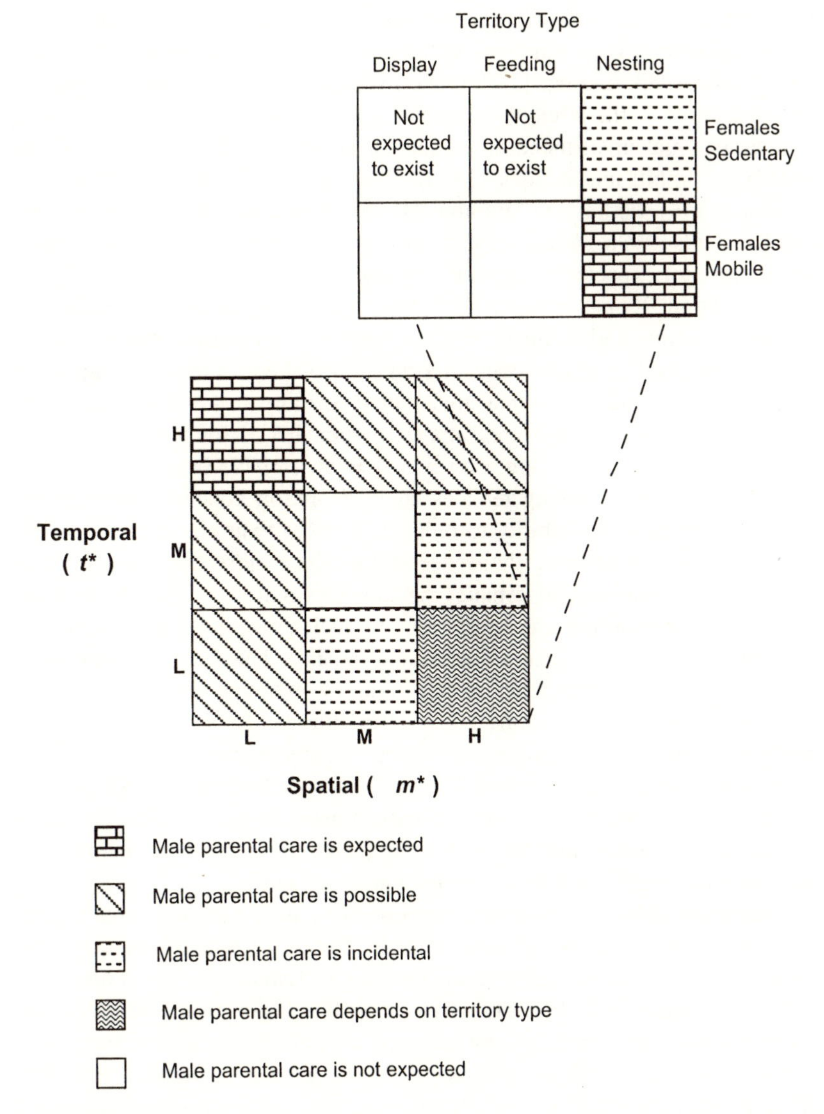

Figure 8.19. The spatial and temporal distributions of females favoring the evolution of male parental care.

planation is that male parental care evolves *secondarily* rather than as a primary response to resource distributions. We suggest that when resources are dispersed, so are females. Males must therefore attend individual females to successfully mate, and since they are unlikely to locate additional mates, males increment their fitness by caring for young. This hypothesis is consistent with Bateman's (1948) argument that the source of sexual selection lies in the variance in mate numbers within each sex (see also Wade 1979; Wade and Arnold 1980; and eq. [2.7] above). However, this hypothesis is *not* expected to hold if the strength of sexual selection depends on differences in relative parental investment, as argued by Trivers (1972), Emlen and Oring (1977), Thornhill and Alcock (1983), Bradbury (1985), Clutton-Brock and Vincent (1991), Parker and Simmons (1996), Kvarnemo and Ahnesjö (1996), and Reynolds (1996). Males in such species, with their smaller investment in sperm, are not expected to become parental.

In situation 2 (m^* = high, t^* = low; fig. 8.19), receptive females are dispersed in time and clumped in space. Such circumstances permit males to become attached to a particular site and, once there, attempt to attract or defend multiple mates. Parental males may be more successful in attracting multiple mates than nonparental males, particularly if females can oviposit and leave nesting sites (Downhower and Brown 1979; Oring et al. 1994). This latter tendency among females could evolve in concert with male parental care because the ability to oviposit in more than one nest is possible *only* if females mate more than once and produce more than one clutch of offspring. Alternatively, if females remain with males and are defended in harems, the only option non-harem-holding males may have to mate may be by contributing parental care to individual females. Thus, parental care could represent an alternative male mating strategy evolving in response to an extreme "skew" (cf. Johnstone 2000) in the number of successfully mating males. Or, stated more quantitatively, as the number of mating males, $1 - p_0$, approaches zero, average harem size, H, becomes large (eq. [2.8]; see also chapter 9). The degree to which female fitness decreases with increased harem size will determine the relative frequency of single-pair matings with male parental care given this combination of m^* and t^*.

Conditions favoring monandry can also favor male parental care. That is, when (3) extreme temporal synchrony in female receptivity occurs, combined with moderate to high spatial clumping of females (m^* = moderate to high, t^* = high; fig 8.19), or (4) extreme spatial dispersal of females occurs, combined with moderate to high temporal asynchrony in female receptivity (m^* = low, t^* = moderate to low; fig. 8.19). Prolonged female receptivity in each situation could serve to reinforce male tendencies to defend their mates, and thus enhance male tendencies to become parental. However, as we explained earlier, prolonged female receptivity is only expected to evolve if covariance arises between male and female tendencies to mate more than once, or if females have a low probability of successful insemination due to their monopolization by guarding males.

Thus, prolonged female receptivity, moderate spatial clumping of receptive females and moderately asynchronous sexual receptivity may allow males who seek extra-pair copulations to invade parental male populations as an alternative male mating strategy. The success of such invaders will depend on the degree to which female receptivity is asynchronous or the degree to which female receptivity permits multiple mating. Depending on the life history characteristics of the species, as well as local variation in the spatial and temporal availability of sexually receptive females, both male parental care and multiple mating attempts by males could persist in the same population, either as discrete strategies among individuals or as more environmentally contingent mating tactics used by individual males (m^* = moderate, t^* = high; m^* = low, t^* = moderate; fig. 8.19; chapter 11).

Male parental care can also evolve incidentally under two additional sets of circumstances, although male care is not likely to be as well developed in these cases as it is in situations 1–4. In the first of these situations (5), females are highly clumped in space and moderately asynchronous in time (m^* = high, t^* = moderate; fig. 8.19). At such times, males are expected to guard small groups of females. By defending these groups, males are in a position to defend females and their young from predators as well as from males attempting to usurp control of the female group. However, the spatial proximity of females whose receptivities do not completely overlap provides males with multiple mating opportunities. When multiple breeding opportunities among males generates greater variance in offspring numbers among males than can be achieved by male parental care, sexual selection will favor males who emphasize mating effort over parental effort.

In situation 6, females are moderately clumped in space and highly asynchronous in time (m^* = moderate, t^* = low; fig. 8.19). At such times, as in situation 5, moderately aggregated females are likely to attract dominant, guarding males. Although female groups are expected to be more loosely associated in space than in 5, asynchrony in female receptivity provides incentive for males to guard these aggregations. As in situation 5, male defense of females and their offspring is incidental to male mating effort. However, such effort may still enhance male, female, and offspring fitness. Again, as in situation 5, if greater fitness variance is likely to arise among males from successful defense of female groups than from male parental care, sexual selection will favor males who emphasize mate-guarding effort over parental effort.

Male care is *unlikely* to evolve when receptive females are moderately dispersed in space and time (m^* = moderate, t^* = moderate; fig. 8.19). Males in such circumstances have the option to pursue and secure several mates in sequence, and so are unlikely to become parental. Male parental care is also unlikely to evolve when, although females are clumped in space and dispersed in time (m^* = high, t^* = low; fig. 8.19), males defend locations that are not used by females as nesting sites. Such locations include feeding territories (Alcock 1979) as well as mating territories in leks (Höglund and Alatalo 1995). In both situations, males are spatially disassociated from nest and young, and thus can enhance their offspring numbers only by successful matings.

Our fundamental hypothesis, then, is that male parental care and mate guarding coevolve, especially when males lack mating opportunities. Some empirical results, not in accord with the general predictions of parental investment theory, are consistent with our explanation (Reynolds 1987; Reynolds and Szekely 1997; Balshine-Earn and Earn 1998; Beck 1998; Tallamy 2001).

Sexual Reversal

Sex role reversal (Smith 1979) describes situations in which males become more parental and females become more aggressive and showy, yet both sexes retain their sexual identities in terms of gamete size and sexual physiology. We define *sexual reversal* as the condition in which male and female phenotypes become so completely reversed that "females" are designated males and vice versa. Sex role reversal is evidenced in a number of animal taxa (Gwynne 1997; Delehanty et al. 1998; Ens and Pixten 2000). Sexual reversal is likely to be difficult to identify but we predict it occurs because of the extreme degree of sex role reversal seen in groups like the sea spiders (Pycnogonida, Bain 1992) and the pipe fish (Syngnathidae, Berglund 1989). Species in which females are dispersed in space, as well as asynchronously distributed in time, in which males provide parental care to young at fixed nest sites, and in which the quality of male care decreases with increasing brood size, are most likely to lead to sex role and sexual reversal ($m^* =$ low, $t^* =$ low; fig. 8.20). I_{mates} in these species will be smaller than in any of the situations considered so far, because in these species, the number of progeny males can successfully rear restricts male mating opportunities.

On the other hand, I_{clutch} is likely to be larger than in any situation yet considered because females are freed of the responsibilities of parental care and thus can move among nesting males, thereby increasing their mate numbers as well as their numbers of clutches. As I_{mates} and I_{clutch} become equal in magnitude, the sex difference in the opportunity for selection will disappear. If this situation remains stable and I_{males} approximately equals $I_{females}$, mate choice criteria that accurately reflect male and female contributions to offspring numbers may evolve (i.e., indicators of mate "quality"). However, particularly in species with male parental care at fixed nesting sites, this situation is unlikely to remain stable because females are likely to have more opportunities for multiple mating than males. When this occurs, all nesting males are likely to obtain mates, but not all roving females may be as uniformly successful. When variance in mating success arises among females, the sex difference in the opportunity for selection will favor the reversal of

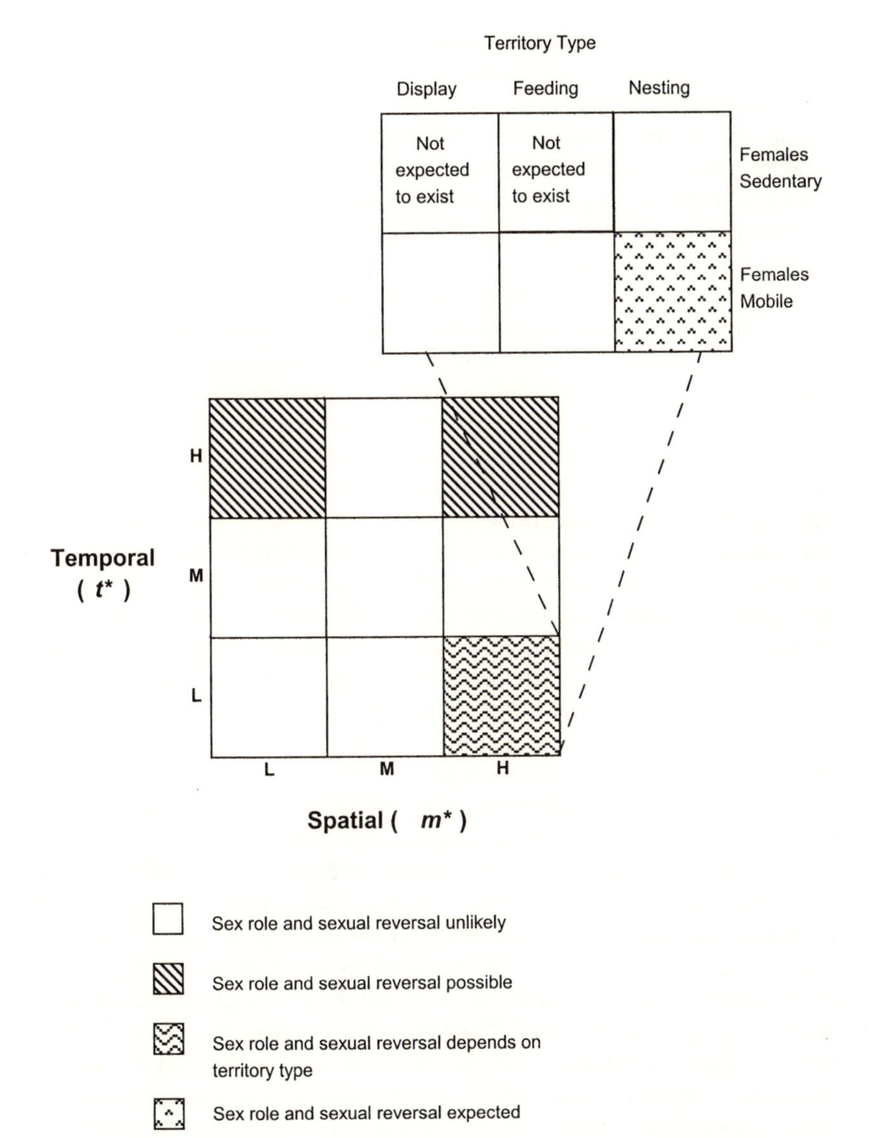

Figure 8.20. Sex role reversal is expected to evolve when male parental care and male mating success coincide, and female iteroparity allows females to seek multiple mates; sexual reversal is expected to occur when, after this initial state, ecological conditions shift such that males become incapable of rearing multiple broods; alternatively, sexual reversal is possible when there is no sex difference in the opportunity for selection and ecological conditions change, favoring increased mate numbers in ancestal females.

parental roles between the sexes and will favor characteristics in females such as aggressiveness, territoriality, and bright coloration that may correlate with increased mate numbers.

Once the process of sex role reversal begins, it may or may not lead to the complete phenotypic reversal of the sexes. Females may become increasingly male like in phenotype, expressing nonparental tendencies and gamete size, and males may become increasingly femalelike in these same characteristics. However, this process will be arrested if ecological factors limit female mate numbers or allow males to secure multiple mates either within one or over several breeding seasons. In such cases, "sex role reversed" species will be observed (cf, Smith 1979; 1984; Gwynne 1984; Delehanty et al. 1998). However, if female mating opportunities do not become immediately limited and if male mate numbers become increasingly restricted, the sexes are expected to *completely reverse* in phenotype. The rate which this process takes place may be so rapid, particularly if it is driven by some runaway process, that transitional species, that is, species in the *process* of sex role reversal, will be rare. Indeed, this appears to be the case based on reviews of existing mating systems (Clutton-Brock and Vincent 1991). A possible clue to which species have undergone this transformation may be found in interspecific patterns of chromosomal sex determination. Examples of female heterogamety are expected to be disproportionately higher in taxa in which polygamy associated with nesting sites as well as in which polyandrogyny occur. The diversity of sperm morphologies exhibited across taxa could also provide clues to when and in which lineages sexual reversal has occurred.

Additional cases of sexual reversal are possible in circumstances in which the sex difference in the opportunity for selection approaches zero, as in species with low mate number promiscuity and/or with biparental care. As we explained in chapter 5, such situations are unstable because slight changes in ecological conditions could allow variance in mate numbers to arise in one or the other sex. If opportunities for multiple mating remain even slightly higher for males, no morphological changes in sexual phenotype are expected. However, if opportunities for multiple mating become slightly higher for females, perhaps due to nest failure and greater female mortality (L. W. Oring, pers. com.), sexual phenotypes may begin to reverse. Such cases include species in which females are aggregated in space and synchronously distributed in time, as well as those in which females are dispersed in space and synchronously distributed in time (m^* = high or low, t^* = high; fig. 8.20). In both situations, male parental care is likely to already exist.

Alternative Mating Strategies

Polymorphisms in male reproductive behavior and/or morphology are thought to evolve when sexual selection is intense (Gadgil 1972). Recent reviews support this expectation (Lank et al. 1995; Gross 1996; Sinervo and Lively 1996; Shuster and Wade 1991a,b; Shuster and Sassaman, 1997; Shuster et al. 2001). As shown in chapter 1, sexual selection arises when some males are disproportionately successful in mating. When this is true, the among-male component of mating success increases and I_{mates} becomes large. As Darwin observed (1874), and as we later demonstrate (chapter 11) when few males control access to most receptive females, large male harems are equivalent to subpopulations with female-biased sex ratios. Moreover, as the number of mating males decreases, the more female-biased local sex ratios become, and the more easily a mutant male strategy, i.e., one that uses unconventional means for acquiring mates, can invade the conventional male population.

Since the values of m^*, t^*, and $I_{cs,sires}$ may each contribute to ΔI in different ways, our method identifies at least three contexts in which male alternative mating strategies may evolve. Diversity in male mating strategies is likely to arise when sexually receptive females are (1) aggregated in space (m^* is large), (2) dispersed in time (t^* is small to moderate), or (3) have prolonged periods of sexual receptivity ($I_{cs,sires}$ is large; fig. 8.21), as well as when these three factors occur in combination.

Two additional factors that may influence the invasion of alternative male strategies include (4) female fecundity and (5) the location of fertilization. As female fecundity increases, unfertilized ova become increasingly spatially clumped and are often synchronous in their "receptivity." Unless male mate guarding prevents multiple insemination, increasing female fecundity increases opportunities for different males to sire offspring with a single female and so can favor alternative mating strategies among males. The effect of mass release of unfertilized ova can convert such mating systems from those in which individual females are spatially dispersed and asynchronous in their sexual receptivity (m^* = low; t^* = low; I_{mates} = low) to those in which individual fertilizable ova are spatially clumped and moderately synchronous in their availability (m^* = high; t^* = moderate; I_{mates} = moderate to high). Similarly, if greater opportunities exist for satellite males to steal fertilizations at the time of spawning, species with external fertilization may be more likely to evolve alternative mating strategies than those with internal fertilization for a given spatial and temporal distribution of receptive females. These issues have been briefly considered within each of the sections described above, and will be considered in greater detail in chapters 10, 11, and 12.

Chapter Summary

We provide a theoretical framework, consistent with that in preceding chapters, for understanding behavioral influences on the sex difference in the opportunity for selection. For example, prolonged receptivity, like syn-

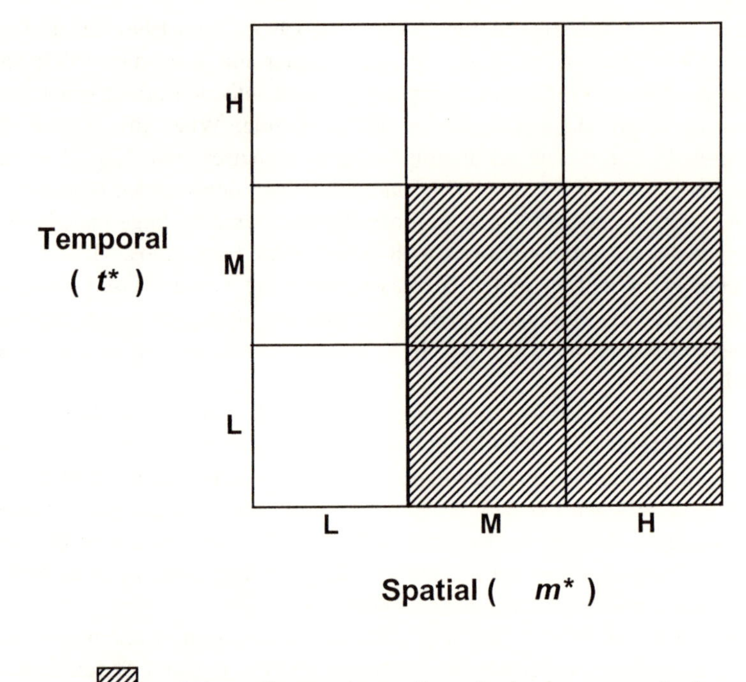

Figure 8.21. Alternative male mating strategies are expected to evolve whenever some males mate more than once and other males are prevented from mating altogether.

chronous receptivity, diminishes I_{mates} by increasing opportunities for satellite males to mate. We expect the duration of female receptivity to increase when females in harems experience decreased insemination success. Further erosion of I_{mates} may occur if increasing harem size decreases average female fecundity. We suggest, therefore, that harem polygyny is self-limiting.

Contrary to most current thought on this issue, multiple mating need not increase the intensity of sexual selection. Instead, multiple mating *decreases* the intensity of sexual selection in all cases except those in which positive covariance exists between male mate-number promiscuity and male fertilization success among females. Only when the above covariance exists will males with the most competitive sperm sire more offspring than those with less competitive sperm. On average, the mates of these males will also be more promiscuous. Thus, not only will promiscuous males leave more offspring, but their sons will have more competitive sperm and their daughters will be more promiscuous. This runaway process arises more easily than others discussed earlier, and it explains a diverse array of sperm phenotypes and mating strategies now presumed to arise by sperm competition alone.

We introduce a method for visualizing the possible distributions of females in space and time using a modification of the diagram we introduced in chapter 6 to describe the ΔI surface. The nine squares of this diagram show the extremes of female spatiotemporal distributions and we also illustrate how a single square can harbor a variety of mating phenotypes. We extend the discussion of male mate guarding begun in chapter 4, noting that when males mate with more than one female, their overall fertilization success equals the harmonic mean in mate-number promiscuity of their mates. This relationship limits the behavior of males because the fitness of roving males must exceed that of guarding males if they are to invade a population. A surprising prediction of this hypothesis is that mate guarding by males in some form will be favored for nearly all spatial and temporal distributions of females. Mate guarding diminishes female tendencies to mate more than once and the effect on male fitness is asymmetric. That is, more is often lost by increasing female mate-number promiscuity than is gained by increasing male harem size. Mate guarding and sperm competition are thus similar in reducing the effects of multiple mating on the sex difference in the opportunity for selection.

Female tendencies to copy the behavior of other females can enhance the opportunity for sexual selection. However, male mate guarding restricts female opportunities to engage in mate copying for the majority of the possible distributions of females in space and time. Runaway processes involving female copying and male guarding behavior can favor large harems of females that are defended by large and pugnacious males, or may lead to explosive breeding aggregations. Alternatively, genetic covariance between female copying behavior and male tendencies to display in groups can lead to runaway sexual selection for leks or breeding hot spots. When female copying increases the variance in mate numbers among mating males, I_{mates} will increase beyond the value measurable from the spatial and temporal distributions of receptive females. However, if female aggregations also attract aggregations of males, the number of mating males may increase and the value of I_{mates} will decrease.

Instead of directly defending females, males may defend feeding sites, nesting sites or display sites, on leks. When males defend food sources, males will remain at these sites and females will become mobile. Nest site defense will depend on the degree to which nest control influences male mate numbers. When male mating effort and male parental effort both enhance offspring number for males, the opportunity for sexual selection on males can become extreme. In locations where females predictably appear, males position themselves and attempt to attract the attention of transient females by visual, auditory, or chemical displays. In general, when males have few options for multiple mating or when a large fraction of males are excluded from mating, monandry will arise and persist. Despite suggestions from recent avian literature to the contrary, such circumstances are surprisingly widespread.

We have defined sexual conflict as a sex difference in the covariance between offspring numbers and mate number promiscuity (chapter 4). Thus, sexual conflict will be greatest when sexual selection is strong, and nonexistent when there is no sex difference in the opportunity for selection. In either case, evidence of sexual conflict may be inconspicuous, either because fitness variance in one sex is so small that an evolutionary response to exploitation is impossible, or because no response is expected when no sex difference in fitness exists. As with other aspects of mating system evolution, the outcome of distinct selective processes may produce similar behavioral results. Conflict resolution, at a populational level, is the result neither of differences in relative parental investment nor of individual cost-benefit analyses. If the average effect of male monopolization imposes costs on average female fecundity, exploiting males must mate with multiple females to overcome the cost to their own fertility. This prediction suggests that sexual exploitation is a self-limiting process rather than an accelerating "arms race."

Sexual selection by female choice can be stronger than that by male-male combat and its potential results are more extreme. If female preferences exist, for traits used by males as weapons, then the intensity of selection on them due to preference can exceed that due to combat. We expect runaway selection to modify these preferred male combat traits rapidly so that their usefulness in combat will be diminished relative to their utility as displays.

We identify four general forms of sexual dimorphism, attachment organs, body size differences, weaponry, and displays. We expect male attachment structures and male guarding of individual females to evolve in concert. When females are synchronous in their receptivity, male competition is relaxed, and if attachment organs arise, regardless of female spatial distributions, they will evolve by natural selection. When females are spatially dispersed, male attachment organs will evolve by sexual selection, particularly as female receptivities become asynchronous. When males guard females, large male body size as well as territoriality and aggressiveness will evolve. When roving behavior or satellite behavior by males is favored, we expect males to be comparatively small and secretive. Extreme female fecundity and large female body size will be favored when juvenile survivorship is low, and if relatively few females breed successfully due to preadult mortality, the opportunity for selection on large female body size can become intense because few females contribute to each generation. If I_{mates} as well as I_{clutch} favor increased body size, sexual dimorphism will not be obvious because large size is favored in both sexes. However, if fecundity selection favors large size in females, and sexual selection favors small size among searching males, sexual size dimorphism can become extreme. When no sex difference in the opportunity for selection exists, the phenotypes of each sex will be most similar.

Sexually dimorphic aggressiveness and weaponry will evolve when fe-

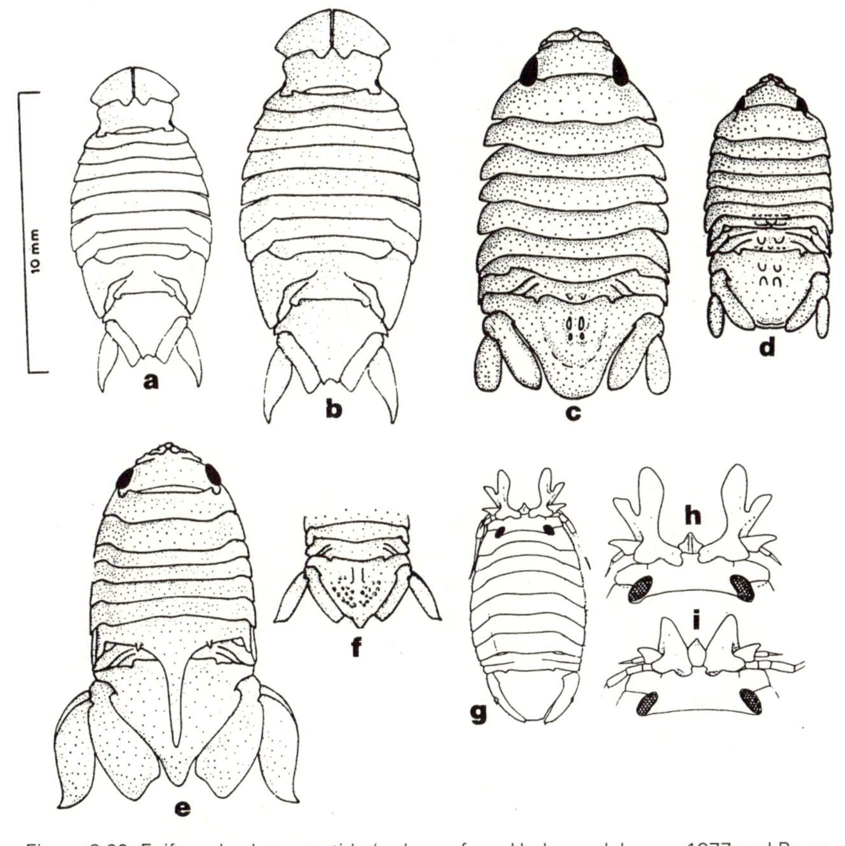

Figure 8.22. Epifaunal sphaeromatids (redrawn from Hurley and Jansen 1977 and Bruce 1994). *Amphoroidea falcifer* males (a) are smaller than females (b) which live dispersed among algae. *Parasphaeroma campbellensis* males (c) are larger than females (d); both sexes aggregate under stones near streams. *Isocladus armatus* males (e) are larger than females (f), and possess a pleonal horn and flattened uropods; both sexes are collected on hard substrates. *Paracassidina cervina* males (g, h) are slightly larger than females (i) and possess enlarged antennal peduncles; congeners inhabit mud and sand substrates; all epifaunal females possess unmetamorphosed mouthparts, i.e., are iteroparous (Harrison 1984; Shuster 1991a).

males become spatially clumped and receptive asynchronously. We expect to observe weapons in highest frequency among cursorial species, in species which breed in cavities, and in species in which males defend fixed sites where receptive females aggregate. Individuals in any species may exchange displays that communicate the locations of food, nests, offspring, or enemies. However, the greater the sex difference in the opportunity for selection, the more exaggerated signals broadcast in the context of mate acquisition signals can become, and the more exaggerated the response of receivers can be. We

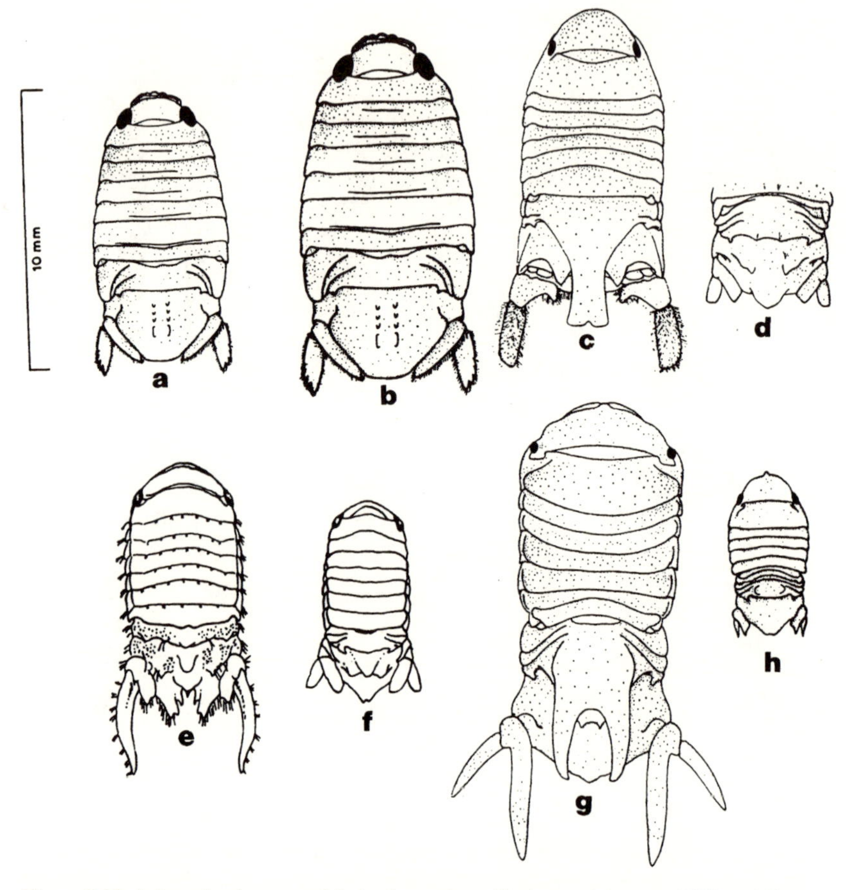

Figure 8.23. Infaunal sphaeromatids (redrawn from Hurley and Jansen 1977 and Brusca 1980). *Sphaeroma quoyanum* males (a) are smaller than females (b) and associate with individual females in mud burrows. *Cilicaea dolorosa* males (c) are larger than females (d) and possess a pleonal horn with elongated, movable uropod exopods; breeding aggregations are found under stones and in algal holdfasts. *Paracerceis sculpta* males (e) are larger than females (f) and possess rugose pleotelsons and elongated uropod exopods; breeding aggregations are found in sponges and tunicates. *Cymodopsis impudica* males (g) are larger than females (h) and possess elongated uropod expods and endopods, as well as bifurcate pleonal horns; breeding aggregations are found within burrows in coral rubble; all infaunal females except *Sphaeroma* possess metamorphosed mouthparts, i.e., are semelparous (Harrison 1984; Shuster 1991a).

therefore expect the intensity of selection on visual, auditory, and chemical displays to follow the ΔI surface, becoming increasingly exaggerated and arbitrary as ΔI becomes large.

We describe conditions in which males may provide care to offspring. Our fundamental hypothesis is that male parental care evolves when males lack mating opportunities and thus enhance their fitness by caring for young, or when mating opportunities for males coincide with male parental care. Empirical results not in accord with the general predictions of parental investment theory are consistent with these explanations. In sex role reversal, males become parental and females become aggressive and showy, yet both sexes retain their sexual identities in terms of gamete size and sexual physiology. In sexual reversal, male and female phenotypes completely reverse. Species in which males provide parental care to young at fixed nest sites, and in which the quality of male care decreases with increasing brood size, are most likely to lead to exhibit these traits. Alternative mating strategies will evolve when variance in fitness exists within one sex. Such variance usually exists in males, but can arise in females as well. In males, diverse mating strategies may arise, particularly in species with high fecundity and/or external fertilization, when females are aggregated in space, dispersed in time, have prolonged periods sexual receptivity or when these three factors occur in combination.

We expect the evolutionary outcome of the diverse selection pressures described above to cause marked divergence in male and female phenotypes among closely related taxa. Examples of such divergence are likely to be widespread, but can be readily seen among isopod crustaceans in the family Sphaeromatidae (figs. 8.22 and 8.23).

9 A Classification of Mating Systems

There are two ways of describing mating systems in the current literature, each of which has different implications for the sex difference in the strength of selection and for runaway sexual selection. In behavioral ecology, mating systems are usually described in terms of the *numbers of mates* per male or per female. For example, monogamy, polygyny, and polyandry are all descriptions of mate numbers per male or per female. In chapter 6 we modified this scheme to more accurately include male and female mate numbers, and as we saw in chapters 1, 2, and 3, these relationships are fundamental to the sex difference in the strength of selection. A second way to describe mating systems is in terms of the *genetic relationships* between mating males and females. For example, random mating, negative assortative mating (outbreeding), and positive assortative mating (inbreeding), are all descriptions of mating systems in terms of the genetic correlation across mating pairs. This type of description is most often found in literature concerning plants, or in discussions of parasite resistance. A combination of these two approaches captures the covariances between male and female promiscuity, between male guarding and female tendency to aggregate, or between male sperm plugs and female detergents. This description of a mating system determines the nature as well as the rates of the "runaway" processes that may derive from it. We believe that a great deal more of the phenotypic variance in sex dimorphism that exists within and among species can be explained if these two classification systems are combined.

In this chapter, we will expand the approach we have introduced in previous chapters to develop a classification scheme for mating systems. We will discuss how specific combinations of the spatial and temporal crowding of receptive females, as well as how female life history, male and female reproductive behavior, and runaway processes in various forms, may interact to influence the sex difference in the opportunity for selection, ΔI. To facilitate discussion, we will illustrate combinations of m^* and t^*, with the diagram shown in fig. 9.1, where m^* varies along the horizontal axis and t^* varies along the vertical axis. This diagram is based on the diagram introduced in chapter 8, wherein each axis is divided into sections, representing low, intermediate, and high values for the parameters m^* and t^*, so the diagram is divided into nine squares. Thus, each highlighted square on the floor of the diagram represents a unique combination of m^* and t^* that corresponds to a different mating system. However, within each square in our diagram, variations in female mating behaviors (monandry or polyandry) as well as in female life history (semelparity or iteroparity) introduce minor

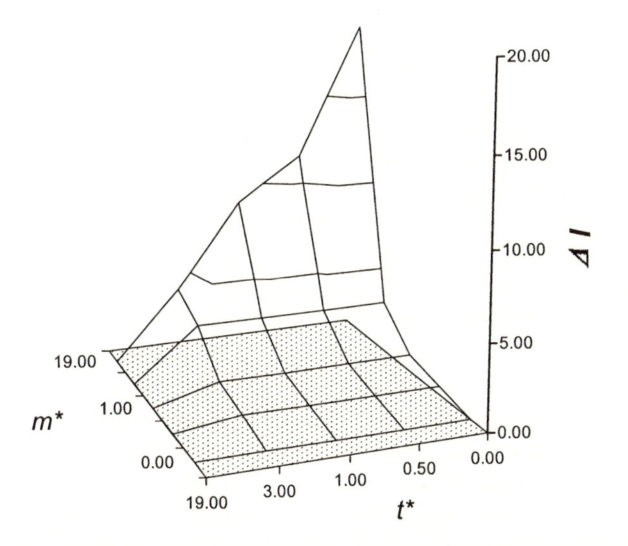

Figure 9.1. The ΔI surface. As explained in chapter 6, this surface represents the adjusted sex difference in the opportunity for selection, I_{mates}, arising from the spatial and temporal distributions of matings; the spatial distribution of matings is measured by the mean spatial crowding of receptive females, m^*; the temporal distribution of matings is measured by the mean temporal crowding of receptive females, t^*; the value of I_{mates} is then reduced by the opportunity for selection arising from the effects of multiple mating by females ($I_{cs,sires}$), the opportunity for selection arising from the effects of multiple reproductive episodes by females ($I_{cs,clutch}$). The surface is further adjusted by the opportunity for selection arising from the effects of sperm competition among male ejaculates as well as the effects of female copying behavior. Note that this figure is rotated 90° counterclockwise relative to the figures in chapter 8 to show the contour of the ΔI surface.

deviations from the overall category mean mating system. Thus, we will add the three-dimensional surface described by ΔI to provide a means for visualizing the approximate sex difference in the opportunity for selection as it incorporates these effects. We will show that the major mating system categories are robust to these variations, but also that the variations are of some interest in their own right. We discuss them to show that some mating systems are closely related to others (i.e., within-square rather than between-square variants) and to show why some minor variants and not others can initiate transitions among the major categories. Runaway processes between male and female traits can lead to rapid transition from one mating system (or square) to another, but we will argue that some transitions are more likely to occur than others.

We will illustrate each mating system as well as its minor variants using examples from natural populations. However, because quantitative estimates of m^* and t^* are not known for most species, and because estimates of the mate number and brood number distributions of females, as well as quantitative estimates of levels of sperm competition and female copying are largely unavailable, we do not expect our assignments of different species to be correct in all instances. Empirical studies should help us refine (or verify) our assignments for particular species, and phylogenetic studies can test our predictions regarding the likelihood of transitions among the major categories.

In our approach, we take the parameters that measure the spatial and temporal distributions of female receptivity as the starting point. We then ask two questions: First, how do these female attributes affect the sex difference in strength of selection? Second, what degree of sex dimorphism should we expect as a consequence? Male and female traits evolve in response to the evolutionary process of sexual selection and its strength relative to natural selection. Male and female traits coevolve because nonrandom patterns of mating success and paternity result in a genetic correlation between them. Weak, direct selection on a trait in females can be enhanced or opposed by strong, indirect selection, derived from the genetic correlation between the female trait and a male trait under strong sexual selection. Reciprocally, the value of some traits in either sex can permit, limit, or prohibit some forms of sexual selection. For example, monandry prohibits sperm competition while female mate-number promiscuity permits it. Similarly, male mate guarding diminishes female mate-number promiscuity, and thus limits sperm competition. Although male and female trait values can signal the underlying selective forces, as we have stated above, our approach is to define mating systems in terms of their *causal processes*, i.e., the selective forces that produced them, rather than as *outcomes*, i.e., observable male and female trait values.

In previous chapters, we have outlined an approach for quantifying the selective contexts in which mating systems evolve. We will now combine all possible extreme and intermediate values of the elements of our model to predict the forms mating systems are likely to assume in nature. We will show that our measures of the opportunity for selection, arising from the spatial and temporal patchiness of females, from differences in mate numbers and breeding episodes among females, as well as from behavioral adjustments by males and females in response to the above conditions, can explain the diversity of animal mating systems now recognized in existing classifications (Emlen and Oring 1977; Thornhill and Alcock 1983; Reynolds 1996). We will also show that our classification system predicts additional, now unrecognized mating systems when the effects of covariances between male and female mating phenotypes are considered.

The variety of possible runaway processes extends beyond the classic Fisherian one between male traits and female preferences. Thus, our approach helps to resolve several current paradoxes in mating system research. These include the diversity of mating systems in which males and females form "monogamous" pairs (Wachtmeister 2001; Cezilly et al. 2000; Bull 2000; Reid 1999; Reavis and Barlow 1998; Komers and Brotherton 1997;

Wickler and Seibt 1981), the high frequency of multiple mating by females in apparently monogamous mating systems (Cooper et al. 1997; reviews in Birkhead and Møller 1998; Ligon 1999; Taylor et al. 2000), sexual dimorphism arising and persisting in pair-forming species (Darwin 1874), the "lek paradox" (Borgia 1979; Reynolds 1996; Kotiaho et al. 2001), differences between "classic" and "exploded" leks (Höglund and Alatalo 1995), the relative rarity of polyandry (Thornhill and Alcock 1983; Ligon 1999; Clutton-Brock and Vincent 1991), as well as the taxonomic distribution of male parental care, sexual conflict, and sex role reversal (Smith 1979, 1984; Gross and Shine 1981; Clutton-Brock and Vincent 1991; Choe and Crespi 1997; Eens and Pinxten 1995, 2000).

Interspecific Comparisons

We will begin by returning to the graphical model introduced in chapter 6 and adding to it the conditions explained in the last chapter (fig. 9.1). Recall that for a given species, within- and between-breeding-season variation in the value of ΔI generates a cloud of points summarized by the ΔI surface. This surface arises by combining values of I_{mates}, determined from the spatial and temporal distributions of females, with the effects of multiple mating and multiparity on $I_{females}$ (see fig. 6.4). If females in related species share life history characters that lead to similar patterns of resource use or to similar temporal patterns of sexual receptivity, related taxa are expected to show greater overlap in their values of ΔI than unrelated taxa (fig. 6.3). Differences in the degree of sexual dimorphism observed among species can be compared against these surfaces of ΔI to determine whether the observed degree of sexual dimorphism is coincident with the current patterns of selection. We showed in the last chapter that, with few exceptions (see below), the larger the value of ΔI, the greater the expected sexual dimorphism. Although ΔI represents the current strength of selection, we assume that some continuity in the strength and direction of selection over time has been necessary to produce the existing degree of sexual dimorphism.

The spatial and temporal distributions of receptive females affect the distribution of mate numbers among males. However, the distribution of paternity is modified by the abilities of males to engage in sperm competition, to search for or guard females, to provide parental care to young, to exploit or cooperate with their mates, or to change mating strategy. Although we expect related species to have similar mating systems, we also expect mating system divergence among such species to be predictable. For example, we expect a derived tendency for females to aggregate spatially to be associated with males who defend female aggregations (fig. 9.2). Compared to more ancestral populations, these males will experience more intense sexual selection (i.e., a larger ΔI) and are expected to become more sexually dimorphic. Such species are expected to exhibit distinct alternative mating strategies,

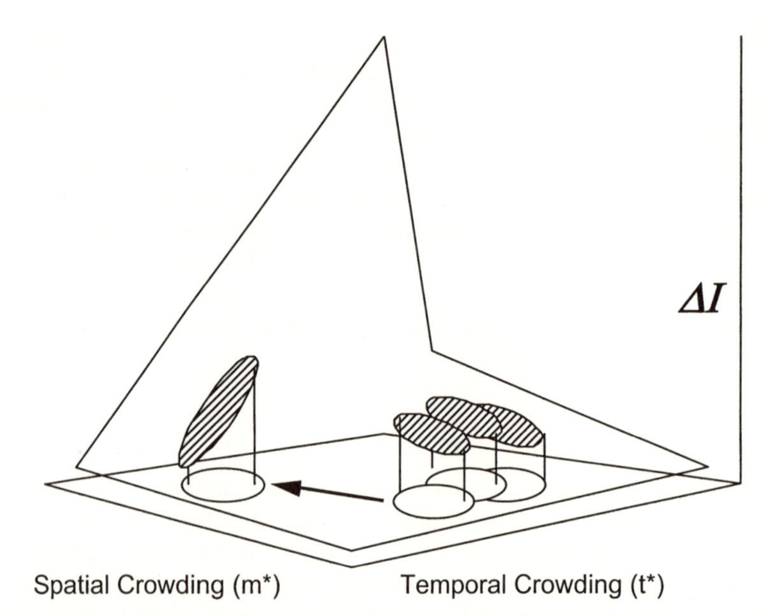

Figure 9.2. Related species tend to show similar mating systems, but divergent species within larger taxa should be predictably divergent. A derived tendency for females to aggregate spatially is expected to be associated with males that defend females; if females are temporally aggregated as well, males are expected to defend individual females, with consequent reductions in the adjusted sex difference in the opportunity for selection, ΔI.

and multiple paternity, especially within the largest harems, is likely to be common. Similarly, a derived tendency for females to synchronize their reproduction is expected to weaken sexual selection (i.e., produce a smaller ΔI), and be associated with diminished tendencies among males to engage in mate searching behavior, and enhanced tendencies to guard individual females. Genetic covariances between male and female characters and their attendant runaway processes are expected to displace some species from their position on the ΔI surface relative to other taxonomically related species (chapter 4). We believe that, by identifying the components of ΔI and the sources of such genetic correlations, causal effects on mating systems evolution can be more clearly identified.

The Combined Effects of Female Life History and the Ecology and Phenology of Female Receptivity on Mating System Evolution

To estimate I_{mates}, our model summarizes the spatial and temporal distributions of sexually receptive females, and thus measures the opportunity for sexual selection on males using two parameters, (A) the mean spatial crowding of receptive females, m^*, and (B) the mean temporal crowding of receptive females, t^* ($= m^*_{obs}$, chapter 3). Values of I_{mates} arising from each *pair* of m^* and t^* coordinates will be unique (fig. 9.1). We summarize the effects of life history characters on the opportunity for sexual selection using two additional parameters, (C) the opportunity for selection due to the effects of multiple mating by females, $I_{cs,sires}$, and (D) the opportunity for selection due the effects of multiple reproductive episodes by females, $I_{cs,clutch}$ (chapter 5). By assembling combinations of extreme and intermediate values of m^*, t^*, $I_{cs,sires}$, and $I_{cs,clutch}$ (tables 9.1–9.12) we identify 12 major mating system categories we consider likely to evolve. Each category appears robust to small variations in the parameters A–D, and runaway processes are the likely cause of transitions from one to another category.

From the possible values of these parameters, we predict specific details in behavior and morphology that allow each combination of traits to be classified within each major category. Our predictions include (1) the degree to which sperm competition may occur, (2) whether female mate copying is likely, and (3) the estimated magnitude of the adjusted opportunity for sexual selection, ΔI, that arises from factors A–D and 1 and 2. We also predict (4) the likely form of male-female associations at breeding sites, (5) the degree and form of sexual dimorphism visible between the sexes, (6) the tendency for males to provide parental care, (7) whether and how sexual conflict will arise between the sexes, and (8) whether as well as in what form alternative mating strategies are likely to exist. We next (9) assign each mating system to a descriptive category that summarizes each suite of male and female adaptations, and lastly, (10), we cite some of the existing literature on mating systems that may allow known mating systems to be classified under our scheme.

We illustrate our predictions using selected references with detailed information on the spatial and temporal distributions of females, as well as female mate numbers and reproductive episodes. In some cases, we extrapolate from descriptions of these characteristics when detailed information is unavailable; we will note such extrapolations as we make them. Our intent is to stimulate future researchers to collect the type of information that will allow specific tests of our model. In some cases, our terminology overlaps with existing ones, but it also classifies mating systems that have previously been difficult to categorize. We discuss the types of information necessary to test our model. We identify convergences wherein apparently different combinations nevertheless give rise to the same mating system. We also identify combinations that are unlikely to generate stable mating systems in nature. This part of our theory predicts which mating systems will arise and why, as well as which will not and why not.

The notion that mating systems will change in response to environmental conditions is not new. It is the foundation of the Emlen and Oring (1977) hypotheses and many have explicitly considered these effects. Davies (1983, 1985, 1989, 1991), in particular, has documented in exquisite detail the degree to which dunnock (*Prunella modularis*) mating systems shift as conditions affecting mating success and parental care change. Our goal in this chapter is not to restate this point, but rather to show how to quantify these shifts and identify those environmental changes that can lead to a major change in mating system. We believe that our method predicts and explains phylogenetic patterns in mating system diversity. We will also show that patterns of prezygotic and postzygotic investment in offspring can shift in response to changes in mating system. Thus, as we have argued, sex differences in initial parental investment, arise from, rather than lead to a sex difference in the variance in mate numbers. Once again, our method focuses on the selective forces that shape mating system evolution, rather than on the outcomes of these causal processes. We will demonstrate that these outcomes are neither necessary nor sufficient, to describe mating system diversity, or to understand its evolution.

A Classification of Mating Systems: Sedentary Pairs

Eumonogamy

When females are spatially dispersed, the mean spatial crowding of females, m^*, approaches zero (column A, row 1, table 9.1). If females are synchronized in their sexual receptivity, the mean temporal crowding of females, t^*, is large (column B, row 1, table 9.1), and I_{mates} is small. If females mate only once and produce only one clutch of offspring, both $I_{cs,sires}$ and $I_{cs,clutch}$ are zero (columns C and D, row 1, table 9.1). Without multiple mating by females, sperm competition is impossible (column 1, row 1, table 9.1), and spatially dispersed females are unlikely to have the opportunity to display copying behavior (column 2, row 1, table 9.1). With I_{mates} and I_{clutch} near zero, ΔI will also nearly equal zero. Here, there will be no net sex difference in the opportunity for selection (M = F; column 3, row 1, table 9.1; fig. 9.3).

In such circumstances, males are expected to associate with and defend individual females at or before the onset of receptivity (Gwynne 1984; Jormalainen 1998; column 4, row 1, table 9.1). Because receptive females are spatially dispersed and temporally synchronous, most males are likely to locate mates. Also, since ΔI is negligible, the sexes are expected to show little if any dimorphism, unless it is relictual from a recent ancestor (column 5, row 1, table 9.1). Since males may enhance their fitness by caring for young, male parental care is expected to occur (column 6, row 1, table 9.1). Moreover, because male and female mate numbers are similar, male and female reproductive interests coincide. Thus, there is little opportunity for

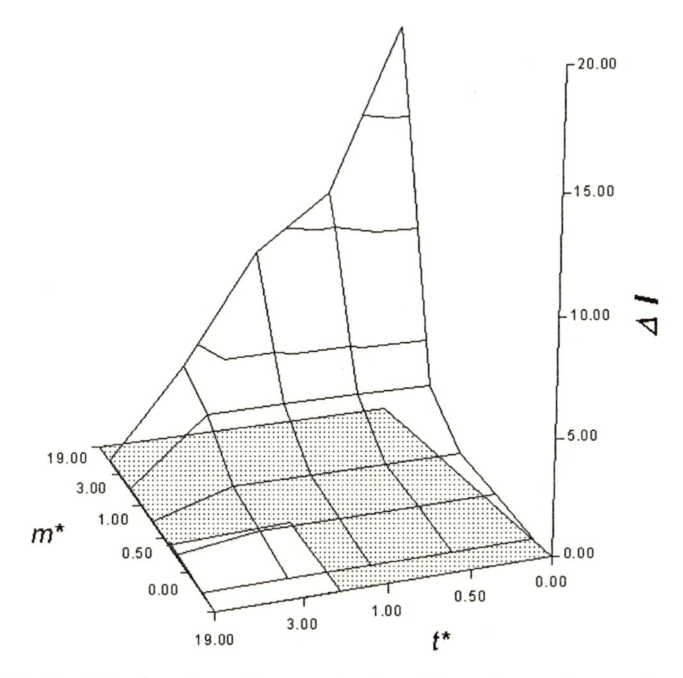

Figure 9.3. The ΔI surface for *sedentary pairs*; the white region on the two-dimensional floor of the diagram identifies circumstances in which mean spatial crowding of matings, m^*, is low, and the mean temporal crowding of matings, t^*, is high.

selection to favor sexual exploitation, and we expect no sexual conflict (column 7, row 1, table 9.1). Without substantial values for I_{mates} or for $I_{\text{cs,sires}}$, alternative male mating strategies are not expected to evolve (column 8, row 1, table 9.1).

Since males and females form exclusive mating pairs, we label such associations *eumonogamy* (column 9, row 1, table 9.1). Examples of this mating system appear to be rare, although, as we describe below, species with somewhat different life histories may converge toward eumonogamy. Univoltine insects or other short-lived species inhabiting harsh environments are possible candidates (column 10, row 1, table 9.1). Biparental care is favored in such species (column 6, row 1, table 9.1). However, short-lived males may be unable to provide direct parental care to females and their young. Instead, males may contribute nutritious secretions to females with their ejaculates, or be eaten by females after mating.

Many slight modifications of this mating system are possible, but none move ΔI significantly away from zero and there is little or no sexual dimorphism. For example, if females mate more than once (column C, row 2, table 9.1), and produce only one clutch of offspring (column D, row 2, table 9.1), sperm competition is possible, but is likely to be greatly weakened by male

Table 9.1

The influences of the spatiotemporal distribution of sexually receptive females as measured by m^*, t^*, $I_{cs,sires}$, and $I_{cs,clutch}$ on the characteristics of animal mating systems; sedentary pairs.

column ->	A	B	C	D	1	2	3	4
	I_{mates}		I_{clutch}					
row	m*	t*	$I_{cs,}$ sires	$I_{cs,}$ clutch	*Sperm competition*	*Female copying*	ΔI	*Expected male-female associations*
1.1	Small (females spatially dispersed)	Large (females temporally synchronous)	0	0	No; females do not mate more than once	No; spatial dispersal of females likely favored by NS	M = F	Prolonged single pairs w/precopulatory guarding of semelparous females
1.2			>0	0	Possible; but prevented by mate guarding	No; females are spatially dispersed	M = F	Mate guarding favored since females are dispersed; should converge to row 9.1.1
1.3			0	>0	No	No	M = F	Mate guarding favored since females are dispersed; should converge to row 9.1.1
1.4			>0	>0	Possible; but prevented by mate guarding	No	M = F	Iteroparous pairs with continual male attendance; death of one pair member results in sterility of the other
1.5			>0	>0	Possible; but unlikely, see row 1	No	M ≤ F	Prolonged iteroparous pairs w/precopulatory and postcopulatory guarding; pairs separate between breeding seasons; I_{within} > than possible in row 9.1.4

mate guarding (column 1, row 2, table 9.1; see below). Female copying is also possible but the opportunity is mitigated by female dispersion (column 2, row 2, table 9.1). This mating system will thus be indistinguishable from eumonogamy (row 1, table 9.1).

If females with low m^* and high t^* tend to mate only once, yet produce

Table 9.1, sedentary pairs (rows 1–5) (*continued*).

5	6	7	8	9	10
Evolutionary responses of males to female spatiotemporal distributions					
Sexual dimorphism	*Male parental care*	*Sexual conflict*	*Alternative mating strategies*	*Mating system*	*Examples*
Negligible $\Delta I = 0$	Yes males benefit by enhancing female fitness	No; male and female fitnesses coincide	No; Imates is negligible	Eumonogamy	univoltine insects in harsh environments?
see row 9.1.1	see row 9.1.1	see row 9.1.1	see row 9.1.1	see row 9.1.1	see row 9.1.1
see row 9.1.1	see row 9.1.1	see row 9.1.1	see row 9.1.1	see row 9.1.1	see row 9.1.1
Large female size due to fecundity selection; small male size to conserve food	Yes; see 1; reduced male size or care of young	No; see 9.1.1	No; see 9.1.1	Persistent pairs	Sponge shrimp; termites; desert isopods; parasitic peracarids; parasitic barnacles.
Large female size due to fecundity selection; large male size enhances female defense; sexes appear nearly monomorphic	Possible; males guard females; may assume territorial defense during nesting; direct care to young uncertain	Possible; over guarding quality or infidelity	Possible; by extrapair matings; mate abandonment	Sequential pairs	Marine copepods; gonodactylid stomatopods; Cirolanid isopods; burying beetles; solitary nesting birds and mammals.

more than one clutch of offspring (columns A–D, row 3, table 9.1), sperm competition and mate copying are again unlikely (columns 1 and 2, row 3, table 9.1), although $I_{cs,clutch}$ now exceeds zero. ΔI remains near zero because males are expected to locate a mate and remain with her until receptivity is complete. Males are not likely to search for additional mates because receptive females mate only once, are widely separated in space, and breed synchronously. If most males die without remating, this mating system too converges to eumonogamy.

PERSISTENT PAIRS

When females are interoparous and exhibit mating-number but not mate-number promiscuity, there is a high covariance in the identity of the sire across clutches (row 4, table 9.1). Males who remain with females through successive reproductive events avoid the costs of searching for females yet continue to sire young. In such cases, $I_{cs,clutch}$ can become extremely large if some females breed and others do not, or if variance in female life span causes variance in the number of breeding episodes. Since males and females exist as functional, permanent pairs, sperm competition does not exist (column 1, row 4, table 9.1), female copying is unlikely (column 2, row 4, table 9.1), and I_{mates} and I_{clutch} are equivalent (column 3, row 4, table 9.1).

In such iteroparous pairs with continual male attendance (column 4, row 4, table 9.1), selection may favor particular male or female morphologies and lead to sexual dimorphism (column 5, row 4, table 9.1). For example, fecundity selection favoring large female body size, may favor males who minimize foraging between clutches. Thus, selection could favor small body size in males, and in some species males may attach themselves permanently to females, living as apparent parasites. Such size reduction in males could be considered a form of male parental care since enhancing offspring production is in both sexes' interest. Thus, a male who is incorporated into the female's body (e.g., in certain rhizocephalan barnacles [Hoeg 1995] and ceratioid anglerfish [Demski 1987]) is not parasitic at all. We call such associations *persistent pairs* (column 9, row 4, table 9.1).

In extreme cases, the death of one member of the breeding pair may result in sterility of the other member. Examples of such species may include sponge shrimp (*Sponigocola* [Hayashi and Ogawa 1987; Saito and Konishi 1999] and *Spongiocaris* [Bruce and Baba 1973], which form pairs within hexactinellid sponges as larvae or juveniles. Young shrimp enter the sponge through its basketlike body wall, but soon grow too large to escape and remain imprisoned for their entire adult lives within the rigid spongocoel of their host (Brusca and Brusca 1990). Termites (Wilson 1971) and desert isopods (*Hemilepistus* sp.; Linsenmair and Linsenmair 1971; Roeder and Linsenmair 1999; Baker et al. 1998) may also belong to this category, since the ecology of these species limit individual opportunities to find secondary mates if one member of a pair dies. Parasitic species of rhizocephal barnacles and epicaridean isopods (Markham 1979, 1992; Bourdon and Bruce 1983; Hoeg 1995; Ghani and Tirmizi 1993) may belong to this group, but only if a single male is associated with each female. Mating systems in which gigantic females associate permanently with more than one dwarf male are better described as a form of polyandrogyny, and are summarized in table 9.6 below.

Sequential Pairs

If breeding becomes somewhat asynchronous, widowed individuals may pair again before reproduction becomes impossible (row 5, table 9.1). Male-male competition for mates and sperm competition are possible in this mating system, which could either enhance or reduce I_{mates} (column 1, row 5, table 9.1). If females with few clutches die earlier than females with more clutches, competition between widowed males and guarding males will tend to equalize the variance in clutch number. We expect males to guard individual females throughout the duration of their sexual receptivity and functionally eliminate sperm competition.

The duration of the interval between female reproductive episodes, relative to adult life span, is expected to influence the persistence of male-female pairs. If this interval is relatively short, spatial dispersion of females and synchronized female reproduction, combined with prolonged male and female life spans, could favor males who lengthen their guarding durations to extend over multiple breeding seasons, if not over entire male and female lifetimes. Mate-tending males sacrifice no fertilizations between breeding seasons due to sperm competition (chapter 8). Mate-tending males also avoid searching and guarding costs that nontending males cannot escape. Such mating systems are functionally indistinguishable from persistent pairs (row 4, table 9.1), except that pair members mate more than once with each other.

If the interval between breeding episodes is long relative to adult life span, pairs may separate between episodes, and females could store sperm from one episode to the next. However, the spatial and temporal distributions of females, as well as the within-season pattern of female receptivity described above, are likely to favor males who guard their mates intensely. If guarding males also copulate repeatedly in response to prolonged female receptivity, they may minimize sperm competition and overwhelm any residual sperm. Thus, sperm storage between breeding episodes may be unnecessary if females gain adequate sperm supplies within seasons. Although females seeking male "genetic quality" could retain sperm for such purposes, the number of males individual females could sample, given their spatial and temporal distribution, is likely to be small, making this explanation unsatisfactory. Thus, although sperm storage between breeding episodes is possible, the likely pattern of male mate guarding within breeding episodes is expected to reduce ejaculate competition to a low level.

If sperm are not stored between reproductive events, or if residual sperm are overwhelmed by the sperm of the guarding male, pairs are expected to form prolonged associations within breeding seasons, to separate between breeding episodes, and then reassemble with the same mate or with another individual in the next interval of female receptivity (column 4, row 5, table 9.1). In such cases, I_{mates} may nearly equal I_{clutch} (column 3, row 5, table 9.1), although variance in both male and female offspring numbers will undoubtedly be larger than is possible with persistent pairs. Sexual dimorphism, such as larger body size or clasping organs in males, may evolve if these characters enhance males' ability to retain their hold on females and resist widowed males (Darwin 1874). However, since I_{mates} is expected to approximately equal I_{clutch}, and since I_{clutch} is expected to favor enlarged female body size, size dimorphism is expected to be slight (column 5, row 5, table 9.1).

Male defense of females has been viewed as a form of parental investment (Thornhill and Alcock 1983). However, since prolonged mate guarding may be costly, males may enhance their fitness more by provisioning themselves between breeding episodes than by caring for young after they are born or hatched. Male parental care, if provided, may be indirect (column 6, row 5, table 9.1). For the same reasons as described above, alternative male mating strategies are unlikely to evolve with small I_{mates} (column 8, row 5, table 9.1; see chapter 11). However, sperm competition and thus opportunistic attempts at multiple mating are possible, albeit at a low level determined by the variance in female longevity. Examples of such *sequential pairs* (column 9, row 5, table 9.1) may include certain marine crustaceans (copepods [Guewuch and Croker 1973]; gonodactylid stomatopods [Caldwell 1986; Shuster and Caldwell 1989]; cirolanid isopods [Shuster, pers. obs]), burying beetles (Scott 1990; Eggert and Sakaluk 1995) as well as birds and mammals with solitary nesting pairs (Møller 1997; Clutton-Brock 1989). In such species, males locate individual females (or females may locate males) and remain with their mates for prolonged periods before and during female sexual receptivity. After oviposition, males abandon their mates to feed until receptive females can again be found (column 10, row 5, table 9.1).

Itinerant Pairs

Attendance Polygyny

When females are spatially dispersed (m^* = low; column A, row 1, table 9.2) but asynchronous in the timing of their receptivity (t^* = moderate to low; column B, row 1, table 9.2), I_{mates} can range from low to intermediate values (column C, row 1, table 9.2), with $I_{cs,sires}$ and $I_{cs,clutch}$ equal to zero (columns C and D, row 1, table 9.2), sperm competition is impossible (column 1, row 1, table 9.2), spatially dispersed females have little opportunity to copy the mate choices of other females (column 2, row 1, table 9.2).

With positive ΔI arising mainly from female breeding asynchrony, males are expected to search for unmated females close to their period of sexual

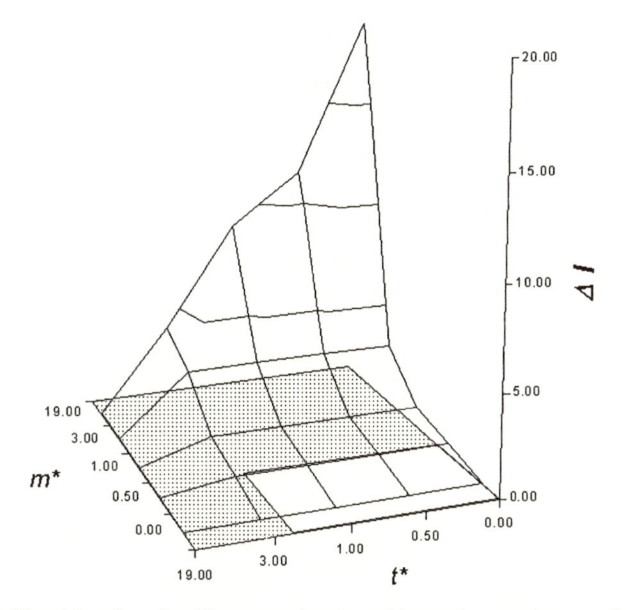

Figure 9.4. The ΔI surface for *itinerant pairs*; the white region on the two-dimensional floor of the diagram identifies circumstances in which mean spatial crowding of matings, m^*, is low, and the mean temporal crowding of matings, t^*, is moderate to low.

receptivity, mate, and then continue searching for other females (column 4, row 1, table 9.2). Since males must locate dispersed females to mate successfully, males are expected to possess modified sensory or locomotory structures that are different from those possessed by females. Males may also possess adaptations, such as size, strength, or attachment structures, associated with holding onto females once they are found. As in the previous mating system, male attachment organs are likely to evolve here, where asynchronously receptive females are spatially dispersed (column 5, row 1, table 9.2; fig. 9.4; chapter 8). The fewer females available for mating at any time, the larger R_O will become and the more intense male-male competition for females will be. However, as explained in chapter 3, the degree to which this condition influences I_{mates} will depend on whether or not some males are dominant across all mating bouts. If male dominance is inconsistent, I_{mates} will erode.

Males are unlikely to engage in parental care given their opportunities to mate with the additional receptive females created by reproductive asynchrony. Thus, after mating, males are expected to abandon females (column 6, row 1, table 9.2). Sexual exploitation by males is likely because I_{mates} is moderate in magnitude while I_{clutch} is zero (column 7, row 1, table 9.2). Although individual variation in condition and size may influence the outcome of contests within pairs (Jormalainen 1998), the searching and guard-

Table 9.2
The influences of the spatiotemporal distribution of sexually receptive females as measured by m^*, t^*, $I_{cs,sires}$, and $I_{cs,clutch}$ on the characteristics of animal mating systems; itinerant pairs.

column ->	*A*	*B*	*C*	*D*	*1*	*2*	*3*	*4*
	I_{mates}		I_{clutch}					
row	m*	t*	$I_{cs,sires}$	$I_{cs,clutch}$	*Sperm competition*	*Female copying*	ΔI	*Expected male-female associations*
2.1	Small (females spatially dispersed)	Small (females temporally asynchronous)	0	0	No; females do not mate more than once	No; spatial dispersal of females likely favored by NS	M > F	Males search for and guard individual dispersed semelparous females; males abandon mates shortly after mating and continue searching
2.2			>0	0	Possible; but minimized by mate guarding	No	M > F	Males search for and mate w/ individual dispersed semelparous females; however male guarding prevents multiple mating and due to female dispersion, likely convergence to attendance polygyny
2.3			0	>0	No	No	M ≥ F	Males search for and guard individual dispersed iteroparous females, mate, then abandon females and continue searching; convergence to attendance polygyny
2.4			>0	>0	Possible; but minimized by mate guarding	No	M ≥ F	Males search for and guard individual dispersed iteroparous and remain with them to reproduce, or abandon females and continue searching if other females become available

Table 9.2, itinerant pairs (rows 1–4) (*continued*).

5	*6*	*7*	*8*	*9*	*10*
Evolutionary responses of males to female spatiotemporal distributions					
Sexual dimorphism	*Male parental care*	*Sexual conflict*	*Alternative mating strategies*	*Mating system*	*Examples*
Male structures for locating and guarding dispersed females	No; males emphasize searching effort	Yes; guarding may inhibit female feeding	Males attempt to usurp guarded females	Attendance polygyny	Semelparous solitary bees; anostracans; notostracans
As above	As above	As above	As above	Attendance polygyny	unknown
As above	As above	As above	As above	Attendance polygyny	Ischnuran damselflies
Male structures for locating, remaining with and protecting dispersed females;	Possible see row 9.2.1; males may physically protect females	Yes; males may enhance female fecundity at females' expense	Males attempt to usurp guarded females	Attendance polygynandry	Desert scorpions; sleepy lizards; schistosomes; parasitic nematodes; ixodid ticks.

Table 9.2, itinerant pairs (rows 5–6) (*continued*).

column ->	A	B	C	D	1	2	3	4
	I_{mates}		I_{clutch}					
row	m*	t*	$I_{cs,}$ sires	$I_{cs,}$ clutch	*Sperm competition*	*Female copying*	ΔI	*Expected male-female associations*
2.5			>0	>0	Possible; but minimized by mate guarding between broods	No	M < F	Males search for and guard individual dispersed iteroparous and remain with them to reproduce, females cannibalize males after mating.
2.6			>0	>0	Possible; but minimized by mate guarding	No	M > F	Males defend, mate with, and then abandon iteroparous females between breeding events/seasons

ing tactics males use in their efforts to mate more than once are likely to exact measurable costs on female fecundity.

With moderate I_{mates}, some taxonomic diversity in male mating behavior among related species is likely (column 8, row 1, table 9.2). For instance, male mate-locating ability and male competitive ability with mate guarding males could generate equivalent I_{mates}. Thus, the degree to which searching, guarding, or usurpation occurmay vary among male populations. Depending on the time scale over which the availability of receptive females changes, males may be relatively flexible in their tendencies to shift between these alternatives (see chapter 12). Since females are dispersed in space and time, and since males must guard individual females, yet also maintain sufficient mobility to obtain multiple mates, we call such mating systems *attendance polygyny* (column 9, row 1, table 9.2). Examples of this mating system appear to exist in semelparous, solitary bees (Alcock 1979; Thornhill and Alcock 1983), as well as in anostracan and notostracan crustaceans (Belk 1991; column 10, row 1, table 9.2).

Table 9.2, itinerant pairs (rows 5–6) (*continued*).

5	6	7	8	9	10
Evolutionary responses of males to female spatiotemporal distributions					
Sexual dimorphism	*Male parental care*	*Sexual conflict*	*Alternative mating strategies*	*Mating system*	*Examples*
Male structure for locating, remaining with and protecting dispersed females; females larger than males due to fecundity selection	Possible see row 9.2.1; males may provide nutrition to females and their young	Possible males may enhance female fecundity at their own expense; but may ensure their own fitness as well.	Males attempt to usurp guarded females	Attendance polyandry	Redbacked spiders; widow spiders; mantids?
Male structures for locating and guarding dispersed females; also for defending mates against other males; trade-offs may exist.	No; males emphasize searching and guarding; care is incidential	Yes; guarding may inhibit female feeding	Males attempt to usurp guarded females; may interfere with matings	Coercive polygynandry	Strongylid nematodes; acanthocephalans; peracarid crustacea; waterstriders; odonates; salamanders; sand lizards; snakes; scorpionflies

When, in addition to the above conditions, sperm competition is possible (column 1, row 2, table 9.2), but mate guarding renders multiple mating rare (Jormalainen et al. 1999), the mating system may be indistinguishable from attendance polygyny, except that males may defend their mates for somewhat longer durations than in species in which females mate only once. When females mate only once, yet produce more than one clutch of offspring (columns C and D, row 3, table 9.2), sperm competition is unlikely, as is mate copying (columns 1 and 2, row 3, table 9.2). However, multiple breeding episodes among females will diminish ΔI and male opportunities to acquire multiple mates may be limited by guarding across breeding episodes. Overall, despite female iteroparity, this mating system will likely be functionally indistinguishable from other forms of attendance polygyny. Examples may exist in certain ischnuran damselflies (Robinson and Allgeyer 1996), as well as in other insects or invertebrates in which monandrous, iteroparous females are widely dispersed in space and become receptive sequentially over a prolonged breeding season.

Attendance Polygynandry

When m^* and t^* are low (columns A and B, row 4, table 9.2), I_{mates} can become moderate in magnitude as described above. If females can mate more than once and have multiple breeding episodes, $I_{cs,sires}$ as well as $I_{cs,clutch}$ can both contribute to I_{clutch} (columns C and D, row 4, table 9.2). The spatial dispersion of females and the degree of reproductive synchrony will determine whether males guard females when they locate them, and thus whether sperm competition will be eliminated (column 1, row 4, table 9.2).

This mating system differs from persistent pairs only in that males are likely to abandon a guarded female if she becomes unreceptive and/or other females become available. Thus, species with persistent pairs may become increasingly polygynandrous if ecological circumstances cause female receptivities to become less synchronous (column 8, row 4, table 9.2). We call such mating systems *attendance polygynandry* because males and females may find other mates. Examples of this mating system may include desert scorpions (*Hadrurus arizonensis*, Tallarovic et al. 2000), desert tenebrionids (*Physadesmia globosa*; Marden 1987), and sleepy lizards (*Tiliqua rugosa*, Bull 2000). If males remain with particular females for extended periods, they may provide their mates with protection or otherwise enhance female fitness. Such examples may include parasitic flatworms such as blood flukes (*Schistosoma*, Southgate et al. 1982; Pica-Mattoccia et al. 2000), in which males hold females within a gonocophoric canal, protecting females not only from other males, but also from the immunological responses of these parasites' hosts. In certain sexually dimorphic strongylid nematodes, males may use their expanded copulatory bursae to retain and repeatedly mate with iteroparous females (Poulin 1997; Roberts and Janovy 2000). Male ixodid mites (*Rhipicephalus appendiculatus*, Wang et al. 1998) are chemically attracted to females feeding on hosts and after mating, by feeding near them, provide salivary secretions that increase female feeding efficiency.

If males who sacrifice their bodies to females after mating also increase female fecundity by removing the need for females to feed after mating or between clutches, then males may prolong mating and transfer more sperm (Lawrence 1992; Andrade 1996, 1998). If males in such species mate only once, whereas females have opportunities to mate more than once, this mating system could be described as attendance polyandry (row 5, table 9.2). We will return to our discussion of such mating systems at the end of the chapter.

Coersive Polygynandry

If reproductive episodes that each constitute a "breeding season" do not occur in close succession, then males are expected to abandon females between seasons. When this occurs, different males will sire the different clutches produced by each female, and $I_{cs,sires}$ as well as $I_{cs,clutch}$ will contribute to $I_{females}$. However, as female breeding asynchrony increases, variance in male mate numbers also increases, and we believe the contribution of I_{mates} to ΔI will often outweigh that of I_{clutch} (column 3, row 6, table 9.2).

When males defend their mates within breeding seasons, but abandon them between breeding seasons (column 4, row 5, table 9.2), they will possess structural adaptations associated with mate location, acquisition, and control similar to those described for attendance polygyny above. However, since males are more often forced to defend their mates from usurpation attempts by other males, owing to prolonged female receptivity, male body size and strength may be favored, in addition to, or as opposed to, male mobility (column 5, row 6, table 9.2). Males are unlikely to provide parental care, although prolonged male defense of receptive females could enhance survivorship by females and their young (column 6, row 6, table 9.2). However, since prolonged male guarding could interfere with female opportunities to feed (Jormalainen 1998; Jormalainen and Shuster 1999), sexual conflict is more likely than sexual cooperation in this mating system (column 7, row 6, table 9.2). As described above, male and female adaptations associated with conflict resolution are expected to evolve to the degree that I_{mates} and I_{clutch} each influence the value of ΔI.

The value of t^* may permit the evolution of alternative male mating strategies where usurpation of females from guarding males is common (column 8, row 6, table 9.2). Trade-offs between rapid mate location and successful mate guarding may arise, taking form as distinct strategies, as ontogenetic switches, or as flexible behavioral tactics (chapter 12). Intense guarding may occur at the expense of female reproductive interests. However, since I_{mates} as well as I_{clutch} are nonzero, females may respond with adaptations that minimize their fitness costs due to sexual conflict. For these reasons, we call this mating system *coercive polygynandry* (column 9, row 6, table 9.2). Examples of this common breeding system (column 10, row 6, table 9.2) may include acanthocephalan worms (Abele and Gilchrist 1977; Poulin and Morand 2000) peracarid crustaceans (reviews in Shuster 1981b; Jormalainen and Merilaita 1993, 1995; Sparkes 1996; Sparkes et al. 1996; Jormalainen 1998; Jormalainen and Shuster 1997, 1999; Jormalainen et al. 2001), waterstriders (Lauer 1996; Lauer et al. 1996), scorpionflies (Thornhill 1981, but see below), as well as amphibians (Verrell 1997), reptiles (Duvall et al. 1985, 1995; Olsson et al. 1994a,b, 1996, 1997) and mammals (Clutton-Brock 1989). As mentioned above, when females exhibit this spatiotemporal distribution of receptivity but release multiple unfertilized ova at once, the mating system shifts to one with greater spatial and temporal crowding of fertilizations. In such cases, while increased temporal crowding of unfertilized ova can decrease I_{mates}, we expect the sudden increase in the spatial crowding of ova that occurs during spawning to have an overriding influence on the opportunity for sexual selection. In such cases, mating systems will rapidly converge toward those described in tables 9.3 and 9.4 in which male competition is more intense and in which individual females may be attended by more than one male.

Scorpionflies may also represent a special case of this type of mating system because males defend individual insect carcasses on which females are allowed to feed during copulation (Thornhill 1981; review in Thornhill and Alcock 1983). Although males capture and defend carcasses, these resources are sufficiently dispersed that males must attract females to them using pheromones. If the carcass defended by a male scorpionfly is large enough, males can attract and mate with more than one female. Scorpionfly mating systems thus resemble feeding site mating systems (see table 9.9 below), in that some males can attract and mate with several females in sequence, generating a large value of I_{mates}. However, because the density of the resource is usually not sufficient to attract more than a few females, males offer a movable feast rather than one that is fixed. This mating system shows characteristics of both coercive polygynandry and *iteroparous feeding site polygyny*, described in more detail below (table 9.9). The somewhat larger value of ΔI possible in resource-based mating systems explains why scorpionfly males differ markedly from females and exhibit unusual variation in their alternative mating behaviors (review in Thornhill and Alcock 1983).

A Summary of Sedentary and Itinerant Pairs

Eumonogamy and *attendance polygyny* occur in monandrous, semelparous species, in which females mate with one male and produce one brood of offspring. *Persistent pairs* occur in monandrous, iteroparous species, in which females mate with one male but produce more than one brood of offspring. *Sequential pairs, attendance polygynandry*, and *coercive polygynandry* occur in polyandrous as well as polygynandrous, iteroparous species, in which females mate more than once and produce more than one brood of offspring. Common to all of these mating systems is the feature that females are spatially dispersed (m^* = low), a condition that favors prolonged mate guarding by males. However, *eumonogamy* differs from *attendance polygyny*, and *persistent pairs* differs from *sequential pairs, attendance polygynandry*, and *coercive polygynandry*, in the degree to which males are able to move between females and mate more than once. The degree of asynchrony in female receptivity (t^* = moderate to low) determines these variations in mating system. Although spatial dispersion favors males who remain with individual females for the duration of their sexual receptivity, the greater the asynchrony in female receptivity, the more opportunities males have to locate multiple mates. Such conditions lead to larger values of ΔI, a result of more intense sexual selection on males (fig. 9.4). Given low m^*, the greatest sexual dimorphism is found in those systems with the smallest t^*.

Mass Mating

Semelparous Mass Mating

When females are aggregated in space, m^* is high (column A, row 1, table 9.3) and t^* is also high if females are synchronized in their sexual receptivity (column B, row 1, table 9.3). Under these conditions, mating systems change markedly. If females mate only once (column C, row 1, table 9.3) and produce one clutch of offspring (column D, row 1, table 9.3), sperm competition is impossible (column 1, row 1, table 9.3), but females are likely to copy the behavior of other females to locate female groups (column 2, row 1, table 9.3; Wade and Pruett-Jones 1990; Dugatkin 1998; chapter 8).

Even a moderate tendency for females to aggregate can result in a runaway process, rapidly evolving toward extreme levels of spatial patchiness. Spatial aggregation of females will attract males (Wade 1995), and if these males enjoy increased mating success relative to solitary males, this condition establishes a genetic covariance between male attraction to female aggregations and female tendencies to aggregate (chapter 4). The coevolution of female tendencies to aggregate and male tendencies to seek aggregations will push such mating systems rapidly toward high values of m^* (fig. 9.5).

Temporal synchrony in female receptivity can mitigate the opportunity for a single male to defend an aggregation of females, despite the presence of numerous receptive females in close spatial proximity (McCurdy et al. 2000). Thus, although females are aggregated in space, I_{mates} will be *reduced* when female receptivities are synchronous (fig. 9.5). Since I_{mates} is small and I_{clutch} is zero, the value of ΔI will be small as well (column 3, row 1, table 9.3). In such circumstances, males are expected to converge *en masse* toward female aggregations, or to aggregate *en masse* in areas in which female assemblages are likely to form (Parker 1974, 1978b).

Little sexual dimorphism is expected to arise because ΔI is near zero, although male clasping organs may be favored if environmental conditions such as wind or wave action prevent matings by individuals unable to remain in copula during the narrow window of female receptivity. Sexual conflict is expected to be minimal because male and female mate numbers will be nearly equal, and male and female reproductive interests coincide (column 7, row 1, table 9.3). Because male and female reproductive interests coincide, male parental care may evolve (column 6, row 1, table 9.3), although the degree to which male care is apparent will depend on the life span of males and females. Because I_{mates} is small, alternative male mating strategies are not expected to evolve (column 8, row 1, table 9.3).

Table 9.3

The influences of the spatiotemporal distribution of sexually receptive females as measured by m^*, t^*, $I_{cs,sires}$, and $I_{cs,clutch}$ on the characteristics of animal mating systems; mass mating.

column ->	*A*	*B*	*C*	*D*	*1*	*2*	*3*	*4*
	I_{mates}		I_{clutch}					
row	m*	t*	$I_{cs,}$ sires	$I_{cs,}$ clutch	*Sperm competition*	*Female copying*	ΔI	*Expected male-female associations*
3.1	Large (females spatially aggregated)	Large (females temporally synchronous)	0	0	No	Yes; copying produces groups of receptive females and prevents multiple mating by males	M = F	Males defend and mate individual semelparous females
3.2			>0	0	Possible; but prevented by mate guarding	Yes	M = F	Males locate and mate individual females; probable convergence to 9.3.1 duo to male mate guarding
3.3			0	>0	No	Yes	M = F	Males defend, mate and remain with individual iteroparous females
3.4			>0	>0	Possible; but prevented by mate guarding	Yes	M = F	Males locate, mate and remain with individual iteroparous females over several reproductive events
3.5			>0	>0	Possible; but prevented by mate guarding	Yes	M ≥ F	Males locate and mate with individual females; pairs separate between breeding events

Table 9.3, mass mating (rows 1–5) (*continued*).

5	*6*	*7*	*8*	*9*	*10*
Evolutionary responses of males to female spatiotemporal distributions					
Sexual dimorphism	*Male parental care*	*Sexual conflict*	*Alternative mating strategies*	*Mating system*	*Examples*
Little dimorphism; possible male clasping structures but only if favored by natural selection	Possible; may not be obvious due to brief life spans	No; male and female interests coincide	No; ΔI is near zero	Semelparous mass mating	Eunicid polychaetes; mysids; cumaceans; Nematomorpha; semelparous swarming insects; Semelparous broadcast spawners; Wind-pollinated annual plants
As above	As above	As above	As above	Semelparous mass mating	
Negligible; I_{mates} and I_{clutch} are nearly equal	Yes; male and female fitnesses coincide	Minimal; except over quality of care	No; Imates is small	Mass mating w/ male parental care	monogamous social hymenoptera
Minimal; possible structures associated with territory establishment and defense	Yes; male and female fitnesses coincide	Minimal; except over quality of care	No; Imates is small	Mass mating w/ male parental care; possible shifts to polyandrogyny; see table 9.12.	Dance flies; group nesting birds and primates
Large body size in either sex; possibly male structures for mate retention	Yes; within breeding seasons	Minimal within breeding seasons	Possible in both sexes	Iteroparous mass mating	Corals, wind pollinated perennials; Mass breeding amphibians polychaetes w/epitokes; land crabs, grunion

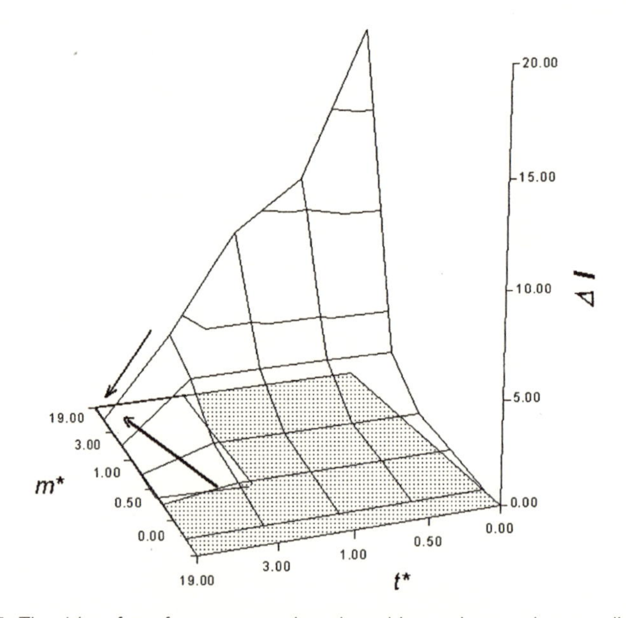

Figure 9.5. The ΔI surface for *mass mating*; the white region on the two-dimensional floor of the diagram identifies circumstances in which mean spatial crowding of matings, m^*, is moderate to high, and the mean temporal crowding of matings, t^*, is high. Female copying is likely to influence the spatial distribution of females as well as favor male tendencies to be attracted to female aggregations. However, synchrony in female receptivity as well as female tendencies to mate more than once will not only enhance the spatial crowding of females, but will also enhance male tendencies to guard individual mates. Increased spatial crowding of females combined with intense male mate guarding of individual females drives the functional temporal synchrony in female receptivity higher (lower arrow) and reduces ΔI (upper arrow).

We call mating systems in which males and females breed once in large assemblages of pairs *semelparous mass mating* (column 9, row 1, table 9.3). Examples of this mating system (column 10, row 1) may exist in marine annelids that breed synchronously (e.g., eunicids, Brusca and Brusca 1990; Porchet and Olive 1988), horsehair worms that form mating balls in fresh water (Nematomorpha, Cochran et al. 1999), crustaceans in which receptive females aggregate in large groups (mysids, cumaceans, Guewuch and Croker 1973; Duncan 1983), insects with synchronous mass emergences of sexually receptive females (reviews in Alcock 1979; Thornhill and Alcock 1983; Sivinski and Petersson 1997), semelparous, broadcast-spawning species (Bartels-Hardege and Zeeck 1990; Desrosiers et al. 1994; Armendáriz 2000), as well as wind-pollinated annual plants (Goodwillie 1999). If, in addition, females mate more than once (column C, row 2, table 9.3), but produce only one clutch of offspring (column D, row 2, table 9.3), sperm competition is possible (column 1, row 2, table 9.3), but is mitigated by male mate guarding. In such cases, I_{mates} will approximately equal I_{clutch}, ΔI will be near zero, and this mating system will be indistinguishable from semelparous mass mating.

Mass Mating with Male Parental Care

When receptive females are spatially clumped (m^* = high; column A, row 3, table 9.3), synchronized in their receptivity (t^* = high; column B, row 3, table 9.3), mate only once ($I_{cs,sires}$ = 0; column C, row 3, table 9.3), yet produce more than once clutch of offspring ($I_{cs,clutch} > 0$; column D, row 3, table 9.3), sperm competition is not possible (column 1, row 3, table 9.3). However, female copying behavior is expected to have the same effects on male and female distributions as described above (column 2, row 3, table 9.3).

Males may also remain with females after mating rather than search for additional, unlikely mates (column 4, row 3, table 9.3). Although females are aggregated, synchrony in female receptivity limits male mating opportunities. With ΔI nearly zero, sexual dimorphism will be inconspicuous (column 5, row 3, table 9.3), male parental care will be favored (column 6, row 3, table 9.3), sexual conflict is likely to be minimal (column 7, row 3, table 9.3), and alternative male mating strategies are not expected to evolve (column 8, row 3, table 9.3).

When male-female pairs form within large aggregations and males provide care to young, we call such mating systems *mass mating with male parental care* (column 9, row 3, table 9.3). Examples of this mating system may exist in monogamous social Hymenoptera (Wilson 1971), or other species that exhibit mass emergences in which females mate only once, but oviposit multiple times (column 10, row 3, table 9.3). Extreme synchrony in female reproduction could exist because conditions favoring reproduction are widely separated in time, and survival between breeding seasons may be low. Short-lived males may provide no direct parental care to females and their young, but may instead contribute nutritious secretions to females with their ejaculates, or allow their bodies, or parts of their bodies, to be eaten by females after mating.

The situation changes little when females mate more than once and have multiple breeding episodes (columns C and D, row 4, table 9.3). Since ΔI is small, sexual dimorphism is expected to be minimal, although fecundity selection could favor larger female body size (column 5, row 4, table 9.3). Because male and female mate numbers are similar, and the fitness of pairs coincides, sexual conflict is reduced as well (column 7, row 4, table 9.3). Moreover, with I_{mates} small due to synchrony in female receptivity and male guarding of individual females within the breeding aggregation, alternative mating strategies are unlikely to evolve (column 8, row 4, table 9.3). Examples of this mating system appear to exist in certain dance flies (Svensson

1997), and may also exist in species of group nesting birds such as sulids (Osorio-Beristain and Drummond 1999) and corvids (Marzluff and Balda 1992), as well as certain group nesting primates (Clutton-Brock 1989; Daly and Wilson 1992). Seahorses (Vincent 1994, 1995; Masonjones and Lewis 2000) might appear to exhibit this mating system. However, as explained below (table 9.12) seahorse and pipefish mating systems appear to have evolved from a mating system involving male defense of the oviposition site (McCoy et al. 2001). Males sequester embryos on their bodies or within abdominal pouches, but restrictions on male brooding space prevent males from mating more than once (Berglund et al. 1986, 1989).

Recent analyses of paternity indicate that multiple paternity is common in group nesting birds and mammals (Westneat and Sherman 1997; Travis et al. 1997; Soukup and Thompson 1998). It should therefore be noted that the persistence of mass mating with male parental care relies on *extreme* synchrony in female receptivity. As we will describe below, this mating system can rapidly shift toward the mating systems described in table 9.6 (*social pairs*), if female receptivities become even slightly asynchronous. This mating system may also shift toward *polyandrogyny* (table 9.12) if altricial young and male-biased sex ratios allow females to leave their brooding mates and seek nest sites again (Oring et al. 1994).

Iteroparous Mass Mating

When both m^* and t^* are high but females mate more than once and have multiple breeding episodes, $I_{cs,sires}$ and $I_{cs,clutch}$ become important ($I_{cs,sires}$, $I_{cs,clutch} > 0$; columns C and D, row 5, table 9.3). As expected, male guarding reduces sperm competition (column 1, row 5, table 9.3), and hence reduces I_{mates} to low levels (column 2, row 5, table 9.3, fig. 9.5). If $I_{cs,sires}$ is also reduced toward zero, then I_{mates} will approximately equal I_{clutch} (column 3, row 5, table 9.3), and ΔI may be relatively small.

If the interval between breeding episodes is long relative to adult life span, pairs may separate between episodes, and females could store sperm from one episode to the next. Males will guard as well as copulate repeatedly in response to prolonged female receptivity. If sperm are not stored between reproductive events, or if displacement of residual sperm is common, pairs will form prolonged associations within breeding seasons, separate between seasons, and then reassemble with the same mate or with another individual in the next interval of receptivity (column 4, row 5, table 9.3).

If pairs separate between breeding episodes, variance in offspring numbers can increase among males as well as among females. In such cases, I_{mates} may nearly equal I_{clutch}, although variance in both male and female offspring numbers will undoubtedly be larger than is possible with semelparous mass mating or with mass mating with male parental care. Since I_{mates} can by itself become large due to differential male abilities to defend mates or to differences in male life span, larger male body size or male structures useful in defending mates are expected to evolve (column 5, row 5, table 9.3). Since I_{clutch} can also become large due to variation in female brood size or to differences in female life span, larger female body size is expected to evolve. When increases in body size are simultaneously favored in both sexes, sexual dimorphism will not be extreme.

Male parental care may exist within breeding seasons (column 6, row 5, table 9.3) and sexual conflict will be minimal because ΔI is nearly zero. However, conflict is possible if long-term reproduction is compromised within breeding seasons (column 7, row 5, table 9.3). As described above, alternative mating strategies are unlikely to evolve in such situations. However, alternative mating strategies remain possible in both sexes because I_{mates} and I_{clutch} are positive and because both sexes tend to mate more than once (column 8, row 5, table 9.3; see chapter 8). When males and females assemble as pairs in large, conspecific aggregations, but also separate between breeding events, we call such associations *iteroparous mass mating* (column 9, row 5, table 9.3). Examples of this mating system appear to exist in land crabs (Seeger 1996), iteroparous, mass-spawning fish such as grunion (Thomson and Muench 1977) and surgeonfish (Myrberg et al. 1988), and iteroparous amphibians that breed in large assemblages (Wells 1979; Sullivan 1992; column 10, row 5, table 9.3). Iteroparous, broadcast-spawning species such as corals (*Dendronephthya*, Dahan and Benayahu 1997; *Pocillopora*, Kruger and Schleyer 1998) and wind-pollinated perennial plants (Lyons et al. 1989; McKone et al. 1998), as well as polychaetes which either broadcast gametes (*Arenicola*, Howie 1982; *Eupolymnia*, McHugh 1993) or produce epitokes (*Perinereis*, Hardege et al. 1994; Cerratulidae, Petersen 1999) may also belong to this group.

Breeding Pairs versus Mass Mating

Eumonogamy and *semelparous mass mating* occur in monandrous, semelparous species, while *persistent pairs* and *mass mating with male parental care* occur in monandrous, iteroparous species. *Sequential pairs* and *iteroparous mass mating* occur in polygamous, iteroparous species. While these mating systems share synchronous female receptivites, they differ in the degree to which females are spatially aggregated (figs. 9.3 and 9.5). Although spatial aggregation of females can increase the opportunity for sexual selection on males, in semelparous species, female breeding synchrony prevents most males from mating more than once. In iteroparous species, long-lived males may have opportunities to mate more than once, but the spatial distribution of females may favor prolonged male guarding between breeding seasons, or the repeated assembly of males in breeding aggregations where only pairs can form. Both circumstances cause the variance in mate numbers, and thus the variance in offspring numbers, to be approximately equal between the sexes. Thus, in species that form large assemblages of synchronously breeding male-female pairs, we expect to see reduced sexual dimorphism and a high frequency of male parental care.

Even slight tendencies for receptive females to aggregate can cause a mating system characterized by spatially isolated male-female pairs to evolve rapidly toward a mating system characterized by male-female pairs that are spatially aggregated (fig. 9.5). One possible explanation for the apparent rarity of *eumonogamy*, as well as other single-pair mating systems, could be the ease with which clumped assemblages of male-female pairs may form. Increasing spatial aggregation tends to intensify male-female pair bonds, provided that synchrony in female receptivity remains high (fig. 9.5). However, even slight decreases in the synchrony of female receptivity can cause spatially aggregated single-pair mating systems to rapidly become more polygynous or polygynandrous (fig. 9.6).

Polygamy

Cursorial Polygyny

When the values of m^* and t^* are both moderate, the opportunity for sexual selection on males, I_{mates}, is also moderately high (columns A and B, row 1, table 9.4). If females mate once and produce one clutch of offspring ($I_{cs,sires}$, $I_{cs,clutch} = 0$; columns C and D, row 1, table 9.4), sperm competition is impossible (column 1, row 1, table 9.4), but female mate copying is likely owing to aggregation among females.

Males in such situations are expected to search for clusters of semelparous females, mate, and mate again within clusters (column 4, row 1, table 9.4). Since I_{mates} is moderately large, male structures that are useful in searching for *as well as* for defending females during mating are likely to be favored (column 5, row 1, table 9.4). Body size trade-offs may exist such that, in contests with rivals, large size provides a competitive advantage, whereas in searching for clusters of females, smaller, more agile males are favored over more cumbersome individuals. These alternatives could be favored in a frequency-dependent manner, or male body size may represent a compromise between these selective pressures. Males are not expected to engage in parental care owing to opportunities for remating (column 6, row 1, table 9.4). Sexual conflict will be conspicuous, and males are expected to exploit female fecundity (column 7, row 1, table 9.4, fig. 9.7).

Since ΔI is moderately large and positive, and since circumstances favoring different male behaviors are likely to change over short time scales, alternative male mating strategies are expected to involve behavioral variation in searching, guarding and usurpation tactics (column 8, row 1, table

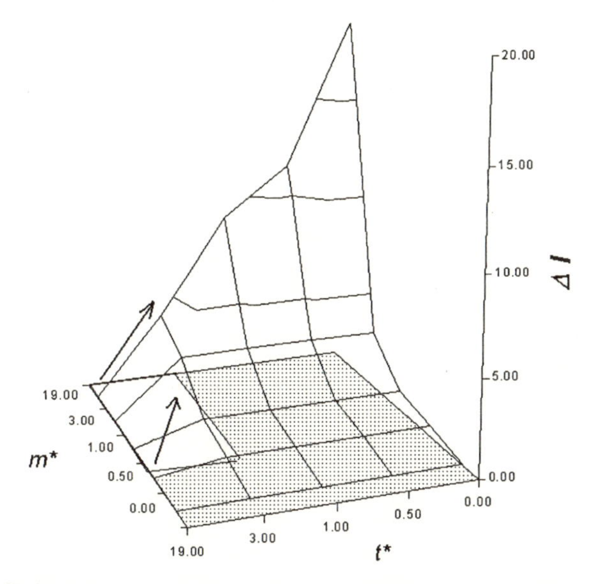

Figure 9.6. Slight decreases in the synchrony of female receptivity (lower arrow) can cause spatially aggregated single-pair mating systems to rapidly become more polygynous or polygynandrous, thereby rapidly increasing ΔI (upper arrow).

9.4). We call such mating systems *cursorial polygyny* because males are highly mobile and male mating success will be variable, whereas females will tend to mate only once (column 9, row 1, table 9.4). Examples may include semelparous insects or other invertebrates in which males rove for mates (column 10, row 1, table 9.4). As described in chapters 4 and 7, multiple insemination allows sperm competition. If males who mate with previously mated females also sire disproportionate numbers of progeny, a genetic covariance between sperm competitiveness in males and tendencies toward multiple mating in females could arise (chapter 4).

Polygamy

When females are moderately clumped and the timing of female receptivity is moderately asynchronous, tendencies among females to copy the mate choices of other females could strongly increase the variance in mating success among roving males (column 2, row 2, table 9.4). Since sperm competition is also possible, I_{mates} as well as $I_{cs,sires}$ can become quite large (chapters 4 and 8). The opportunity for sexual selection on males (I_{mates}) arising from sperm competition and female copying could easily exceed $I_{females}$. Thus, although somewhat reduced compared to row 1, table 9.4, ΔI is still likely to be moderate to large and positive (column 3, row 2, table 9.4). This is true even if the relative sperm competitive ability of males varies among females

Table 9.4
The influences of the spatiotemporal distribution of sexually receptive females as measured by m^*, t^*, $I_{cs,sires}$, and $I_{cs,clutch}$ on the characteristics of animal mating systems; polygamy.

column ->	*A*	*B*	*C*	*D*	*1*	*2*	*3*	*4*
	I_{mates}		I_{clutch}					
row	m*	t*	$I_{cs,}$ sires	$I_{cs,}$ clutch	*Sperm competition*	*Female copying*	ΔI	*Expected male-female associations*
4.1	Moderate (females moderately aggregated)	Moderate (females moderately asynchronous)	0	0	No	Possible, but will not accelerate because males stay mobile searching for receptive females	M > F	Males search for moderately clumped, semelparous females, mate once and move on
4.2			>0	0	Yes	Possible as above	M > F	Males search for moderately clumped, semelparous females, mate and move on
4.3			0	>0	No	Possible as above	M > F	Convergence to cursorial polygyny or to polygamy
4.4			>0	>0	Yes	Possible as above	M ≥ F	Convergence to polygamy
4.5	Large (females spatially aggregated in matings with certain males)	Small (females temporally asynchronous while waiting to mate with preferred males	>0	>0	Yes	Possible as above	M > > F	Females aggregate around particular males and mate in sequence; males perform behaviors that attract female attention

(Clark and Begun 1998). Although multiple mating could reduce I_{mates}, this effect may be offset by female copying.

Males are expected to search for moderately clumped, semelparous females, to mate with any females that are receptive, and then to move on in search of other receptive females, as in cursorial polygyny (column 4, row 2,

Table 9.4, polygamy (rows 1–5) (*continued*).

5	*6*	*7*	*8*	*9*	*10*
Evolutionary responses of males to female spatiotemporal distributions					
Sexual dimorphism	*Male parental care*	*Sexual conflict*	*Alternative mating strategies*	*Mating system*	*Examples*
Yes, associated w/searching and fending off rivals	No; energy devoted to searching and fighting	Yes; males exploit females	Yes; trade offs in searching and guarding	Cursorial polygyny	Semelparous insects with roving males?
Yes, as above, also sperm competition and attendant changes in sperm morphology	No; energy devoted to searching, fighting and repeated mating	Yes; males exploit females; possible fem. response to sperm competition	Yes; variation in sperm morphology, secretions	Polygamy	Uniparous bivalves with multiple mating; annual plants with pollen mixing
Yes, but reduced compared to above due to female iteroparity	As above	Yes, but females may evolve countermeasures	As above	Cursorial polygyny or polygamy	Univoltine insects with multiparity
Yes but reduced compared to above due to female iteroparity	As above	Yes; males exploit females; possible fem. response to sperm competition	As above	Polygamy, possible convergence to leks	ground squirrels; drosophilids
Yes; males exhibit conspicuous phenotypes and behavioral displays	No; male energy devoted to display	Yes; males exploit females; possible fem. response to sperm competition	Yes; satellite males associate with successful displaying males	Classic leks	Picture-winged drosophila; Manakins; Uganda kob

table 9.4). As in the above mating system, males are expected to exhibit modifications suited for searching for mates and fending off rivals. However, with greater opportunities for sperm competition, we also expect modifications such as large testis size, diversity in sperm morphology, and chemical complexity in prostatic secretions to evolve. Male signals that exploit female

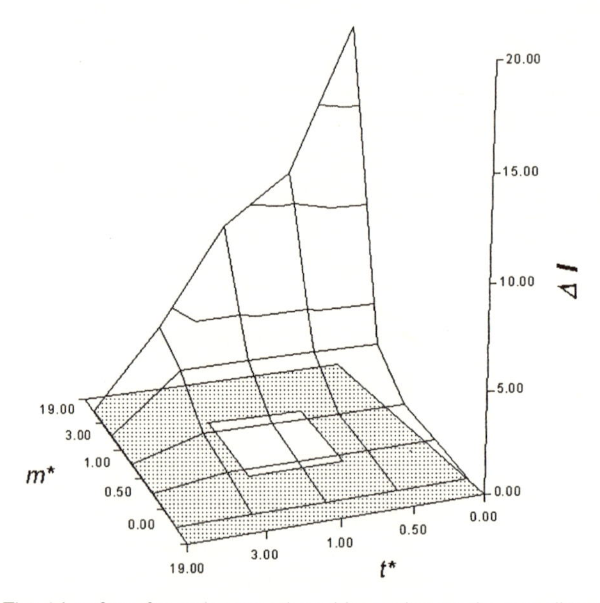

Figure 9.7. The ΔI surface for *polygamy*; the white region on the two-dimensional floor of the diagram identifies circumstances in which mean spatial crowding of matings, m^*, is moderate, and the mean temporal crowding of matings, t^*, moderate.

sensory biases could also evolve if transmission of such signals enhances the abilities of certain males to attract the attention of receptive females. Because of covariance between male sperm competitive ability and female tendencies to mate more than once, this mating system is expected to enhance displays and elaborate morphologies associated with sperm competition (column 5, row 2, table 9.4).

Since receptive females are moderately dispersed in space and time, males are not expected to become parental (column 6, row 2, table 9.4). The large and positive value of ΔI will lead to sexual exploitation, and conflict resolution will come at a cost to female fecundity (column 7, row 2, table 9.4). Since both males and females mate with more than one individual, following Darwin (1874) and our table 6.1, we call such mating systems simply *polygamy* (column 9, row 2, table 9.4). Cursorial polygyny converges to polygamy if roving males inseminate already mated females and sperm competition ensues. For this reason, examples of mating systems in which females mate more than once, but produce only one clutch of offspring, are likely to be common. Examples include invertebrate species in which females mate more than once (or receive sperm from multiple sources), yet produce only one clutch of offspring, as do certain bivalves (Morton 1985; Araujo et al. 1999), as well as annual plants whose flowers receive pollen from several stamens before setting seed (Delph et al. 1998) (column 10, row 2, table 9.4).

When females mate once but produce more than one clutch of offspring

($I_{cs,sires} = 0$, $I_{cs,clutch} > 0$; columns A–D, row 3, table 9.4), sperm competition is impossible (column 1, row 3, table 9.4). I_{mates} is moderately large due to the spatial and temporal distributions of females, and although multiparity may increase the variation in female fitness, because females mate only once, this variability does not decrease the sex difference in the opportunity for selection. Thus, ΔI is expected to remain moderate and positive in value (column 3, row 3, table 9.4). Although females are multiparous, this mating system will be indistinguishable from cursorial polygyny. Again, as described above, mating systems in which males mate more than once may become polygamous if females are fortuitously inseminated more than once (row 2, table 9.4). Thus, regardless of the number of clutches that females produce, this mating system too may to converge to polygamy.

When iteroparous females mate more than once and produce more than one clutch of offspring ($I_{cs,sires}$, $I_{cs,clutch} > 0$, columns A–D, row 4, table 9.4), I_{mates} can be large due to multiple mating and sperm competition, as can I_{clutch} due to multiple mating and multiparity in females. We expect ΔI to be positive, because I_{mates} arising from the distribution of females will exceed I_{clutch} (column 3, row 4, table 9.4). However, the reduction of ΔI by female iteroparity and mate number promiscuity is expected to diminish sexual dimorphism (Schwagmeyer and Wootner 1985). This mating system is similar to polygamy in semelparous species, but the resulting sexual dimorphism is reduced to the degree that females are iteroparous (column 5, row 4, table 9.4).

Because males and females mate more than once, sexual conflict is expected in this mating system, and unlike mating systems in which sexual selection on males overwhelms viability selection on females (chapters 1, 4, and 5), the evolutionary responses of males and females to intersexual conflict may be similar in complexity (column 7, row 4, table 9.4). Examples include odonates (Waage 1979), drosophilids (Pitnick 1996; Snook 1997), and rodents (Schwagmeyer 1986; Schwagmeyer and Foltz 1990) in which adaptations for sperm competition ability are well known (column 10, row 4, table 9.4). Internally fertilizing poecellid fish such as guppies and swordtails, which engage in mate copying and multiple insemination, also belong in this group (Ryan et al. 1992; Dugatkin 1996).

Many dioecious annual and perennial plants produce multiple inflorescences and disperse pollen to numerous other plants. However, the resemblance of these mating systems to animal polygamy is only superficial due to pollen limitation. Male plants can compete with one another through pollen production and pollen tube growth on the female stigmatic surface in a manner analogous to sperm competition between males within multiply inseminated females in insects. However, this kind of male-male competition through pollen does not necessarily result in greater variance in male than in female reproductive success. Just as in female animals that vary in the number of clutches they can produce, female plants differ from one another in the numbers of flowers and fruits they produce. A male plant, successful in pollinating one flower on a plant, will not necessarily be successful pollinating other flowers on the same plant. Thus, variation among female plants in the numbers of flowers affects the variance in female reproductive success in the same way that variance in the number of clutches of eggs laid affects the variance in reproductive success of polyandrous, iteroparous female animals. Because the effects are common to males and females the sex difference in reproductive variance necessary for sexual selection does not develop.

Pollen limitation is analogous in its effects on plants to multiple mating and multiple reproductive episodes in polyandrous, iteroparous female animals, particularly in species that are broadcast spawners (chapter 5). Females with more mates (i.e., pollen or sperm) have an increased reproductive success, and those females with fewer mates have a diminished reproductive success. This too prevents the sex difference in reproductive variance necessary for sexual selection. Mate-number promiscuity increases in both sexes and, mediated by pollinators, involves the same phenotypes. The females with more mates leave more offspring and, on average, their daughters as well as their sons are more promiscuous. This principle also applies to hermaphroditic animals. Thus, unless individuals have differential abilities to reproduce as one sex or the other, as may exist in certain hermaphroditic turbellarians (Michiels and Newman 1998), sexual selection in hermaphrodites will be relatively weak.

Classic Leks

In polygamous mating systems, females are moderately aggregated in space and moderately asynchronous in the timing of their receptivity (m^*, t^* = moderate; columns A and B, rows 1–5, table 9.4). Since males do not physically guard females under such circumstances, females can move between potential mates and have opportunities to mate as well as to copy the mating behavior of other females (chapter 8). We suggested above that female copying will be unimportant in most of these mating systems. This is true because roving males are unlikely to remain with small aggregations of females, leaving little opportunity for genetic covariance to become established between female tendencies to aggregate and male tendencies to attend these aggregations. However, if certain characteristics of males tend to *attract* the attention of females in such a way that these conspicuous males secure a disproportionate number of mates, then genetic covariance could arise between female aggregation behavior and male conspicuousness (chapters 4 and 7).

When females begin to prefer particular males as mates, the spatial distribution of females changes, such that a disproportionate number of matings occurs in locations surrounding preferred males. Since females must wait to gain access to preferred males, the temporal distribution of receptive females

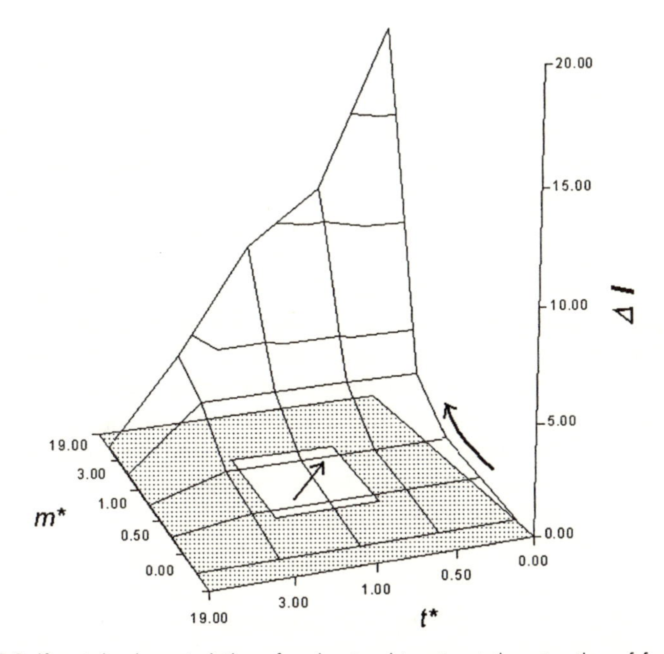

Figure 9.8. If certain characteristics of males tend to attract the attention of females in such a way that these conspicuous males secure a disproportionate number of mates, then the spatial distribution of matings becomes more clumped and the temporal distribution of mating becomes more dispersed (left arrow); the value of ΔI will increase (right arrow) and may increase dramatically if genetic covariance between female mating behavior and male conspicuousness arises.

becomes functionally asynchronous (fig. 9.8). Such circumstances favor female mate copying. Furthermore, a genetic covariance may also become established between female tendencies to copy the mate selections of other females and the conspicuous phenotypes of males (chapter 8; column 3, row 5, table 9.4). Under these circumstances, the opportunity for sexual selection can shift, from moderate to extremely high values (see table 9.8 below). Because females become attracted to males solely on the basis of their conspicuous phenotypes, mating is expected to occur independently of feeding or oviposition sites, and may bear no relationship to landmarks or other environmental characteristics that influence male or female fitness (column 4, row 5, table 9.4).

Males are expected to develop extremely conspicuous displays and external morphologies as described in chapters 1 and 7 (column 5, row 5, table 9.4). Males are not expected to become parental because of their large prezygotic investment in display (column 6, row 5, table 9.4). Sexual conflict is likely because variance in male mating success may far exceed that of female fitness (column 7, row 5, table 9.4). As certain males become increasingly successful in mating, males who aggregate around them become more successful in obtaining matings than males who attempt to display to females in isolation, especially with an initial predisposition to usurpation. Presumably, the success of satellite males will decrease as their numbers increase (column 8, row 5, table 9.4). We call such mating systems *classic leks* (column 9, row 5, table 9.4) because (1) males display with minimal aggression directed at other males in locations that contain no resources required by females, (2) display sites can shift from year to year and do not represent landmarks by which females orient themselves, and (3) females mate with males and then leave display sites to rear offspring alone (Bradbury 1981; Höglund and Alatalo 1995). Examples of such mating systems appear to include picture-winged *Drosophila* (Spieth 1981, 1984; Hoikkala and Kaneshiro 1997), white bearded manakins (Lill 1974a,b), and Uganda kob (Deutsch 1994). Other mating systems resembling this polygamous mating system, as well as other circumstances that can lead to the evolution of classic leks, are described below (table 9.8).

Male Dominance

Dominance Polygyny

When females are moderately clumped in space, (m^* = moderate; column A, row 1, table 9.5) and asynchronous in the timing of their sexual receptivity (t^* = low; column B, row 1, table 9.5), the opportunity for sexual selection on males, I_{mates}, will be moderately large. If females mate once and produce only one clutch of offspring, I_{clutch} will be zero ($I_{cs,sires}$, $I_{cs,clutch} = 0$; columns C and D, row 1, table 9.5), and sperm competition will be impossible (column 1, row 1, table 9.5). Female copying could exaggerate aggregation size, and males could become attracted to groups of females. However, asynchrony in female receptivity will allow one or a few males to defend clusters of females as they become receptive. Thus, with asynchronous breeding, males guarding clusters are more likely to mate repeatedly than roving males. A genetic covariance arises between female tendencies to aggregate and male guarding ability so that ΔI will remain large and positive (column 3, row 1, table 9.5; fig. 9.9).

Males in such circumstances are expected to defend small groups of females and to mate with individual females as each one becomes sexually receptive (column 4, row 1, table 9.5). The moderately large value of ΔI will favor male structures and behaviors useful in asserting dominance over other males attracted to female groups, although male modification may not be as extreme as in other systems with larger values of I_{mates} (column 5, row 1, table 9.5). Prolonged association of dominant males with groups of females could favor the evolution of male parental care. Indeed, male defense of

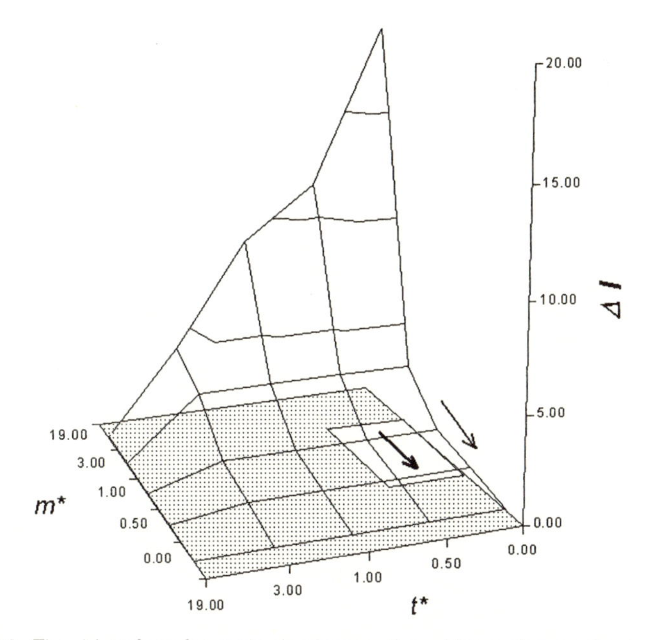

Figure 9.9. The ΔI surface for *male dominance*; the white region on the two-dimensional floor of the diagram identifies circumstances in which mean spatial crowding of matings, m^*, is moderate, and the mean temporal crowding of matings, t^*, low. Limitations on ΔI imposed by male dominance: although females are moderately aggregated in space, allowing dominant males to mate with the females in their group, dominant males prevent each other from mating with large numbers of females (left arrow). Although a runaway process between multiple mating in females and sperm competitiveness in males could transport this mating system toward polygamy, the spatiotemporal distribution of females favoring male dominance will oppose the transformation of this mating system to one that is more polygynous (right arrow).

females and their young could enhance variance in offspring numbers among males. However, since mate defense by males may be expensive, and since this activity most directly influences male mating success, most male parental care is expected to be incidental to male mate guarding activities (column 6, row 1, table 9.5). Sexual conflict is likely to exist between males and females, and conflict may also arise among overaggregated females. Given the magnitude and sign of ΔI sexual conflict is likely to be resolved in favor of male interests (column 7, row 1, table 9.5). If cluster size is small, male mating strategies are more likely to involve forceful takeovers of female groups by other males than harem invasion.

Since males faced with such selection are likely to become quite fierce, successful takeovers may require males to form coalitions, which could persist after takeovers are accomplished. Dominant status could thus become transferred among male group members and enhance cooperation among male coalitions attending groups of females (column 8, row 1, table 9.5).

Table 9.5
The influences of the spatiotemporal distribution of sexually receptive females as measured by m^*, t^*, $I_{cs,sires}$, and $I_{cs,clutch}$ on the characteristics of animal mating systems; male dominance.

column ->	*A*	*B*	*C*	*D*	*1*	*2*	*3*	*4*
	I_{mates}		I_{clutch}					
row	m*	t*	$I_{cs,}$ sires	$I_{cs,}$ clutch	*Sperm competition*	*Female copying*	ΔI	*Expected male-female associations*
5.1	Moderate (female moderately aggregated)	Small (female receptivity asynchronous)	0	0	No	Possible, could enhance male guarding abilities	M > F	Males defend small groups of females and mate with females in sequence; males guard mates until receptivity is complete
5.2			>0	0	Possible, but male guarding minimizes multiple mating	Possible as above	M > > F	Convergence to dominance polygyny or dominance polygynandry
5.3			0	>0	No	Possible as above	M > F	Convergence to dominance polygyny
5.4			>0	>0	Possible, but male guarding minimizes multiple mating	Possible as above	M ≥ F	Convergence to dominance polygynandry

Since males defend and mate with groups of females who mate only once, we call such associations *dominance polygyny* (column 9, row 1, table 9.5). Examples of this mating system are unknown in semelparous species, perhaps because sufficient time does not exist within the lifetimes of individuals in semelparous species for the complexity of social dynamics associated with this mating system to arise. However, complex social dynamics are likely in the mating systems described in row 4, table 9.5 below.

Table 9.5, male dominance (rows 1–5) (*continued*).

5	*6*	*7*	*8*	*9*	*10*
Evolutionary responses of males to female spatiotemporal distributions					
Sexual dimorphism	*Male parental care*	*Sexual conflict*	*Alternative mating strategies*	*Mating system*	*Examples*
Yes, associated with dominance and aggression	Possible, since males remain with females, but most male care may be indirect	Yes, males exploit females	Yes, satellite males attempt to take over group control by force or stealth	Dominance polygyny	Unknown
As above	As above	As above	As above	Dominance polygyny or dominance polygynandry	Peracarid crustaceans
As above	As above	As above	As above	Dominance polygyny or dominance polygynandry	Unknown
As above	As above	As above	As above	Dominance polygynandry	Horned beetles; reef fish; side-blotched lizards; redwing blackbirds; marmots bighorn sheep; horses, ungulates wolves; lions; primates

Dominance Polygynandry

If guarding males experience sperm depletion, or if female receptivities begin to overlap, females may need to mate more than once simply to acquire enough sperm to fertilize their ova (see chapter 8). Thus, alternative male mating strategies in which males assume morphologies or behaviors that allow them to invade female aggregations guarded by dominant males may arise (column 8, row 2, table 9.5). Since variance in male mate numbers exceeds that of females, we call this mating system *dominance polygynan-*

dry. Examples of this mating system may include isopod crustaceans in which males defend clusters of polyandrous, semelparous females (*Paracerceis sculpta*, Shuster 1989a,b, 1992; *Elaphognathia cornigera*, Tanaka and Aoki 1999, 2000).

If females are multiparous, $I_{cs,clutch}$ will contribute to $I_{females}$, and erode I_{mates}. Although ΔI will be smaller than described for other mating systems in table 9.5, this mating system will be indistinguishable from dominance polygyny. Once again, however, the possibility of sperm depletion in dominant males could favor females who mate more than once and allow satellite males to invade the population, causing this mating system to converge toward dominance polygynandry or even to polygamy (table 9.4). While roving males may persist in the population at low frequencies, the success of male dominance as a mating strategy is likely to prevent such mating systems from converging entirely toward polygamy.

When iteroparous females are long lived, despite male guarding, females are likely to produce offspring with more than one male, increasing I_{clutch} and decreasing I_{mates} (column 3, row 4, table 9.5). Reductions in ΔI due to multiple mating and multiple female broods will produce concomitant reductions in sexual dimorphism among the sexes compared to other forms of dominance polygyny (column 5, row 4, table 9.5). Persistence of dominance and family relationships within groups could favor male parental care (column 6, row 4, table 9.5), and although sexual conflict will exist, it will be mitigated somewhat by the decrease in ΔI compared to dominance polygyny (column 7, row 4, table 9.5), as well as by the abilities of females to respond to selection in the context of male exploitation (columns C and D, row 4, table 9.5).

Long life spans could also allow subordinate males to steal copulations from dominant males, or trade sexual access to certain females for assistance in group defense. Although the number of offspring produced by such matings may be small, subordinate males are likely to reproduce with greater success than males excluded from male coalitions entirely. Subordinate males may also have a greater probability of eventually inheriting dominant status than if they leave the group (column 8, row 4, table 9.5). Since variance in female offspring numbers may exist within groups, females may also exhibit alternative mating strategies, and for this reason, linear dominance hierarchies may arise among females as well. Depending on the mechanism by which sex is determined, these strategies could intergrade.

This mating system will be similar in most respects to dominance polygyny, but long life spans will also allow complex social interactions. Thus, females as well as males may have opportunities to mate with more than one partner, but since variance in male mating success is likely to be greater than variance in female mating success, we label this mating system dominance polygynandry as well. Although a runaway process between multiple mating in females and sperm competitiveness in males could transport this mating system toward polygamy, the spatiotemporal distribution of females favoring male dominance will prevent complete transformation of this mating system. That is, dominant males will be prevented by other dominant males from mating with sufficient numbers of females to allow this process to accelerate (fig. 9.9). Examples of this mating system appear to exist in long-lived species that form mixed-sex groups, in which males may cooperate to defend small groups of females, and linear dominance hierarchies exist in both sexes. Such species may include protogynous reef fish (Warner 1984, Robertson and Warner 1978; Warner and Robertson 1976), certain rodents and birds in which males defend small aggregations of females (yellow-bellied marmots, Downhower and Armitage 1971; redwing blackbirds (*Agelaeus phoenicus*, Weatherhead and Robertson 1979), as well as group dwelling ungulates, canids, felids, and primates with multimale bands (Packer 1986; Packer and Pusey 1987; Clutton-Brock 1989; Hogg and Forbes 1997).

Social Pairs

Polygynous Social Pairs

When m^* is large and t^* is moderate in magnitude (columns A and B, row 1, table 9.6), I_{mates} will be large. With semelparity, $I_{cs,sires}$, $I_{cs,clutch} = 0$ and sperm competition is impossible (columns C–1, row 1, table 9.6). Female copying is likely if natural selection favors persistent associations of females.

Like mass mating, nonrandom mating can establish a genetic covariance between the tendency for females to aggregate and the attraction of males to female assemblages, intensifying these characters in both sexes. When females mate only once, slight asynchrony in female receptivity will favor intensified male mate guarding (column 3, row 1, table 9.6). Males are expected to locate and defend females before and during their periods of sexual receptivity, but with moderate breeding asynchrony males may attempt to locate other, still receptive females after securing a primary mate. Since most such females will already be guarded by males, roving males are expected to attempt takeovers. The distinction between guarding and roving is thus reduced, and encounters between these strategies are likely to be more common than when t^* is high (column 4, row 1, table 9.6). Male structures associated with retaining females once they are located, as well as those that are useful in dislodging guarding males from receptive females, will both be favored (column 5, row 1, table 9.6). Asynchrony in female receptivity, as well as the degree to which females are spatially aggregated, determine the degree to which this happens (column 6, row 1, table 9.6). Greater breeding asynchrony combined with greater spatial clumping of receptive females results in larger values of I_{mates} (fig. 9.10). However, guarding reduces the

Table 9.6
The influences of the spatiotemporal distribution of sexually receptive females as measured by m^*, t^*, $I_{cs,sires}$, and $I_{cs,clutch}$ on the characteristics of animal mating systems; social pairs.

column ->	*A*	*B*	*C*	*D*	*1*	*2*	*3*	*4*
	I_{mates}		I_{clutch}					
row	m*	t*	$I_{cs,}$ sires	$I_{cs,}$ clutch	*Sperm competition*	*Female copying*	ΔI	*Expected male-female associations*
6.1	Large (females spatially aggregated)	Moderate (female receptivity moderately asynchronous)	0	0	No	Yes; causes males to associate with individual females within female groups	M > F	Males guard individual females and mate; also attempt to usurp mates from other males
6.2			>0	0	Yes	Yes; causes males to associate with individual females within female groups	M > F	Males guard individual females and mate; also attempt to mate other females as they become available.
6.3			0	>0	No	Yes; causes males to associate with individual females within female groups	M ≥ F	Convergence to polygynous social pairs
6.4			>0	>0	Yes	Yes; as above	M > > F	Convergence to polygamous social pairs but possibility of extrapair mating may enhance sexual dimorphism

number of available females and thus decreases the functional asynchrony in female receptivity. Moreover, if males attempt to sequester the females they guard away from aggregations, the functional spatial distribution of females will become less crowded as well (fig. 9.11).

Male parental care in the form of protection of the guarded female could evolve from prolonged mate guarding. However, male care may be compromised by the availability of extra-pair copulations. The degree to which female receptivity is asynchronous also determines the degree of sexual conflict, since variance in male mating success is directly proportional to female asynchrony (column 7, row 1, table 9.6). Alternative mating strategies

Table 9.6, social pairs (rows 1–4) (*continued*).

5	*6*	*7*	*8*	*9*	*10*
Evolutionary responses of males to female spatiotemporal distributions					
Sexual dimorphism	*Male parental care*	*Sexual conflict*	*Alternative mating strategies*	*Mating system*	*Examples*
Moderate; associated with guarding and usurping females	Yes, but is compromised by extrapair mating opps.	Yes, males exploit females	Yes, roving vs guarding	Polygynous social pairs	Unknown
Moderate to high depending on degree of sperm competition	Unlikely due to availability of extrapair mating opps.	Yes; males exploit females; females may counter male exploitation if $I_{cs,sires}$ is large	Yes, searching vs guarding and sperm competition	Polygamous social pairs	Lovebugs
As above	As above	Yes; males exploit females but females may coevolve	As above	Polygynous social pairs	Insects with small moderately asynchronous emergences of females
As above	Yes, but is compromised by extra-pair mating opps.	Yes; possible mutual exploitation due to opportunities for extra-pair matings	Yes, searching vs guarding; also egg dumping	Polygynandrous social pairs	Group nesting birds and primates; salmonids

among males are likely to involve alternation between roving and guarding tactics (column 8, row 1, table 9.6). Such mating systems will resemble semelparous mass mating, but with greater sexual dimorphism than with synchronous female receptivity. The combination of stable male-female pairs, with occasional usurpation by roving males, leads us to label such mating systems *polygynous social pairs* (column 9, row 1, table 9.6). Examples are common among short-lived insects with mass emergences from clustered resources, spread over several days or weeks (Choe and Crespi 1997), as well as among crustaceans and other invertebrates with similar population patterns of female receptivity (column 10, row 1, table 9.6).

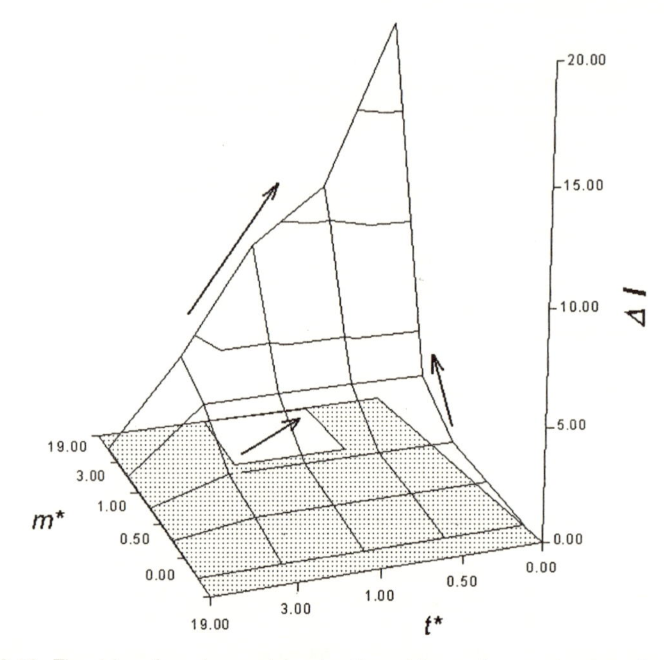

Figure 9.10. The ΔI surface for *social pairs*; the white region on the two-dimensional floor of the diagram identifies circumstances in which mean spatial crowding of matings, m^*, is high, and the mean temporal crowding of matings, t^*, is moderate. Increases in ΔI in species with social pairs: the degree to which such male modifications evolve will depend on the relative asynchrony in female receptivity as well as the degree to which females are spatially aggregated. Greater breeding asynchrony combined with greater spatial clumping of receptive females (lower arrow) will lead to larger values of ΔI (upper arrows).

Polygynandrous Social Pairs

Female mate promiscuity permits sperm competition and thus modifies polygynous social pairs (column 2, row 2, table 9.6). Since females may mate more than once, the intensity of male mate guarding may erode, and favor roving for receptive females (column 4, row 2, table 9.6). Whereas male guarding of individual females reduces I_{mates}, roving with sperm competition could increase I_{mates} (column 3, row 2, table 9.6). Thus, sexual dimorphism is expected to increase, and, depending on the frequency of extra-pair copulation, males may also exhibit variable sperm morphologies or displays that attract the attention of receptive females (column 5, row 2, table 9.6).

Male parental care is unlikely to evolve due to the availability of multiple mating opportunities (column 6, row 2, table 9.6), and sexual conflict is likely to be conspicuous, although $I_{cs,sires}$ may permit females to evolve countermeasures to male exploitation (column 7, row 2, table 9.6). Males are

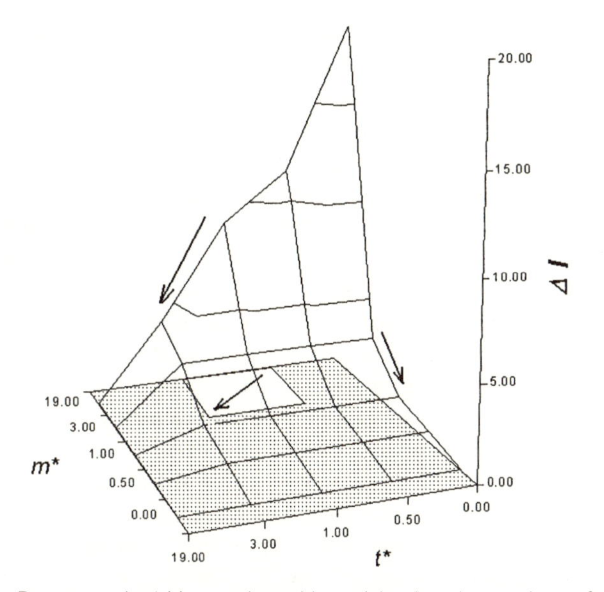

Figure 9.11. Decreases in ΔI in species with social pairs: the tendency for males to closely guard females is expected to reduce the availability of females and thus decrease the functional asynchrony in female receptivity. Moreover, if males attempt to sequester the females they guard, the functional spatial distribution of females will become less crowded as well (lower arrow); such changes will lead to smaller values of ΔI (upper arrows).

expected to show variable mating behaviors that include mate guarding as well as mate searching, usurpation, and displays (column 8, row 2, table 9.6). Because both members of breeding pairs are likely to mate more than once, and because roving males may displace guarding males, we call such mating systems *polygamous social pairs* (column 9, row 2, table 9.6). Examples of this mating system include insects that breed in swarms but which remain in copula for prolonged durations, such as lovebugs (*Plecia nearctica*, Thornhill 1976) and other swarming insects with semelparous females.

With iteroparity, but no female promiscuity, the tendency for genetic covariance to form between female aggregation and male attraction to female assemblages is enhanced. With female promiscuity, slight asynchrony in female receptivity favors intensified male mate guarding, and reduces I_{mates}. However, since females produce more than one clutch of offspring, I_{clutch} as well as I_{mates} are somewhat enhanced beyond these values for polygynous social pairs with semelparous females (column 3, row 3, table 9.6). Males are expected to locate and defend females before and during periods of sexual receptivity. Males are also expected to attempt to locate other, still receptive females if they are available, a condition that will make encounters between roving and guarding males common (column 4, row 3, table 9.6).

Selection could favor male structures for retaining females and dislodging guarding males (column 5, row 3, table 9.6), but the degree to which such modifications evolve will depend on the relative asynchrony in female receptivity. Since females produce more than one clutch of offspring, males successfully guarding mates for multiple clutches could remain with females to assist in offspring care. However, male care is likely to be compromised by the availability of extra-pair copulations (column 6, row 3, table 9.6). Female asynchrony also determines the degree of sexual conflict because asynchrony and male mating success will vary in direct proportion (column 7, row 3, table 9.6). The increased fitness variance of females, owing to $I_{cs,clutch}$, permits the evolution of countermeasures to male exploitation.

Alternative mating strategies among males are likely to involve reciprocal fighting versus searching and guarding tactics (column 8, row 3, table 9.6). This mating system resembles iteroparous mass mating, although with greater sexual dimorphism than is observed when t^* is small. Since most individuals form pairs, and only some males mate more than once by usurping mates from other guarding males, this mating system will converge to polygynous social pairs (column 9, row 3, table 9.6). Examples of this mating system are likely to be common among insects with mass emergences that occur over several days or weeks as mentioned above (column 10, row 3, table 9.6).

When females are both promiscuous and iteroparous, ($I_{cs,sires}$, $I_{cs,clutch} > 0$; columns A–D, row 4, table 9.6), sperm competition is possible (column 2, row 4, table 9.6), and the advantage of mate guarding relative to roving is reduced (column 4, row 4, table 9.6). ΔI can become larger via sperm competition because females are spatially clumped. With multiple insemination common, males are expected to abandon females after mating and seek additional mates rather than perform parental duties. Thus, sexual dimorphism is likely to involve male morphologies used in aggression and dominance. Moreover, if sperm competition is possible, sperm morphologies enhancing postopulatory, prezygotic competitive ability as well as male characteristics that attract the attention of still receptive females may also arise (column 4, row 4, table 9.6).

In this way, sexual dimorphism may evolve in species in which many males and females appear to form lasting pair bonds (column 5, row 4, table 9.6). Since females produce more than one clutch of offspring, males may elect to remain with their mates through several oviposition cycles. However, the degree to which males attempt to be parental will depend on the proximity and asynchrony of other sexually receptive females (column 6, row 4, table 9.6). Sexual conflict is likely to be more conspicuous in this mating system than in other mating systems in this section (table 9.6) because both sexes mate more than once with more than one partner, and because females can evolve countermeasures to male sexual exploitation (column 7, row 4, table 9.6). For this reason, alternative mating strategies are also likely to be favored in both sexes. Thus, males as well as females are likely to solicit extra-pair copulations, and males as well as females may attempt to induce other individuals to provide parental care to offspring (column 8, row 4, table 9.6).

If male fitness is increased more by guarding than by roving, this mating system could converge to mass mating with male parental care (row 4, table 9.3). However, as described above, rapid sequential mating attempts by males may be the form of alternative mating strategy that invades this last mating system. Thus, both schemes are possible in this mating system. We call this mating system *polygynandrous social pairs* (column 9, row 4, table 9.6). Examples (column 10, row 4, table 9.6) of this combined system include many species of birds, rodents, and primates that breed in groups, but form "monogamous" pair bonds, that is, "socially monogamous" species (Gowaty 1997; Liker and Szekely 1997; Blumstein and Arnold 1998; Soukup and Thompson 1998). Not only is moderate sexual dimorphism common in such species, but evidence of widespread sperm competition and egg dumping is increasingly reported (Westneat and Sargent 1996; Webster et al. 1995; Baker and Belis 1995). Salmonids may also exhibit this mating system, although greater mating frequency by females in these species than in birds and mammals may enhance the mating success of particular males, increasing I_{mates}, and lead to greater observed degrees of sexual dimorphism (Willson 1997) as well as more conspicuous, i.e., behaviorally or morphologically distinct, alternative male mating strategies (Gross 1985; Montgomery et al. 1987; Hutchings and Myers 1994). When females in this group spawn large egg masses or allow multiple matings, the spatial and temporal crowding of fertilizable ova simultaneously increase. Because each female's clutch may be more easily subdivided at this time, more than one male may associate with each female, as is observed in horseshoe crabs (Brockmann 1996; Brockmann et al. 2000).

Diversity in Polygamous Mating Systems

Combinations of mate guarding and roving by males are expected to coexist when the spatial crowding of receptive females is moderate to high, and when breeding synchrony is moderate to low. Breeding asynchrony combined with moderate clustering of females places a premium on harem defense and leads to dominance polygyny or dominance polygynandry. When the balance between guarding and roving favors roving, polygynous social pairs or polygynandrous social pairs are likely outcomes. The isolation and defensibility of clusters in conjunction with the degree of asynchrony and iteroparity determine this balance and the diversity among polygamous mating systems.

Fixed Sites

Eumonogamy and Persistent Pairs

When females reach sexual maturity in particular locations, or when females travel certain routes in search of food, shelterm or mates, the mean spatial crowding of receptive females (*m**) can become very large (column A, row 1, table 9.7). If females emerge from natal sites asynchronously or travel preferred routes in sequence, the mean temporal crowding of receptive females (*t**) will become very small (column B, row 1, table 9.7). When females mate only once and produce only one clutch of offspring ($I_{cs,sires}$, $I_{cs,clutch}$ = 0; columns C and D, row 1, table 9.7), sperm competition is prohibited (column 1, row 1, table 9.7), yet conspecific cueing by females could lead disproportionate numbers of females to travel toward particular landmarks or to follow the same paths toward resources, thereby increasing the spatial concentration of females (column 2, row 1, table 9.7).

Although I_{mates} has the potential to become very large, the opportunity for sexual selection depends on the behavior of males at breeding aggregations. If males tend to pair with individual females, leave the breeding aggregation, and never mate again, I_{mates} is small, whereas if males pair with individual females, leave the aggregation to mate, but later return to the breeding aggregation, I_{mates} is larger. I_{mates} is larger still when males remain in a particular location with an aggregation of females and mate with different females in sequence. Differences in female life history under each of these circumstances can produce further diversity in mating system character as follows (fig. 9.12).

The emigration of mated males from breeding aggregations can decrease I_{mates} to the point that I_{mates} equals I_{clutch}, so that the mating system converges toward eumonogamy or persistent pairs depending on whether monandrous females produce one or more than one clutch of offspring, respectively (rows 1–5, table 9.7). Examples of this mating system include species of monogamous swarming insects such as termites (Shelley and Whittier 1997; Sivinski and Petersson 1997).

Polyandrous Mating Swarms

When the spatial and temporal distributions of females match those described for row 4, table 9.7 above, but females outlive their monogynous mates, they could return to breeding aggregations repeatedly. This situation could allow I_{clutch} to exceed I_{mates} by a factor equal to the average number of mates per female. ΔI would thus become extremely large but negative in sign (column 3, row 5, table 9.7). Sexual dimorphism in such circumstances is expected to favor characteristics that enhance female fecundity such as

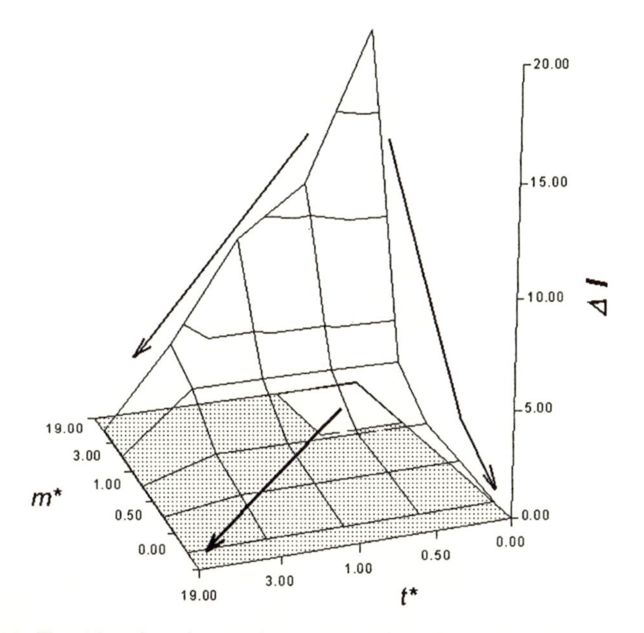

Figure 9.12. The ΔI surface for *mating swarms*; the white region on the two-dimensional floor of the diagram identifies circumstances in which mean spatial crowding of matings, m^*, is high, and the mean temporal crowding of matings, t^*, low. Reductions in ΔI due to emigration of pairs from mating swarms: when mated males leave the breeding aggregations altogether, as more females arrive, the fraction of mating males within the breeding aggregation will steadily increase, creating a situation similar to that which occurs when female sexual receptivity is synchronous. The emigration of mated males from breeding aggregations will decrease I_{mates} to the point that I_{mates} equals $I_{clutches}$ (lower arrow), thus reducing ΔI to zero (upper arrows).

large body size (column 5, row 5, table 9.7). Male parental care is expected to arise, but if male life spans are short, such male contributions may be indirect and far from apparent (column 6, row 5, table 9.7). Sexual conflict is also expected to exist given the magnitude of ΔI. However, unlike mating systems discussed above, in this case, males rather than females are likely to be exploited. Multiple mating by females at the expense of male fecundity is consistent with this expectation (column 7, row 5, table 9.7).

Although the large, negative value of ΔI makes alternative mating strategies among males unlikely, female alternative strategies are expected to evolve, with females aggressive toward other females and undermining the efforts of reproductive competitors whenever possible (column 8, row 5, table 9.7). Since mate acquisition takes place in large aggregations of males, and since variance in mate numbers exists among females (but not among males) and affects their fecundity, we call such mating systems *polyandrous mating swarms* (column 9, row 5, table 9.7). Examples of this mating system

Table 9.7
The influences of the spatiotemporal distribution of sexually receptive females as measured by m^*, t^*, $I_{cs,sires}$, and $I_{cs,clutch}$ on the characteristics of animal mating systems; mating swarms.

column ->	*A*	*B*	*C*	*D*	*1*	*2*	*3*	*4*
	I_{mates}		I_{clutch}					
row	m*	t*	$I_{cs,}$ sires	$I_{cs,}$ clutch	*Sperm competition*	*Female copying*	ΔI	*Expected male-female associations*
7.1	Large (females spatially clumped) But emigration by pairs reduces *m**	Large (females temporally asynchronous) But emigration by pairs increases *t**	0	0	No	Yes, females may copy attraction of females to swarms	M = F	Semelparous females emerge or travel through particular locations males swarm at these sites and attempt to mate with individual emerging females; pairs emigrate from aggregations, causing I_{mates} to decrease.
7.2			>0	0	Possible but prevented by male guarding	As above	M = F	Convergence to eu-monogamy
7.3			0	>0	No	As above	M = F	Males remove females from swarms to mate and assist females in building a nest and rearing young
7.4			>0	>0	No; male guarding prevents multiple mating	Yes, as above	M = F	Convergence to persistent pairs
7.5			>0	>0	Yes, males are unable to guard mates between breeding episodes	Yes, females may copy attracton of females to swarms	M < F	Iteroparous females emerge or travel through particular locations; monogynous males grasp females, emigrate, mate once and do not return to breeding aggregations; females return to breeding aggregations

Table 9.7, mating swarms (rows 1–5) (*continued*).

5	*6*	*7*	*8*	*9*	*10*
Evolutionary responses of males to female spatiotemporal distributions					
Sexual dimorphism	*Male parental care*	*Sexual conflict*	*Alternative mating strategies*	*Mating system*	*Examples*
Unlikely, unless large male size allows males access to early arriving females possible structures for clasping and retaining females	Yes, if males do not attempt to reenter the breeding aggregation	Unlikely although male mating attempts may damage females	Unlikely, if males leave breeding aggregation	Convergence to eumonogamy	Semelparous swarming insects; semelparous anurans
As above	As above	As above	As above	As above	As above
Minimal as breeding adults, however, females could become extremely large	Yes, male and female reproductive interests coincide	No, ΔI equals zero	No, I_{mates} is small	Convergence to persistent pairs	Termites
As above	As above	As above	As above	As above	As above
Large body size in females	Possible, but males do not live long enough to provide for young	Yes; males are exploited by females; sperm are taken but may not be used	Yes, but among females; since offspring numbers contribute more to female fitness than mate numbers female strategies involve aggression toward other breeding females.	Polyandrous mating swarms	Social hymenoptera

Table 9.7, mating swarms (rows 6–9) (*continued*).

column ->	*A*	*B*	*C*	*D*	*1*	*2*	*3*	*4*
	I_{mates}		I_{clutch}					
row	m*	t*	$I_{cs,}$ sires	$I_{cs,}$ clutch	*Sperm competition*	*Female copying*	ΔI	*Expected male-female associations*
7.6	Large (females spatially clumped) But emigration by pairs reduces *m**	Small (females temporally asynchronous) But emigration by pairs increases t*	0	0	No	As above	M > F	Males grasp and guard individual semelparous females; pairs leave aggregation to mate, after mating males abandon females and return to aggregation
7.7			>0	0	Possible but prevented by male guarding	As above	M > F	Convergence to polygynous mating swarms
7.8			0	>0	No	As above	M > F	Convergence to polygynous mating swarms
7.9			>	>0	No	As above	M ≥ F	Males grasp and guard individual iteroparous females; pairs leave aggregation to mate, after mating males abandon females and both sexes return to aggregations

exist in many social Hymenoptera (column 10, row 5, table 9.7) in which multiple mating by queens occurs in mating swarms and reproductive competition among foundresses is common (Wilson 1971, 1975). If competition is sufficiently intense, then the social organization of females at the founding of colonies will resemble the formation of male coalitions described earlier.

Table 9.7, mating swarms (rows 6–9) (*continued*).

5	6	7	8	9	10
Evolutionary responses of males to female spatiotemporal distributions					
Sexual dimorphism	*Male parental care*	*Sexual conflict*	*Alternative mating strategies*	*Mating system*	*Examples*
Male structures for grasping and guarding females in aggregations	No; males emphasize swarming and guarding effort	Yes; male grasping may damage females	Males attempt to usurp guarded females	Polygynous mating swarms	Semelparous group signaling Insects and anurans
As above	As above	As above	As above	As above	As above
As above	As above	As above	As above	As above	As above
Male structures for grasping and guarding females in aggregations also for attracting females Large female size	No; males emphasize swarming, displays and guarding effort	Yes; male grasping may damage females; but females fight back	Males attempt to usurp guarded females	Polygynandrous mating swarms or polygamy	Iteroparous group signaling Insects and anurans

Polygynous Mating Swarms

A male tendency to grasp receptive females and sequester them before and after mating amounts to male mate guarding. Thus, when female lifetimes are short, male mate guarding is likely to prevent females from mating with more than one male (chapter 5; column 1, row 6, table 9.7). Such mating systems are likely to resemble attendance polygyny, wherein individual males guard and mate with spatially dispersed receptive females (column 4, row 6, table 9.7). However, in these mating systems, pairs form when receptive females move toward aggregations of males rather than when assemblages of males form around individual receptive females (row 6, table 9.7). Thus, since pairs form within breeding aggregations, and since variance in mate numbers is negligible among females but can be large among males, we call such mating systems *polygynous mating swarms* (column 9, row 6, table 9.7).

Examples of this mating system (column 10, row 6, table 9.7) include swarming species with moderate sexual dimorphism, usually involving structures in males that assist in retaining their grasp on females. This mating system also includes species in which males perform visual, auditory, or chemical displays that attract females to male aggregations (Wells 1979; Lloyd 1966, 1979; Sullivan 1992; Snedden and Greenfield 1998; Grafe 1999). The argument that the intensity of such displays does not increase linearly with aggregation size has been used by some researchers to argue that these male displays do not function as mechanisms to attract females (Bradbury 1981; Höglund and Alatalo 1995). However, the intensity of male signals need not increase linearly with aggregation size to be favored by sexual selection. Male signals need only increase the probability that receptive females will be more attracted to signaling males than to nonsignaling males (Christy 1995).

When both sexes return to breeding assemblages to locate mates (column 4, row 9, table 9.7), males can mitigate low fertilization success by increased mate numbers. If males can encounter multiple females by returning to swarms, males may abandon rather than guard females after mating. Because sperm competition is possible (column 1, row 9, table 9.7), a genetic covariance can arise between female tendencies toward mate-number promiscuity and competitive sperm in males (chapters 4 and 7). Multiple mating opportunities and iteroparity among females will decrease the magnitude and possibly reverse the sign of ΔI (column 3, row 9, table 9.7). Moreover, if males cease to defend females after mating, females can move between potential mates within a male aggregation and exercise mate choice. Such mating systems are expected to resemble attendance polygynandry (table 9.2) if males guard females before returning to the breeding aggregation, or polygamy (table 9.4) if they do not. Large testis size, diversity in sperm morphology, and chemical complexity in prostatic secretions are expected to evolve, as are male displays that enhance male mating success in polygynous mating swarms. Male signals that exploit female sensory biases are also expected to be common.

Leks

Semelparous Exploded Leks

When females complete their development in patchily distributed locations, or when females tend to travel particular routes to reach food, nesting sites or shelter, the mean spatial crowding of sexually receptive females (m^*) is large (column A, row 1, table 9.8). If receptive females appear in particular locations asynchronously, or travel certain routes in sequence, the mean temporal crowding of receptive females (t^*) can become small (column B, row 1, table 9.8). Monandry and semelparity prevent sperm competition (columns C–1, row 1, table 9.8). However, if genetic covariance arises between female tendencies to aggregate and male tendencies to assemble at or defend locations where females are found, I_{mates} may be greatly enhanced (column 2, row 1, table 9.8). With I_{mates} large (m^* = high, t^* = low), ΔI will remain large (column 3, row 1, table 9.8), particularly if females are attracted to and mate with certain site-fixed males (column 4, row 1, table 9.8; fig. 9.13).

As is well known (Bradbury 1985; Höglund and Alatalo 1995), males able to attract the attention of females via display are likely to experience greater mating success than nondisplaying males. Also, the more females tend to mate with particular males, the greater the reward is in mate numbers for males able to retain their position by aggression (column 5, row 1, table 9.8). Male courtship displays and male aggressiveness are a common behavioral combination in many insect and vertebrate species (reviews in Höglund and Alatalo 1995; Shelley and Whittier 1997; Borgia and Presgraves 1998). However, male aggression tends to disperse the spatial distribution of males. This spatial dispersion allows more males to successfully mate because the ability of females to make direct comparisons among males as well as to copy the mate choices of other females is reduced (fig. 9.13).

Males are not expected to become parental when male aggressiveness and display enhance male mating success (column 6, row 1, table 9.8). Given the sign and magnitude of ΔI, sexual conflict is expected to exist and be resolved in favor of males. That is, females may be damaged by male copulatory attempts, and are not expected to evolve much in the way of countermeasures to prevent male exploitation (column 7, row 1, table 9.8). Large ΔI also favors alternative male mating strategies of two kinds: (1) displaying, site guarding males and (2) nondisplaying, satellite males who congregate around display sites (column 8, row 1, table 9.8). Because males remain at particular sites where they are likely to encounter semelparous females, and because males aggressively defend territories in these locations, we call such mating systems *semelparous exploded leks* (column 9, row 1, table 9.8). Examples of this mating system may include univoltine lekking insects in which aggression occurs between displaying males.

Table 9.8

The influences of the spatiotemporal distribution of sexually receptive females as measured by m^*, t^*, $I_{cs,sires}$, and $I_{cs,clutch}$ on the characteristics of animal mating systems; leks.

column ->	*A*	*B*	*C*	*D*	*1*	*2*	*3*	*4*
	I_{mates}		I_{clutch}					
row	m*	t*	$I_{cs,}$ sires	$I_{cs,}$ clutch	*Sperm competition*	*Female copying*	ΔI	*Expected male-female associations*
9.1	Large (females spatially clumped)	Small (females temporally asynchronous)	0	0	No	Yes, females may copy, causing attraction of females to particular sites	M > > F	Males defend sites where semelparous females are likely to be found; females mate once and leave to oviposit; if males leave with females this mating system converges toward polygynous mating swarms (row 9.7.6)
9.2			>0	0	Yes; males do not defend females after mating	As above	M > > F or M > > > F	Males display at sites where semelparous females are likely to be found; females mate more than once and leave to oviposit.
9.3			0	>0	No	Yes, females may copy attracton of females to swarms	M > > F	Males defend sites where iteroparous females are likely to be found; females mate once and leave to oviposit several times
9.4			>0	>0	Yes; males do not defend females after mating	Yes, as above	M > > > F	Males display at sites where iteroparous females are likely to be found; females mate more than once and leave to oviposit.

Table 9.8, leks (rows 1–4) (*continued*).

5	*6*	*7*	*8*	*9*	*10*
Evolutionary responses of males to female spatiotemporal distributions					
Sexual dimorphism	*Male parental care*	*Sexual conflict*	*Alternative mating strategies*	*Mating system*	*Examples*
Male structures for aggression and defense of mating sites; also males displays to attract female attention	No; males emphasize guarding of mating site	Yes; male mating attempts may damage females	Site guarders vs site satellites	Semelparous exploded leks	Semelparous lekking insects with aggression among males
Males display to attract females; also, adaptations of sperm and seminal fluids to enhance fertilization success.	No, males emphasize mating effort	Yes, but exploitation is balanced somewhat by female countermeasures	Yes, satellite males attempt to intercept females	Semelparous classic leks	Semelparous lekking insects with reduced aggression among males
Male structures for aggression and defense of mating sites; also males displays to attract female attention	No; males emphasize guarding of mating site	Yes; male mating attempts may damage females	Site guarders vs site satellites	Iteroparous exploded leks	Bowerbirds; black grouse
Males display to attract females.	No, males emphasize mating effort	Yes, but exploitation need not negatively influence female fecundity	Yes, satellite males attempt to intercept females	Iteroparous classic leks	Prairie chickens; Sage grouse

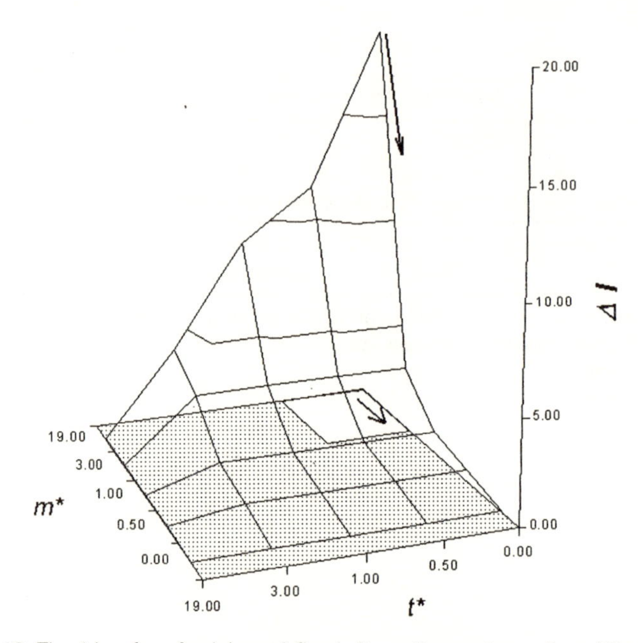

Figure 9.13. The ΔI surface for *leks* and *fixed site* mating systems; the white region on the two dimensional floor of the diagram identifies circumstances in which mean spatial crowding of matings, m^*, is high, and the mean temporal crowding of matings, t^*, low. Aggression among males tends to disperse the spatial distribution of males, which allows more males to successfully mate because the ability of females to make direct comparisons among males as well as to copy the mate choices of other females is reduced; this causes the spatial distribution of females to become less aggregated (lower arrow) and causes ΔI to decrease (upper arrow).

If males in the above circumstances associate with individual females and leave the breeding aggregation to mate, ΔI will be sharply reduced (column 4, row 1, table 9.8). This tendency causes receptive females to become more spatially dispersed, and since mating males leave the breeding aggregation, the number of mating males will increase. Both activities decrease I_{mates} and thereby decrease the magnitude of ΔI. As explained above, males are unlikely to become territorial in such situations unless females approach male aggregations from particular directions where they can be intercepted (Shelley and Whittier 1997). The smaller the sex difference in the opportunity for selection, the less likely elaborate male behaviors and morphologies are to evolve. When males escort semelparous females away from display sites, we expect the mating system to converge toward polygynous mating swarms (row 7, table 9.6).

Semelparous Classic Leks

When m^* is high and t^* is low (columns A and B, row 2, table 9.8; fig. 9.13), and when females are promiscuous but semelparous ($I_{\text{cs,ires}} > 0$, $I_{\text{cs,sclutch}} = 0$;

columns C and D, row 2, table 9.8), ΔI will be large. Sperm competition is possible (column 1, row 2, table 9.8) and female mate copying will also arise for reasons described above (column 2, row 2, table 9.8). Although males could guard individual females after mating to mitigate or prevent sperm competition, the large number of available females attracted to male display sites can make mate guarding less profitable for males than attempting to mate with as many receptive females as possible (chapter 8). Thus, the correlation between male mating success and fertilization success among females (Webster et al. 1995), determines whether ΔI is reduced or enhanced (column 3, row 2, table 9.8). Males are expected to exhibit elaborate behaviors and morphologies that are attractive to females, as well as modifications of their spermatozoa or seminal secretions that enhance the success of sperm in competition with other ejaculates (column 5, row 2, table 9.8).

With female promiscuity, males are not expected either to defend their location on the lek or to exhibit parental care (column 6, row 2, table 9.8). Although both sexes mate more than once, variance in male mate numbers is likely to exceed that of females. Thus, although $I_{cs,sires}$ may affect female fecundity, a large and positive ΔI will still favor conspicuous male exploitation of female fecundity compared to female countermeasures (column 7, row 2, table 9.8). Large ΔI also favors the evolution of alternative male mating strategies, similar to those described in row 1, table 9.8 (column 8, row 2, table 9.8). Because males display to females in particular locations, but exhibit reduced aggressiveness toward females and toward other males (that is, males are more closely spaced than in exploded leks), we call such mating systems *semelparous classic leks* (column 9, row 2, table 9.8). Examples of this mating system may exist in univoltine lekking insects, as described above.

Iteroparous Exploded Leks

When receptive females are aggregated in space, appear asynchronously in time (m^* = high, t^* = low, columns A and B, row 3, table 9.8), tend to mate once, but are iteroparous ($I_{cs,sires} = 0$; $I_{cs,clutch} > 0$; columns C and D, row 2, table 9.8), sperm competition is prevented (column 1, row 3, table 9.8), but female copying is expected to occur for reasons described above (column 2, row 3, table 9.8). Variance in offspring numbers produced by iteroparous females diminishes ΔI, but it remains positive owing to the initially large value of I_{mates} (column 3, row 3, table 9.8). Males are expected to aggressively defend locations where iteroparous females congregate. Females are expected to mate once with the most attractive male in this location and leave to oviposit elsewhere (column 4, row 3, table 9.8). As explained above, male aggression will decrease the spatial clumping of matings and thus decrease I_{mates}.

Males with structures and behaviors that confer an advantage in aggressive mating site defense (column 5, row 3, table 9.8) are unlikely to exhibit parental behavior (column 6, row 3, table 9.8). If female tenure at mating sites is long, the covariance of mate numbers and offspring numbers among males will exceed that of females; thus sexual conflict is expected to exist. Male mating attempts that occasionally injure females as well as outright male aggression toward females are expected to be common. Females are not expected to evolve conspicuous countermeasures to male exploitation (column 7, row 3, table 9.8). The large and positive value of ΔI also favors the evolution of alternative male mating strategies in the form of site guarding versus satellite males (column 8, row 3, table 9.8). Because males aggressively defend display sites visited by iteroparous females, who mate and oviposit at distant locations, we call such mating systems *iteroparous exploded leks* (column 9, row 3, table 9.8). Examples of such mating systems are abundant in birds. Particularly well-known examples include bowerbirds (Borgia 1979; Borgia and Presgraves 1998; Borgia and Coleman 2000; Pruett-Jones and Pruett-Jones 1994) and black grouse (Höglund and Alatalo 1995; column 10, row 3, table 9.8). In all of these species, males aggressively defend display sites whose positions are geographically similar but seldom identical from year to year. Iteroparous females visit male display sites, mate, and depart to oviposit elsewhere.

Iteroparous Classic Leks

If females are promiscuous and iteroparous, then sperm competition is likely (column 1, row 4, table 9.8), to co-occur with female mate copying (column 2, row 4, table 9.8). Like semelparous classic leks (row 2, table 9.8), males could guard individual females after mating to prevent sperm competition. However, the large number of available females at male display sites is expected to make mate guarding less profitable than attempting to mate repeatedly (chapter 8). If male mating success correlates with fertilization success (Webster et al. 1995), ΔI will either be reduced compared to row 3, table 9.8 or may become markedly greater (column 3, row 4, table 9.8). Since females in this mating system tend to mate with more than one male, as in semelparous classic leks, males are expected neither to aggressively defend their location on the display site, nor to exhibit parental care (column 6, row 4, table 9.8).

Because both sexes are promiscuous, sexual conflict exists, and the variance in female fecundity arising from $I_{cs,sires}$ as well as from female iteroparity ($I_{cs,clutch}$) provides females with an opportunity to mitigate this conflict. The greater value of $I_{females}$ resulting from female iteroparity will diminish sexual dimorphism more than in other lekking species. If a high degree of female mate copying or intense sperm competition keeps ΔI large (column 7, row 4, table 9.8), then alternative male mating strategies, including diversity in male sperm and seminal secretions, are also expected (column 8, row 4, table 9.8). Because males display to females in particular locations, but are less aggressive toward females and toward other males, we call such mating systems *iteroparous classic leks* (column 9, row 4, table 9.8). Examples of this mating system are also well known in birds and evidently include prairie chickens and sage grouse (Bradbury and Gibson 1983; Cannon and Knopf 1981; Schroeder and White 1993; Ryan et al. 1998). As described above for semelparous classic leks, we expect intermale distances to be less in this mating system than in iteroparous exploded leks.

Feeding Sites

Semelparous Feeding Site Polygyny

As explained in chapter 8, there are several causes for female aggregation. Female fecundity correlates with body size in many animal species, so the clustering of females at nutritious resources is not surprising. When females assemble en masse at feeding sites, males too are expected to find these aggregations en masse. With extreme female breeding synchrony there can be little opportunity for sexual selection (table 9.3). However, if females arrive at feeding sites *asynchronously*, so that m^* is high and t^* is low (columns A and B, row 1, table 9.9), I_{mates} can become very large. In such situations, if females mate once and produce only one clutch of offspring ($I_{cs,sires}$, $I_{cs,clutch}$ = 0; columns C and D, row 1, table 9.9; fig. 9.13), sperm competition is prohibited (column 1, row 1, table 9.9), and female copying enhances the ability of females to locate patchy feeding sites. This tendency toward female aggregation may become genetically correlated with male aggressiveness (chapters 4 and 7), enhancing I_{mates}. In such circumstances, individual males are expected to defend feeding sites and attempt to mate with each female that arrives there to feed, mate once, and leave to oviposit elsewhere (column 4, row 1, table 9.9; Wade 1995). The intensity of sexual selection in this context will lead to considerable modification of male structures associated with aggression, dominance, and fighting ability (column 5, row 1, table 9.9).

Since male mating success depends primarily on successful resource monopolization, male parental care is unlikely to evolve. As explained in chapter 8, in fixed site mating systems in which females are attracted to food sources, males may even drive females away after they have fed and mated. Moreover, if females nest away from food sources, males will be unable to interact with young (column 6, row 1, table 9.9). Although sexual conflicts of interest are likely to be extreme due to large variance in male mate numbers, females are not expected to respond evolutionarily to male exploitation since I_{clutch} is near zero (column 7, row 1, table 9.9). Thus, male aggression and mating attempts that impede females' ability to feed are likely to be observed, and females may appear to submit to such aggression.

Table 9.9
The influences of the spatiotemporal distribution of sexually receptive females as measured by m^*, t^*, $I_{cs,sires}$, and $I_{cs,clutch}$ on the characteristics of animal mating systems; feeding sites.

column ->	*A*	*B*	*C*	*D*	*1*	*2*	*3*	*4*
	I_{mates}		I_{clutch}					
row	m*	t*	$I_{cs,}$ sires	$I_{cs,}$ clutch	*Sperm competition*	*Female copying*	ΔI	*Expected male-female associations*
9.1	Large (females spatially clumped)	Small (females temporally asynchronous)	0	0	No	Yes, females attracted to other females at feeding sites	M > > F	Males defend feeding sites where semelparous females aggregate to feed and mate; females mate once and leave to oviposit
9.2			>0	0	Yes; males do not defend females after mating	Yes as above	M > > > F	Males defend feeding sites where sem. females aggregate; multiple mating favors conspicuous males, feeding context may eventually be lost
9.3			0	>0	No	Yes as above	M > > F	Males defend feeding sites where iteroparous females aggregate
9.4			>0	>0	Yes as above	Yes as above	M > > F	Males defend feeding sites where iteroparous females aggregate; multiple mating favors conspicuous males, feeding context is eventually lost

Table 9.9, feeding sites (rows 1–4) (*continued*).

5	6	7	8	9	10
Evolutionary responses of males to female spatiotemporal distributions					
Sexual dimorphism	*Male parental care*	*Sexual conflict*	*Alternative mating strategies*	*Mating system*	*Examples*
Male structures for aggression and defense of feeding sites	No; males emphasize guarding of feeding site	Yes; male mating attempts may impede female feeding	Site guarders vs site satellites	Semelparous feeding site polygyny	Solitary bees
Male displays; rapid mating sperm competition	No; males emphasize rapid and frequent mating	Yes; males may enhance female fecundity at female expense	Display repertoire sperm morphology satellite males	Semelparous classic leks	Swallowtail butterflies
Male structures for aggression and fighting	No; males emphasize rapid and frequent mating	Yes; male mating attempts may impede female feeding	Site guarders vs site satellites female mimicry may allow satellites to avoid aggression by guarding males	Iteroparous feeding site polygyny	Neriid flies; sunbirds; hummingbirds
Males displays; rapid mating sperm competition	No; males emphasize rapid and frequent mating	Yes; males may enhance female fecundity at females' expense	Display repertoire sperm morphology satellite males; also female mimics since copying is important to territorial male	Iteroparous classic leks	Lekking Drosphilia; bellbirds; ruffs; epauletted bats

Male alternative mating strategies involving satellite male behaviors are likely to arise, in which smaller or younger males assemble near feeding sites, where they attempt to steal matings from larger or older males (column 8, row 1, table 9.9). The success of these alternative behaviors will be in proportion to the synchrony with which females arrive at feeding sites. As we will explain in chapter 11, the more females are attracted to particular site-guarding males, the smaller the mating success of satellite males needs to be for this strategy to invade and persist. Because semelparous females are attracted to feeding sites that are fixed in space, because females mate only once, and because resource-guarding males can mate with several females in succession, we call such mating systems *semelparous feeding site polygyny* (column 9, row 1, table 9.9). Examples of this mating system appear to include solitary bees (column 10, row 1, table 9.9; Alcock 1979).

If females share the above traits and tend to mate more than once (columns A–D, row 2, table 9.9), sperm competition is possible, and female copying is likely because such behavior enhances the ability of females to locate patchy feeding sites (columns 1 and 2, row 2, table 9.9). A genetic covariance between female tendencies to aggregate and male abilities to defend aggregation sites will also arise. However, when females are promiscuous, males who shift their activities, away from resource defense and toward direct attraction of females, could be favored by sexual selection. Female tendencies to mate more than once can lead to covariance between multiple mating in females and competitive sperm in males (chapters 4 and 7). Moreover, as explained above, male displays that enhance female tendencies to mate can also lead to covariance between female mate preferences and male mating displays (chapters 4 and 7).

The degree to which females are able to exercise mate choice depends on how successfully males can defend groups of females assembled at resources. If male defense prevents females from mating with non-resource-holding males, this mating system will converge toward semelparous feeding site polygyny (row 1, table 9.9). However, if multiple mating by females favors conspicuous males, rather than resource-defending males, the original context for this mating system, that is, the tendency for females to aggregate at food sources, could be lost. When this occurs, the tendency for females to mate with particular displaying males, combined with asynchrony in female receptivity (a situation that allows preferred males to mate with multiple females in sequence), could cause I_{mates} to become extremely large. Although some variance in offspring numbers may arise among females from their matings with different sires, We expect the magnitude of I_{mates} to exceed $I_{\text{cs,sires}}$, so that ΔI remains very large and positive (column 3, row 2, table 9.9).

As females become increasingly attracted to particular male characters, males are expected to cease to defend feeding sites and may begin to form aggregations around males with the most attractive displays (column 4, row

2, table 9.9); that is, around males who draw the most reliable streams of females. As explained in chapters 1, 4, and 7, when the fraction of males who obtain no mates at all becomes large, males who behave as satellites around extremely successful displaying males are favored. At such times, the average success of satellites exceeds the average success of males attempting to attract males by displaying. If females prefer to mate with certain displaying males over others, many displaying males will not mate. Satellite males may shift their allegiances depending on the rate at which females mate with particular males, and if the time scale becomes sufficiently short, males able to rapidly change their mating tactics may be favored by sexual selection (chapter 11). Sexual dimorphism favoring male aggressiveness and dominance can then shift toward conspicuous male displays that covary with female mating preferences. Since multiple mating is likely, adaptations among males associated with sperm competition are also expected to arise (columns 5 and 8, row 2, table 9.9).

With male mating success dependent on display rate (Höglund and Alatalo 1995), male fertilization success dependent on mate numbers (chapter 8), and nesting sites located apart from mating sites, males are not expected to evolve parental care (column 6, row 2, table 9.9). When females approach displaying males to mate and male mating success is not limited by resource depletion, no apparent exploitation of females by males may be observed. However, with multiple insemination, sperm competition could occur at some expense to female fitness (column 7, row 2, table 9.9). Alternative male mating strategies are likely to evolve as described above. Satellite behavior around successful displaying males as well as variation in sperm morphology or seminal secretions that influence fertilization success will be common (column 8, row 2, table 9.9).

Since males display to females in aggregations that do not serve as feeding or nesting sites, and since females produce only one clutch of offspring, such mating systems are likely to converge toward semelparous classic leks (column 9, row 2, table 9.9). Examples of this mating system may include semelparous insects in which males form swarms or display on open surfaces such as leaves, tree limbs, light gaps, or other open areas (e.g., swallowtail butterflies, *Papilio*, Lederhouse 1981; Krebs 1987; column 10, row 2, table 9.9). This mating system differs from semelparous mass mating (table 9.3) in which females and males arrive at breeding sites en masse, thus preventing males from mating with several females in immediate succession.

Iteroparous Feeding Site Polygyny

When receptive females aggregate at feeding sites and are asynchronous in their sexual receptivity, m^* is large, t^* is small, and the opportunity for sexual selection on males is large, as described above (columns A and B, row 3, table 9.9). Monandry prohibits sperm competition, even if females produce more than one clutch of offspring. Female copying could aid females in locating patchy feeding sites, and, as described above, this tendency will enhance the abilities of certain males to defend locations where females aggregate to feed. Although iteroparity decreases I_{mates} (column 2, row 3, table 9.9), the arrival of females at patchy feeding sites in sequence allows ΔI to remain large (column 3, row 3, table 9.9). Males are expected to defend sites to which females are attracted, and to mate with females as they arrive (column 4, row 3, table 9.9). The intensity of sexual selection in this context will lead to considerable modification of male structures associated with aggression, dominance and fighting ability as described in row 1, table 9.4, above (column 5, row 3, table 9.9; fig. 9.13).

Since male mating success depends primarily on successful resource monopolization, male parental care is unlikely to evolve in this mating system. Males are expected to drive females away from resources after females have been mated (column 6, row 3, table 9.9). Sexual conflicts of interest are expected to arise due to the magnitude of ΔI as well as from male attempts to drive females away from feeding sites. Moreover, since I_{clutch} is positive, females are expected to respond evolutionarily to, and mitigate, the deleterious effects of male exploitation (column 7, row 3, table 9.9).

Alternative male mating tactics involving satellite male behaviors are likely to arise, in which smaller or younger males assemble near feeding sites defended by larger or older males, and attempt to steal matings (column 8, row 3, table 9.9). However, as described above, these activities are not expected to be successful unless the synchrony with which females arrive at feeding sites increases. Since iteroparous females are attracted to feeding sites that are fixed in space, since resource-guarding males can mate with several females in succession, and since females mate only once, we call such mating systems *iteroparous feeding site polygyny* (column 9, row 3, table 9.9). Examples of this mating system appear to include iteroparous insects in which males defend inflorescences or other food sources (Alcock 1979; Preston-Mafham 2001), as well as sunbirds and hummingbirds that do not form obvious leks (Evans and Hatchwell 1992; Höglund and Alatalo 1995; Dearborn 1998; column 10, row 3, table 9.9).

When females are promiscuous, the degree to which females are able to exercise mate choice will depend on how poorly males defend females assembled at resources. If male defense prevents females from mating with non-resource-holding males, this mating system will converge toward iteroparous feeding site polygyny. However, with female promiscuity, and a runaway process with conspicuous males, the original feeding site context of such mating systems could be lost, as described above for semelparous feeding site polygyny. The net result is a smaller ΔI than in semelparous leks, with a concomitant decrease in sexual dimorphism (column 3, row 4, table 9.9).

Alternative male mating strategies are likely to evolve and are expected to include satellite behaviors around successful, displaying males as well as variation in sperm morphology or seminal secretions that influence fertilization success (column 8, row 4, table 9.9). Since males display to females in aggregations that do not serve as feeding or nesting sites, and since females produce more than one clutch of offspring, such mating systems are expected to converge toward iteroparous classic leks (column 9, row 4, table 9.9). Examples of this mating system may include iteroparous insects such as *Drosophila* in which males display and are joined by females in succession (Droney 1992) as well as lekking hummingbirds (Atwood et al. 1991; Höglund and Alatalo 1995), bellbirds (Sagar 1985), epauletted bats (Bradbury 1981), and perhaps ruffs (Höglund et al. 1998; Widemo and Owens 1995; column 10, row 4, table 9.9). The difference between this mating system and iteroparous mass mating lies in a much higher degree of female breeding synchrony in the latter mating system.

Nesting Sites—Female Care

Harem Polygynandry

When nesting sites are patchy in their distribution and mating immediately precedes oviposition, the spatial distribution of receptive females can become large (m^* = large; column A, row 1, table 9.10). As in mating systems associated with feeding sites, if females assemble at nesting sites en masse, males too are expected to find these aggregations en masse. Pairs are likely to form and the opportunity for sexual selection will be small. However, if females arrive at nesting sites asynchronously, their temporal distribution becomes dispersed (t^* = low; column B, row 1, table 9.10) and the opportunity for sexual selection on males can become large, especially if females are semelparous and not promiscuous ($I_{cs,sires}$, $I_{cs,clutch} = 0$; columns C-1, row 1, table 9.10).

As explained above (chapter 8), female copying may be favored by natural selection because such behavior can enhance the ability of females to locate patchy nest sites or swamp egg or nest predators. If males defend locations where females aggregate, genetic covariance can arise between female tendencies to aggregate and male abilities to defend aggregation sites, and enhance I_{mates}. The runaway between male defense of nesting sites and female aggregation can be halted in either of two ways. The first is sperm depletion. As more females arrive at a particular nesting site, males may become increasingly less capable of providing sufficient sperm to each female. Second, unfertilized, otherwise monandrous females may seek additional matings to realize their fecundity, a situation which is tantamount to prolonging the duration of female receptivity. In either case, roving males have a potential mating niche (chapter 7). Mating opportunities permit not

Table 9.10

The influences of the spatiotemporal distribution of sexually receptive females as measured by m^*, t^*, $I_{cs,sires}$, and $I_{cs,clutch}$ on the characteristics of animal mating systems; nesting sites: female care.

column ->	A	B	C	D	1	2	3	4
	I_{mates}		I_{clutch}					
row	m*	t*	$I_{cs,}$ sires	$I_{cs,}$ clutch	Sperm competition	Female copying	ΔI	Expected male-female associations
10.1	Large (females spatially clumped)	Small (females temporally asynchronous)	0	0	No	Yes	M > > > F	Males defend nesting sites where semelparous females oviposit and remain to brood young; extreme female clumping and sperm depletion in guarding males compromises effective mate guarding; convergence to semelparous harem polygynandry (row 9.10.2)
10.2			>0	0	Yes	Yes	M > > F	Males defend nesting sites where semelparous females oviposit and remain to brood young; females mate with satellites when they are present
10.3			0	>0	No	Yes	M > > F	Males defend nesting sites where iteroparous females mate once and remain to brood multiple clutches of young; extreme female clumping and sperm depletion in guarding males compromises effective mate guarding; convergence to iteroparous harem polygynandry (row 9.10.4)

Table 9.10, nesting sites: female care (rows 1–3) (*continued*).

5	*6*	*7*	*8*	*9*	*10*
Evolutionary responses of males to female spatiotemporal distributions					
Sexual dimorphism	*Male parental care*	*Sexual conflict*	*Alternative mating strategies*	*Mating system*	*Examples*
see row 9.10.2	see row 9.10.2	see row 9.10.2	see row 9.10.2	see row 9.10.2	see row 9.10.2
Male structures for site defense; large male body size	Possible; male defense of nesting sites may constitute care	Yes; male mate numbers exceed those of females; little ability for females to evolve countermeasures	satellite males; also female mimics since copying is important to territorial male mating success	Semelparous harem polygynandry	Cavity dwelling peracarids
see row 9.10.4	see row 9.10.4	see row 9.10.4	see row 9.10.4	see row 9.10.4	see row 9.10.4

Table 9.10, nesting sites: female care (row 4) (*continued*).

column ->	*A*	*B*	*C*	*D*	*1*	*2*	*3*	*4*
	I_{mates}		I_{clutch}					
row	m*	t*	$I_{cs,}$ sires	$I_{cs,}$ clutch	*Sperm competition*	*Female copying*	ΔI	*Expected male-female associations*
10.4			>0	>0	Yes	Yes	M > F	Males defend nesting sites where iteroparous females oviposit and remain to brood young; multiple mating and sperm storage occur

only satellite males but also guarding males to rove outside their own territories.

If females are promiscuous, "winner-fertilize-all" sperm competition could enhance the variance in offspring number among males and increase the value of ΔI. The magnitude of ΔI is expected to produce structures in males that are useful in nesting territory defense, such as large body size. Since nesting sites are fixed in space, males may also develop structures that are useful in displacing rivals, such has horns, antlers, or bosses (chapter 8; column 5, row 2, table 9.10).

Indirect male parental care could arise since males and offspring are closely associated and males may drive predators as well as rivals away from nesting sites. However, parental care by territorial males may be incidental to their mating activities (column 6, row 2, table 9.10). Sexual conflicts of interest are expected to exist in such species given that males attempt to maximize mate numbers, and, although females also may mate more than once, the magnitude of ΔI will permit considerable male exploitation of their mates (column 7, row 2, table 9.10).

As with other mating systems described in this section, the large, positive value of ΔI favors the evolution of alternative mating strategies. Smaller or younger males are expected to congregate at the margins of nesting territories and attempt to mate with still receptive females. Since female copying is an important source of attraction of females to nesting sites, satellite male

Table 9.10, resting sites: female care (row 4) (*continued*).

5	*6*	*7*	*8*	*9*	*10*
Evolutionary responses of males to female spatiotemporal distributions					
Sexual dimorphism	*Male parental care*	*Sexual conflict*	*Alternative mating strategies*	*Mating system*	*Examples*
Males structures for site defence large body size, weapons	No; males emphasize territory defense and mating effort	Yes; males attempt to mate may decrease female longevity	Satellite males; also female mimics since copying is important to territorial male mating success	Iteroparous herem polygynandry	Marine iguanas pinnepeds

behavior may also involve female mimicry. Such behavior may allow satellite males not only to invade territories, but also to get closer to receptive females within nesting territories. Indeed, female mimics may even enhance the attractiveness of the territories they occupy to additional females, thereby making their presence tolerable by territorial males and increasing average harem size. As suggested in chapter 8, another type of alternative mating strategy, involving males who focus not on mating effort but instead on parental effort, may also appear in such mating systems (column 8, row 2, table 9.10).

Because semelparous females are attracted to nesting sites, where they mate primarily with a male who defends his site against other males, but may also mate with satellite males, we call this mating system *semelparous harem polygynandry* (column 9, row 2, table 9.10). Examples of this mating system appear to exist in certain gnathiid isopods (*Paragnathia*; Upton 1987) and perhaps in some sphaeromatids (e.g., *Cilicaea*; Hurley and Jansen 1977) in which large aggregations of semelparous females brood their young in cavities that are defended by aggressive, weapon-bearing males (column 10, row 2, table 9.10). This mating system differs from male dominance polygynandry (table 9.6) in that the size of breeding aggregations is extremely large, that is, females are *highly*, rather than moderately, clumped in space.

Multiple mating and iteroparity reduces I_{mates} since $I_{\text{cs,sires}}$ as well as $I_{\text{cs,clutch}}$

enhances $I_{females}$. Sexual dimorphism is expected to be decreased in species with iteroparous females, compared to those in which females produce a single clutch of young, unless a positive correlation exists between male mate number and sperm competitive ability. For the same reason, structures useful in displacing rivals such has horns, antlers or bosses (chapter 8; column 5, row 4, table 9.10) are expected to be less well developed in iteroparous species than in semelparous species, although sperm competitive structures may also characterize males in such mating systems. Male parental care, incidental to male territoriality (column 6, row 4, column 9.10) and sexual conflict will exist because, despite female iteroparity, the magnitude of ΔI will permit considerable male exploitation of their mates (column 7, row 4, table 9.10). Females may also exhibit alternative breeding strategies of their own depending on the magnitude of the variance in offspring numbers among females.

In this mating system, iteroparous females are attracted to nesting sites, where they mate primarily with the male who defends his site against other males. However, since females may also mate with satellite males, we call this mating system *iteroparous, harem polygynandry* (column 9, row 4, table 9.10). Examples of this mating system may include marine iguanas (*Amblyrhynchus cristatus*, Trillmich and Trillmich 1984; although this species may also form iteroparous exploded leks; Wikelski et al. 1996) and pinnipeds (Mesnick 1997). In each of these species, iteroparous females congregate on or near rocks, or on beaches to brood their young. Large, pugnacious males defend territories that include the numerous females, and satellite males attend the peripheries of harems and attempt to force or sneak copulations with unguarded females (column 10, row 2, table 9.10). Again, this mating system differs from *dominance polygynandry* (table 9.6) in that the size of breeding aggregations is extremely large, with females highly, rather than moderately, clumped in space. Whereas in dominance polygynandry females may or may not be associated with nesting sites (table 9.5), in harem polygynandry females release or rear young within aggregations.

Nesting Sites—Male Care

Semelparous Nest Site Polygynandry

When females are attracted to patchily distributed nest sites, mate, deposit offspring, and immediately leave, the spatial distribution of females can become highly dispersed. However, as described above for leks, the spatial distribution of *matings* becomes highly aggregated around preferred nest sites (m^* = high, column A, row 1, table 9.11). When females become receptive asynchronously, or, as is more likely, when females must mate with males at preferred nest sites *in sequence*, the temporal distribution of receptive females, and thus of matings, becomes highly dispersed (t^* = low, column B, row 1, table 9.11). If females mate once and produce only one clutch of offspring ($I_{cs,sires}$, $I_{cs,clutch}$ = 0, columns 1 and 2, row 1, table 9.11), sperm competition, at least initially, will be prevented (column 3, row 1, table 9.11). However, as explained for all fixed-site mating systems, female copying is expected to cause females to aggregate around preferred nesting sites. Female copying behavior thus enhances the spatial clumping of receptive females (column 4, row 1, table 9.11).

A genetic covariance between female tendencies to copy each other's nest choices and male attraction to assembled females will result in congregations of males around nest sites as well as around females. As in mating systems involving harems (table 9.10), the attraction of many receptive females to nest sites defended by particular males can lead to sperm depletion. If (Cov[P,m^*] < 0 (chapter 4), initially monandrous females may prolong their periods of receptivity and begin multiple mating. Such conditions diminish ΔI, and will cause this mating system to converge toward the mating system described below (column 5, row 1, table 9.11).

With female promiscuity and semelparity ($I_{cs,sires} > 0$, $I_{cs,clutch}$ = 0; columns A–D, row 2), sperm competition is possible (column 1, row 2, table 9.11), but I_{mates} and ΔI are expected to remain very large because I_{clutch} is small or zero (column 3, row 2, table 9.11). Sperm depletion in territorial males favors the success of satellite males, who steal copulations or mix their ejaculates with those of territorial males attempting to spawn (column 4, row 2, table 9.11). Males are expected to evolve conspicuous structures that attract females to nesting sites, as well as structures and behaviors that are useful in nest site defense. Indirect paternal care as a correlate of aggression toward mating competitors as well as toward predators is the expected result (column 5, row 2, table 9.11).

Although territorial males may lose some fertilizations to satellite or roving males, a large accumulation of zygotes or offspring within a male's territory may serve to stimulate other mate-copying females to oviposit there as well (column 6, row 2, table 9.11). Far from representing the "cruel bind" proposed by Trivers (1972), the more females tend to mate, oviposit, and leave a male guarding a particular nest site, the greater that parental male's fitness will become (Tallamy 2001). Sexual conflict may exist, attendant on the large and positive value of ΔI. However, since females can oviposit and leave nest sites where matings occur, they may suffer little physical damage from mating attempts by males. Although males tending nest sites are in a position to cannibalize zygotes and developing young, they reduce their own fitness by doing so (column 7, row 2, table 9.11).

As described above, alternative male mating strategies in such mating systems are likely to consist of satellite males who congregate at territory edges and specialize as sneak copulators. Female mimics may also gain access to territories by stealth or by their resemblance to receptive females (column 8,

Table 9.11

The influences of the spatiotemporal distribution of sexually receptive females as measured by m^*, t^*, $I_{cs,sires}$, and $I_{cs,clutch}$ on the characteristics of animal mating systems; nesting sites: male care.

column ->	A	B	C	D	1	2	3	4
	I_{mates}		I_{clutch}					
row	m*	t*	$I_{cs,}$ sires	$I_{cs,}$ clutch	Sperm competition	Female copying	ΔI	Expected male-female associations
11.1	Large (females spatially clumped)	Small (females temporally asynchronous)	0	0	No	Yes	M > F	Males defend nesting sites where semelparous females oviposit and leave; female clumping and sperm depletion prevent effective male guarding; convergence to row 9.11.2
11.2			>0	0	Yes	Yes	M > > F	Males defend nesting sites where semelparous females oviposit and leave; sperm depletion in territorial males and persistent satellites lead to multiple paternity; female copying maintains mating success of guarding male at nest site.
11.3			0	>0	No	Yes		Males defend nesting sites where iteroparous females oviposit and leave; female clumping and breeding asynchrony prevent effective male guarding; convergence to row 9.11.4

Table 9.11, nesting sites: male care (row 1) (*continued*).

5	*6*	*7*	*8*	*9*	*10*
Evolutionary responses of males to female spatiotemporal distributions					
Sexual dimorphism	*Male parental care*	*Sexual conflict*	*Alternative mating strategies*	*Mating system*	*Examples*
see row 9.11.2	see row 9.11.2	see row 9.11.2	see row 9.11.2	see row 9.11.2	see row 9.11.2
Male structures for aggressive defense of nesting site; male indicators of parental care;	Yes; males defend nesting sites and their offspring within	Yes; male mate numbers exceed female mate numbers	Satellite males; also female mimics since copying is important to territorial male mating success	Semelparous nest site polygynandry	unknown
see row 9.11.4	see row 9.11.4	see row 9.11.4	see row 9.11.4	see row 9.11.4	see row 9.11.4

Table 9.11, nesting sites: male care (row 4) (*continued*).

column ->	*A*	*B*	*C*	*D*	*1*	*2*	*3*	*4*
	I_{mates}		I_{clutch}					
row	m*	t*	$I_{cs,}$ sires	$I_{cs,}$ clutch	*Sperm competition*	*Female copying*	ΔI	*Expected male-female associations*
11.4			>0	>0	Yes	Yes	M > F	Males defend nesting sites where iteroparous females oviposit and leave; sperm depletion in territorial males and persistent satellites lead to multiple paternity; female copying maintains mating success of guarding male at nest site.

row 2, table 9.11). Since semelparous females are attracted to nest sites where multiple mating occurs, since males are likely to be more variable in mate numbers than females, and since females leave nest sites after mating, we call such mating systems *semelparous nest site polygynandry* (column 9, row 2, table 9.11). Examples of this mating system are unknown at this time, but could exist in univoltine, hemipterous insects with male parental care, or in semelparous species within the Opiliones or Pycnogonida (Bain 1992; Machado and Raimundo 2001; column 10, row 2, table 9.11).

Iteroparous Nest Site Polygynandry

When females are monandrous and iteroparous (columns A–D, row 3, table 9.11), the results are similar to those described above. Since females vary in clutch numbers, $I_{cs,clutch}$ will significantly erode I_{mates}. Although ΔI may still remain positive, the degree of sexual dimorphism in such mating systems is expected to diminish in proportion to $V_{cs,clutch}$ (column 3, row 4. table 9.11). Sexual conflict may exist, although there may be little obvious aggression between the sexes since fitness in both sexes depends on numbers of matings (column 7, row 4, table 9.11). Nest-defending males are unlikely to abandon nests containing zygotes fertilized by other males since their mating success depends on accumulated egg masses. In addition to tolerating matings by

Table 9.11, nesting sites: male care (row 4) (*continued*).

5	6	7	8	9	10
Evolutionary responses of males to female spatiotemporal distributions					
Sexual dimorphism	*Male parental care*	*Sexual conflict*	*Alternative mating strategies*	*Mating system*	*Examples*
Male indicators of parental care; Male structures for aggression and fighting	Yes; males defend nesting sites and their offspring within	Yes; male mate numbers exceed female mate numbers; females may eat eggs	Satellite males; also female mimics since copying is important to territorial male mating success	Iteroparous nest site polygynandry	Belostomatids; certain opiliones; sea spiders; sculpins; sticklebacks; blennies; carcinids; plethodontid salamanders; leptodactylid frogs; sandpipers, stilts.

sneaker males, nesting males may steal eggs from other nests and place them in their own to lure females.

Because iteroparous females are attracted to nest sites where multiple mating occurs, because males are likely to be more variable in mate numbers than females, and because females leave nest sites after mating, we call such mating systems *iteroparous nest site polygynandry* (column 9, row 2, table 9.11). Examples of this mating system appear to exist among nereid polychaetes (Ridley 1978) as well as insects in the family Belostomatidae (Smith 1979, 1984, 1997; Ichilkwa 1995). They also exist among opilione spiders (Mora 1990; Martens 1993; Machado and Raimundo 2001), in sea spiders (Pycnogonida; Bain 1992), and in bottom dwelling or coastal fish species such as sculpins (Downhower and Brown 1979), sticklebacks (Oestlund and Ahnesjö 1998; Kuenzler and Bakker 2000), and carcinids (Dominey 1981; Gross 1982) (column 10, row 4, table 9.11). A significant fraction of amphibians (14%) may also exhibit this mating system, including plethodontid salamanders and leptodactylid frogs (Ridley 1978). Several birds may also exhibit this mating system, including sandpipers and stilts (Delehanty et al. 1998; Oring et al. 1994). External fertilization in many of these species is consistent with the expectation that mating systems with male parental care should have low probability of sperm competition. In species that copulate, female oviposition occurs immediately after mating (Oring et al. 1994;

Smith 1997). However, as explained by Gross and Shine (1981) as well as by Tallamy (2001), external fertilization alone is not sufficient to favor the evolution of male parental care in these species. As these authors argue, male opportunities to mate, rather than displaced male parental investment, seems the best explanation for the evolution of this form of male parental care. However, as we will see, confidence of paternity *is* important when males are parental *and* I_{mates} is reduced.

Limited Male Fertility

In iteroparous nest site polygynandry, the greater the asynchrony in female receptivity (i.e., as t^* becomes small), the more some males may attract mates and the more other males will be excluded from mating. When constraints on brooding space arise, the quality of male parental care may decrease with increasing clutch numbers, or male mate choice may be more apparent. Examples include species in which males brood embryos internally or on their bodies (Smith 1979; Berglund et al. 1989). Alternatively, ecological circumstances, such as increased predation or short breeding seasons (Oring et al. 1994), may reduce the ability of males to care for large numbers of progeny. If such circumstances arise after nest site polygynandry evolves, increased demands on male care or increasing male discrimination could limit the number of females ovipositing within a single male's territory.

These limitations, in turn, reduce the variance in offspring numbers for males while increasing it for females. If males can rear only one or at most a few clutches, then the influence of the spatial distribution of males on female mating success, will exceed the influence of the spatial distribution of females on male mating success. Similarly, if males can tend but a limited number of young, the timing of female receptivity may constrain male fertility less than female fecundity is constrained by the temporal availability of parental males. When the availability of males limits female fecundity, females are likely to respond by moving between males to locate individuals that are still able to accommodate their young. This is expected to occur until females bearing ova locate all of the available parental males. The more dispersed nesting males are in space, the more individual females will tend to associate with individual males. In contrast, the more spatially aggregated nesting males are, the more individual females will tend to monopolize male groups. The time scale over which females search may be variable, but, as described above, increasing synchrony in the availability of nesting males will allow a larger proportion of females to successfully locate nests. Decreasing synchrony in the availability of nesting males increases the opportunity for certain females to oviposit in several nests in sequence. Since the ability of each male to tend young is limited, and since the ability of females to locate males will vary, some females will be unable to locate any nesting males. Such females will be prevented from breeding altogether.

This set of circumstances *decreases* the variance in mate numbers among males, and simultaneously *increases* the variance in mate numbers among females. Male mating success may depend less on the spatial and temporal distribution of females than on the number and size of the clutches of offspring they rear (chapter 5). When female mating success depends on the availability of nesting males, variance in offspring numbers among females depends on the spatial and temporal distributions of parental males. In order to understand how selection operates between the sexes in such mating systems, it is still necessary to estimate the value of ΔI. However, we must first reconsider what to measure.

When a male mates with only one female, he is monogynous (chapter 6). When a female mates with more than one male, she is polyandrous. Earlier in this chapter, we described two polyandrous mating systems, *attendance polyandry* (table 9.2), as may occur in certain mantids (Lawrence 1992) and spiders (Andrade 1996), as well as polyandrous mating swarms (table 9.7), as occurs in many social hymenopterans (Wilson 1971). In attendance polyandry, males search for spatially dispersed females, mate, and after mating, are often devoured. Females usually oviposit after mating, but may cannibalize additional mates, and oviposit again. In polyandrous mating swarms, males join large aggregations of conspecific males, where they compete for access to females, who are attracted to and enter the swarm. In the case of social hymenopterans, successful males engage females with their genitalia and transfer sperm, but after copulation is complete, females, in their attempts to detach themselves from males, tear open the male's abdomen. Although males are not consumed entirely, females may eat some of their mate's body when they remove his residual genitalia. Males die after a single mating and females often return to the mating swarm to mate again (Franck et al. 2000), evidently because the long reproductive tenure of females makes reproduction sperm limited.

In both of these mating systems, males devote considerable energy toward *prezygotic* mating effort. Not only do males search for females or travel to locations where potential mates are found, but they also contribute part or all of their somatic mass to females for the privilege of mating. This pattern of male behavior stands in sharp contrast to that of males in mating systems involving direct male parental care. While males in these latter species may devote some energy to prezygotic mating effort, the amount they need to expend to successfully mate is considerably less.

There are three classes of mating systems in which males primarily emphasize postzygotic mating effort. In nest site polygynandry (table 9.11), females are attracted to nest sites defended by males. Mating takes place at nests, females deposit their young within nests and leave, often visiting the nests of other males. In such situations, males mate with more than one female and the variance in mate numbers among males is large, often larger than that of females. Two other mating systems with exclusive male parental care also exist, although they differ from the first in the relative degree to which male fecundity diminishes with increasing mate numbers. In each of these cases, ecological, energetic, or brooding space constraints place an upper limit on the number of offspring that a male can successfully rear in each nest.

In the first case, the upper limit is *greater than* the average clutch size. Thus, certain males can attract more than one mate although the variance in mate numbers among males is *exceeded* by that among females. We label such mating systems as *nest site polyandrogyny.*

In the second type of mating system with exclusive and constrained male care, the upper limit on offspring numbers per nest is *less than or equal to* the average clutch size. In this case, variance in male mate numbers *does not exist.* Moreover, if in such situations, females can mate and oviposit in the nests of multiple males, variance in female mate numbers can become extremely large. Since variance in male mate numbers is zero, but variance in female mate numbers can become extreme, we can identify such mating systems as *nest site polyandry.* Thus, there are two additional classes of mating systems in which males defend nest sites that are visited by females; polyandrogynous and polyandrous.

Unlike nest site polygynandry, in both of these latter classes of mating systems, the variance in offspring numbers among males *is less than* variance in offspring numbers among females. In species exhibiting these mating systems, therefore, the sex difference in the opportunity for selection, ($I_{clutch} - I_{mates} > 0$) arises from differences in mate numbers *among females*, rather than among males. Since female mating success depends on the spatial and temporal distribution of nesting males, we can measure I_{clutch} for females by measuring the mean spatial crowding, m^*, and temporal crowding, t^*, of *nesting males.* It is interesting to note that, although we may measure I_{mates} among females in polyandrogynous and polyandrous mating systems with male parental care, differences in sire numbers ($I_{cs,sires}$) and clutch numbers (I_{clutch}) among females still have a powerful influence on the magnitude of I_{mates} because they determine the variance in offspring numbers among males. Thus, in polyandrogynous as well as in polyandrous mating systems, ΔI will usually be *negative.* As we will see, when ΔI remains at zero, the mating system is likely to converge toward some form of monogamy.

Polyandrogyny

Eumonogamy

We will now consider *polyandrogynous* mating systems in which the sex difference in the intensity of selection favors females (columns A–2, row 1, table 9.12). Limitations on male parental care, the spatial dispersion of nests, the synchrony with which nests become available, female monandry, and female semelparity all reduce the value of ΔI (column 3, row 1, table 9.12). Females are expected to search for males occupying dispersed nests, mate, and oviposit within nests. Moreover, monandrous, females are expected to remain with their mates and assist in rearing offspring; thus, such mating systems will converge toward eumonogamy (table 9.1).

When nesting males are spatially dispersed and become available to females synchronously in time (m^* = small, t^* = large, columns A and B, row 2, table 9.12), the opportunity for sexual selection on either sex, I_{mates}, is limited. If females mate more than once but have only one clutch of offspring ($I_{\text{cs,sires}} > 0$, $I_{\text{cs,clutch}} = 0$; columns C and D, row 1, table 9.12), sperm competition is possible (column 1, row 1, table 9.12). However, if the ability of males to provide care for offspring is also limited, then males may be unreceptive to previously mated females. With I_{mates} near zero, this mating system too is expected to converge toward eumonogamy (table 9.1).

Similar results are expected for all mating systems with constraints on the spatial and temporal distribution of nesting, care-giving males, regardless of female promiscuity (columns A–D, row 3, table 9.12), or iteroparity (columns A–D, row 4, table 9.12). Dispersed nests and male breeding synchrony, combined with male paternity-assurance behavior, are expected to prevent females as well as males from securing multiple mates. Convergence toward eumonogamous mating systems will be a common result (tables 9.1 and 9.12).

Cursorial Polyandrogyny

When nesting males are spatially dispersed and become available to females asynchronously in time (m^* = small, t^* = large, columns A and B, row 5, table 9.12), I_{clutch}, can become moderately large for females. As described above, males with limited brooding space are expected to prevent previously mated females from ovipositing in their nests. When females are capable of mating only once (column C, rows 5 and 7, table 9.12), these limitations are expected to cause females to remain with the first male they locate and jointly care for young, perhaps reducing their fecundity if iteroparous and investing the residual energy in parental care (column D, row 7, table 9.12), or, perhaps, since care of the young involves both parents, allowing females to oviposit repeatedly with the same male. Such mating systems also converge toward eumonogamy (table 9.1).

Parental ability, while well developed in males, may be lacking in females (column 6, row 8, table 9.12). Given a negative ΔI, sexual conflict affecting males rather than females may occur (column 7, row 8, table 9.12). Given limited male brooding space, females may attempt to destroy or replace existing young to produce multiple clutches of progeny (Ichikawa 1995). Such

Table 9.12

The influences of the spatiotemporal distribution of nesting males as measured by m^*, t^*, $I_{cs,sires}$, and $I_{cs,clutch}$ on the characteristics of animal mating systems; polyandrogyny.

column ->	*A*	*B*	*C*	*D*	*1*	*2*	*3*	*4*
	I_{mates}		I_{clutch}					
row	m*	t*	$I_{cs,}$ sires	$I_{cs,}$ clutch	*Sperm competition*	*Female copying*	ΔI	*Expected male-female associations*
12.1	Small (nesting males spatially dispersed)	Large (male availability temporally synchronous)	0	0	No	No	M = F	Females search for dispersed nest sites; mate with males and remain to rear young; convergence to eumonogamy.
12.2			>0	0	Possible, but males unlikely to allow multiple mating w/constrained male care; convergence to row 9.12.1	No		
12.3			0	>0	No	No	M = F	Females search for dispersed nest sites; mate with males and remain to rear young; convergence to eumonogamy.
12.4			>0	>0	Possible, but males unlikely to allow multiple mating w/constrained male care; convergence to row 12.3	No		

Table 9.12, polyandrogyny (rows 1–4) (*continued*).

5	*6*	*7*	*8*	*9*	*10*
Evolutionary responses of males to female spatiotemporal distributions					
Sexual dimorphism	*Male parental care*	*Sexual conflict*	*Alternative mating strategies*	*Mating system*	*Examples*
See table 9.1	See table 9.1	See table 9.1	See table 9.1	See table 9.1	See table 9.1
See table 9.1	See table 9.1	See table 9.1	See table 9.1	See table 9.1	See table 9.1

Table 9.12, polyandrogyny (*continued*).

column ->	A	B	C	D	1	2	3	4
	I_{mates}		I_{clutch}					
row	m*	t*	$I_{cs,}$ sires	$I_{cs,}$ clutch	*Sperm competition*	*Female copying*	ΔI	*Expected male-female associations*
12.5	Small (nesting males spatially dispersed)	Small (male availability temporally asynchronous)	0	0	No	No	M = F	Females search for dispersed nest sites; mate with males and remain to rear young; convergence to eumonogamy.
12.6			>0	0	Possible, but males unlikely to allow multiple mating w/constrained male care; convergence to row 9.12.5	No		
12.7			0	>0	No	No	M = F	Females search for dispersed nest sites; mate with males and remain to rear young; convergence to eumonogamy.
12.8			>0	>0	Possible, but males unlikely to allow multiple mating w/constrained male care;	No	M ≤ F	Females search for dispersed nest sites; mate with males and move on to find others

Table 9.12, polyandrogyny (rows 5–8) (*continued*).

5	*6*	*7*	*8*	*9*	*10*
Evolutionary responses of males to female spatiotemporal distributions					
Sexual dimorphism	*Male parental care*	*Sexual conflict*	*Alternative mating strategies*	*Mating system*	*Examples*
See table 9.1	See table 9.1	See table 9.1	See table 9.1	See table 9.1	See table 9.1
See table 9.1	See table 9.1	See table 9.1	See table 9.1	See table 9.1	See table 9.1
Female attributes that enhance movement between nests; allow competition for limited nest sites; attract the attention of males.	Well developed; in males, no parental care by females.	Yes, females may damage males in mating attempts may also destroy existing young	Females may compete for access to limited nests, may destroy existing young	Cursorial polyandrogyny	Sea spiders; asian belostomatids; broad-nosed pipefish; cardinalfish; tidewater gobies.

Table 9.12, polyandrogyny (*continued*).

column ->	*A*	*B*	*C*	*D*	*1*	*2*	*3*	*4*
	I_{mates}		I_{clutch}					
row	m*	t*	$I_{cs,}$ sires	$I_{cs,}$ clutch	*Sperm competition*	*Female copying*	ΔI	*Expected male-female associations*
12.9	Large (nesting males spatially clumped)	Large (male availability temporally synchronous)	0	0	No	Yes; but only enhances tendencies for social breeding	M = F	Females search for aggregated nest sites; mate with males and remain to rear young; convergence to semelparous mass mating with male parental care
12.10			>0	0	Possible, but males unlikely to allow multiple mating w/constrained male care; convergence to row 12.9	Yes; but only enhances tendencies for social breeding		
12.11			0	>0	No	Yes; but only enhances tendencies for social breeding	M = F	Females search for aggregated nest sites; mate with males and remain to rear young; convergence to iteroparous mass mating with male parental care

Table 9.12, polyandrogyny (rows 9–11) (*continued*).

5	*6*	*7*	*8*	*9*	*10*
Evolutionary responses of males to female spatiotemporal distributions					
Sexual dimorphism	*Male parental care*	*Sexual conflict*	*Alternative mating strategies*	*Mating system*	*Examples*
See table 9.3	See table 9.3	See table 9.3	See table 9.3	See table 9.3	See table 9.3
See table 9.3	See table 9.3	See table 9.3	See table 9.3	See table 9.3	See table 9.3

Table 9.12, polyandrogyny (*continued*).

column ->	*A*	*B*	*C*	*D*	*1*	*2*	*3*	*4*
	I_{mates}		I_{clutch}					
row	m*	t*	$I_{cs,}$ sires	$I_{cs,}$ clutch	*Sperm competition*	*Female copying*	*ΔI*	*Expected male-female associations*
12.12			>0	>0	Possible, but males unlikely to allow multiple mating w/ constrained male care; nesting synchrony prevents females from mating w/ multiple males convergence to mass mating w/ male care.	Yes; but only enhances tendencies for social breeding	M = F	See table 9.3
12.13	Moderate (nesting males moderately aggregated)	Moderate (males availability moderately asynchronous)	0	0	No	Yes; but only enhances tendencies for social breeding	M = F	Females search for moderately aggregated nest sites; mate with males and remain to rear young; possible convergence to mass mating with male parental care; groups moderate in size.

Table 9.12, polyandrogyny (rows 12–13) (*continued*).

5	*6*	*7*	*8*	*9*	*10*
Evolutionary responses of males to female spatiotemporal distributions					
Sexual dimorphism	*Male parental care*	*Sexual conflict*	*Alternative mating strategies*	*Mating system*	*Examples*
See table 9.3	See table 9.3	See table 9.3	See table 9.3	See table 9.3	Seahorses
See table 9.3	See table 9.3	See table 9.3	See table 9.3	See table 9.3	See table 9.3

Table 9.12, polyandrogyny (*continued*).

column ->	A	B	C	D	1	2	3	4
	I_{mates}		I_{clutch}					
row	m*	t*	$I_{cs, sires}$	$I_{cs, clutch}$	*Sperm competition*	*Female copying*	ΔI	*Expected male-female associations*
12.14			>0	0	Possible, but males unlikely to allow multiple mating w/constrained male care; convergence to row 9.12.13	Yes; but only enhances tendencies for social breeding		
12.15			0	>0	No	Yes; but only enhances tendencies for social breeding	M = F	Females search for moderately aggregated nest sites; mate with males and remain to rear young; possible convergence to mass mating with male parental care; groups moderate in size.
12.16			>0	>0	Possible, but males unlikely to allow multiple mating w/constrained male care.	Yes; but only enhances tendencies for social breeding	M < F	Females search for moderately aggregated nest sites; mate with males (after males gain confidence of paternity) and may attempt to locate other nests; convergence to cursorial polyandrogyny.

Table 9.12, polyandrogyny (rows 14–16) (*continued*).

5	6	7	8	9	10
Evolutionary responses of males to female spatiotemporal distributions					
Sexual dimorphism	*Male parental care*	*Sexual conflict*	*Alternative mating strategies*	*Mating system*	*Examples*
See table 9.3	See table 9.3	See table 9.3	See table 9.3	See table 9.3	See table 9.3
See row 12.8	See row 12.8	See row 12.8	See row 12.8	See row 12.8	See row 12.8

Table 9.12, polyandrogyny (*continued*).

column ->	*A*	*B*	*C*	*D*	*1*	*2*	*3*	*4*
	I_{mates}		I_{clutch}					
row	m*	t*	$I_{cs,}$ sires	$I_{cs,}$ clutch	*Sperm competition*	*Female copying*	ΔI	*Expected male-female associations*
12.17	Moderate (nesting males moderately aggregated)	Small (male availability temporally asynchronous)	0	0	No	Yes; but only enhances tendencies for social breeding	M = F	Females search for moderately aggregated nest sites; mate with males and remain to rear young; possible convergence to mass mating with male parental care; groups moderate in size.
12.18			>0	0	Possible, but males unlikely to allow multiple mating w/constrained male care; convergence to row 9.12.17	Yes; but only enhances tendencies for social breeding		
12.19			0	>0	No	Yes; but only enhances tendencies for social breeding	M = F	Females search for moderately aggregated nest sites; mate with males and remain to rear young; possible convergence to mass mating with male parental care; groups moderate in size.

Table 9.12, polyandrogyny (rows 17–19) (*continued*).

5	*6*	*7*	*8*	*9*	*10*
Evolutionary responses of males to female spatiotemporal distributions					
Sexual dimorphism	*Male parental care*	*Sexual conflict*	*Alternative mating strategies*	*Mating system*	*Examples*
See table 9.6	See table 9.6	See table 9.6	See table 9.6	See table 9.6	See table 9.6
See table 9.6	See table 9.6	See table 9.6	See table 9.6	See table 9.6	See table 9.6

Table 9.12, polyandrogyny (*continued*).

column ->	*A*	*B*	*C*	*D*	*1*	*2*	*3*	*4*
	I_{mates}		I_{clutch}					
row	m*	t*	$I_{cs, sires}$	$I_{cs, clutch}$	*Sperm competition*	*Female copying*	ΔI	*Expected male-female associations*
12.20			>0	>0	Possible, but males unlikely to allow multiple mating w/constrained male care; males expected to reject non-virgins or females who do not permit male to gain confidence of paternity	Yes; enhances tendencies for social breeding also permits alternative females mating strategies	M < < F	Females defend small groups of nesting males mate and oviposit as males become available
12.21	Large (nesting males spatially clumped)	Moderate (male availability moderately asynchronous)	0	0	No	Yes; but only enhances tendencies for social breeding	M = F	Females search for aggregated nest sites; mate with males and remain to rear young; convergence to mass mating with male parental care

Table 9.12, polyandrogyny (rows 20–21) (*continued*).

5	*6*	*7*	*8*	*9*	*10*
Evolutionary responses of males to female spatiotemporal distributions					
Sexual dimorphism	*Male parental care*	*Sexual conflict*	*Alternative mating strategies*	*Mating system*	*Examples*
Yes; females aggressive, larger than males, possibly with weapons	Well developed; in males, no parental care by females; mating effort emphasized.	Yes, females exploit males	Yes, satellite females attempt to take over control of nests; possible infanticide to allow oviposition.	Dominance polyandrogyny	Jacanas; phalaropes; narrow-nosed pipefish
See table 9.3	See table 9.3	See table 9.3	See table 9.3	See table 9.3	See table 9.3

Table 9.12, polyandrogyny (*continued*).

column ->	*A*	*B*	*C*	*D*	*1*	*2*	*3*	*4*
	I_{mates}		I_{clutch}					
row	m*	t*	$I_{cs,}$ sires	$I_{cs,}$ clutch	*Sperm competition*	*Female copying*	ΔI	*Expected male-female associations*
12.22			>0	0	Possible, but males unlikely to allow multiple mating w/constrained male care; convergence to row 9.12.9	Yes; but only enhances tendencies for social breeding		See row 9.12.9
12.23			0	>0	No	Yes; but only enhances tendencies for social breeding	M = F	Females search for aggregated nest sites; mate with males and remain to rear young; convergence to mass mating with male parental care
12.24			>0	>0	Possible, but males unlikely to allow multiple mating w/ constrained male care	Yes; enhances tendencies for social breeding	M < F	Females guard individual nesting males until they become available; attempt extrapair matings as other males become available

Table 9.12, polyandrogyny (rows 22–24) (*continued*).

5	*6*	*7*	*8*	*9*	*10*
Evolutionary responses of males to female spatiotemporal distributions					
Sexual dimorphism	*Male parental care*	*Sexual conflict*	*Alternative mating strategies*	*Mating system*	*Examples*
See table 9.3	See table 9.3	See table 9.3	See table 9.3	See table 9.3	See table 9.3
Moderate; associated with guarding mates but also possibly to attract attention of available males.	Males parental; females maybe but care is compromised by extrapair mating opportunities	Yes; females exploit males	Yes, searching vs guarding	polyandrogynous social pairs	Group nesting birds

Table 9.12, polyandrogyny (*continued*).

column ->	*A*	*B*	*C*	*D*	*1*	*2*	*3*	*4*
	I_{mates}		I_{clutch}					
row	m*	t*	$I_{cs,}$ sires	$I_{cs,}$ clutch	*Sperm competition*	*Female copying*	ΔI	*Expected male-female associations*
12.25	Large (nesting males spatially clumped)	Small (male availability temporally asynchronous)	0	0	No	Yes; enhances tendencies for social breeding	M = F	Females search for aggregated nest sites; mate with males and remain to rear young; convergence to mass mating with male parental care
12.26			>0	0	Possible, but males unlikely to allow multiple mating w/constrained male care; convergence to row 9.12.25	Yes; enhances tendencies for social breeding	M = F	
12.27			0	>0	No	Yes; enhances tendencies for social breeding	M = F	Females search for aggregated nest sites; mate with males and remain to rear young; convergence to mass mating with male parental care

infanticidal activities may constitute an alternative mating strategy among females (column 8, row 8, table 9.12). Since nesting males can rear the young of one or only a few females, whereas females are capable of moving between and mating with many males, we call such mating systems *cursorial polyandrogyny* (column 9, row 8, table 9.12). Although the term *poly-*

Table 9.12, polyandrogyny (rows 25–27) (*continued*).

5	*6*	*7*	*8*	*9*	*10*
Evolutionary responses of males to female spatiotemporal distributions					
Sexual dimorphism	*Male parental care*	*Sexual conflict*	*Alternative mating strategies*	*Mating system*	*Examples*
See table 9.3	See table 9.3	See table 9.3	See table 9.3	See table 9.3	See table 9.3
See table 9.3	See table 9.3	See table 9.3	See table 9.3	See table 9.3	See table 9.3

andry is often used to describe mating systems superficially similar to this one, in most described cases, males mate with and brood the young of *more than one female*, either within or between seasons (Ichikawa 1995; Smith 1997). We consider these mating systems *polyandrogynous* because both sexes mate more than once, but variance in mate numbers among females

Table 9.12, polyandrogyny (*continued*).

column ->	A	B	C	D	1	2	3	4
	I_{mates}		I_{clutch}					
row	m*	t*	$I_{cs,}$ sires	$I_{cs,}$ clutch	*Sperm competition*	*Female copying*	*ΔI*	*Expected male-female associations*
12.28			>0	>0	Possible, but males unlikely to allow multiple mating w/constrained male care	Yes; enhances tendencies for social breeding	M << F	Females defend sites where nesting males are aggregated; mate with males and oviposit in their nests sequentially

exceeds that of males (see table 6.1). Examples may include derived pycnogonids (Bain 1992), asian belostomatid bugs (Ichikawa 1995), pipefish (*Syngnathus*, 1986, Berglund et al. 1989), tidewater gobies (Swenson 1997), and certain cardinalfish (Okuda 1999). We know of no existing examples of cursorial polyandry, *sensu stricto*.

Mass Mating with Male Parental Care

When nesting males are spatially aggregated and become available to females synchronously in time (m^* = large, t^* = large, columns A and B, row 9, table 9.12), the opportunity for sexual selection on females can potentially become large. With monandry and semelparity (column C, rows 9 and 11, table 9.12), females are expected to remain with the first male they locate and jointly care for young, perhaps reducing their fecundity if iteroparous and investing residual energy in parental care (column D, row 11, table 9.12). When females mate more than once (column C, rows 10 and 12, table 9.12), males are expected to reject nonvirgins or attempt to mate repeatedly to enhance their confidence of paternity. The aggregated spatial distribution of parental males combined with male and female tendencies to copy the behavior of other individuals in these groups could cause groups of nesting pairs to become large (column 2, rows 9–12, table 9.12).

This spatial distribution of breeding adults could allow females to move between males and attempt to mate repeatedly. However, male breeding synchrony would limit such opportunities for females. Thus, regardless of the life history characteristics of females, such mating systems are expected to

Table 9.12, polyandrogyny (row 28) (*continued*).

5	6	7	8	9	10
Evolutionary responses of males to female spatiotemporal distributions					
Sexual dimorphism	*Male parental care*	*Sexual conflict*	*Alternative mating strategies*	*Mating system*	*Examples*
Female structures for site defence large body size, weapons	Yes, however females emphasize territory defense and mating effort	Yes; female attempts to mate may decrease male fertility	Satellite females	Harem polyandrogyny	Unknown

converge toward mass mating with male parental care (table 9.3). Examples of this mating system appear to include many seahorses (*Hippocampus*, Vincent 1994; Vincent and Sadler 1995; Masonjones and Lewis 2000). In these species, males provide considerable parental care, but the sexes appear monomorphic and exhibit conventional sex roles. The sex difference in the opportunity for selection calculated from existing data on seahorse reproduction shows, as this hypothesis predicts, that ΔI is small and positive ($\Delta I = 0.23$, Shuster and Wade, unpubl. data). Similar results are expected when nesting males are moderately aggregated in space and moderately asynchronous in the timing of their availability for ovipositing females (rows 13–16, table 9.12), although when females are polyandrous and iteroparous, the mating system could converge toward a form of cursorial polyandrogyny (row 16, table 9.12).

Dominance Polyandrogyny

When nesting males are moderately aggregated in space and become available to females asynchronously in time (columns A and B, row 17, table 9.12), females are expected to defend small groups of nesting males, mating and ovipositing in nests as they become available (column 4, row 20, table 9.12). In such classic cases of sex role reversal, females are expected to become aggressive, larger in body size, and more brightly colored to attract the attention of nesting males (column 5, row 20, table 9.12). Parental ability, while well developed in males, is expected to be lacking in females (column 6, row 20, table 9.12). Given the large and negative ΔI, sexual conflict is expected to exist, and males, rather than females, are expected to suffer damage or exploitation from females attempting to mate more than once (column 7, row 20, table 9.12). With limited male brooding space, females may attempt to destroy existing young in their attempts to produce multiple clutches of progeny as described above (column 8, row 20, table 9.12). Females may become satellites around aggregations of nests, or become subordinate to dominant females, helping to defend nest sites in exchange for matings and opportunities to lay eggs in established nests.

Since nesting males can rear the young of one or only a few females, whereas females who aggressively defend small groups of nesting males are capable of moving among and mating with many males, we call such mating systems *dominance polyandrogyny* (column 9, row 20, table 9.12). Whereas the term *polyandry* is often used to describe mating systems superficially similar to this one, in most described cases, males in such mating systems mate with and brood the young of more than one female (see below). We consider these mating systems *polyandrogynous* rather than *polyandrous* because both sexes mate more than once, and because the variance in mate numbers among females exceeds that of males (see table 6.1). Examples appear to include straight nosed pipefish (*Nerophis*, Berglund et al. 1989; Rosenqvist, 1993), as well as phalaropes and jacanas (Delehanty et al. 1998; Butchart 2000). We know of no existing examples of dominance polyandry, *sensu stricto*.

Polyandrogynous Social Pairs

When nesting males are aggregated in space and become available to females with moderate asynchrony (columns A and B, row 21, table 9.12), but brooding space is limited and females mate only once (column C, rows 21 and 23, table 9.12), females may remain with the first male they locate and jointly care for young (column D, row 23, table 9.12). Such mating systems may also converge to mass mating with male parental care (table 9.3).

Polyandrous, iteroparous females are expected to mate with and defend individual nesting males within a large group of similar pairs and remain vigilant for opportunities to mate and oviposit in nests of other males when they become available (column 4, row 24, table 9.12). Females are expected to become only moderately dimorphic compared to males, although they may develop bright pigmentation to attract the attention of males that become available after females have initially paired (column 5, row 24, table 9.12). Parental ability, while apparent in both sexes, is expected to be less well developed in females (column 6, row 24, table 9.12). Given that most females will acquire mates, alternative mating tactics among females are expected to consist primarily of attempts to secure extra-pair copulations (column 8, row 24, table 9.12).

Since nesting males are attended by individual females who cooperate to rear young but also may attempt to secure extra-pair copulations with adjacent nesting males, we call such mating systems *polyandrogynandrous social pairs* (column 9, row 24, table 9.12). This mating system is very similar to polygynandrous social pairs except that females rather than males are the primary seekers of extra-pair matings. It is possible that these two mating systems could persist within the same species with the sign of ΔI varying among generations or localities. Moreover, with joint parental care, it may be possible for males to abandon their mates and mate with other females, making polygynandrous social pairs and polyandrogynous social pairs indistinguishable within the same species. In such cases, these mating systems are better described as polygamous social pairs (table 9.6). As explained in the discussion of table 9.6, many species of group nesting birds may exhibit this mating system.

Harem Polyandrogyny

When nesting males are highly aggregated in space and are available to females asynchronously (columns A and B, row 25, table 9.12), the opportunity for sexual selection on females, I_{clutch}, is extremely large. With large and negative ΔI, females are expected to defend extremely large groups of nesting males, mating and ovipositing in nests as they become available (column 4, row 20, table 9.12). Females are also expected to become extremely aggressive, larger in body size and more brightly colored than males, and possibly exhibit weapons to exclude other females from breeding aggregations (column 5, row 28, table 9.12). Parental ability, while well developed in males, is expected to be lacking in females (column 6, row 28, table 9.12). Moreover, sexual conflict is expected to exist, with males and their nests suffering damage or exploitation by females attempting to mate or oviposit more than once (column 7, row 28, table 9.12). Females may become satellites around aggregations of nests, or become subordinate to dominant females, helping to defend nest sites in exchange for occasional matings (column 8, row 28, table 9.12).

Since nesting males can rear the young of one or only a few females, whereas females aggressively defend large groups of nesting males, we call such mating systems *harem polyandrogyny* (column 9, row 28, table 9.12). We know of no actual examples of this mating system, nor do we know of any examples of harem polyandry. In all cases in which females defend groups of males, these groups tend to be relatively small. Thus, it appears that examples mating systems in which females tend large aggregations of nesting males do not exist in nature.

Polyandry

We next consider *polyandrous* mating systems. That is, mating systems in which m^* is large and t^* is small for nesting males, and in which the total number of offspring each nesting male can rear is severely limited. Here, the maximum number of offspring a male can successfully rear is equal to or less than the average female clutch size. In such mating systems, males are unable to mate with more than one female, although females may have multiple mates. Thus, the variance in female mating success may far exceed that of males.

Truly polyandrous mating systems with male parental care *could* conceivably exist in nature, but evidently they do not. A possible exception may occur in situations in which males die after mating (tables 9.3 and 9.6) or after nearing a single brood perhaps as in polyandrous social pairs. However, as we have argued, this latter mating system is likely to coexist with polyandrogynous and polynandrous social pairs, or converge toward polygamous social pairs, wherein joint parental care is likely to favor mutual exploitation by males and females. We have argued that most apparently polyandrous mating systems are actually *polyandrogynous*. That is, these mating systems represent situations in which *both sexes* mate more than once, but variance in female mating success is greater than that of males. It is our position that the reason polyandrogynous mating systems persist, whereas polyandrous mating systems usually do not, is because males are able to mate more than once.

We have argued that when males provide parental care to young and brood care is limited, males may exert little prezygotic reproductive effort. In fact, nearly all male reproductive effort in such species is postzygotic. In contrast, females in such mating systems must move between males to locate oviposition sites, and female reproductive effort becomes more conspicuously prezygotic. We suggest that increasing emphasis on parental care by males, combined with increasing emphasis on mate numbers by females, conspire to shift patterns of initial parental investment. As male parental care increases and mate numbers decrease, sperm production as well as structures used to defend nests from other males become energetically expensive. Similarly, females must reduce gamete size if they are to increase mate numbers, and structures useful in locating and/or defending locations containing nesting males may be favored.

Changes in the physical characteristics of males and females under such circumstances are well known as "sex role reversal" (Smith 1981, 1984, 1997; Gwynne 1984; Alcock 1995; Svensson 1997; Delehanty et al. 1998; Ritchie et al. 1998). This condition describes what we have called cursorial polyandrogyny, dominance polyandrogyny, or mass mating with male parental care, depending on the degree to which females move among or defend aggregations of nesting males (table 9.12), or females rather than males, form mating swarms. However, we further suggest that when males become parental and are prevented from multiple mating altogether, the process of reversed sexual dimorphism may continue unchecked. Moreover, patterns of initial parental investment between the sexes (i.e., gamete size) may follow this same course until sexual reversal is complete. That is, individuals who were ancestrally male are expected to become indistinguishable from females in other, nonreversed species and vice versa. Subsequently, depending on the spatial and temporal distributions of parental "males," who are now indistinguishable from females in most nonreversed species, the mating system may converge toward any of the mating systems previously described in this chapter.

Such complete transformations are not expected if initial parental investment is considered fixed, as implied by Williams (1966), Trivers (1972), and Emlen and Oring (1977). In these discussions, relative gamete size is used to define the sexes; that is, males are the producers of small motile gametes and females are the producers of large, nonmotile gametes. We suggest that sexual transformation may occur over evolutionary time when ΔI is high first for one-sex and subsequently for the other.

Examples of so-called "sex role reversal" (see the above references) consistently show anisogamy of an ancestral form, although the degree to which this condition exists may be variable compared to fixed-sex species (Shuster and Bain, in prep.). Such cases may represent circumstances in which significant opportunities for sexual selection persist for both sexes. While circumstances arise in which I_{clutch} for females exceeds I_{mates} for males, periodic increases in I_{mates} due to variation in resource availability or sperm competition may prevent complete transformation of the pattern of primary parental investment. Taxa exhibiting nest site polygynandry, and particularly those exhibiting nest site polyandrogyny, are predisposed toward partial or compete sex role reversals. It is not surprising therefore that ratios of maximum reproductive rate between the sexes (Clutton-Brock and Vincent 1991) differ in these species compared to species with other mating systems. Moreover, within these susceptible taxa, sexual reversals could explain diverse patterns of chromosomal sex determination (Jones and Singh 1985; Green et al. 1993; Kraak and De Looze 1993; Baroiller et al. 1996; Hayes 1997; Quintana-Murci et al. 2001), as well as diversity in sperm morphology within apparently monophyletic taxa (Sivinski 1980; Pitnick 1996; Snook 1997). In particular, we expect examples of female heterogamety to be disproportionately higher in taxa in which nest site polygynandry and nest site polyandrogyny occur.

Chapter Summary

We have described twelve major categories of mating systems: (1) sedentary pairs, (2) itinerant pairs, (3) mass mating, (4) polygamy, (5) dominance polygyny, (6) social pairs, (7) mating swarms, (8) leks, (9) feeding sites, (10) nesting sites with female care, (11) nesting sites with male care, and (12) polyandrogyny (table 9.13). Within each of these categories, one or more

Table 9.13
Major categories and subcategories of mating systems.

Major Category	*Subcategory*
Sedentary pairs	Eumonogamy
	Persistent pairs
	Sequential pairs
Itinerant pairs	Attendance polygyny
	Attendance polygynandry
	Attendance polyandry
	Coercive polygynandry
Mass mating	Semelparous mass mating
	Mass mating with male parental care
	Iteroparous mass mating
Polygamy	Cursorial polygyny
	Polygamy
	Iteroparous classic leks
Male dominance	Dominance polygyny
	Dominance polygynandry
Social pairs	Polygynous social pairs
	Polygamous social pairs
	Polygynandrous social pairs
Mating swarms	Eumonogamy
	Persistent pairs
	Polyandrous mating swarms
	Polygynous mating swarms
	Polygynandrous mating swarms
Leks	Semelparous exploded leks
	Semelparous classic leks
	Iteroparous exploded leks
	Iteroparous classic leks
Feeding sites	Semelparous feeding site polygyny
	Semelparous classic leks
	Iteroparous feeding site polygyny
	Iteroparous classic leks
Nesting sites with female care	Semelparous harem polygynandry
	Iteroparous harem polygynandry
Nesting sites with male care	Semelparous nest site polygynandry
	Iteroparous nest site polygynandry
Polyandrogyny	Eumonogamy
	Cursorial polyandrogyny
	Mass mating with male parental care
	Dominance polyandrogyny
	Polyandrogynous social pairs
	Harem polyandrogyny (?)

subcategories exist. All 12 major categories have apparent representatives in nature and are robust to slight variation in m^* and t^*. However, these designations are based on existing literature in which the spatial and temporal distributions of receptive females are often documented anecdotally rather than quantified as application of our framework requires. Thus, further study of each mating system is warranted to obtain more precise values for m^*, t^*, and I_{mates}, the details of female life history, patterns of sperm competition, and of male and female copying behavior. Such analyses will test the power of our model, possibly correct our preliminary classifications, and very likely change our current impression of the relative abundance of the different mating systems. We predict that future research will define all mating systems fundamentally by the spatial and temporal distributions of receptive mates, as Emlen and Oring (1977) originally proposed.

10 A Darwinian Perspective on Alternative Mating Strategies

The Descent of Man and Selection in Relation to Sex (Darwin 1871, 1874) was Charles Darwin's most explicit statement on human evolution, as well as his most detailed account of how sexual selection can occur through male-male combat and female mate choice. In his lifetime, this book earned Darwin fame and notoriety equal to if not exceeding that generated by his revolutionary first volume (Darwin 1859). Darwin's expanded reputation resulted primarily from his hypothesis that humans arose from a "lower" anthropoid form (Darwin 1874, p. 613), a perspective that was eventually accepted by most biologists. However, the bulk of Darwin's 1871 thesis was devoted, not to human evolution, but rather to sexual selection, and unlike the first subject, the existence and importance of sexual selection was debated long after Darwin described its mechanisms (Andersson 1994). In one of his "letters" in volume 2 of his collected papers, Darwin (1882) states:

> Most of the naturalists who admit that natural selection has been effective in the formation of species, likewise admit that the weapons of male animals are the result of sexual selection—that is, of the best-armed males obtaining most females and transmitting their masculine superiority to their male offspring. But many naturalists doubt, or deny, that female animals ever exert any choice, so as to select certain males in preference to others.

He goes on to say (see chapter 7), "It would, however, be more correct to speak of the females as being excited or attracted in an especial degree by the appearance, voice, & c. of certain males, rather than of deliberately selecting them."

Major contributions by Fisher (1930), Bateman (1948), and O' Donald (1962) notwithstanding, Darwin's insights on sexual selection were not generally accepted until the late 1970s, when research documenting its effects became simply too voluminous to ignore (review in Andersson 1994). Although some debate continues over when natural and sexual selection are truly distinct (Darwin himself noted that this distinction is occasionally blurred; see Endler 1986; also chapter 1), the recent surge in sexual selection research has turned the tide of opinion. Selection "in relation to sex" is now widely recognized as a genuine phenomenon, and has become a fundamental area of theoretical and empirical research in evolutionary and behavioral biology (Andersson 1994).

The scope of sexual selection research has become vast, as we have discussed (chapter 1). Yet most studies citing sexual selection as their raison d'être simply document the mechanisms Darwin described, and speculate on the degree to which certain traits are shaped by the relative action of male-male competition and/or female mate choice (reviews in Campbell 1972; Thornhill and Alcock 1983; Andersson 1994). Ironically, Darwin's own descriptions of these processes were almost cursory compared to the other subjects he addressed. Whereas Darwin (1874) devoted over one-third (252 pp.) of his text to the subject of human evolution, he completed his outline of how sexual selection works in a mere 55 pages. He then committed the remaining several hundred pages, well over half of the volume, to detailed descriptions of phenotypic differences among males and females in animal taxa.

At first glance, the final half of *The Descent of Man* appears to be a systematic inventory of sexual dimorphism in the Animal Kingdom including humans; an excellent example of Darwin's fondness for singularly "long arguments." However, Darwin's comment that "whole chapters could be filled with details on the differences between the sexes in their sensory, locomotive and prehensile organs" (1874, p. 206), suggests that Darwin considered the book's last half, not only minimally concise, but also much more than a catalog of animal sex differences. Indeed, closer examination reveals that Darwin had a very specific purpose for these last chapters.

"Secondary sexual characters" were widely documented in Darwin's day, but the tendency for the sexes to appear distinct in most species bore little explanation other than that which Darwin himself provided (Darwin 1859). However, Darwin's discussion of sexual dimorphism stimulated a debate with Alfred Russell Wallace that continued long after both men had died, and influenced research in such diverse areas as sex-limited inheritance, sex role reversal, and gene expression (see below). Wallace (1868) agreed with Darwin (1859) that male combat could favor weaponry or pigmentation that made males conspicuous. Yet Wallace considered natural selection necessary for males and females to become truly distinct. Wallace's hypothesis had two parts. Using birds for his examples, Wallace suggested first that hereditary factors causing males to become conspicuous were transmitted to offspring of both sexes. Wallace next proposed that since conspicuous characters would most likely be disadvantageous to females caring for young, natural selection must be responsible for the suppression of conspicuous traits in females.

While Darwin agreed that natural selection could favor the suppression of certain traits in both sexes, he evidently considered Wallace's model unnecessarily complex, and used the latter half of his 1871 book (revised in 1874) as a forum for refuting Wallace's ideas. Darwin seemed to have been determined not simply to show that sexual selection alone could account for the evolution of male-female differences. He was evidently also resolute in his position that selection acting in the context of reproductive competition was unique in its ability to rapidly increase, decrease, maintain, or extirpate traits within populations. Darwin asserted, moreover, that sexual selection's enormous power to generate evolutionary change was due to the fact that males in most species were considerably more variable than females in their ability to successfully mate.

What significance do these details have for the study of mating systems and alternative mating strategies? There are two important points: The first is that, despite his assertion that sexual selection was weaker or less rigorous than natural selection (1874, p. 225; see also chapter 1), Darwin recognized sexual selection's power to rapidly change populations, and, in particular, to generate polymorphisms among members of the sexually selected sex, usually males. In his descriptions of male polymorphisms, moreover, Darwin used concepts that are now familiar from population genetic and game theory models (these approaches are conceptually similar, though not always equivalent; chapters 7, 8, and 11; Maynard Smith 1982). Darwin's explanations of the stability, invasibility, and persistence of different male mating "strategies" are models of simplicity and eloquence. That Darwin himself used such reasoning is significant because more recent discussions on the evolution of alternative mating strategies have violated assumptions inherent in this methodology (see chapters 11 and 12). We hope that the weight of Darwin's insight will support our view that a population genetics approach to these problems is indeed appropriate.

The second important point relating particularly to alternative mating strategies is that, while Darwin recognized polymorphisms in which individuals assumed different but distinctly male forms, he failed to recognize the now well-documented phenomenon of female mimicry, a form of intraspecific "deception" that allows males to acquire mates by imitating females. Female mimicry may be accomplished either by imitation of female reproductive behavior by males (e.g., Thornhill 1980) or by morphological convergence by males on mature female phenotypes (e.g., Dominey 1981; Gross 1982; Shuster 1987) or both. This latter form of mimicry is usually accomplished by accelerated maturation in males without corresponding development of their secondary sexual characteristics (Borowsky 1985; Shuster and Wade 1991a).

Darwin knew that precociously mature males occur in varied and diverse taxa. He regarded such examples as important for understanding "species that undergo great modification of character" (1874, p. 480). The context in which such males had arisen, however, was difficult for Darwin to explain given (1) his conclusion that sexual selection can act only on sexually mature individuals, (2) his understanding of sex-limited inheritance, and (3) his conviction that natural selection was unnecessary to explain most cases of sexual dimorphism. Darwin skillfully used these arguments to demonstrate the existence and importance of sexual selection. However, as we will illustrate below using Darwin's own words, this conceptual framework also prevented Darwin from recognizing accelerated maturation and female mimicry as male alternative mating strategies.

Darwin on Male Polymorphism

As is widely known, Darwin described male combat in exquisite detail (1874, p. 497):

> All male animals which are furnished with special weapons for fighting are well known to engage in fierce battles. The courage and the desperate conflicts of stags have often been described; their skeletons have been found in various parts of the world, with their horns inextricably locked together, showing how miserably the victor and vanquished had perished. No animal in the world is so dangerous as an elephant in must. Lord Tankerville has given me a graphic description of the battles between the wild bulls in Chillingham Park, the descendants, degenerated in size but not in courage from the gigantic *Bos primigenius*. In 1861, several contended for mastery; and it was observed that two of the younger bulls attacked in concert the old leader of the herd, overthrew and disabled him, so that he was believed by the keepers to be lying mortally wounded in the neighboring wood. But a few days afterwards one of the young bulls approached the wood alone; and then the "monarch of the chase" who had been lashing himself up for vengeance, came out and in a short time, killed his antagonist. He then quietly joined the herd, and long held undisputed sway.

Darwin noted too that certain male attributes can provide mating advantages in the complete absence of other male competitors (1874, p. 270):

> The Rev. O. P. Cambridge accounts in the following manner for the extreme smallness of the male in the genus *Nephila*: "M. Vinson gives a graphic account of the agile way in which the diminutive male escapes from the ferocity of the female, by gliding about and playing hide and seek over her body and along her gigantic limbs: In such a pursuit it is evident that the chances of escape would be in favor of the smallest males while the larger ones would fall early victims; thus gradually a diminutive race of males would be selected, until at last they would dwindle to the smallest possible size compatible with the exercise of their generative functions,—in fact probably to the size we now see them, i.e., so small as to be a sort of parasite upon the female and either beneath her notice, or too agile and too small for her to catch without great difficulty."

Darwin was also aware of the fact that males in many species are more brightly pigmented than females, although the cases usually cited seldom include examples from one of Darwin's true loves, the Crustacea (1874, p. 268):

> I am informed by Fritz Muller, that in the female of a Brazilian species of *Gelasimus* (an amphipod crustacean), the whole body is of a nearly uniform grayish-brown. In the male, the posterior part of the cephalothorax is pure white, with the anterior part a rich green, shading into dark brown; and it is remarkable that these colors are liable to change in the course of a few minutes—the white becoming dirty gray or even black, the green, "losing much of its brilliancy." It deserves especial notice that the males do not acquire their bright colors until they become mature.

In point of fact, it was among the Crustacea that Darwin first noted examples of polymorphisms restricted to members of one sex, particularly among sexually mature males (1874, p. 262):

> In various crustaceans, belonging to distinct families, the anterior antennae are furnished with peculiar thread-like bodies, which are believed to act as smelling organs, and these are much more numerous in males than in females. As the males, without any unusual development of their olfactory organs, would almost certainly be able sooner or later to find females, the increased number of the smelling-threads has probably been acquired through sexual selection, by the better-provided males having been more successful in finding partners and in producing offspring. Fritz Muller has described a remarkable dimorphic species of *Tanais*, in which the male is represented by two distinct forms, which never graduate into each other. In the one form the male is furnished with more numerous smelling-threads, and in the other form with more powerful and more elongated chelae or pincers, which serve to hold the female. Fritz Muller suggests that these differences between the two male forms of the same species may have originated in certain individuals having varied in the number of the smelling threads, whilst other individuals varied in the shape and size of their chelae; so that of the former, those which were best able to find the female, and of the latter, those which were best able to hold her, have left the greatest number of progeny to inherit their respective advantages.

Darwin continued several pages later (1874, p. 275):

> Another Brazilian amphipod [*Orchestia darwinii*, fig. 10.1] presents a case of dimorphism, like that of *Tanais*; for there are two male forms, which differ in the structure of their chelae. As either chela would certainly suffice to hold the female,—for both are now used for this purpose,—the two male forms probably originated by some having varied in one manner and some in another; both forms having derived certain special, but nearly equal advantages, from their differently shaped organs.

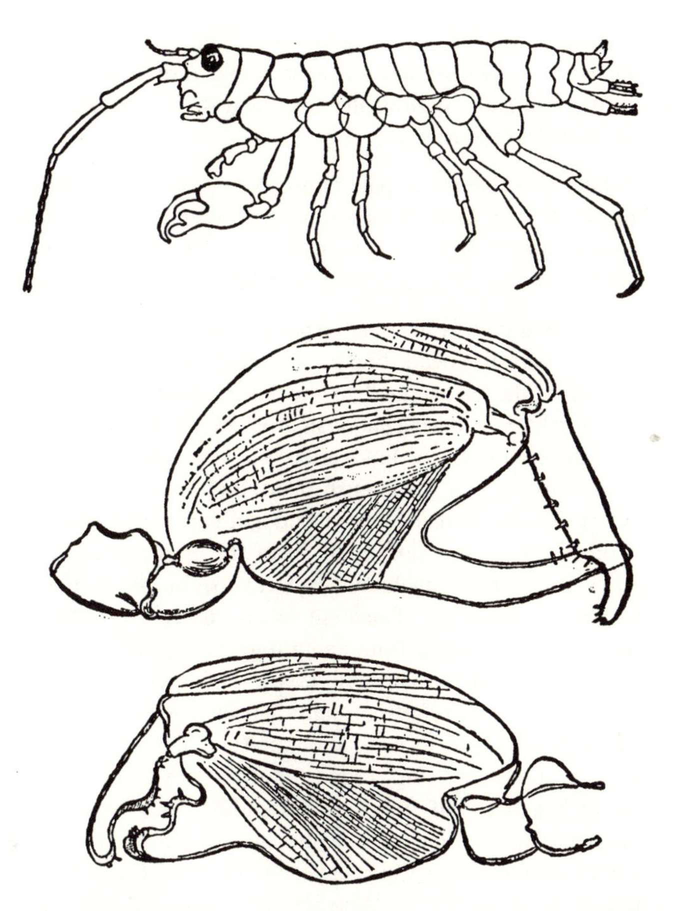

Figure 10.1. *Orchestia darwinii* and the comparative construction of the chelae of the two adult male morphs (from Darwin 1874, p. 266).

Darwin, as well as Muller, clearly recognized that male dimorphism, like sexual dimorphism, can evolve via sexual selection. Moreover, both researchers acknowledged that the mating success of the different males *must be equal* for the distinct morphs to persist in the same population. Darwin evidently found this point to be of particular importance, for he provided in following pages yet another example of male dimorphism, this time in the Arachnida (spiders, p. 269):

> We may well admit with some confidence that the well-marked differences in color between the sexes of certain species are the results of sexual selection; though we have not here the best kind of evidence,—the display by the male of his ornaments. From the extreme variability of color in the male of some species, for instance of *Theridion lineatum*, it would appear that these sexual characters of the males have not as yet

become well fixed. Canestrini draws the same conclusion from the fact that the males of certain species present two forms, differing from each other in the size and length of the jaws; and this reminds us of the above cases of dimorphic crustaceans.

Darwin later explained that male dimorphisms are not simply restricted to invertebrates (1874, p. 418). After discussing the great variability in plumage characteristics among birds of the same species, noting specifically that males of these species are much more variable than females, Darwin wrote:

> The following case is in some respects more interesting. A pied variety of the raven with the head, breast, abdomen and parts of the wings and tail-feathers white, is confined to the Feroe Islands. It is not very rare there, for Graba saw during his visit from eight to ten living specimens. Although the characters of this variety are not quite constant, . . . it has been named by several distinguished ornithologists as a distinct species. The fact of the pied birds being pursued and persecuted with much clamor by the other ravens (possibly males?) of the island was the chief cause which led Brunnich to conclude they were specifically distinct; but this is now known to be in error. This case seems analogous to that lately given of albino birds not pairing from being rejected by their comrades.
>
> In various parts of the northern seas a remarkable variety of the common Guillemot (*Uria troile*) is found; and in Feroe one out of every five birds, according to Graba's estimation, presents this variation. It is characterized by a pure white ring 'round the eye, with a curved narrow white line, an inch and a half in length, extending back from the ring. This conspicuous character has caused the bird to be ranked by several ornithologists as a distinct species under the name of *U. lacrymans*, but it is now known to be merely a variety. It often pairs with the common kind, yet intermediate gradations have never been seen; nor is this surprising, for variations which appear suddenly, are often, as I have elsewhere shown, transmitted unaltered or not at all. We thus see that two distinct forms of the same species may coexist in the same district, and we cannot doubt that if the one had possessed any advantage over the other, it would have been multiplied to the exclusion of the latter. If, for instance, the male pied ravens, instead of being persecuted by their comrades, had been highly attractive (like the above pied peacock) to the black female ravens, their numbers would have rapidly increased. And this would have been a case of sexual selection.

Darwin also described male polymorphism in North American deer (p. 508), and in this case outlined circumstances in which a secondary morph had not simply invaded a population, but appeared to be sweeping through it:

> An interesting case has lately been published, from which it appears that the horns of a deer in one district in the United States are now being

> modified through sexual and natural selection. A writer in an excellent American Journal (the *American Naturalist* Dec. 1869, p. 552) says that he has hunted for the last twenty-one years in the Adirondacks, where the *Cervus virginianus* abounds. About fourteen years ago he first heard of spike-horn bucks. These became from year to year more common; about five years ago he shot one, and afterwards another, and now they are frequently killed. 'The spike-horn differs greatly from the common antler of the *C. virginianus*. It consists of a single spike, more slender than the antler and scarcely so long, projecting forward from the brow, and terminating in a very sharp point. It gives a considerable advantage to its possessor over the common buck. Besides enabling him to run more swiftly through the thick woods and underbrush (every hunter knows that does and yearling bucks run much more rapidly than the large bucks when armed with their cumbrous antlers), the spike-horn is a more effective weapon than the common antler. With this advantage the spike-horn bucks are gaining upon the common bucks, and may, in time, entirely supersede them in the Adirondacks. Undoubtedly the first spike-horn buck was merely an accidental freak of nature. But his spike-horns gave him an advantage and enabled him to propagate his peculiarity. His descendants having a like advantage, have propagated the peculiarity in a constantly increasing ratio, 'till they are slowly crowding the antlered deer from the region they inhabit.

In the next paragraph Darwin noted (1874, p. 509):

> A critic has well objected to this account by asking, why, if the simple horns are now so advantageous, were the branched antlers of the parent-form ever developed? To this I can only answer by remarking, that a new mode of attack with new weapons might be a great advantage, as shown by the case of the *Ovis cycloceros*, who thus conquered a domestic ram famous for his fighting power (p. 504). Though the branched antlers of a stag are well adapted for fighting his rivals, and though it might be an advantage to the prong-horned variety slowly to acquire long and branched horns if he had to fight only with others of the same kind, yet it by no means follows that the branched horns would be the best fitted for conquering a foe differently armed. In the forgoing case of the *Oryx leucoryx* (p. 507), it is almost certain that the victory would rest with an antelope having short horns, and who therefore did not have to kneel down (to fight), though an oryx might profit by having still longer horns if he fought only with his proper rivals.

Darwin recognized not only that male polymorphisms can persist in nature if different male types have equal fitnesses, but also that, to invade an established male population, a novel male type must employ a mode of combat that is significantly advantageous over the conventional fighting style. This point especially appears to have been omitted in modern discussions of alternative male strategies, where the minority strategy is labeled "making the best of a bad situation" or "fitness satisficing" (see chapter 11).

For different male types to persist within a population, Darwin succinctly described how the fitnesses of male types relative to each other must be equal over time, otherwise the type with greater fitness, "however slight," will eventually supercede the type with lesser fitness. Male fitness satisficing may avoid zero mating success, but by definition, it does not achieve fitness comparable to, let alone exceeding, that of the dominant male strategy. Darwin was also aware of the interaction between success in battle and the relative frequency among the different male types, since males may only profit by their weapons when fighting against their "proper rivals." His arguments here forecast the "scissors-paper-rock" dynamical interplay of male strategies in male lizards (*Uta stansburiana*) discovered by Sinervo (2001a; Sinervo and Lively 1996).

It is clear that Darwin included in his discussion of male polymorphism all of the central components of modern population genetics and evolutionary (as opposed to "behavioral"; see chapters 7 and 11) game theory models (Maynard Smith 1982). These components are (a) the tendency for a successful phenotype to become widespread and remain so within a population, (b) the requirement that to increase in frequency within a population, the fitness of a novel phenotype must exceed that of the established phenotype, and (c) that to persist within a population the fitnesses of distinct phenotypes relative to each other must be equal over time. Moreover, if this last condition is not met, (d) the male reproductive phenotype with greater fitness will become fixed within the population.

Darwin on Female Mimicry

Having argued persuasively on the power of sexual selection to rapidly change populations, and, in particular, to generate polymorphisms among members of the sexually selected sex, Darwin next asserted that natural selection was unnecessary or unable to account for the observed variation in sexually dimorphic characters as suggested by Wallace (1868). Darwin supported this view with three observations: (1) sexual selection can act only on sexually mature individuals, (2) sex-limited characters are inherited only by members of one sex and are expressed only at maturity, and (3) patterns of sexual dimorphism can be classified into six mutually exclusive groups whose existence can be explained by sexual rather than natural selection. The first two of these arguments were specifically responsible for Darwin's rejection of precocious male maturation and female mimicry as viable male mating strategies. As explained above, these points are best understood by considering Darwin's own statements (p. 235):

> Sexual selection can never act on any animal before the age for reproduction arrives. From the great eagerness of the male it (is clear that selection) has generally acted on this sex and not on the females. The males have thus become provided with weapons for fighting with their rivals, with organs for discovering and securely holding the female, and for exciting or charming her. When the sexes differ in these respects, it is also, as we have seen, an extremely general law that the adult male differs more or less from the young male; and we may conclude from this fact that the successive variations by which the adult male became modified, did not generally occur much before the age for reproduction.

Darwin continued:

> It is probable that young male animals have often tended to vary in a manner which would not only have been of no use to them at an early age, but would have been actually injurious—as by acquiring bright colors, which would render them conspicuous to their enemies, or by acquiring structures, such as great horns, which would expend much vital force in their development. Variations of this kind occurring in the young males would almost certainly be eliminated through natural selection. With the adult and experienced males, on the other hand, the advantages derived from the acquisition of such characters, would more than counterbalance some exposure to danger, and some loss of vital force.

Darwin thus considered sexual selection, acting primarily among males and only after males had become sexually mature, both necessary and sufficient to account for sexual dimorphism. Darwin's discussion rendered natural selection's role in producing sex differences superfluous, except as Darwin noted, in certain unusual cases in which male characters had indeed been transmitted to females (see below). Darwin thus was willing to state that natural selection could, in certain circumstances, act on individuals after sexual selection had occurred. However, he was unwilling to allow, as Wallace did, that natural selection had been responsible for generating male-female differences. Darwin's error appears in our hindsight to be that he assumed traits to be sex-limited ab initio, rather than granting, as Wallace did, that sex-limited expression is a secondary adaptation. Darwin recognized that the logic of his argument addressed one part of Wallace's hypothesis, namely, that sexual dimorphism is refined by natural selection. However, Darwin also recognized that his argument did not account for the other part of Wallace's hypothesis, i.e., conspicuous male characters were inherited by individuals of both sexes. To address this latter point, Darwin drew on his considerable studies of sex-limited inheritance, particularly in pigeons and livestock (1874, p. 225).

Through these observations, Darwin concluded that there were two types of traits: (1) characters expressed early in life were transmitted to members of both sexes, whereas (2) characters expressed late in life were transmitted only to members of one sex. As described above, Darwin observed that secondary sexual characters were expressed only after sexual maturity (i.e., "later" in life). He thus concluded that such characters were transmitted only to members of the sexually selected sex. These observations rendered modification of secondary sexual characters in females by natural selection unnecessary to account for their exclusive appearance in males, leaving sexual selection alone to explain the existence of sexual dimorphism. Darwin did not specifically discuss how a trait might make the evolutionary transition to a sex-limited adult character from a non-sex-limited juvenile character, although he suggested this process was likely to be difficult and would take considerable time (see below). As far as sexually selected traits in adults are concerned, this transition is tantamount to assuming sex-limited expression.

To cement his argument on the primacy of sexual selection in generating sexual dimorphism, Darwin recognized six categories (1874, p. 460) of sexual differences and resemblance in adult and juvenile morphology among members of the same species. Here, Darwin provided his most detailed evidence that sexual selection acts mainly on males and only at particular life stages. Here too, Darwin extended his argument that sex-limited inheritance prevents the evolution of precocial male maturation as a viable male mating strategy. It is interesting that Darwin chose birds to illustrate these points because birds show more pronounced and varied cases of sexual dimorphism than any other animal group. It seems likely that Darwin wished to use the very same taxa to refute Wallace that Wallace himself used to argue the importance of natural selection for the evolution of sexual dimorphism. Darwin's six categories of sexual dimorphism are as follows.

Class 1. When the adult male is more beautiful or conspicuous than the adult female, the young of both sexes resemble the adult female. Darwin noted that this category included most sexually dimorphic species and represented his strongest argument for the evolution of sexual dimorphism though sexual selection. To make this point, Darwin restated his earlier (1859) observation that comparison of character similarities among groups of organisms was an effective way to establish evolutionary relationships among them (1874, pp. 458–460). Darwin argued that because new characters arise rarely, are subsequently shaped by local selection, and may persist within lineages for generations, the ancestral condition of a character within related groups of organisms is usually the one that is most widespread. Conversely, unique characters, or characters shared by few taxa, are most likely to have arisen later. Darwin noted, for example, that stripes in the plumage of thrushes occur in the young of all members of this family and persist in adult plumage in certain species, whereas bright feathers or patches of white occur in fewer species and are seen only in adults (1874, p. 459).

Darwin extended his review of similarities between juvenile forms among related species to include a variety of birds and mammals, and noted that adult females in most species closely resembled juveniles of both sexes. Since females and young exhibited the same patterns of cryptic coloration in most cases, and since these patterns were widespread among taxa, Darwin considered such characters representative of these species' ancestral form. Moreover, since adult males in sexually dimorphic species usually exhibited morphologies different from females and juveniles of both sexes, Darwin concluded that the distinctive male phenotypes must have arisen as a result of selection acting specifically on *adult* males and that this selection had given rise to male limited traits. Darwin saw no reason to suggest that natural selection, acting after males had become conspicuous, was responsible for making females inconspicuous, as proposed by Wallace (1867).

Class 2. When the adult female is more beautiful or conspicuous than the adult male (as rarely occurs) the young of both sexes in their first plumage resemble the adult male. Darwin saw this pattern as similar to that of the first class, but in reverse. Acknowledging that Wallace (1867, 1868) was the first to document such cases of "sexual role reversal," he then argued that here too was evidence of sexual selection's work. In this case, females varied in their abilities to acquire more than one mate and therefore sexual selection had favored pugnacity, brighter coloration, and larger size in this sex, leaving the males to retain their ancestral juvenile plumage. Here again, Darwin cited evidence of his two principles. The characteristics exhibited by females in Class 2 appeared only in mature individuals. That is, they were never observed to appear in the young of either sex. Instead, the plumage of young birds resembled that of the mature male. Darwin provided a detailed list of species included in this class, including members of the gallinaceous genus *Turnix* (p. 469), three species of painted snipes (*Rhynchaea*, p. 470), two species of phalaropes (*Phalaropus*, p. 471), several waders (including *Limosa lapponica*, p. 472), cassowaries (*Casuarius*, p. 472), the Falklands carrion hawk (*Milvago*, p. 473), as well as Australian tree creepers and night jars (*Climacteris* and *Eurosopodus*, p. 473). Darwin asserted again that in these species, sexual selection was "steadily adding to the attractions of the females; the males and the young being left not at all or but little modified" (1874, p. 474).

Class 3. When the adult male resembles the adult female, the young of both sexes have a peculiar plumage of their own. This pattern appeared to Darwin to result from selection acting on individuals of both sexes to acquire a particular plumage pattern. Darwin assembled evidence from diverse bird species (robins, swans, ibis, and herons) to show that the plumage of young birds in these taxa resembled ancestral forms, thus substantiating his second principle, i.e., plumage modifications in this class exist only in adults. These modifications, Darwin asserted, are likely to have arisen from sexual selection acting on males. However, unlike the previous two classes, in which the transmission of modifications is sex-limited, in this class, although limited by age, modified characters were expressed by both sexes.

Darwin noted that while the inheritance of sex-limited characters was poorly known, certain principles were clear in his time. The first was that, while it was difficult to induce sex-limited inheritance in a character initially observed in both sexes, when characters were observed as initially sex-limited, this expression could be easily enhanced in that sex by intensifying selection on individuals in which the character is expressed. Thus, characters favored in males by sexual selection, were predicted to appear only in mature individuals according to Darwin's second rule. However, if that character existed initially in members of both sexes, then that character during its modification under sexual selection was transmissible to females, leading adults to resemble each other and become distinct from a presumably ancestral juvenile condition.

4. When the adult male resembles the adult female, the young of both sexes resemble the adults. Darwin was uncertain about this pattern. However, after some discussion, he asserted that this class is extremely widespread and again confirmed his rules about the inheritance of sexually selected characters by females and young. Darwin noted that the plumage of adult and juvenile members of this class were typically brilliant, conspicuous, and thus of little service in providing protection from predators. This observation allowed Darwin to conclude that such bright plumage was gained in males via sexual selection and later transmitted to females and young. Still unresolved, however, was the question of whether the circumstances favoring the evolution of bright plumage occurred while males were young or mature. The second of Darwin's two rules regarding age-related inheritance stated: "variations occurring late in life are transmitted to one and the same sex." Darwin noted that it is sometimes difficult to determine when "early" and "late" actually are within individual lifetimes. Moreover, Darwin had reported earlier that, when the law of inheritance at corresponding ages occasionally failed, offspring can inherit the characters in question at ages earlier than when they first appeared in their parents (Darwin 1868). This uncertainty allowed Darwin to overlook possible exceptions of his rule regarding the ability of subadult individuals to breed and thus for sexual selection to influence their behavior and morphology. Moreover, after noting as "remarkable" that many birds do breed bearing apparently immature plumage, Darwin stated (1874, p. 479):

> The fact of birds breeding in their immature plumage seems opposed to the belief that sexual selection has played as important a part, as I believe it has, in giving ornamental colors, plumes, etc., to the males, and, by means of equal transmission, to the females of many species. The objection would be a valid one, if the younger and less ornamented males were as successful in winning females and propagating their kind, as the older and more beautiful males. But we have no reason to suppose this is the case . . . If males with immature plumage were more successful in win-

> ning partners than the adults, the adult plumage would probably soon be lost, as the males would prevail which retained their immature dress for the longest period, and thus the character of the species would be changed. If, on the other hand the young never succeeded in obtaining a female the habit of early reproduction would perhaps be sooner or later eliminated from being superfluous and entailing waste of power.

Darwin thus reasoned that if unmodified males achieved significant mating success, secondary sex characters among males would change, such that elaborated characters would become less pronounced. Since secondary sexual characters did not appear to erode in species in which precocially mature males were reported, Darwin concluded that these males never achieved significant mating success. Darwin did not consider the possibility, as he did in other forms of male polymorphism, that precocially mature males or males resembling females might persist in populations with males not so modified, by obtaining "certain special, but nearly equal advantages," from their different approaches to mate acquisition.

And yet Darwin seemed bothered by observations contrary to his assertion that combat with mature males must preclude the evolution of precocial male maturation. Darwin noted that, in some orthopterans, earlier male instars resemble females, and only in later molts acquire their distinctive masculine characters, adding that "strictly analogous cases occur at the successive moults of certain male crustaceans" (1874, p. 232). In ducks, distinctive male plumage develops later in life among males. In particular, in *Mergus cuculiatus*, males and females differ in the plumage of the speculum, "which is pure white in the male and grayish-white in the female . . . the young males at first entirely resemble the females and have a grayish-white speculum, which becomes pure white at an earlier age than at which the adult male acquires his other more strongly marked sexual differences." There was no comment by Darwin about whether males are sexually mature at this time (p. 231).

Darwin cited Audubon and Swinhoe at some length for their descriptions of "premature breeding," occurring in ibis, redstarts, blue herons, harlequin ducks, bald eagles, and orioles (1874, p. 479). He mentioned salmon, in which habitually or occasionally young males breed, and he cited Muller's work showing that males in several amphipod crustaceans become sexually mature while young. Yet Darwin regarded the appearance of such cases, not as evidence that alternative mating strategies involving precocial maturation could exist, but instead as "one means by which species may undergo great modifications of character" (p. 480). Thus, Darwin never considered seriously the issue of female mimicry as a male mating strategy.

Class 5. When the adults of both sexes have a distinct winter and summer plumage, whether or not the male differs from the female, the young resemble the adults of both sexes in their winter dress, or much more rarely in their summer dress, or they resemble the females alone. Darwin again considered this confusing class as confirmation of his primary rules. Modifications between the sexes arise first in males as a result of sexual selection, and these characters are variously limited in their transmission to females and young according to the rules of age, season, and sex. Darwin did not allow that distinct winter plumage serves to provide protection as Wallace (1867) proposed. Rather, Darwin suggested that an ancient plumage pattern, partially modified by the transference of characteristics from summer plumage, is retained as winter plumage by the adults. Once again, this idea is consistent with Darwin's assertion that, although environmental conditions experienced during summer and winter could impose divergent pressures on plumage, the widespread retention of ancestral plumage patterns by young and female adults birds provides a simpler explanation than Wallace's hypothesis.

Class 6. When the young in their first plumage differ from each other according to sex, young most closely resemble adult individuals of their own sex. Darwin noted that this last class is the rarest of all, but also suggested that this class, by itself, illustrated several of the above principles. Darwin recommended that, if the plumage of the young of these species is considered first, as in previous examples, it becomes clear that the sexes have become modified independently, in ways similar to the males mentioned in Class 1 and the females in Class 2. How both processes could occur in the same species puzzled Darwin, but he supposed that extreme sex ratio biases could produce conditions analogous to sexual selection, first in one sex and then in the other. With both sexes so modified, Darwin suggested that these characters must have been transmitted to young at an age earlier than they were first acquired. If this were not so, the characters unique to each sex would be lost in their transmission to offspring of both sexes.

Darwin was convinced, not only by the simplicity of his arguments compared to those of Wallace (1868), but also by the power of sexual selection to modify individuals of each sex. Darwin appears to have been correct about the primacy of sexual selection in producing sexual dimorphism (reviews in Slatkin 1984; Jormalainen and Tuomi 1989), and despite his frequent 1874 references to pangenesis, as well as his unwillingness to consider sex-limited expression as a secondary adaptation, Darwin's observations on the expression of sex-limited characters are difficult to improve on to this day. Darwin showed that sexual selection acts *only* on sexually mature individuals. Although his persistence on this point prevented him from seeing female mimics for what they are, he demonstrated without a doubt that age specificity in the effects of sexual selection explains much about the pattern of morphological differences between the sexes. In particular, Darwin showed the limits of natural selection in producing sex differences in males and females, despite the greater intuitive appeal of Wallace's (1868) hypothesis, and the fact that modification in the expression of sex-limited characters requires the influence of natural selection (chapter 1). Whether Darwin consciously focused the second part of his 1874 book on resolving this particular issue may never be confirmed. What is clear is that Darwin's detailed explanation of how sexual selection works, when in the lives of individual organisms it is most likely to act, which characters sexual selection is capable of modifying, and how these characters are ultimately expressed, will stand as one of his most elegant contributions.

Chapter Summary

Darwin's 1871 book, *The Descent of Man and Selection in Relation to Sex*, is well known for its explicit treatments of human evolution and sexual selection. However, three aspects of this volume have gone unrecognized. We suggest first that the conceptual thrust of this book was not human evolution, or even sexual selection per se, but rather Darwin's response to Wallace's hypothesis that the expression of sexual dimorphism required natural selection. Darwin evidently considered Wallace's model unnecessarily complex, and devoted over half of his book toward making this point.

We next show how Darwin, in developing his argument, demonstrated a sophisticated grasp of population genetic theory. Well in advance of game-theoretical discussions of evolutionary stable strategies, Darwin identified the conditions necessary for the invasion, spread, and persistence of novel phenotypes. We consider it noteworthy that Darwin used male alternative mating strategies to illustrate these principles, because more recent discussions of the evolution of these traits have violated assumptions inherent in this methodology (e.g., "making the best of a bad job").

Lastly, we show that, while Darwin clearly recognized polymorphisms in which males assume distinct forms, he failed to recognize female mimicry as an actual phenomenon. Despite his masterful understanding of the expression of sex-limited characters, Darwin's incomplete understanding of sex-limited inheritance prevented him from recognizing female mimicry for what it was, as well as from seeing the components of Wallace's hypothesis that were in fact correct. Darwin's argument was detailed and complex. We use it to illustrate, not only his exquisite understanding of plumage variation in birds, but also why, despite his superior logical and observational skills, Darwin neglected a now well-documented form of male polymorphism.

11 Sexual Selection and Alternative Mating Strategies

There can be little doubt that Darwin recognized the circumstances necessary for alternative mating strategies to evolve and persist (chapter 10). Game theory, evolutionary genetic models, and laboratory selection experiments since 1874 have mainly added precision to Darwin's original observations (Gadgil 1972; Maynard Smith 1982; Hartl and Clark 1989). General evolutionary theory now describes four conditions that are necessary for the invasion and persistence of *any* phenotype in a population, regardless of its association with mating behavior.

First and most importantly, genetic variation for the traits under selection must exist. If genetic variation is lacking, that is, if a population is genetically monomorphic for a given trait, *no response to selection is possible*. Second, to *invade* a population, the average fitness of an alternative phenotype, when rare, must be greater than the average fitness of the conventional phenotype. Only if the average fitness of the invading phenotype exceeds that of the conventional phenotype will the invading phenotype increase in frequency. Third, to *persist* in a population at stable frequency, the average fitness of the alternative phenotype at equilibrium must equal the average fitness of the conventional phenotype. Phenotypes with inferior mean fitness are relentlessly culled from natural populations. With frequency-dependent selection, equal fitness does not mean equal frequency of the different phenotypes. One phenotype may be much rarer than the other and, thus, may experience an increased risk of loss by random genetic drift (Eckert and Barrett 1992; Agren and Ericson 1996); indeed, the combination of frequency-dependent selection and random genetic drift can lead to a faster rate of loss than by drift alone. Fourth, to be *modified by selection*, that is, for the function of a phenotype to be refined, heritable fitness variance must exist among those individuals expressing the alternative phenotype.

Despite theoretical unanimity on these principles, controversy continues over how and why alternative mating strategies evolve and persist in natural populations. Consensus remains elusive on three issues in particular. The first concerns the relative role of genes and environment in the origins of polymorphic male mating behaviors: That is, are male genetic polymorphisms ancestral to condition-dependent phenotypic plasticity, or vice versa (Slatkin 1979; Lively 1986; Moran 1992; Gross 1996)? The second issue is whether or not the average fitnesses of different male strategies must be equal to maintain variation in male reproductive behavior and morphology. That is, can a polymorphism of male mating strategies persist in a population when the fitnesses of different male morphs are *unequal* (Maynard Smith 1982; Shuster and Wade 1991a; Gross 1996; Shuster and Sassaman 1997; Repka and Gross 1995; Gross and Repka 1995, 1998)? The third issue concerns whether existing differences among males in mating strategy are heritable and, if so, to what degree. That is, are all males genetically the same, sharing a common *sensitivity* to an environmental signal that elicits different reproductive behaviors in different males, or are males, which display different reproductive behaviors, also genetically distinct (Gadgil 1972; Dawkins 1980; Eberhard 1982; Thornhill and Alcock 1983; Shuster and Wade 1991a; Gross and Repka 1998)?

This last question is clearly related to the debate over whether the ancestral condition of a particular trait was genetically polymorphic or phenotypically plastic. This question and the second are confounded, because if an existing male polymorphism is not genetic, then there is no necessity that the fitnesses of the male morphs be equal. Nongenetic male mating polymorphisms might appear to have equal fitnesses today because they are derived from an ancestral condition of genetic polymorphism that required equal fitnesses. However, if male polymorphisms were nongenetic and condition dependent ab initio, then these male phenotypic polymorphisms would never be required to have equal fitnesses.

Part of the tenor of these controversies, in our opinion, stems from fundamental differences between evolutionary biology and behavioral biology as we have discussed in previous chapters. In evolutionary theory, most phenotypes are assumed to be influenced simultaneously by genes *and* environments, while in behavioral studies, the focus appears more often to be genes *versus* environments, as though the two were alternative (rather than simultaneously acting) causes of phenotypic variation. In addition, a very high level of adaptation in the face of environmental variation is assumed in behavioral biology while, in evolutionary theory, it is not. In the words of Scheiner and Berrigan (1998, p. 368): "The world is heterogeneous. It varies from place to place. As a consequence of this variation, the optimal phenotype of an individual changes. In an ideal world, an individual would alter it phenotype to always match the optimum. In the real world, however, organisms do not always do this." In our opinion, discussions of the evolution of male mating behavior often sound more like prescriptions for an "ideal *male* world" than descriptions of the "real world."

To address these issues, we first review their discussion in the literature. We next consider hypotheses used to explain the evolution and maintenance of male polymorphisms, in particular, those hypotheses assuming genetic uniformity and unequal fitnesses among alternative male phenotypes. We will then show that the invasion of a novel, mutant mating strategy may occur whenever some males are excluded from mating and not only when sexual selection is intense. Similarly, the invasion of an alternative female strategy, such as helping or egg dumping, may occur whenever some females are excluded from nesting sites. Using examples, we will illustrate that the fitness achieved by an alternative male mating strategy is not inferior to that of the dominant strategy. Indeed, we show that the ease of invasion of an alternative mating strategy is proportional to the degree to which males practicing the dominant strategy are unsuccessful. Differently put, it is the fact that mates are distributed *unevenly* among males of the dominant strategy that allows invasion by an alternative mating strategy. This alternative strategy need not result in fitness superior to that of males *successfully* using the dominant mating strategy. The alternative strategy invades as long as its relative fitness is greater than the *average* male using the dominant strategy.

We discuss our concept of the "*mating niche*" (Shuster and Wade 1991a) available to an alternative mating strategy and show how its size can be determined from the observed distribution of mates among males. Thus, we show that the existence as well as the frequency of alternative male morphs in a particular species may be predicted from available field data. When the distribution of mates of a secondary male phenotype can be identified, we argue that the existence and the frequency of tertiary, or perhaps even quaternary, male phenotypes are also predictable from field data. We show that even slight deviations from strict monogamy allow alternative male morphs to invade and persist in natural populations, according to established evolutionary genetic principles. Thus, we assert that *genetic* polymorphisms in male mating behavior are more common than is now recognized, and the hypothesis that satellite or alternative males have lower average fitness or "make the best of a bad job" (Dawkins 1980; Eberhard 1982; Gross 1996) is unnecessary, and, in many cases, inaccurate.

Lastly, we explain how condition-dependent polymorphisms in male mating behavior may evolve from genetic polymorphisms for these traits. Genetically distinct alternative mating strategies clearly exist, but they can, in theory, be invaded and replaced by plastic phenotypes with greater ability to adapt to changing environments, provided that the environment is common or that cues predictive of environmental change are available and reliable (Levins 1968; Slatkin 1978, 1979; Lloyd 1984). Although such flexibility can evolve if the average success of males with flexible phenotypes exceeds that of males with fixed genetic strategies, the fitness effects can be sufficiently weak that pleiotropic effects, linkage, or random genetic drift would greatly slow or impede their spread. With weak selection, the equilibrium heterozygosity at mutation-selection balance would be relatively high for flexible strategies. As a result, phenotypic flexibility can, in turn, be replaced by genetic polymorphism if cues predicting environmental change become unavailable.

Regardless of whether male polymorphisms are controlled by a few loci of major effect, or represent conditional phenotypes, no evolutionary change in male phenotype is possible if males are genetically identical in their ability to express a conditional phenotype (e.g., Gross 1996; Repka and Gross 1995; Gross and Repka 1995, 1998). We expect genetic variation to exist in the threshold for expression of condition-dependent male mating behaviors just as it does for the evolution of plasticity in more ordinary phenotypes, like those affecting dispersal, dormancy, or germination (Levins 1968; Via and Lande 1985; Lively 1986; Schlichting and Pigliucci 1998). Differently put, no special rules for inheritance or expressivity are necessary to explain the evolution and persistence of alternative mating strategies (but see Gross and Repka 1998). Our scheme fits well with current mechanistic explanations for the form of alternative mating strategies appearing at different life stages (Moore 1991; Hews et al. 1994).

Current Issues

Males in many animal species exhibit discontinuous variation in behavior and morphology associated with reproduction (reviews in Parker 1970b; Gadgil 1972; Perrill et al. 1978; Hamilton 1979; Dominey 1980, 1984; Cade 1981; Eberhard 1979, 1982; Emlen 1996; Gross 1982, 1985, 1991b, 1996; Clutton-Brock et al. 1982; Thornhill and Alcock 1983; Austad 1984; Waltz and Wolf 1984; Howard 1984; Hedrick 1988; Travis and Woodward 1989; Warner 1984; Ryan et al. 1990, 1992; Morris et al. 1995; Lank et al. 1995, 1999; Sinervo 2001a,b; Sinervo and Lively 1996; Shuster and Wade 1991a; Shuster and Sassaman 1997). These variations are called *alternative mating strategies*, or *male polymorphisms*, and they occur primarily in polygynous, polygamous, or polygynandrous species, in which variance in male mating success is high and thus sexual selection is strong (Gadgil 1972; chapter 1). Depending on the species, some male mating strategies represent discrete genetic differences among males, some are environmentally induced shifts in development, and some involve a combination of environmental and genetic factors that together produce distinct male morphs.

Although many variations in male reproduction behavior are discussed within this framework, some behaviors commonly identified as alternative male mating strategies may not merit such distinction. For example, when breeding territories are established by male-male contests, it is not unusual for the larger males to win and hold territories, while the smaller males lose and do not. Many contests may end with the defeated, smaller males hiding at territory borders to prevent further attacks. Indeed, experimentally trapping and removing a dominant, territory-holding male often results in a nearby and previously defeated male taking his place. If a female enters the territory, then the resident dominant male, adjacent dominant males, and the hiding males may all attempt copulation. In this scenario, is a small, hiding male "playing" a separate "mating strategy"? In this case, the behavioral phenotype "hiding at the margins between territories of dominant males and waiting for a female" need not be a property of the small male. Rather, it is the outcome of a series of behavioral interactions between the small male with both dominant males as well as the behavioral interactions between the dominant males, which are necessary to establish the territory boundaries whether or not a small male is present. When social context determines a component of fitness as in this case, a different approach is employed in evolutionary genetics, the theory of indirect effects (Wolf et al. 1999). In this framework, group structure is of critical importance to the evolutionary dynamic (Agrawal et al. 2001a), whereas it plays no role at all in most discussions of alternative male strategies. Thus, inappropriately assigning the label "alternative male strategy" to a behavior can place evolutionary discussion in an uncertain (and possibly inadequate) theoretical framework.

When male phenotypes *are* genetically distinct, the average fitness of the different male types must be equal over time. Genes with consistently deleterious effects on fitness, i.e., with lower than average fitness, are removed from populations by selection. When one genetic male mating strategy has consistently lower fitness than another, it will not persist (see also Gadgil 1972; Brockmann et al. 1979; Gross and Charnov 1980; Rubenstein 1980; Maynard Smith 1982). When differences among male morphs are genetic, then, like other phenotypic polymorphisms, we expect them to be recognizable across a broad range of environmental conditions (Levins 1968; McKenzie et al. 1983; Morris et al. 1992; Roff 1992; Roff et al. 1998). Although genetic alternative strategies are expected in theory to be rare owing to the selective advantage of genotypes with conditional phenotypes (Levins 1968; Slatkin 1978, 1979; Dawkins 1980; Eberhard 1982; Gross 1996), genetic polymorphisms in male reproductive morphology in nature *are not rare* (see reviews in Shuster and Sassaman 1997; chapter 12; see also assertions to the contrary in Gross 1996). Indeed, examples of genetically different male strategies have steadily increased in number over the last two decades (Cade 1981; Kallman 1983; Zimmerer and Kallman 1989; Travis and Woodward 1989; Shuster and Wade 1991a; Ryan et al. 1992; Emlen 1996; Lank et al. 1995; Sinervo 2001; Sinervo and Lively 1996; Shuster and Sassaman 1997). Although considered relatively rare and uncommon at one time (Austad 1984; Dominey 1984; Gross 1996), genetically distinct polymorphisms in male reproductive behavior and morphology can no longer be considered unusual (chapter 12).

Despite the increasing number of documented, genetic male polymorphisms, far more common are reports of species in which males appear to adjust their reproductive behavior or morphology in response to local environmental conditions (reviews in Austad 1984; Waltz and Wolf 1984; Gross 1996). Many authors now argue that, if males employ such "condition-dependent" reproductive alternatives, no genetic polymorphism need exist (Dawkins 1980; Eberhard 1982; Alcock 1995; Emlen 1996; Lucas and Howard 1995; Gross and Repka 1998; Lank et al. 1999). Despite male phenotypic polymorphisms, all males are considered "genetically monomorphic" with respect to their ability to exhibit a "conditional strategy," allowing each male to respond adaptively to local variation in his special mating environments (Gross 1996). That is, males are considered genetically identical in their ability to adjust their reproductive phenotypes in response to local conditions.

According to this hypothesis, each male makes the best of his reproductive options given his body size, physical condition, or relative social standing. Because all males are presumed to be genetically similar in their ability to adjust to changing environments, and because the frequencies of different male phenotypes reflect or match the frequencies of the environmental contexts that elicit the different strategies, the average fitnesses of the different male phenotypes are not expected to be equal (Dawkins 1980; Eberhard 1982; Repka and Gross 1995; Gross and Repka 1998). Indeed, if all males are genetically identical, their fitness differences, if any, are of little evolutionary consequence. Such reasoning provides an explanation for the widespread observation that average fitnesses among males employing distinct mating "tactics" are seldom equivalent (Gross 1996; Gross and Repka 1998). We find it paradoxical that many discussions of female mate choice invoke extremely subtle and indirect genetic differences among phenotypically indistinguishable males, while, at the same time, studies of alternative male mating strategies discard entirely the notion of any genetic difference among conspicuously different male phenotypes.

The theoretical framework for investigating male alternative strategies is similar to life history models in which individuals adjust their growth or maturation schedules to maximize fitness in variable environments (Bradshaw 1965; Levins 1968; Schaffer 1974; Stearns 1976). The evolutionary theory of phenotypic plasticity in behavior or morphology requires environmental cues that predict an individual's success when expressing particular behavioral or morphological alternatives (Levins 1968; Via and Lande 1985; Lively 1986; Moran 1992; Zera and Denno 1997). The evidence for such developmental flexibility is rich (reviews in Travis 1983; Reznick 1983; Reznick and Bryga 1996; Schlicting and Pigliucci 1998), especially with respect to reproductive alternatives involving sex change (review in Warner 1984; Charnov 1989), variable maturation rates (Gross and Charnov 1980; Gross 1985), migration polyphenisms (Dingle 1991), and germination strategies (Gutterman 1994; Olff et al. 1994). However, in many discussions of male mating strategies, males are accorded a degree of sensitivity and responsiveness to local conditions that is unparalleled relative to these other examples of phenotypic plasticity. This is in contrast with standard theory, where "Infrequent or temporally unpredictable disturbances should have little effect on the evolution of life-history strategies, even though they may cause high mortality" (Lytle 2001). Adaptive developmental flexibility is associated with predictable environmental change experienced by most (or a large fraction) of the population. It is not expected to evolve in response to rare environments experienced by only a few members of only one sex.

In the quantitative genetic theory of reaction norm evolution, a single trait manifest in two environments is considered as two distinct but genetically correlated traits, one expressed in each environment (e.g., Falconer and MacKay 1996). Evolution of the trait in response to selection in one environment leads to a correlated response of the unexpressed trait in the other environment via the genetic correlation. In reaction norm evolution, trait expression is often assumed to involve trade-offs, such as large size in one environment but small in the other, ornaments present in one environment but absent in the other, or disease resistance in one but susceptibility in the other. Such trade-offs are common in discussions of the "cost of resistance" or the "cost of plasticity" (e.g., West-Eberhard 1989; Relya 2002), which specifically indicate that high fitness in one environment is associated with lowered fitness in the other. With such trade-offs, the genetic correlation between the traits expressed in different environments is negative and the population has "crossing-type" $G \times E$. The evolution of "adaptive plasticity" in this framework means selection for a flexible reaction norm producing the fittest phenotype in each environment. If only a few individuals experience one environment, while the majority experience the other, then the commonly experienced environment is weighted more heavily in calculating mean fitness than is the rare environment. As a result, the direct response to selection in the rare environment can be overwhelmed in the averaging by the opposing indirect (or correlated) response to selection in the common environment. These kinds of considerations, though standard in evolutionary theory, are rarely raised in behavioral discussions of the evolution of alternative male strategies.

There is an additional difficulty with discussions of conditional male mating strategies. In life history models, plastic mutants become established because their fitness *exceeds* that of existing nonplastic individuals (Levins 1968; Via and Lande 1985; Lively 1986). In models of conditional male mating strategies, on the other hand, it is routine to assign *inferior* fitness to satellite or sneaker males (Lucas and Howard 1995; Gross and Repka 1998). This fitness satisficing approach to the study of alternative mating strategies provides an explanation for the inequality of fitnesses among males employing alternative mating tactics. However, as we explain below, it cannot explain why individuals with inferior fitness invade existing male populations in the first place, nor can it explain how the phenotypes of satellite males become specialized in their reproductive morphology.

Game Theory and Alternative Mating Strategies

Since its formal introduction to biological systems by Maynard Smith and Price (1973), game theory has revolutionized the study of animal behavior (see, for example, chapters in Dugatkin and Reeve 1998). The source of this revolution is a simple set of procedures that permits (or sometimes forces) behavioral researchers to visualize animal interactions within a consistent evolutionary framework. This approach usually includes five steps:

1. Identification of all possible competing phenotypes in a population (I, J, K, . . . , N).
2. Placing each phenotype into a matrix that identifies all possible pairwise interactions.
3. Identifying each of the frequencies, P_j, with which pairwise interactions between two phenotypes occur.
4. Identifying the fitness, W_{ij}, gained or lost by the jth participant in the ith interaction.
5. Summing the various products of fitness and phenotype frequency across all possible interactions to obtain the average fitness of each phenotype.

This methodology identifies the conditions that permit a mutant phenotype to invade an existing population, as well as the conditions of evolutionary stability, i.e., the "evolutionary stable strategy," or ESS (Maynard Smith 1982; Dugatkin and Reeve 1998).

The form of these conditions is now familiar. To invade a population, a mutant phenotype J, when rare, must have fitness $W(J)$ that *exceeds* the average fitness $W(I)$, of the dominant phenotype I, i.e., $W(J) > W(I)$. To persist in a population at equilibrium, the mean fitness of the mutant phenotype must *equal* the mean fitness of the dominant phenotype, or algebraically $W^{\sim}(J) = W^{\sim}(I)$, where "$^{\sim}$" signifies the equilibrium mean fitnesses of each phenotype. Under these conditions, the rate at which an ESS is obtained depends on the heritability of the character in question. In most cases, heritability is implicitly assumed to equal 1, as though each strategy were equivalent to a new allele in a haploid, asexual species. As Maynard Smith (1982) took pains to note, the standard game-theoretic procedure generates analytical models that are *isomorphic* to haploid population genetic models in their conclusions. Note also that the conditions for invasion and stability of a polymorphic equilibrium are identical to those discussed by Darwin (1871; see our discussion in chapter 10).

Like haploid population genetic models, a genetically polymorphic ESS, with one strategy in frequency p and the other in frequency $(1 - p)$, requires the fitness of each strategy to be frequency dependent and mean fitnesses to be equal at equilibrium. Alternatively, a monomorphic population consisting of a single hybrid or "mixed" strategy may also be an ESS. Here, individuals play one tactic with probability p and the other tactic with probability $(1 - p)$. (Note that "strategy" refers to characteristics of a genetic type while the different behaviors displayed by the mixed type or flexible strategy are called "tactics." Thus, in behavioral discussions, strategy is a genotypic description and tactic is a phenotypic description.) However, ESS models are somewhat limited in their conclusions compared to population genetic models that consider the average *as well as* the variance in fitness and consider several evolutionary forces in addition to natural selection. In particular, the distribution of relative fitness within a population determines the intensity of selection, as well as the population's expected response to selection (Crow 1958; 1962; Wade 1979; Wade and Arnold 1980; chapter 1). We suggest that the focus of game theory models on average fitness, without considerations of fitness *variance*, has caused them to overlook an important aspect of alternative mating strategies (Maynard Smith 1982; Austad 1984; Dominey 1984; Gross and Repka 1998). This oversight neither renders game theory models incorrect, nor invalidates their application to the study of alternative mating strategies. But it does mean that additional insights can be gained by considering the variance in fitness, especially that which, in males, arises from variations in the numbers of mates.

A quantitative genetic approach to male alternative mating strategies would assume that male behaviors or phenotypes consisting of two (or more) discrete phenotypes, are "threshold traits," whose expression is controlled by many loci of small effect (see chapter 12). Individuals with some genotypes exceed the threshold and have an enhanced tendency to display one mating morph, while individuals below it display the alternative morph. Differences among individuals or among populations in the expression of the alternative phenotypes are the result of differences in the sensitivity of the threshold to environmental variation. Theory shows that threshold traits closely connected to fitness, like male mating strategies, can have high heritabilities owing to the balance between frequency dependent selection and mutation. As one strategy becomes predominant, selection between strategies becomes weak and mutational variance within the predominant strategy accumulates (Roff 1998a,b). Differently put, there is special relationship between mutation and frequency-dependent selection acting on threshold traits that tends to increase or at least maintain genetic variation. It is believed that this variation facilitates the evolution of the threshold in response to environmental gradients.

As explained above, theoretical discussions of alternative mating strategies for over 100 years after Darwin (1871), assumed that the fitnesses of morphologically distinct males in the same species were equal (Gadgil 1972; Gross and Charnov 1980; Maynard Smith 1981, 1982). However, as field researchers began to document the fitnesses of males expressing distinct mating phenotypes in nature, four unexpected patterns emerged. First, in most cases, alternative strategies existed only at *low frequency* in natural populations. Second, compared to males exhibiting the dominant phenotype, alternative male phenotypes were relatively *unsuccessful* at mating. Third, in spite of their apparent conjugal failures, alternative morphs *persisted* in populations over long periods of time. Fourth, despite their apparently inferior fitness, alternative male phenotypes often were *highly specialized* in mating behavior and morphology.

These observations led Dawkins (1980) to erect a remarkably resilient concept. His hypothesis stated that, if sexual selection favored the most combative, showy, and vigorous males, then males unable to compete in this arena might adopt alternative sets of behaviors or morphologies that would still allow them to mate. These inferior males might be less successful than the dominant males in the population, but they would still "do better" than if they had secured no mates at all. That is, they could "make the best of a bad job." Following the form of game theory models, Dawkins (1980) proposed that if all males possessed such flexibility, that is, if this form of plasticity were *fixed* in the population, there would be *no need* for fitness of alternative variants of male mating behavior to be equal among all male phenotypes. When all males are genetically identical, fitness differences among them of any kind are of no evolutionary consequence.

Eberhard (1979, 1982) championed Dawkin's view and postulated a developmental mechanism for switching among male morphs, similar to the developmental "thresholds" known for other developmentally flexible phenotypes. The switch would be activated when insufficient resources prevented an individual from achieving the size, strength, or brilliance of the dominant male morph. This developmental switch would allow underfed or otherwise impaired males to adopt a phenotype that, while competitively inferior, might still allow males expressing it to achieve some mating success.

It is relatively common in holometabolous insects for individuals as larvae to delete one or more instars in response to nutritional stress or to prolong development by postponing pupation under social stress (Tschinkel 1993; Bell 1994; see also Fischer and Fiedler 2001 in the context of sexual selection in a butterfly). These developmental polymorphisms can reduce the risk of mortality from starvation and cannibalism, respectively, and *increase relative fitness*. In some species, such as the bean weevil, *Callosobruchus maculatus* (Messina and Renwick 1985), a developmental switch triggers a winged dispersal morph instead of an apterous, flightless morph in both sexes. Conditions of temperature and humidity are predictive of local population density and, thus, the future availability of local resources. The metabolic activity of many feeding and growing larvae increases average temperature and humidity. Thus, certain combinations of temperature and humidity activate the developmental switch within individuals that permits them to grow wings and escape intense resource competition. Because there is always some variation in the way individuals experience the environment and because there is genetic variation among individuals in the value of the developmental switch, there are wing polymorphisms in *all* populations. Like our other examples, the flexibility in producing winged versus apterous forms is adaptive in relation to resource availability. The majority of such phenotypic polymorphisms involve escape from deteriorating environments in time (e.g., germination or resting egg polymorphisms) or in space (e.g., winged versus apterous polymorphisms), conditions that are experienced by the *entire population*.

It is reasonable to suggest that a similar developmental switch might underlie alternative male mating morphs to ensure access to limited reproductive resources. Like a deteriorating environment, in any harem polygynous system, some dominant males will consume more than the average amount of available reproductive resource ($H > R$) while many others ($p_0 = 1 - [R/H]$) will have none. Thus, the advent of intense male-male competition for reproductive resources is a certainty. Furthermore, size at some stage during development may be an excellent predictor of an individual's adult ability to compete against other dominant individuals. (Indeed, maternal condition may be such a predictor.) However, it is not clear why a special evolutionary scenario is necessary to explain this situation nor why the switch points for alternative male strategies are not themselves genetically variable as is common for all other known developmental thresholds.

Researchers have extended the developmental switch hypothesis to include male behavioral variation (Thornhill and Alcock 1983; Gross 1996; Gross and Repka 1998), and behavioral and developmental flexibility appear to be widespread across taxa. When the phenotype is behavioral, such as fight or flight, the time between activating the switch and the phenotypic outcome is greatly reduced relative to morphological polymorphisms, in which the time between activation and outcome is much greater. However, a propensity to fight and a propensity to flee can also have genetic components and pose no special evolutionary problem relative to the standard scenario.

"Condition-dependent" reproductive phenotypes, behavioral and morphological, exist in a large number of taxa (reviews in Thornhill and Alcock 1983; Moran 1992; Andersson 1994; Gross 1996; Choe and Crespi 1997). The explanation offered by "making the best of a bad job" appears to explain the paradoxical persistence, at low population frequencies, of noncombative male phenotypes with inferior mating success versus dominant males. The "best of a bad job" hypothesis also seems to fit the empirical observations from field studies and laboratory arenas so much better than more quantitative models, that some (e.g., Austad 1984; Dominey 1984; Gross 1996; Gross and Repka 1998) have advocated removal of game theory models and their population genetic equivalents from the study of alternative mating behaviors altogether. The essence of this perspective exists in the "status dependent selection" (SDS) model of Gross (1996; see below), who argues that genetic polymorphisms underlying male mating behavior are rare or nonexistent. Thus, the theoretical framework of the SDS model claims to explain how conditional phenotypes persist within populations, despite unequal fitnesses among males differing in social status.

Despite the intuitive appeal of the SDS approach, we find nothing so unusual about the biology of alternative mating strategies that warrants placing them outside the realm of conventional evolutionary principles. We also believe that existing data are better explained by standard evolutionary theory than by the "best of a bad job" hypothesis. In the following sections, we show that hypotheses assuming unequal fitnesses among alternative mating strategies are neither necessary nor sufficient to explain observed variation in male phenotypes. Instead, we will demonstrate that simpler and well-established evolutionary genetic principles, in particular, those that examine the variance in fitness as well as its average, can be used to explain the relative frequency, fitness, persistence and complexity of alternative mating strategies in nature. First, however, we consider the SDS model in more detail because of its current popularity.

Status-Dependent Selection

The current theoretical framework explaining the evolution of "almost all alternative reproductive phenotypes" is the SDS model of Gross (1996). According to this approach, variation in mating phenotypes within species is primarily condition dependent, and evolves by a process called *status-dependent selection*. When the fitness consequences of alternative behavioral phenotypes, X and Y, depend on the relative competitive ability of the interactants, that is, on their *status* relative to one another, then selection is said to be status dependent. Status is assumed to be a trait that varies among individuals, primarily due to external environmental influences, such as variation in nutrition, injury, and disease, and secondarily due to genetic variation and developmental state. Overall, the distribution of status among individuals is expected to be normal within most populations (fig. 11.1a). Unlike the mixed strategy of ESS models, where individuals play one or the other behavioral tactic at random a given proportion of the time, in SDS theory, individuals, throughout life, always play the strategy that maximizes their fitness in the population given the particular local conditions of relative status. (As noted above, the theory of indirect effects [Wolf et al. 1999; Agrawal et al. 2001a] provides a well-established population genetic approach for understanding the coevolution of social context traits, like status, and more "ordinary" phenotypes, like body size.)

In SDS theory, the fitnesses of two alternative phenotypes, X and Y, are mapped linearly onto the normal distribution of status but with different slopes (fig. 11.1b). The slopes of these functions differ such that fitness rises faster for a given change in status for phenotype X than it does for phenotype Y. (Note that there is nothing relative about the linear relationships between fitness and status. Although one might expect that the relative status of both interactants would determine the tactic [X or Y] displayed at an

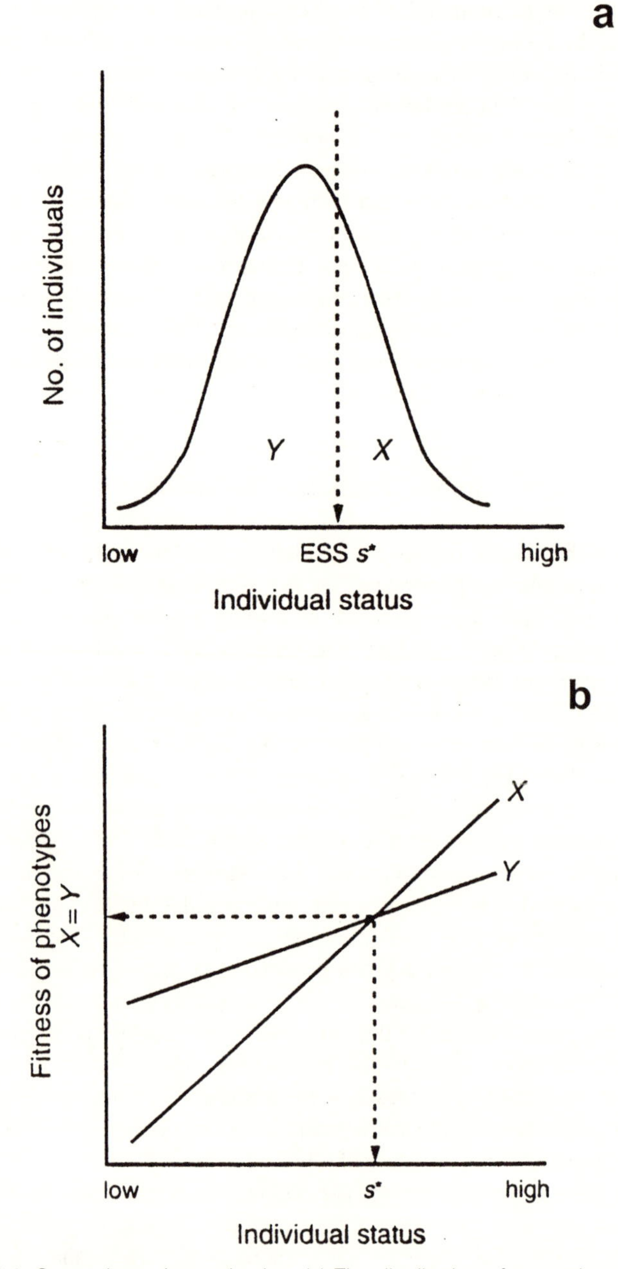

Figure 11.1. Status-dependent selection: (a) The distribution of status in a population; (b) the fitness functions for two morphs in a population, *Y* and *X*. Note that the slope of the fitness function for *X* is steeper than for *Y*; the intersection of these points is the switchpoint *s* (from Gross 1996).

encounter, nothing conditional or relative is evident in the mapping of fitness onto the status.) Nevertheless, at some point within the range of variation in status, the two functions intersect. This intersection is identified as the "switchpoint," s^*, that is, the point where the fitnesses of the two phenotypes are equal. At values of individual status below the switchpoint, the fitness function for phenotype *Y* exceeds that of phenotype *X*. The opposite condition holds for values of status above the switchpoint. Up to this point, the description is very similar (if not identical) to the typical genetical studies, mapping a dichotomous trait onto another, underlying, normally distributed trait (see, for example, Wright 1969; Roff 1994). The only evident difference lies in the argument that none of the variation in status is genetic; rather, it is entirely environmental. (Note: this is a situation explicitly considered in the theory of indirect effects.)

According to the SDS model of phenotypic plasticity, individuals of low status, that is, those with status below the switchpoint, are expected to assume phenotype *Y* because phenotype *Y* has greater fitness than phenotype *X* when individual status is low. Conversely, individuals with high status, that is, with status above the switchpoint, are expected to assume phenotype *X*, because phenotype *X* experiences greater fitness than phenotype *Y* when individual status is high. Again, the relative status of a pair of interactors does not affect the individual's decision to display *X* or *Y*. Given the verbal emphasis on maximizing fitness at every moment, an individual with status just below the switchpoint should display tactic *X* against an individual with status much farther below the switchpoint. Regardless whether that low-ranking individual displays *X* or *Y*, the higher-ranking individual will achieve higher fitness by displaying *X* (fig. 11.1b). Nevertheless, under the SDS model, both individuals display *Y*. In this way, the status-dependent selection model allows each individual to incorporate information about its "relative" competitive ability and subsequently express the phenotype that allows it to maximize its fitness under particular environmental conditions. That is, *all* individuals in the population are able to respond to their individual environmental circumstances in an adaptive way.

The ability to perform this switch is presumed to be identical among all individuals in the population. Differently put, all individuals are presumed to be *genetically monomorphic* with respect to their ability to develop the appropriate conditional strategy (Gross 1996, p. 93). Genetic monomorphism for this conditional strategy also precludes heritable variation in the location of the switchpoint, thereby ensuring that all individuals perform the appropriate mating "tactics" under the appropriate circumstances. The absence of heritable variation of any sort is an important assumption of the status-dependent model. Since all individuals are genetically identical with respect to ability to express the conditional strategy, equal average fitnesses are *not* expected or necessary among the competing phenotypes in the population. Also for this reason, unequal fitnesses among the different phenotypes cannot result in shifts in the phenotypic distribution of individual status or the switchpoint (Gross 1996; Gross and Repka 1998). As noted above, since all individuals, regardless of phenotype, are presumed to be genetically identical with respect to the conditional strategy, distinct phenotypes are described as "tactics," to distinguish them from genetically distinct alternative "strategies" (Austad 1984; Gross 1996). Although the chosen tactic results in higher mean fitness, it is not clear what evolutionary significance can be attached to this owing to the assumed absence of heritable variation.

As Gross (1996, p. 93) states, the five key components of a conditional strategy are:

> (1) tactics involve a "choice" or a "decision" by the individual; (2) the decision is made relative to some aspect of the individual's status; (3) individuals are genetically monomorphic for the decision; (4) the average fitnesses of the tactics are not equal but the fitnesses of the alternatives at the switchpoint are equal; and (5) the chosen tactic results in higher fitness for the individual.

As explained above, this model owes much of its popularity to its ability to "explain" almost any observed variation in mating behavior and do so without having to conduct controlled mating studies and test for evidence of a genetic variation. In many species, individuals *do* appear to make a choice with respect to the phenotype they assume in mating interactions. In many cases, the behavior exhibited by an individual *appears* to be made on the basis of its status (often body size) relative to the status of the prospective interactor (although, as we noted above, this feature is not a part of the formal SDS model). In some species, such choices *appear* as though the focal individual is selecting one or another component from a repertoire of possible behaviors (Thornhill 1981; Forsyth and Alcock 1990; Slagsvold and Saetre 1991). In other species, choices *appear* to be made that move individuals from one to another possible developmental trajectory (Eberhard 1979, 1982; Borowsky 1985; Emlen 1996; Clark 1997; Bass and Grober 2001; Bass et al. 1946). In most cases, although the alternative mating tactics yield lower average fitness, the chosen tactic *appears* to result in higher fitness for the observed individual than if no change in phenotype had been made. That is, the individual's fitness is not as low as it could have been.

Although Gross and Repka (Repka and Gross 1995; Gross and Repka 1998) later acknowledged the existence of hereditary factors that may influence the expression of alternative mating strategies, the significance of these effects is discounted or ignored. In particular, the authors note that the conditional nature of trait expression may shield inferior alleles from selection and prevent their elimination. Genetic monomorphism is still presumed to exist at the switchpoint, a condition that reduces its responsiveness to selection. Despite mentioning the effects of heritable variance in tactic structures in trait expression, these factors are eliminated from their model. The conclusions reached under these conditions reinforce the hypothesis that, despite their experience of unequal fitnesses, conditional strategies can persist in natural populations. Thus, they differ little in practice from those originally identified by Gross (1996) and continue to emphasize the preponderance of conditional strategies in nature.

We believe that the explanatory power of the status-dependent selection model may be more apparent than real (see below). The assumption of absent or even minimal effects of heritable variation for either strategy or threshold makes understanding the origins and refinement of male alternative strategies and the variation among populations in their frequency particularly problematic. Although all individuals in the population appear to choose one phenotype or another, we argue that, in light of existing theory and data on phenotypic plasticity, it is simpler to assume that only *part* of the population responds to environmental cues. Moreover, these cues may or may not involve differences in relative individual condition, but any theory that attempts to explain the influence of social context on individual behaviors must provide an independent description of the distribution of that social context. Without it, it is not possible to properly weight the fitness consequences of different social contexts by their frequency. By analogy, it would be difficult to adequately model the evolution of predator or pathogen resistance in the absence of information about how often predators and pathogens are encountered.

We show below why *genetic variability* rather than genetic uniformity should characterize males competing for reproductive resources and that conditions allowing the persistence of genetic variation for male mating behavior occur *whenever variance in male mating success exists.* That is, we expect genetic polymorphism in male mating strategies *whenever* sexual selection occurs. The observation of equal fitness among the phenotypes at the fitness switchpoint (i.e., threshold) and unequal fitnesses among phenotypes away from the switchpoint is a consequence of how the fitnesses of the different phenotypes are defined (fig. 11.1). They do not describe relative fitness among male phenotypes within the population. The common observation of unequal average fitness between conventional and alternative mating strategies is based upon empirical observations of direct male-male combat between the two types of males in the field or in laboratory arenas. These observations do not include the large fraction of conventional males with zero fitness and, thus, are inadequate for calculating the mean fitness of the dominant strategy. Indeed, by omitting the large fraction of conventional males without mates, these studies typically overestimate the fitness of this male morph. We will show that the apparent tendency for alternative mating strategies to experience inferior mating success compared to a conventional mating strategy vanishes when the unsuccessful males employing the conventional strategy are properly accounted (Shuster and Wade 1991a).

The final feature of the SDS model, that the status-dependent tactic chosen by each individual results in a higher fitness for that individual, is surprising. Chopping off the lower tail of the fitness distribution before it is even manifested reduces the variance in fitness among individuals. This would slow the rate of adaptive evolution, except that the absence of heritable variation obviates this consequence. In the next chapter, we return to this point in the context of among-population variation in the location of the status-dependent switchpoint (Travis and Woodward 1979; Hutchings et al. 1999), when we consider the genetic architectures likely to underlie the expression of alternative mating strategies. We will then show that there is little evidence to support the assumption that individuals exhibiting conditional strategies are "genetically monomorphic" with respect to their ability to express a variable phenotype.

Jack of All Harems Is the Master of Some

In populations with separate sexes, the average mate numbers of males must equal the average mate numbers of females (chapter 4). This result not only constrains the evolutionary potential for apparently "unlimited" gamete production by males (Trivers 1972), it also reinforces the principle that when some males mate more than once, other males are excluded from mating altogether. This observation suggests an important relationship between the population sex ratio and the intensity of sexual selection, thereby providing the key to understanding the evolution of alternative mating strategies. As we explained in chapter 1, the population sex ratio, R, can be conveniently expressed as the ratio of the total number of females to the total number of males, or

$$R = N_{\text{females}}/N_{\text{males}}. \qquad [11.1]$$

Let us now suppose, as we did in chapter 1, that males can be divided into a series of mating classes, each defined by the number of mates each male secures (table 1.3). If i equals the number of mates per male, and p_i equals the frequency of males in each mating class, then in a population of males with variable mating success, p_0 equals the frequency of nonmating males, and $1 - p_0 = p_m$, where p_m equals the frequency of mating males (fig. 11.2).

The average number of mates per male, m, is equal to the number of mates per male times the frequency of each male class, summed over all males. Thus,

$$m = \Sigma\, i\, p_i. \qquad [11.2]$$

This expression gives the average number of mates per male because the total numbers of females in the population are distributed as mates among

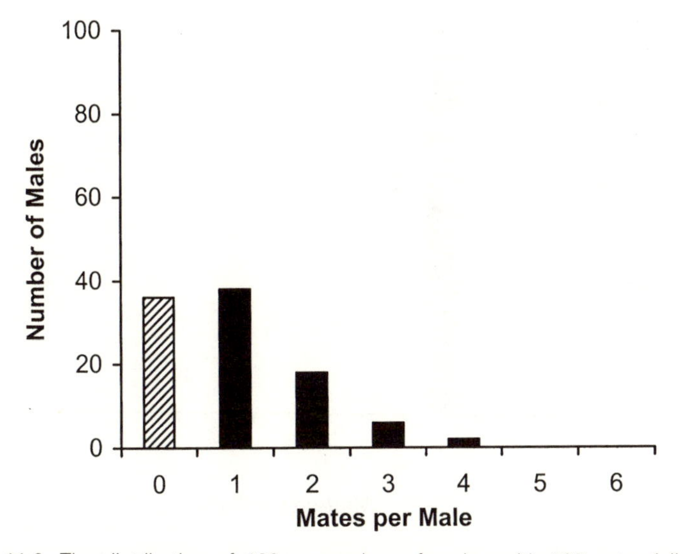

Figure 11.2. The distribution of 100 monandrous females with 100 potentially polygynous males under random mating; the hatched area represents the frequency of nonmating males, p_0; the shaded area represents the frequency of mating males, $1 - p_0 = p_m$; note that the sex ratio R is equal to the average number of mates per male, m, but less than the average harem size, H.

the total number of males. Thus, the sex ratio, R, and the average number of mates per male, m, are equivalent,

$$R = m. \qquad [11.3]$$

In these terms, the sex ratio also equals the frequency of males in each ith mating class, multiplied by the number of mates the males in each class secure, i, summed over all classes of males, or

$$R = \Sigma\, i\, p_i. \qquad [11.4]$$

Note too that the total number of females is equal to the number of males in each mating class, multiplied by the number of mates secured by males in each mating class; therefore,

$$N_{\text{females}} = \Sigma\, i\, (p_i\, N_{\text{males}}). \qquad [11.5]$$

The average harem size, H, expressed as the number of mates per *mating* male, is usually larger than R (fig. 11.2). This is true even when mating occurs at random. As Darwin (1874) noted, whenever one male secures two or more females, even if by chance some other males are excluded from mating at all (chapter 1). Thus, *only* when *all* males are successful in securing a mate is R equal to H (fig. 11.3). In *all* other cases, the smaller the fraction of males who mate successfully, the larger H will become, and, concomitantly, the larger the variance in mating success will be (fig. 11.4).

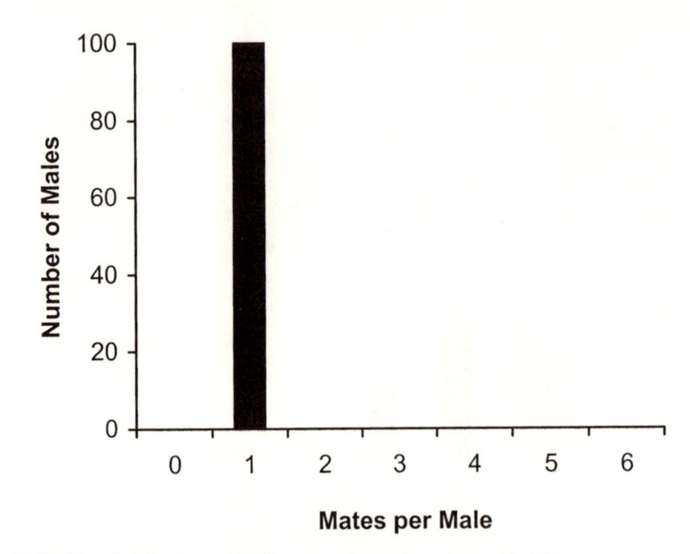

Figure 11.3. The distribution of 100 monandrous females with 100 monogynous males; note that the sex ratio R is equal to the average number of mates per male, m, which also equals the average harem size, H.

The average harem size can be expressed as the total number of females divided by the number of mating males,

$$H = \Sigma\, ip_i/(1 - p_0). \quad [11.6]$$

By substitution, using eq. [11.4] above, we can now express the average harem size, H, in terms of the sex ratio and the fraction of nonmating males, thus,

$$H = R/(1 - p_0). \quad [11.7]$$

By rearranging eq. [11.7], the fraction of nonmating males in the population, p_0, can also be expressed in terms of the sex ratio and the average harem size,

$$p_0 = 1 - (R/H). \quad [11.8]$$

It is now easy to see, especially when R is unity, that as the average harem size H increases, the fraction of nonmating males in the population increases as well (fig. 11.5). For example, when the average harem size for mating males equals 1, *all* males are able to acquire mates and the frequency of nonmating males is zero. If $H = 2$, the frequency of nonmating males rises quickly to 0.50. If $H = 4$, the frequency of nonmating males equals 0.75. If $H = 6$, the frequency of nonmating males equals 0.83, and so on. Two things are obvious from this example. First, as females become concentrated among fewer and fewer mating males, the frequency of nonmating males *must* increase. The second and more useful observation is that when H, the

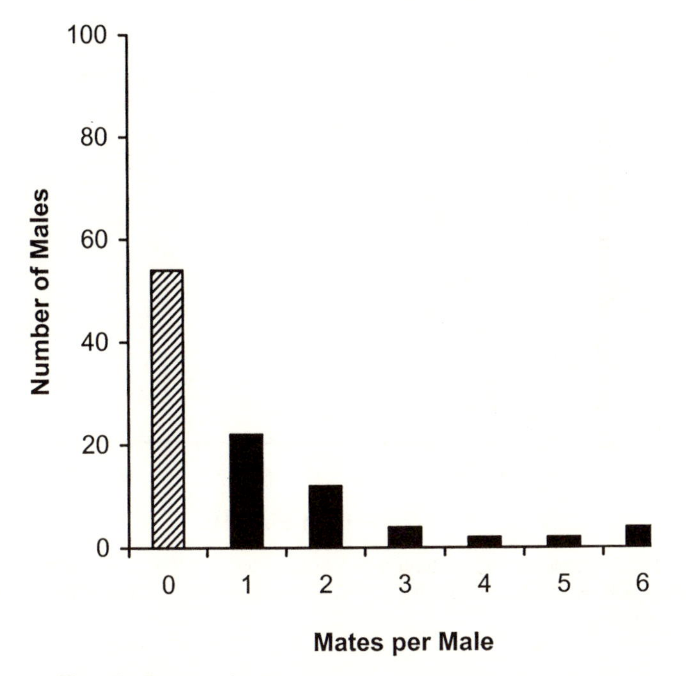

Figure 11.4. The distribution of 100 monandrous females with 100 potentially polygynous males under nonrandom mating, that is, some males acquire a disproportionate number of mates; the hatched area represents the frequency of nonmating males, p_0; the shaded area represents the frequency of mating males, $1 - p_0 = p_m$; note that the sex ratio R is equal to the average number of mates per male, m, but both of these values are much less than the average harem size, H.

average number of mates per mating male, is known and the primary R is observed, the frequency of nonmating males, p_0, can be calculated explicitly.

Let us suppose that all of the males we have described in the above example exhibit the same general phenotype. For example, males in this population could be larger than females and possess horns, which males may use to displace other males from breeding territories (chapter 9). Let us also assume that females tend to aggregate at breeding territories, and, since large, horned males are more successful at displacing smaller, horned rivals, size equals success in acquiring mates among horned males. Let us next assume that another type of male possesses a mutation that induces sexual maturity at a young age. Being precocially mature, such males lack horns, are much smaller in size than territory-defending males, avoid territorial confrontations, yet are attracted to aggregations of females.

We now have two types of males and can refer to horned males as α-males, and to nonhorned males as β-males. If β-males are successful in mating with some of the females in the harems of α-males, we can express their success as a fraction of that achieved by harem-holding α-males. Thus,

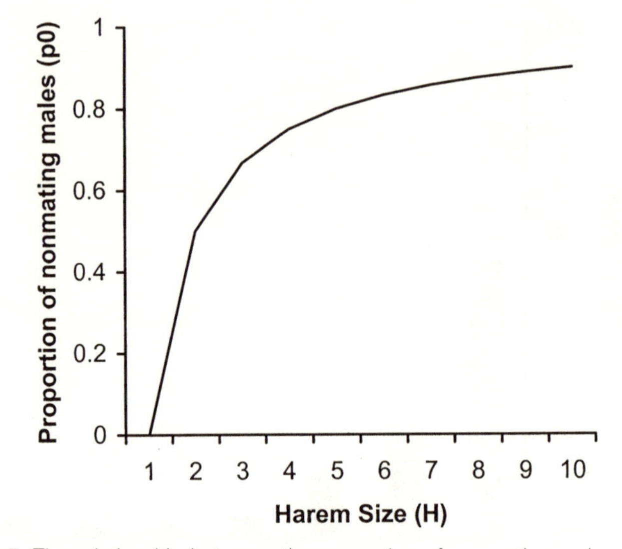

Figure 11.5. The relationship between the proportion of nonmating males (p_0) in the population and *H*; note that, as harem size increases, the proportion of nonmating males increases.

if s equals the success rate of β-males invading harems, the fitness of β-males, W_β, is equal to

$$W_\beta = sH. \qquad [11.9]$$

In most mating systems, the average number of mates per mating male, H, is easier to estimate than the proportion of males with no mates, p_0, because nonmating males disperse from or hide within the vicinity of breeding males. As we have explained above, in species in which male polymorphism occurs, the most common observation is that β-males obtain only *some* mating success from females within aggregations defended by α-males. Most of the observed fitnesses of alternative male mating strategies come from these sorts of pairwise encounters. Thus, the success of β-males appears to be only a fraction of that achieved by harem-holding α-males. Stated differently, the success of α-males appears to exceed that of β-males. We can express this relationship as

$$H > sH \qquad [11.10]$$

and therefore

$$W_\alpha > W_\beta. \qquad [11.11]$$

From this result, it is difficult *not* to conclude, as many have, that the fitness of β-males is less than that of α-males (reviews in Dawkins 1980; Eberhard 1982; Austad 1984; Dominey 1984; Gross 1996). Differently put, β-males certainly appear to "make the best of a bad job," losing in every head-to-head contest with α-males, but getting more mates than by not trying at all.

However, this approach considers only the fitness of α-males that *actually mate*. The average fitness of *all* α-males, $W_{\alpha(\text{all})}$, is equal to R because this value includes the average mate numbers of mating *as well as* of nonmating α-males. To invade a population, the average fitness of a mutant strategy must *exceed* the average fitness of the conventional strategy (Darwin 1874; Maynard Smith 1982). Thus, for β-males to invade a population of α-males, the average fitness of β-males must exceed the average fitness of α-males in the absence of β-males. We can express this condition as

$$W_\beta > W_{\alpha(\text{all})} \qquad [11.12]$$

or by substitution

$$sH > R. \qquad [11.13]$$

By rearrangement, this equation becomes

$$s > R/H \qquad [11.14]$$

and by substitution from eq. [11.7], the inequality becomes

$$s > (1 - p_0). \qquad [11.15]$$

We can see that the β-strategy invades whenever the average proportion of the fertilizations β-males obtain within the harems of α-males exceeds the fraction of mating α-males. This means that the greater the concentration of females within the harems of a few α-males (i.e., as more α-males are *excluded* from mating), the *easier* the invasion by β-males becomes (fig. 11.6)! In chapters 9 and 10, we saw that alternative male mating strategies are more common in species where H is large and $(1 - p_0)$ is small.

Consider a simple example. Suppose that the population sex ratio, R, equals 1. Next suppose that the average harem size among mating males, H, equals 4. Substituting these values into eq. [11.14], gives s a value of 0.25. This means that β-males can invade this population when their fitness is only 1/4 that of the α-males who defend harems (fig. 11.6)! Now consider what happens if females mate with α-males *at random* (fig. 11.2) Sexual selection is usually thought to be nonexistent when such conditions apply (Downhower et al. 1987; Sutherland 1985). However, as shown in fig. 11.2, random mating by females causes some α-males to mate with more than one female, and causes other α-males to secure no mates at all. In a population of 100 α-males with 100 randomly mating, monandrous females, the average harem size, H, is 1.6 (fig. 11.2). Substituting these values into eq. [11.14] gives s a value of 0.63. We conclude that, even when females mate with α-males at random, β-males can *still* invade a population of α-males. Moreover, this invasion will occur even when the fitness of β-males is only 63% of that obtained by harem-holding α-males! Invasion occurs because the

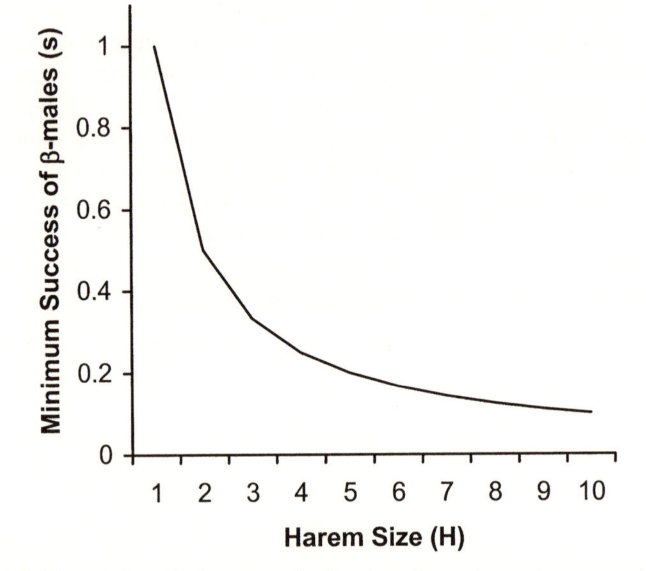

Figure 11.6. The relationship between the fraction of α-male mating success acquired by β-males, s, and $1/H = 1 - p_0$, where p_0 is the proportion of the α-male population that is unsuccessful in acquiring mates.

average fitness of β-males *exceeds* the average fitness of all α-males. It is also interesting to note that when we solve eq. [11.9] using the values for s and H identified in the last example, we find that W_β is approximately equal to 1. Thus, after invasion occurs, β-males, who for all practical purposes are monogamous (i.e., they have the equivalent of $[1/H]$ mates), can persist within a population of α-males in which polygyny is the dominant mating strategy. This means that the average fitness of monogamous males is equivalent to the average fitness of all males attempting to be polygynous, because a large fraction of these latter males are unsuccessful at mating. Because the average fitnesses of α- and β-males are equal, there is no "bad job" for β-males to endure.

The invasion of β-males can be easier still if they are more prone to invade larger harems or if larger harems are more easily invaded. This phenomenon is often reported in studies of alternative male mating behaviors, wherein territories containing large harems are found to be easier for satellite males to invade. Alternatively, territories containing large harems may be more attractive to satellite males than territories containing few or no females (Shuster 1987; Shuster and Wade 1991a; Sinervo and Lively 1996; Cooch et al. 1997; Mesnick 1997). This important covariance is seldom if ever, addressed in laboratory studies, where most male-male contests occur in the presence of a single female. We can express the relationship between harem size and the ability of β-males to invade harems as the covariance between H and s, or

$$\mathrm{Cov}(s, H). \tag{11.16}$$

If this relationship is positive, the fitness of invading β-males is increased by this amount, or

$$W_\beta = sH + \mathrm{Cov}(s, H). \tag{11.17}$$

Thus, when the aggregation of females around particular males enhances the mating success of satellite males, the invasion of an α-male population by β-males becomes easier still. If the covariance between harem size and the success of β-males is negative, invasion will be more difficult. However, as described in chapter 9, the sign of eq. [11.16] will usually be positive. Thus, the simple observation that the mating success of satellite males is less than that of harem-holding males is neither a *necessary* nor a *sufficient* reason to suggest that satellites are "making the best of a bad job." Our results show that, when satellites obtain only a fraction of the total fertilizations within the harems of territory-defending males, as is commonly reported (reviews in Lucas and Howard 1995; Gross 1996; Gross and Repka 1998), they are mating as successfully, or *more* successfully, than the average territorial male. In fact, because the invaded α-males are those with larger than average harems, $H_{\text{invaded}} > H$, the inequality between β- and α-male fitness might be even greater than described above. Consistent with this expectation, Fu et al. (2001, p. 1105) show in bluegill sunfish (*Lepomis macrochirus*) that "sneaks fertilize more eggs than guards during sperm competition," and "that sneakers are superior to satellites in sperm competition."

These same considerations can also be applied to the evolution of alternative female nesting strategies. For example, whenever nesting sites are scarce, some fraction of females, p_{nesting}, may be successful in acquiring them and raise O_{nesting} offspring, while another fraction of females, $p_{\text{nonnesting}}$, may be excluded from resources and not raise any offspring at all. Consider the invasion of a "nest-joining" strategy, wherein females join established nests, with probability f, and thereby obtain resources for a small number of offspring, $O_{\text{joiner}} < O_{\text{nesting}}$ offspring. Specifically, let O_{joiner} equal $(1 - s)$ O_{nesting}, where s is positive. Although joiners are never as fecund as successful nest-site holders because $O_{\text{joiner}} < O_{\text{nesting}}$ in every case, and although joining females are successful only a fraction of the time, f, they will invade if $\{f(1 - s) > p_{\text{nonnesting}}\}$. The smaller the fraction of successful females, $p_{\text{nonnesting}}$, the easier it will be for a joining strategy to invade. However, because the fitness of *successful* dominant males, HO, exceeds that of successful females, O, male alternative mating strategies may be much more likely to evolve than alternative female nesting strategies. We predict that the maintenance of alternative female nesting strategies by frequency-dependent selection should result in an increased mitochondrial haplotype diversity, providing a molecular genetic method for delecting the existence of alternative mating strategies in females and for verifying behavioral observations in the field.

The Size of the Mating Niche

Let us assume that the invasion of our hypothetical α-male population by β-males has been successful. As other authors have explained (Darwin 1874; Maynard Smith 1982), to persist within the invaded population, the average fitness of the mutant strategy must *equal* the average fitness of the conventional strategy. Thus, for α- and β-males to persist in our hypothetical population, their fitnesses must be equivalent, or

$$W_{\beta} = W_{\alpha(\text{all})}. \qquad [11.18]$$

By substitution from eqs. [11.9] and [11.17], we obtain this condition more explicitly:

$$W_{\beta} = sH + \text{Cov}(s,H) = R. \qquad (11.19)$$

If the sex ratio equals 1 and if, as shown in eq. [11.7], $H = 1/(1 - p_0)$, then, by substitution,

$$s\,[1/(1 - p_0)] + \text{Cov}(s,H) = 1, \qquad [11.20]$$

or, by rearranging terms,

$$p_0 = 1 - (s/[1 - \text{Cov}(H,s)]). \qquad [11.21]$$

We can now see that, if the frequency of nonmating α-males is known, we can predict the mating success of β-males necessary for the persistence of this alternative mating strategy within the population of α-males. Alternatively, if we know the mating success of β-males, we can predict the proportion of the α-male population that must be unsuccessful in acquiring mates in order for β-males to persist in the population. Furthermore, when we know the harem size of mating males, as is often the case in many field studies, we can predict the mating success of satellite males necessary for these individuals to invade as well as to persist in the population. In general, when the Cov(H,s) is assumed to be zero, the relative success of satellite males is equivalent to the relative failure of conventional males. This latter relationship may become more complicated if the success of satellite males, like that of territorial males, varies with harem size.

Two things can happen after β-males invade. First, variance can arise in the mating success of β-males. If some β-males obtain multiple mates and others do not, ΔI can again become large and may allow the average fitness of α-males to exceed that of β-males. This will cause the frequency of

α-males to increase at the expense of β-males until variance in mate numbers among α-males again becomes high, whereupon β-males can "reinvade" the population and the frequency of β-males again increases. The second possible consequence of a β-male invasion is that the variance in mating success among β-males does not occur at the expense of variance in α-male fitness, but rather a new mating niche arises and allows a third male type, "γ-males," to invade the population.

As explained above, the size of this mating niche will depend on the value of the average H for α- and β-males relative to R as well as the Cov(H,s). Once a third morph arises, it may invade the mating niches of α- and β-males, or specialize only on the niche provided by one of the male types. The extent to which invasion by each strategy is unidirectional or intransitive not will determine the dynamics of fluctuations in the frequencies of each morph. Unidirectional invasions are expected to create stable limit cycles as observed in "rock-paper-scissors" games (Maynard Smith 1982; Sinervo and Lively 1996). For example, α-males may be vulnerable to invasion by β-males, but not to invasion by γ-males, and, by invading the mating niche of β-males, γ-males are vulnerable to invasion by α-males. If so, the frequencies of the morphs will oscillate regularly. Such cycles are not necessarily expected when invasions are bidirectional or if some morphs are mating niche specialists and some are not.

The magnitude of the difference between H and R will depend on the degree to which females tend to mate with relatively few males. That is, the difference between H and R will depend on the magnitude of the variance in mate numbers among males. When few males mate, the variance in mate numbers per male is high because $(1 - p_0)$ is small. Thus, the larger the difference between H and R, the more easily an alternative mating strategy can invade. A mutation allowing males to mate under such circumstances will be favored and will increase in frequency at a rate proportional to the value of ΔI. Because invasion of a population by an alternative strategy causes the number of mating males to increase, the opportunity for sexual selection will erode as the frequency of β-males increases. If the success of β-males remains constant, the decay in ΔI will be proportional to the frequency of β-males until the mating niche no longer exists. However, the greater the relative success of β-males, the faster this erosion will take place, and the lower the frequency of β-males will be when the reproductive resources that constitute the mating niche are exhausted. This mechanism provides an explanation for the widespread observation that the frequency of males employing alternative mating strategies is generally smaller than the frequency of males employing the conventional mating strategy (Gross and Charnov 1981; Shuster and Wade 1991a; Sinervo and Lively 1996; Lank et al. 1999).

Paracerceis Sculpta: A Worked Example

Can examples of multiple male morphs with repeated invasions exist in nature? Two possible examples include in the marine isopod, *Paracerceis sculpta* (fig. 11.7) and the side-blotched lizard, *Uta stansburiana* (Sinervo and Lively 1996; Zamudio and Sinervo 2000). As explained in previous chapters, *P. sculpta* is native to the eastern Pacific Ocean (Holmes 1904; Brusca 1980), although it now appears cosmopolitan in its distribution as a result of its transport by ships to other locations throughout the world (Rodriguez et al. 1992). The population inhabiting intertidal and subtidal zones in the northern Gulf of California is the most studied population to date (review in Shuster and Guthrie 1999). In our previous examples using this species (chapters 2 and 3) we have focused on males who defend breeding territories in intertidal sponges, that is, on α-males. However, not one, but three, discrete male phenotypes coexist in this population (Shuster 1987, 1992). Each morph possesses mature external genitalia and sperm producing organs, and although males' relative allocations of energy toward somatic and gonadal structures differ significantly (Shuster 1987, 1989a), males of each type do not differ in their ability to sire young (Shuster 1989a, b). Shuster and Sassaman (1997) demonstrated that a major gene (*Ams* = Alternative mating strategy) causes phenotypic differences among males. Inheritance of *Ams* is Mendelian and alleles at this locus exhibit directional dominance ($Ams^{\beta} > Ams^{\gamma} > Ams^{\alpha}$). These results are consistent with an earlier population genetic model proposed by Shuster and Wade (1991a).

As explained in chapters 2 and 3, in monthly samples between October 1983 and November 1985, Shuster (1987, 1991a) collected sponges containing isopod breeding aggregations from permanent tide pools in the intertidal zone located 1.5 km SW of Puerto Peñasco, Sonora, México. Within 15 randomly selected 0.25 m plots along a 100 m transect, Shuster removed all sponges from each plot and examined each spongocoel for isopods, which were removed and identified by adult phenotype. Shuster and Wade (1991a) pooled these monthly samples to calculate the frequencies of males and females within spongocoels (table 11.1). Previous experiments using genetic markers (Shuster 1989) showed that α-males sire a majority of offspring within spongocoels containing a single female and one β-male or one γ-male. However, these results also showed that when more than one female is present in a spongocoel, β-males tended to sire about 60% of all offspring, regardless of female density. The fertilization success of γ-males in spongocoels increased linearly with harem size, but generally represented only a fraction of the total offspring produced by each female (Shuster 1989b; Embry and Shuster 2001; Shuster and Sassaman, unpubl. data). Shuster and Wade (1991a) used these results to identify 14 rules for assigning paternity within spongocoels (table 11.2). When these rules were applied to the aggre-

Figure 11.7. The three male morphs in the marine isopod, *Paracerceis sculpta* (from Shuster and Wade 1991a).

Table 11.1
Distribution of *P. sculpta* α-, β-, and γ-males and females in *Leucetta losangelensis* spongocoels, 1983–85.

Rule ->	*1*	*2*	*3*	*4*	*5*	*6*	*7*	*8*	*9*	*10*	*11*	*12*	*13*	*14*
							1α +						*2α +*	
Harem size	*1α*	*1β*	*1γ*	*2α*	*2γ*	*3α*	*1β*	*1β + 4γ*	*2β + 3γ*	*1γ*	*2γ*	*3γ*	*1γ*	*3γ*
0	121	4	5	1	1	0	2	0	0	5	1	0	0	0
1	101	0	5	2	1	0	4	0	0	8	1	1	0	0
2	55	0	2	2	0	1	4	0	0	7	1	0	0	0
3	33	0	1	1	1	0	2	0	0	3	1	0	1	0
4	22	0	0	2	0	0	0	0	0	6	1	0	1	0
5	17	0	0	0	0	0	1	0	0	2	0	0	0	0
6	6	0	0	1	0	0	0	0	0	2	0	0	0	0
7	0	0	0	2	0	0	0	0	1	0	0	0	0	0
8	2	0	0	0	0	0	0	0	0	0	0	0	0	0
9	1	0	0	1	0	0	0	0	0	0	0	0	0	0
10	1	0	0	0	0	0	0	0	0	0	0	0	0	0
11	1	0	0	0	0	0	0	0	0	1	0	0	0	0
12	0	0	0	0	0	0	0	0	0	0	0	0	0	1
13	0	0	0	0	0	0	0	0	0	1	0	0	0	0
14	0	0	0	0	0	0	0	0	0	0	0	0	0	0
15	0	0	0	0	0	0	0	1	0	0	0	0	0	0
Σ	360	4	13	12	3	1	13	1	1	35	5	1	2	1

Table 11.2
Rules for assigning male fertilization success in spongocoels (from Shuster and Wade 1991a).

Case	Rule
1.	α-male alone; sires all.
2.	β-male alone; sire all.
3.	γ-male alone; sires all.
4.	2 α-males; each sires half.
5.	2 γ-males; each sires half.
6.	3 α-males; each sires one third.
7.	1α- + 1β-male; β sires 0.60, α sires 0.40.[a,b]
8.	1α- + 1β- + 4γ-males; β sires 0.60, α sires none, the remaining progeny are divided among γ-males.
9.	1α- + 2β- + 3γ-males; 0.60 of the progeny are divided between the β-males and 0.40 among the γ-males, the α-male sires none.
10.	1α- + 1γ-male; γ sires 0.08, α-male sires 0.92.[a,b]
11.	1α- + 2γ-males; each γ follows rule 10.[c]
12.	1α- + 3γ-males; each γ sires 0.08, α-male sires the rest.
13.	2α- + 1γ-male; γ follows rule 10,[c] the remaining offspring are divided between the α-males.
14.	2α- + 3γ-males; offspring are evenly divided among the γ-males.

[a]Laboratory tests (Shuster 1989b).
[b]Field data (Shuster and Sassaman, unpub. data).
[c]Linear regression based on above references: $y = -0.047 + 0.113x$.

gate collection of isopods from spongocoels in the 1983–85 transect samples, there were no differences in the fertilization success of the three male morphs over a two-year period (fig. 11.8).

We will now use these same results to test the theoretical framework we have illustrated above. We have suggested that, if the frequency of nonmating α-males is known, we can predict the mating success of β-males necessary for the persistence of this alternative mating strategy within the population of α-males. We also proposed that if we know the mating success of β-males we can predict the proportion of the α-male population that must be unsuccessful in acquiring mates in order for β-males to persist in the population. When we know the harem size of mating males, as is often the case in many field studies, we claimed above that it is possible to predict the mating success of satellite males necessary for these individuals to invade as well as to persist in the population. We also proposed that, if sufficient variance in the mating success of β-males exists, either α-males can reinvade the population, or a tertiary mating strategy, that we might call γ-males, may invade the population.

Let us begin our example by examining the composition of the isopod population summarized in table 11.1. This two-year sample consisted of 555

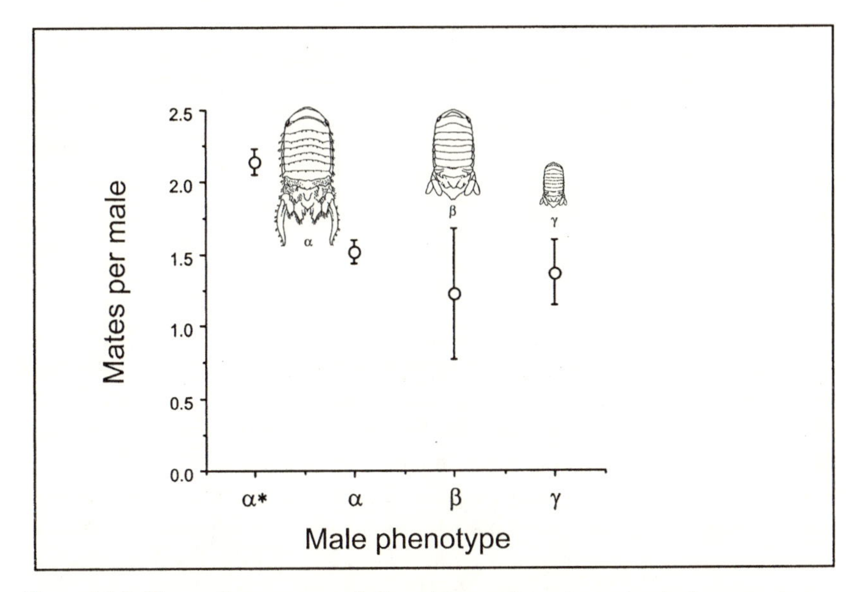

Figure 11.8. The mating success of *P. sculpta* α-, β-, and γ-males in *Leucetta lonangelensis* spongocoels between 1983 and 1985; "α*" represents the average harem size of mating α-males; "α" represents the average harem size of all α-males.

males and 824 females (825 females were incorrectly reported in Shuster and Wade 1991a). Of the 555 males, 452 were α-males (0.81), 20 were β-males (0.04), and 83 were γ-males (0.15). This population of α-males has clearly been invaded by two alternative male mating strategies. However, we can make a few simple adjustments to see what the population might have looked like before β- and γ-males appeared. Note in table 11.1 that column 1 summarizes the contents of spongocoels that contained only α-males and females. Spongocoels summarized in columns 2–14 contained combinations of α-males, β-males, γ-males, and females. We can visualize an uninvaded α-male population by imagining that β- and γ-males do not exist. This allows us to collapse the contents of columns 4 and 6–14 into column 1, and assume that the α-male tending these breeding aggregations was in fact the only male to mate with the females in his harem. Note that columns 4, 13, and 14 summarize the contents of spongocoels that contained two α-males. We will assume, as we did in Shuster and Wade (1991a), that the mating success of these males was equal within these spongocoels, that is, α-males divided the females within these spongocoels equally among themselves as mates. Columns 3 and 5 contained breeding aggregations consisting only of γ-males and females. For the purposes of this example, we will ignore the females mated by these males. This removes only 16 of the 824 females from the total data set (< 2%), allowing us to include over 98% of the total number of females collected in the original two-year sample.

Table 11.3
Summary table of females with α-males, excluding β- and γ-males.

Male type	N_{total}	N_{mating}	$N_{nonmating}$	$N_{females}$
α	452	321	131	808
β	0	0	0	0
γ	0	0	0	0
	555	404	151	808

We now have a population consisting of 452 α-males and 808 females, distributed within the spongocoels summarized in table 11.3. From this distribution, we can calculate the population sex ratio, R, as shown in eq. [11.1], by dividing the number of females by the number of α-males, $808/452 = 1.788$. We can also see in table 11.3 that the number of males in spongocoels containing females equals 321, and the number of α-males in spongocoels by themselves equals 131. Thus, we can explicitly calculate the frequency of mating α-males, $p_{m\alpha}$, as $321/452 = 0.710$. The frequency of nonmating α-males, $p_{0\alpha}$, equals $131/452 = 0.290$. Thus, $1 - p_{0\alpha} = p_{m\alpha}$. We can further calculate the harem size of *mating* α-males, $H_{m\alpha}$, as shown in eq. [11.6], as $(808/321) = 2.517$. We now have the means to examine our model.

Recall that eq. [11.8] shows the relationship between the frequency of nonmating males and harem size of mating males. Rewriting this expression for α-males gives

$$p_{0\alpha} = 1 - (R/H_{m\alpha}). \quad [11.22]$$

Substituting the values obtained from table 11.3, we see that

$$p_{0\alpha} = 1 - (1.788/2.517) \quad [11.23]$$

which equals $1 - 0.170$ or $p_{0\alpha} = 0.290$, a value identical to our observed frequency of nonmating α-males. This expression shows two things. First, it shows that it is possible to explicitly determine the frequency of nonmating males from the distribution of females with mating males. It also provides us with a means for determining the vulnerability of this α-male population to invasion by alternative mating strategies. Recall from eq. [11.16] that the conditions necessary for the invasion of an alternative mating strategy, whose practitioners seek harems at random with respect to harem size are that

$$s_\beta > (1 - p_{0\alpha}) \quad [11.24]$$

where s_β equals the minimum fitness required of β-males for successful invasion to occur. Using the value for $p_{0\alpha}$ identified above, s_β equals $1 - 0.290 = 0.710$. This means that the fitness of β-males must be greater than 71% of that of α-males for a successful β-male invasion to occur. Although in this example, we know the frequency of nonmating α-males explicitly, in many species, nonmating males may disperse from the vicinity of breeding individuals, or be otherwise absent from samples. If for some reason we suspected that we were unable to account for all of the unmated males in our sample, we might assume that $R = 1$. This is a reasonable assumption because the primary sex determination mechanism in this species involves female heterogamy, and although autosomal sex ratio distorters are known to exist most at-birth sex ratios are 1:1 (Shuster and Sassaman 1997; Shuster and Levy 1999; Shuster et al. 2001).

If $R = 1$, the value of $p_{0\alpha}$ becomes 0.533 and the success of β-males necessary for their successful invasion only needs to be $1 - 0.533$ or 0.467. That is, the average fitness of β-males would only have to exceed 47% of the average fitness of α-males for successful invasion to occur. Indeed, our observed estimate of unmated α-males may underestimate the actual frequency of α-males in the population. Thus, our first estimates of $1 - p_0$ and s (= 0.710) represent conservative estimates of the conditions necessary for a β-male invasion.

The above example is admittedly somewhat artificial. We have ignored the existence of satellite males that clearly exist in the population. So now let us return to our original data set, in which we know the distributions of all of the males in spongocoels, as well as their relative fertilization success within spongocoels. Using the rules for assigning fertilization success among males as described in Shuster and Wade (1991a; table 11.2), we can summarize the numbers of α-, β-, and γ-males that secure mates as well as the numbers of individuals bearing these phenotypes who are excluded from mating altogether (table 11.4). We see that the number of α-males who mate, as well as the number of α-males excluded from mating, are unchanged from the previous example. However, given that β- and γ-males obtain their mating success by fertilizing ova *within* the harems of α-males, the number of mates that can be assigned to α-males is decreased from 808 to 687.

We must now recalculate the value of $H_{m\alpha}$, the average harem size of mating α-males. We see from table 11.4 and eq. [11.6] that $H_{m\alpha}$ is equal to $687/321 = 2.140$. We can also calculate the value of R_α, that is, the sex ratio experienced by α-males, as the ratio of females within the harems of

Table 11.4
Summary table of females with α-, β-, and γ-males.

Male type	N_{total}	N_{mating}	$N_{nonmating}$	$N_{females}$
α	452	321	131	687
β	20	14	6	25
γ	83	69	14	112
	555	404	151	824

α-males divided by the total number of α-males, or 687/452 = 1.520. If we substitute the values of R_{α} and $H_{m\alpha}$ into eq. [11.22] we obtain

$$p_0 = 1 - (1.520/2.140) = 0.290. \qquad [11.25]$$

Thus, the predicted frequency of nonmating α-males in the intact sample (0.290) is equal to the observed frequency of nonmating α-males (0.290), as well as equal to the value we identified for $p_{0\alpha}$ in the example excluding β- and γ-males. If we assume that $R = 1$, that is, if we consider the possibility that our observed frequency of unmated α-males underestimated the actual frequency of nonmating α-males, we obtain

$$p_0 = 1 - (1/2.140) = 0.533, \qquad [11.26]$$

a value identical to that determined in the example in which $R = 1$ and which excluded β- and γ-males above. By substituting these two values for $p_{0\alpha}$ into eq. [11.22], we can see that the fitness of β-males necessary for their successful invasion of the α-male population is, more conservatively, $1 - 0.290 = 0.710$, and, less conservatively, $1 - 0.533 = 0.467$.

How can we determine the success of β-males? From table 11.4, we can see that the average harem size of mating β-males, $H_{m\beta}$, is equal to the number of females mated by β-males, divided by the number of mating β-males, or 25/14 = 1.786. The observed frequency of nonmating β-males is equal to the number of nonmating β-males divided by the total number of β-males in the sample, or 6/20 = 0.300. We can calculate the value of R_{β}, that is, the sex ratio as it appears to be experienced by β-males, as the ratio of females mated by β-males divided by the total number of β-males, or 25/20 = 1.250. If we substitute the values of R_{β} and $H_{m\beta}$ into eq. [11.22] we obtain

$$p_{0\beta} = 1 - (1.250/1.786) = 0.300. \qquad [11.27]$$

Thus, the predicted frequency of nonmating β-males in the intact sample (0.300) is equal to the observed frequency of nonmating β-males (0.300). If we assume that $R = 1$, that is, if we consider the possibility that our observed frequency of unmated β-males underestimated the actual frequency of nonmating β-males, we obtain

$$p_{0\beta} = 1 - (1/1.786) = 0.440. \qquad [11.28]$$

The fraction of α-male success experienced by β-males is equal to the ratio of the average harem sizes of mating β-males to that of mating α-males, or

$$H_{m\beta}/H_{m\alpha} = 1.786/2.140 = 0.835. \qquad [11.29]$$

This means that the fitness of mating β-males is about 84% of that achieved by mating α-males. Our more conservative estimate of the fertilization success necessary for β-males to invade the population of α-males was 71%.

Our less conservative estimate of the minimum fertilization success necessary for a β-male invasion was 44%. Clearly, the minimum conditions necessary for invasion of the α-male population by β-males are met.

We can now consider the mating success of γ-males. We have shown that the observed frequency of unmated β-males, $p_{0\beta}$, equals 0.300, a value equivalent to that predicted from the distribution of females within the "harems" of mating β-males. Using this value and eq. [10.22], we can show that the fertilization success necessary for γ-males to invade the β-male population is $1 - p_{0\beta} = 1 - 0.300 = 0.700$. Differently put, in order to invade the β-male population, the fertilization success achieved by mating γ-males must exceed 70% of the fertilization success of β-males. From table 11.4, we can see that the average harem size of mating γ-males, $H_m\gamma$, is equal to the number of females mated by γ-males divided by the number of mating γ-males, or 112/69 = 1.623. The observed frequency of nonmating γ-males is equal to the number of nonmating γ-males divided by the total number of γ-males in the sample, or 14/83 = 0.169. We can calculate the value of $R\gamma$, that is, the sex ratio as it is experienced by γ-males, as the ratio of females mated by γ-males divided by the total number of γ-males, or 25/20 = 1.250. If we substitute the values of $R\gamma$ and $H_m\gamma$ into eq. [10.22] we obtain

$$p_{0\gamma} = 1 - (1.349/1.623) = 0.169. \qquad [11.30]$$

Thus, the predicted frequency of nonmating γ-males in the intact sample (0.169) is equal to the observed frequency of nonmating γ-males (0.169). If we assume that $R = 1$, that is, if we consider the possibility that our observed frequency of unmated γ-males underestimated the actual frequency of nonmating γ-males, we obtain

$$p_{0\gamma} = 1 - (1/1.623) = 0.384. \qquad [11.31]$$

The fraction of β-male success experienced by γ-males is equal to the ratio of the average harem sizes of mating γ-males to that of mating β-males, or

$$H_{m\gamma}/H_{m\beta} = 1.623/1.786 = 0.919. \qquad [11.32]$$

This means that the fitness of mating γ-males is about 92% of that achieved by mating β-males. Our conservative estimate of the fertilization success necessary for γ-males to invade the population of β-males was 70%. Clearly, the minimum condition necessary for invasion of the β-male population by γ-males is met.

Our data also show, however, numerous situations in which γ-males appear to have invaded breeding aggregations in spongocoels that do not contain β-males. Thus, we must also identify the conditions necessary for invasion of the α-male population by γ-males. We showed above that an alternative mating strategy, whether consisting of β-males or of γ-males, must obtain fitness greater than 71% of that achieved by mating α-males.

The fraction of α-male success experienced by γ-males is equal to the ratio of the average harem sizes of mating γ-males to that of mating α-males, or

$$H_{m\gamma}/H_{m\alpha} = 1.623/2.140 = 0.758. \quad [11.33]$$

This means that the fitness of mating γ-males is about 76% of that achieved by mating α-males. Our conservative estimate of the fertilization success necessary for γ-males to invade the population of α-males was 71%. Thus, the minimum conditions necessary for invasion of the α-male population by γ-males are also met.

The above calculations are based primarily on the observed distributions of females with α-, β-, and γ-males. When we calculate the average mating success of *all* males of each type, that is, the number of females mated by each type of male divided by the *total* number of males of each type, mating and nonmating, we see that the average harem sizes for *all* α-, β-, and γ-males are 687/452 = 1.52, 25/20 = 1.25, and 112/83 = 1.35, respectively. These average harem sizes are all quite similar. When we consider the variance in mate numbers around each of these averages, as shown in Shuster and Wade (1991b; 1.520 ± 0.08, 1.25 ± 0.44, 1.37 ± 0.23), we see that there is no significant difference in the average fitnesses of α-, β-, and γ-males. In this species, it is possible to identify mating as well as nonmating males. When we examine only the harem size of mating α-males, α*, we see that it is significantly larger than the harem sizes of β- and γ-males (fig. 11.8). However, when we examine the average number of mates per α-male, that is, including mating *as well as* nonmating α-males, we see that there is no difference in the number of mates obtained by males of each type. This example provides a graphic illustration of why satellite males in many animal species *appear* to "make the best of a bad job," but in reality experience fitness equal to that of territorial males.

What else can be observed in this example? Note first the relative frequencies of the different male morphs. The relative population frequencies of α-, β-, and γ-males are 0.81, 0.04, and 0.15, respectively (N = 555; Shuster and Wade 1991b). Although the fitnesses of the three male morphs are equal, their relative frequencies are not. We have suggested above that the relative size of the mating niche for each morph determines the population frequency of males who invade that niche. The greater the variance in mating success among males of each type, that is, the more certain males in each class are unable to secure mates, the greater the possibility that an alternative mating strategy will invade that population of males. Note in table 11.4 the relative magnitude of the zero class (p_{0i}) of males in α-, β-, and γ-males (0.290, 0.300, and 0.169, respectively). This observation suggests that the relative invasibility of α- and β-males by γ-males is similar. Indeed, the dynamics of the frequencies of the three male morphs in *P. sculpta* indicate that there may be two distinct periods per year in which γ-male frequencies become relatively high. The first of these annual events occurs in the winter when β-males are rare, and the second occurs in the late summer when β-males are more common (Shuster 1989; unpubl. data). The apparent ability of γ-males to invade the harems of α-males, as well as breeding aggregations already invaded by β-males, may explain why γ-males are more abundant than β-males, despite their apparently small mating niche. Beta-males appear to specialize on the mating niche provided by α-males; γ-males may successfully invade the mating niches of α- or β-males.

Our results above show that the conditions favoring the invasion of an alternative mating strategy into a natural population can be remarkably easy to achieve. Equal fitness among phenotypes within a population is the condition considered necessary and sufficient for the maintenance of genetic polymorphism underlying phenotypic differences (Slatkin 1978, 1979). We have shown that this condition can arise *whenever* variance in male mating success exists. Variance in mating success is certain when sexual selection is strong. However, we have also shown that variance in mating success will exist whenever some males mate and others do not, a condition that readily occurs with random mating. Thus, we predict that circumstances in which alternative mating strategies achieve equal fitnesses will be common in nature. If this is indeed the case, we predict that genetic polymorphisms underlying male mating strategies are likely to be more common than is now recognized.

Chapter Summary

We illustrate how a population can be invaded by a novel male mating strategy *whenever* mates are distributed unevenly among males of the dominant strategy. Indeed, conditions for invasion of an alternative mating strategy weaken in proportion to the degree to which males practicing the *dominant strategy* are unsuccessful, i.e., in proportion to I_{mates}. Similar considerations apply to the invasion of alternative *female* nesting strategies, which invade most readily when a large fraction of females are excluded from nesting sites. Despite the use of terms like "fitness satisficing" and "making the best of a bad job," we show that the fitness achieved by an invading alternative mating strategy is not inferior to that of the average dominant male morph although it may be less than the fitness of the *successful* males of the dominant strategy: the alternative strategy invades as long as its relative fitness is greater than the average. We apply these same principles to illustrate how alternative *female* nesting strategies might evolve.

We criticize the casual use of the concept of "alternative mating strategy" and argue that the theoretical framework of indirect effects (chapter 12) might be better suited for investigating the evolution of some of the more questionable examples. In particular, we suggest that status-dependent models might be particularly well suited to exploration with the theory of indirect effects. We introduce the concept of the "*mating niche*," the reproductive resources available to an alternative mating strategy, and show how its size is affected by I_{mates} and the distribution of mates among males. We note that the tendency of females to aggregate together plays a central role in determining the patchiness of the mating niche, and that the ability of alternative male types to differentially exploit the larger aggregations of females has permitted a third male morph to invade and persist in the marine isopod, *P. sculpta*. Although there are other important examples of multiple male strategies, most notably in the side-blotched lizard, *U. stansburiana* (Sinervo and Lively 1996; Zamudio and Sinervo 2000), *P. sculpta* is our choice for a worked example because of our special familiarity with this system.

12 The Forms of Alternative Mating Strategies

In this chapter, we review conditions likely to favor the evolution of genetic architectures underlying alternative mating strategies, as well as taxonomic diversity in the expression of these traits. By genetic architectures, we mean the nature of the genetic factors in influencing whether male behavioral variation is conditional or Mendelian in it expression and whether the male behaviors result from the direct expression of genes in the male genome, from the indirect effects of genes in the genomes of their mothers (i.e., maternal effect genes), or from the interaction of both. Whether a gene acts in the genome of a male or in that of his mother has profound effects on the nature of the selective forces acting on the gene (Wade 1998; Wolf and Wade 2001). In the metaphorical language of behavioral research, it is the difference between "making the best of a bad job" (direct effect in the male genome) and "making the best of a bad batch" (indirect effect in maternal genome).

We will apply the approach developed in the previous chapters to both kinds of gene action in an attempt to explain the biodiversity of mating systems and mating strategies as functions of sexual selection. Because directly acting genes in the male genome affecting the development of male mating strategies have received the most attention in the behavioral literature and are the easiest to treat conceptually, we will address this case in the most depth. Our approach combines the reaction norm approach of Schlicting and Pigliucci (1998) with Levins' fitness sets (Levins 1968) to identify the environmental circumstances in which male genetic polymorphism should give way to plastic responses and vice versa. In short, we propose an alternative to the hypothesis of status-dependent selection (SDS) (see also Hutchings and Myers 1994) for the frequency and appearance of phenotypic plasticity in male mating behavior. As we have throughout this book, our explanation for the taxonomic diversity of male alternative mating strategies is based upon variation in the spatial and temporal distribution of sexually receptive females, as well as variation in female life history. We believe that our approach allows a better consideration of the selective context in which different male mating strategies have evolved than current, status-dependent approaches. We will illustrate the utility of our fitness set plus reaction norm approach for understanding the evolution of precocial male maturation, sequential hermaphroditism, maternal effects on sex ratio and sex ratio, evolution. We will conclude this chapter by proposing a new scheme for classifying alternative mating strategies.

Phenotypic Plasticity and Alternative Mating Strategies

The conventional wisdom with respect to alternative mating strategies is that most variation in male mating behavior represents "conditional" polymorphism, determined by the different expression patterns of genes in the male genome induced by the local environment. According to this view (cf. the status-dependent selection model of Gross [1996]), males are presumed to be genetically similar if not identical in their ability to adjust their reproductive phenotypes appropriately in response to variation in the mating environment. To determine the degree to which this approach may be appropriate, we will begin by reviewing existing theory concerning the evolution of phenotypic plasticity. We will first clarify Levins' (1968) terminology and approach as it relates to alternative strategies. That is, we will describe the environmental circumstances likely to favor genetic architectures leading to Mendelian inheritance or phenotypic flexibility in mating behavior. We will then use a reaction norm approach to illustrate the forms alternative mating strategies are likely to assume, depending on whether they represent genetic polymorphism, developmental conversion, or behavioral polyphenism. Although Levins' method coincides more closely with our attempts to explain why alternative mating strategies evolve rather than how these phenotype are expressed, we consider both of these approaches important for understanding alternative mating strategies.

Evolution in Socially Changing Environments

Levins (1968) proposed a theoretical framework to explain the evolution of plastic phenotypes in response to changing environments. His approach classifies environments with respect to their ability to cause variation in fitness, and examines the phenotypes likely to evolve under circumstances ranging from constant environments, with little or no variation, to variations on the scale of one generation, to variations on the scale of several generations. Levins' (1968) approach has the advantage of allowing simultaneous visualization of the intensity of selection, the domains of phenotypic equilibria, and the range of genetic architectures likely to evolve. Reaction norms (cf. the review in Schlichting and Pigliucci 1998) are a standard approach for illustrating phenotypic plasticity, because they detail the phenotypic patterns and processes of genotypic response across different or changing environments. Because the reaction norm method (Schlichting and Pigliucci 1998) predicts specific developmental trajectories and genetic architectures, it allows more detailed tests of hypotheses relating patterns of phenotypic expression to patterns of environmental variation. We will consider each approach in turn as it relates to the evolution of alternative mating strategies, which are often considered adaptations to a changing environment of male reproductive competition. When the environment itself evolves, as is possible when social context affects fitness (Wolf et al. 2002), it becomes difficult to graphically represent the norm of reaction approach. (Although we emphasize the importance of the theory of indirect effects, this approach is itself relatively new in its application to the evolution of alternative male strategies. Thus, we will not consider indirect effects here.)

The Fitness Set

Levins (1968) suggested that the fitness of each phenotype in a population changes as environmental conditions change. This view stands in marked contrast to the "good genes" views in behavioral studies, in which fitness effects are assigned as constants to specific alleles (e.g., Wade 2002). Levins imagined that the distribution of fitness for a specific phenotype was characterized by two properties: (1) the mean fitness; and (2) the variance about the mean. Different phenotypes were characterized in terms of both properties. For example, environmentally "tolerant" phenotypes manifest a broader distribution of fitness in the face of environmental change, while less tolerant phenotypes have narrower fitness distributions (fig. 12.1). The variance in fitness of a tolerant phenotype, V_W(tolerant), exceeds that of a less tolerant or environmentally sensitive phenotype, V_W(sensitive). Thus, Levins (1968) also assumed that tolerance had a fitness cost, manifested as a lower mean fitness for the tolerant type, W_{tolerant}, than for the sensitive type, $W_{\text{sensitive}}$. (This is similar to the discussions of a cost to plasticity mentioned in chapter 11.) Thus, a phenotype that maintains some fitness in marginal environments will be unable to achieve the highest fitness in the more common environment, whereas a phenotype that achieves low fitness in marginal environments will achieve higher fitness in the environment to which it is specialized. When viewed from a reaction norm perspective, Levins' fitness set approach has two important features: (1) the fitnesses of tolerant and sensitive phenotypes change rank with change in the environment, so there is "crossing-type" genotype × environment interaction (G × E); and (2) the change in fitness is greater for the sensitive phenotype than it is for the tolerant one, so the slopes of the reaction norms differ between tolerant (shallow) and sensitive (steep) phenotypes. In the notation of reaction norms, Levins' "tolerant" phenotype is Schlichting and Pigliucci's "plastic" phenotype.

Levins (1968) devised a graphical means for identifying the optimal phenotype for a particular environment. By holding the distribution of environmental conditions fixed and examining performance as a function of phenotype, a curve results that describes the distribution of performance (i.e., fitness) of different phenotypes across a particular range of environments.

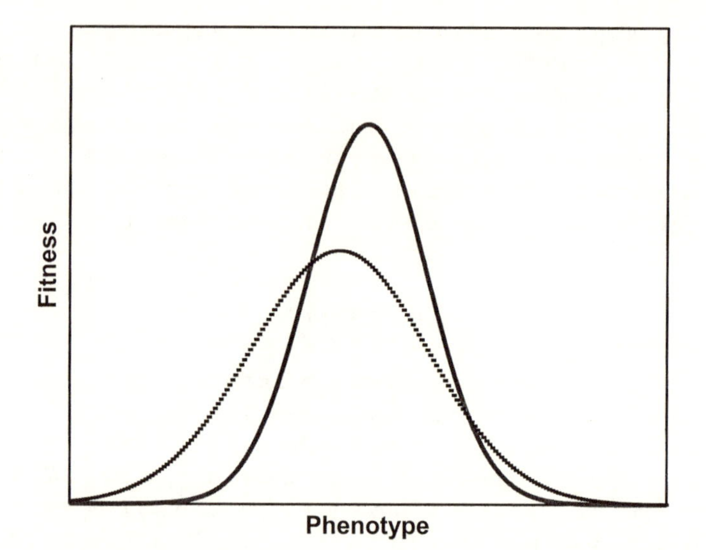

Figure 12.1. The distribution of fitness for phenotypes exposed to environmental variation (after Levins 1968); a phenotype that obtains some fitness in marginal environments (dotted line) will be unable to achieve high fitness in a particular environment, whereas a phenotype that achieves low fitness in marginal environments (solid line) will achieve higher fitness in a particular one.

Although Levins (1968) generated only two curves representing the two most common environmental extremes, many curves are possible. The peak of each curve identifies the optimal phenotype in each environment. Phenotypic deviations from the optimum have lowered fitness with performance decreasing symmetrically toward zero away from the phenotypic optimum (fig. 12.2). When the performance curves generated by the two most common environments are considered together, following Levins (1968), the phenotypic *tolerance* of a population can be quantified. Specifically, tolerance (T) is equal to $2d$, where d is the distance, in phenotypic performance units, from the peak of the distribution to its point of inflection.

The environmental range, E, is defined as the difference in the average phenotypic performances in each environment ($s_2 - s_1$), as in fig 12.2. Approximately overlapping curves produced by each environment indicate a tolerant phenotype; that is, one in which $T > E$ or in which the optimal phenotype is similar in each environment. Nonoverlapping curves, that is, those in which $T < E$, indicate intolerant phenotypes, or, more simply, that different phenotypes are optimal in each environment. When the fitness values of each performance curve are plotted against each other, the familiar shapes of Levins' *fitness sets* are generated. Similar performance curves generate *convex* fitness sets, whereas nonoverlapping performance curves generate *concave* fitness sets (fig. 12.3).

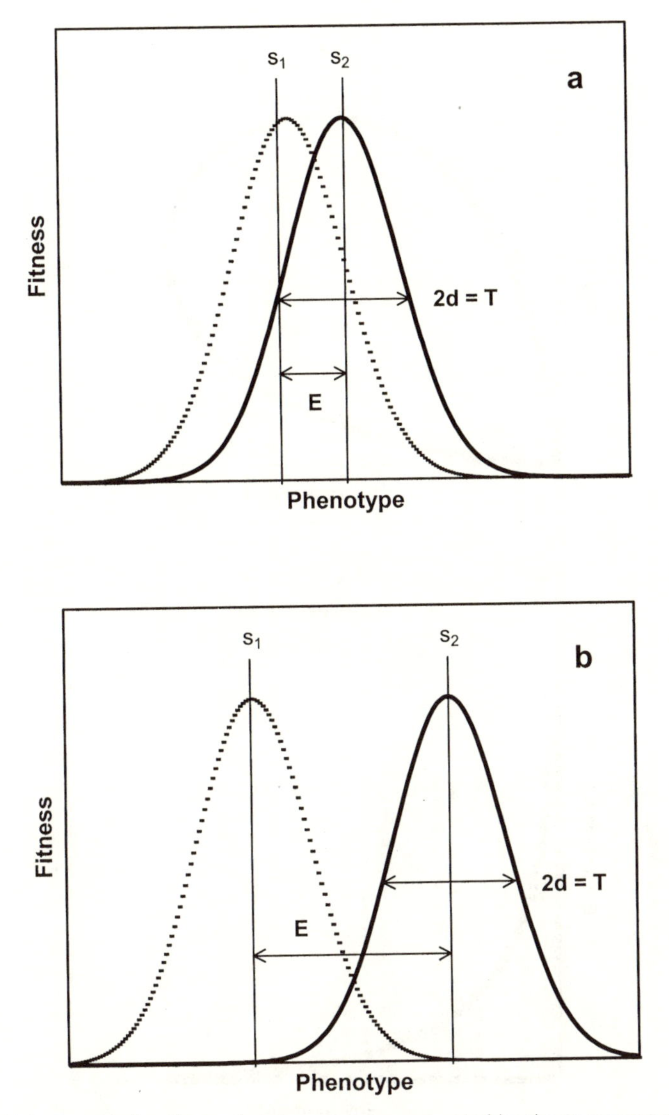

Figure 12.2. Considering the performance curves generated by the two most common environments together (dotted line and solid line) allows the phenotypic *tolerance* of the population to be identified. Tolerance (T) is equal to $2d$, where d is the distance, in phenotypic performance units, from the peak of the distribution to its point of inflection. Environmental range, E, is the difference in the average phenotypic performances in each environment ($s_2 - s_1$). Heavily overlapping curves (a) produced by each environment indicate a tolerant phenotype, one in which $T > E$ or in which the optimal phenotype is similar in each environment. Minimally overlapping curves (b), in which $T < E$, indicate intolerant phenotypes, or, more simply, that different phenotypes are optimal in each environment (after Levins 1968).

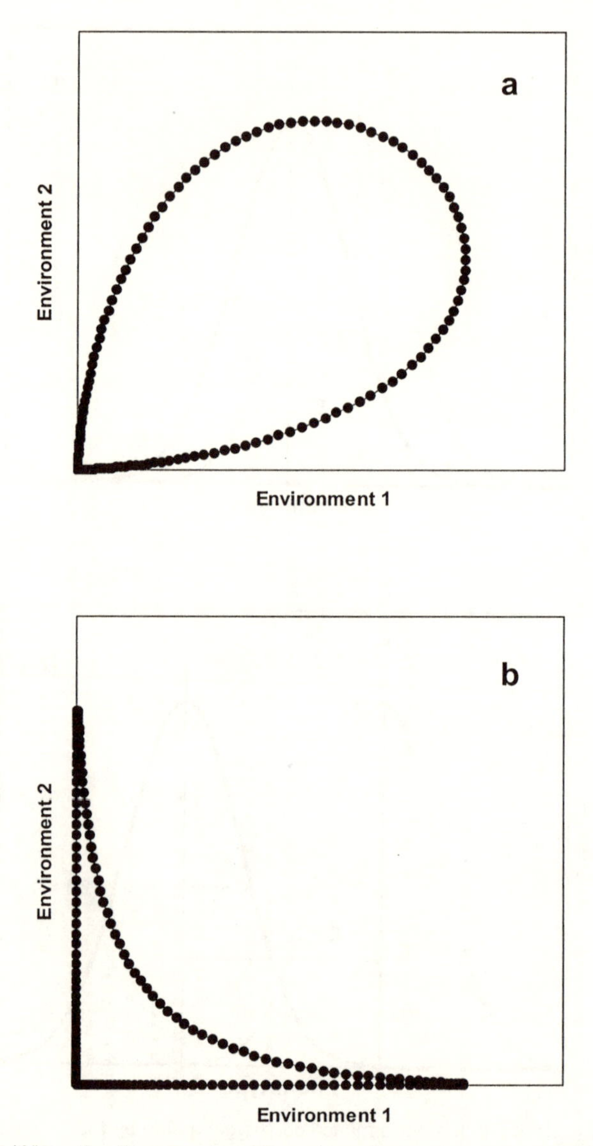

Figure 12.3. When the fitness values of each performance curve are plotted against each other, the now familiar shapes of Levins' *fitness sets* are generated; similar performance curves generate (a) *convex* fitness sets, whereas nonoverlapping performance curves generate (b) *concave* fitness sets (after Levins 1968).

Tolerant phenotypes can persist despite rapid changes in the environment, provided the range of environmental variation, *E*, is small in magnitude. These phenotypes experience the environmental variation as an average of environment types (Bradshaw 1965; Levins 1968; Lloyd 1984). However, increasing the range of environmental fluctuation makes generalization difficult. When the environmental range, *E*, becomes large, tolerant phenotypes may go extinct and be replaced by specialists, who achieve an overall higher average fitness via their enhanced success in a particular environment, compared to tolerate phenotypes who achieve more modest success across all environments. Thus, increasing the range of environmental variation intensifies selection in favor of phenotypes with performance curves specialized to a specific niche. Given the postulated trade-off between fitness mean and variance, as selection intensity increases, specialization is favored and performance distributions must become narrower and more distinct. When these distributions are plotted using Levins' (1968) methods, concave fitness sets appear. Thus, when environments become nonuniform, more specialized phenotypes, with higher average fitness, are expected to invade populations consisting of tolerant, generalist phenotypes.

The range of environmental fluctuations alone can favor phenotypic specialization. However, the more strongly selection acts in favor of a particular phenotype, the more narrow the distribution of performance must be in a particular environment. Thus, if selection becomes intense, a concave fitness set will arise even if the range of environmental fluctuation remains small. This occurs because, under intense selection, the performance distributions within each environmental extreme *contract*, i.e., the variance is reduced. Thus, the stronger selection is, that is, as more specialized phenotypes become favored in particular environments, the more likely concave fitness sets will be observed, and the more likely it is that phenotypes will become specialized. As reviewed in Schlicting and Pigliucci (1998), the results of Bradshaw (1965), Lloyd (1984), Via and Lande (1985), Lively (1986), Moran (1992), and Winn (1996) are in accord with these general predictions. Below, we apply Levins' framework to the invasion of alternative male mating strategies, especially in cases where sexual selection is unusually intense.

Environmental Grain and the Patterns of Environmental Change

Levins (1968) noted that the shape of the fitness set alone does not define the optimum strategy. Rather, it is the *pattern* of environmental change impinging on each fitness set that determines whether polymorphism will evolve or not. When environmental changes occur rapidly within the lifetime of the individual, they cause linear increases or decreases in the fitness of individuals, because the organism experiences the environment as though it were a succession of different developmental conditions with fitness averaged over them. When environmental changes occur more slowly, on the scale of more than one lifetime, then individuals experience their environments as alternative conditions with proportionately larger, and possibly nonlinear, effects on fitness. The spatial and temporal scale of environmental change is the basis of Levins' (1968) concept of environmental *grain*.

The frequency with which the environment changes within the lifetime of a mobile organism determines its environmental grain. A spatially coarse-grained environment will become less so if organisms can move between patches. Thus, mobile organisms experience environmental grain primarily on a temporal scale. Few or no changes within an individual's lifetime constitute *coarse* environmental grain. Rapid changes within an individual's lifetime cause the environment to be experienced as an average, and thereby constitute *fine* environmental grain. Coarse-grained environments favor individuals who specialize. Fine-grained environments favor generalists (Levins 1968; Schlichting and Pigliucci 1998). In the context of sexual selection, we argue that the spatial and temporal distribution of receptive females not only determines the environmental grain and the reproductive fitness consequences for males, but also determines the conditions favoring the evolution of alternative male mating strategies (Waltz and Wolf [1984], Schlichting and Pigliucci [1998], and Sinervo [2001a] have suggested similar relationships to those described below).

Levins' (1968) model has occasionally been misunderstood with respect to the effects of environmental predictability on phenotypic plasticity (Waltz and Wolf 1984). A coarse-grained environment may *seem* more predictable than a fine-grained one because single individuals in a coarse-grained environment may not experience *any* changes within their lifetimes. However, when change does come, it is abrupt and unpredictable. When environments change frequently, the environment may become more or less predictable depending upon cues predicting the change. The presence of predictive cues can turn an otherwise coarse-grained environment into a fine-grained one. In Levins' framework, scale of change and predictability together determine grain.

Evolutionary Response to Environmental Variation: Polymorphism or Polyphenism?

As has been discussed in detail elsewhere (review in Schlichting and Pigliucci 1998), particular phenotypes are expected to evolve within coarse- and fine-grained environments. Monomorphism is predicted, with little debate, to evolve in "uniform" environments, that is, when environmental fluctuation is small in magnitude, and when spatial and temporal variations are fine grained (Bradshaw 1965; Levins 1968). When fitness sets become concave, either as a result of increased environmental range, or due to intensified selection, polymorphic phenotypes of several kinds are expected to evolve, depending on how organisms perceive their environmental grain. If the environmental grain is coarse, that is, if the arrival of change is stochastic and unpredictable, then genetic polymorphisms are expected to evolve. Under these conditions, the frequencies of genetically distinct phenotypes will depend on the probability with which each environment occurs, and the relative fitness that each phenotype obtains therein. Distinct genotypes persist when their fitnesses, averaged across the environmental grain, are equal (Bradshaw 1965; Levins 1968; Maynard Smith 1982; Lively 1986).

If environmental cues predict when change will occur, a coarse-grained environment can be experienced as a fine-grained one by individuals who adjust their developmental trajectories appropriately (Bradshaw 1965; Levins 1968; Lively 1986; Moran 1992; Roff 1992; Winn 1996). If environmental grain is perceived as fine, selection is expected to favor individuals who can express more than one phenotype (i.e., *polyphenism*; Lloyd 1984). This variation differs from simple environmental tolerance because the fitness set is concave. Thus, selection is likely to favor individuals developmentally capable of generating more than one phenotype over individuals developing only a single phenotype with broader tolerance (Bradshaw 1965; Levins 1968; Lively 1986; Moran 1992; Roff 1992). Indeed, polyphenism is a mechanism for tolerance of environmental variation and is evidence of "adaptive plasticity."

Genotype × environment interaction represents a genetic constraint on evolution or, in other words, a genetic trade-off in the ability to respond to a change in one's environment. The equilibrium distribution of genotypes in a population depends upon the distribution of reaction norms, the distribution of environments, and the distribution of fitness of the different possible phenotypes. The level of adaptive plasticity is the average efficiency with which different individuals in the population respond to environmental change. As for any genetic polymorphism, stable phenotypic distributions are expected to persist when the fitnesses of their underlying genotypes are equal.

Fitness Sets and Alternative Mating Strategies

How do these principles apply to alternative mating strategies? We have argued that, in general, sexual selection is more intense than natural selection (chapter 1). Levins argues that intense selection is more likely to generate concave rather than convex fitness sets. For this reason, we suggest that concave fitness sets are more likely to exist under sexual selection than under natural selection. Although extreme environments *can* generate concave fitness sets via natural selection, we have argued that natural selection, overall, is weaker than sexual selection in its ability to rapidly cause changes in phenotype (chapter 1). Thus, whereas natural selection in changing environments is expected to generate convex fitness sets and favor phenotypic plasticity, sexual selection acting in variable environments will often generate concave fitness sets. It is this kind of fitness set that favors genetic polymorphism.

If concave fitness sets are expected under sexual selection, what is the nature of environmental grain as it is perceived by males? In some species, the circumstances a male must negotiate to successfully mate differ with each mating opportunity. In other species, different mating opportunities may be essentially similar to one another. It is not simply the temporal scale on which receptive females become available to males that determines whether males will exhibit genetic or condition-dependent mating strategies. Rather, it is the duration of circumstances in which mating opportunities occur relative to the life span of individual males that determines the grain of the environment (Levins 1968) in which alternative mating strategies evolve.

With respect to the evolution of alternative mating strategies, the grain of the environment is *coarse* if males have few opportunities to breed within their lifetimes and there are no cues predicting mating success by a particular male phenotype. When males are presented with multiple mating opportunities, environmental grain may also be coarse if the circumstances favoring mating success are more or less constant for each male, but differ among males. Under such conditions, specialists are favored and male mating behaviors are expected to represent Mendelian alternatives (Roff 1992; see also below).

The environment will be perceived as *fine grained* if males have few opportunities to breed within their lifetimes, but environmental cues predict the type of mating opportunities that will be available. Environmental grain will also be perceived as fine if multiple opportunities to breed occur within male lifetimes, and the circumstances favoring mating success change from one breeding opportunity to the next. Under both of these circumstances, "phenotypic plasticity" in male mating behavior is expected to evolve. However, the nature of the polyphenism is expected to differ considerably among taxa. The availability of environmental cues is critical to allowing males to developmentally tune their specialized, plastic morphologies to take maximum advantage of limited opportunities.

Whenever reproductive fitness sets are concave and the temporal grain of the environment is fine, so that environmental changes influencing mating success occur often within individual lifetimes, behavioral polyphenism is expected to evolve. Since male mobility combined with male longevity can make many environments fine grained, mobile, long-lived species are more likely to exhibit multiple male behavioral strategies than sedentary short-lived species. Such behavioral polyphenism is consistent with the notion that strong sexual selection, combined with rapidly changing mating opportunities, generates considerable flexibility in individual mating repertoires.

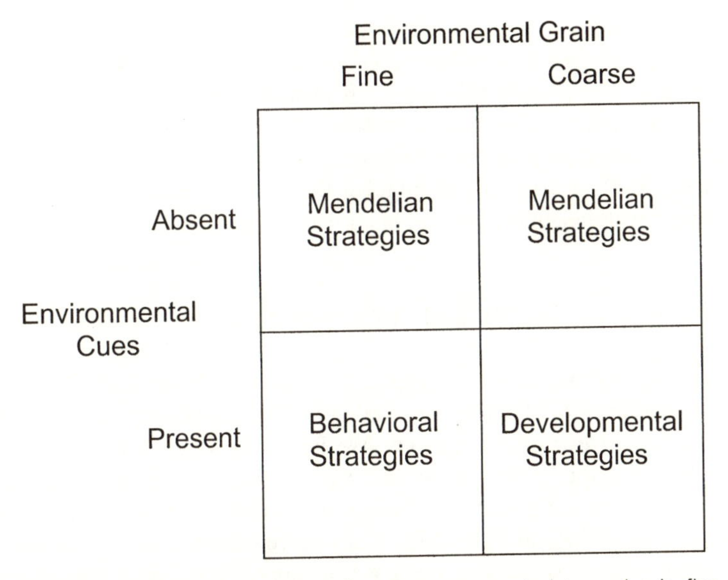

Figure 12.4. The type of alternative mating strategy expected to evolve in fine- and coarse-grained environments with environmental cues present or absent.

However, the grain of an environment can be perceived as coarse by individuals if mating opportunities occur often, but no environmental cues predict fitness in each situation. In such circumstances, Mendelian alternatives, rather than behavioral polyphenism, are expected to persist.

Thus, we predict, using Levins' approach, that polymorphism in male mating behavior will evolve when sexual selection is strong, as is widely observed (Gadgil 1972; Gross 1996; Sinervo 2001a). We also predict, using this approach, that depending on the temporal availability of mating opportunities, relative to the lifetime of individual males, male alternative mating strategies may take one of three forms. Mendelian alternatives are expected to arise when opportunities to breed appear infrequently within individual male lifetimes, are unpredictable in character, and/or are unlikely to change in character when they appear. Developmental alternatives are expected to arise when mating opportunities appear infrequently within male lifetimes, but environmental cues reliably predict the type of mating opportunities (the nature of the contest) that will be available. Behavioral alternatives are expected to arise when environmental changes influencing mating success occur often within individual lifetimes (fig 12.4).

Our perspective is similar to but distinct from that of Sinervo and his coauthors (Sinervo 2001a; Zamudio and Sinervo, in press), who examine the effects of environmental grain on the evolution of alternative mating strategies in lizards (*Uta stansburiana*). This latter approach describes environmental grain in terms of the likelihood of change (fine = changeable;

coarse = constant), but does not specifically address how male mobility or the presence of cues may influence *environmental* grain on spatial and temporal scales. Thus, while this approach identifies the importance of intense selection (concave fitness sets) as well as differences in environmental grain for the evolution of phenotypic plasticity (as reviewed in Schlichting and Pigliucci 1998), it does not distinguish the circumstances in which behavioral and developmental polymorphism may evolve.

Reaction Norms and Alternative Mating Strategies

We can combine the fitness set approach of Levins (1968) with the reaction norm approach of Schlichting and Pigliucci (1998) and propose a classification of alternative mating strategies. We hope this scheme may prevent the facile identification of any behavioral variant as a mating tactic, and assist more detailed investigations of the genetic architecture underlying behavioral and morphological traits. Like many current approaches to the study of alternative mating strategies, the reaction norm approach to the study of phenotypic evolution assumes that adaptive phenotypic plasticity is widespread (Schlichting and Pigliucci 1998). However, the reaction norm approach also views phenotypic plasticity as a developmental threshold character *under genetic control* (Hazel et al. 1990). Thus, the totality of a phenotypic response to environmental change is considered as the object of selection. The application of this approach to the evolution of phenotypic plasticity requires study of the patterns of genetic variation, assessed as genetic correlations among traits expressed in different environments. The constraining effects of genotype × environment interaction are captured in the genetic correlations and relationships among traits across environments (Stearns 1989; Scheiner 1993). All individuals in a population may possess the ability to respond to environmental cues, but heritable variation in cue sensitivity, in the threshold of responsiveness to cues, as well as in the magnitude of individual responses, are also presumed to exist within the population (chapter 11). Genetic monomorphism has no place in this theoretical framework.

The reaction norm approach differs too from a character-state approach for the study of phenotypic plasticity (Via and Lande 1985; Via et al. 1995; Wagner 1996; Wagner and Altenberg 1996). The character-state approach requires statistical investigation of the putative plastic phenotype before it is classified as such. Thus, the character-state approach suggests that portions of a phenotypic response to different environments could arise simply as by-products of selection in other contexts. Alternatively, modified phenotypes in novel environments may represent physiological constraints on developmental conversion, rather than adaptations arising from selection (Scheiner 1993; Nijhout and Paulsen 1997). Although the character-state approach and the reaction norm approach are distinct, their explanatory goals toward providing an explanation for phenotypic plasticity are identical (see Schlichting and Pigliucci 1998).

We have argued that Mendelian alternatives in male mating strategies are likely to be more common than is now recognized. However, phenotypic plasticity in male mating tactics is undeniably widespread. Within the range of behavioral diversity in most animal species, how can we avoid identifying *any* variant in male mating behavior as an alternative mating strategy? As we have argued above, the form of phenotypic plasticity arises from particular selective pressures (chapters 7 and 11). Thus, lacking the ability to directly investigate the inheritance of variation in male mating behavior and morphology, we suggest that the fitness set approach to alternative mating strategies, combined with a reaction norm approach to their classification, can provide a means for distinguishing developmental constraint from phenotypic plasticity. This combination not only identifies the selective pressures under which certain types of plastic reproductive phenotypes may have evolved, it also predicts the form these phenotypes may assume. With this information in hand, the feasibility of direct genetic investigations may be more easily determined.

Phenotypic plasticity has five elements (Schlichting and Pigliucci 1998). These elements include (1) *pattern*, the form of the phenotypic shift that individuals undergo; (2) *amount*, the degree to which a phenotype may shift; (3) *rapidity*, the speed with which the phenotype undergoes transformation; (4) *reversibility*, the degree to which the phenotype can revert to the form it possessed before change was induced; and (5) *competence*, the degree to which transformation of the phenotype is limited to a particular stage or window within an individual lifetime. We will discuss the *pattern* of alternative mating strategies in more detail below (table 12.1). The circumstances influencing mating success arise from the spatial and temporal distributions of females, (chapters 8 and 9) and as well as from the means by which most males in the population secure mates. Usurpers, sneakers, monogynists, and sperm competitors (table 12.1) are distinctive male morphologies that appear in specific selective contexts. The degree to which these phenotypes may appear plastically as alternative tactics *within individuals* will depend on the nature of the remaining four elements.

We have argued that, when fitness sets are concave and the temporal grain of the environment is fine, two types of plasticity exist (fig. 12.4). If the environment is temporally coarse grained, but is perceived as fine grained because cues predict changes in the environment, *developmental* plasticity is likely to arise. However, when the environment is fine grained with respect to the temporal distribution of mating opportunities, *behavioral* flexibility is likely to arise. Each of these situations differs in the *amount* of phenotypic change that occurs (fig. 12.5). Although considerable individual behavioral variation may be possible, an individual performing different behaviors

Table 12.1
The contexts in which alternative mating strategies evolve.

Conventional Morph (context)	*Alternative Morph (response)*
1. Guard individual females in sequence	Usurper Insinuator Cuckolder
2. Guard female groups	Sneaker (satellite; subordinate) Female mimic Monogynist
3. Guard resources required by females	Sneaker (satellite; subordinate) Female mimic Monogynist
4. Mate with many females in sequence	Sneaker (satellite; subordinate) Female mimic Monogynist Sperm competitor[a]

[a]Favored whenever females mate more than once.

undergoes less physical transformation than an individual who adopts a developmental trajectory that changes its body form. Thus, the amount of phenotypic plasticity that exists provides some information about the circumstances in which males may successfully adopt an alternative mating strategy.

The *rapidity* and *reversibility* with which phenotypic change occurs are the other side of the coin relative to the amount of phenotypic change. As shown in fig. 12.5, how much physical change phenotypes undergo depends on the speed and flexibility of transformation. Thus, rapidity and reversibility describe the amount of phenotypic change that is possible. Four possibilities exist: (A) Low rapidity and low reversibility imply that phenotypic change occurs over a prolonged period and is unidirectional. Prolonged, unidirectional change means that considerable phenotypic transformation occurs during development. (B) High rapidity and low reversibility imply only an intermediate amount of phenotypic change. Although low reversibility implies unidirectional change, the faster such change occurs, the less complete the transformation can be. However, since phenotypic transformation does occur, this process too is developmental. (C) Low rapidity and high reversibility describe changes that take time to manifest themselves as well as create prolonged behavior effects, such as occur due to hormonal influences on adult organisms in which physical transformation of the individual is minimal (Moore 1991; see below). Lastly, (D) high rapidity and high reversibility imply behavioral plasticity that occurs over brief periods of time. Greater reversibility implies little lasting transformation, which is possible

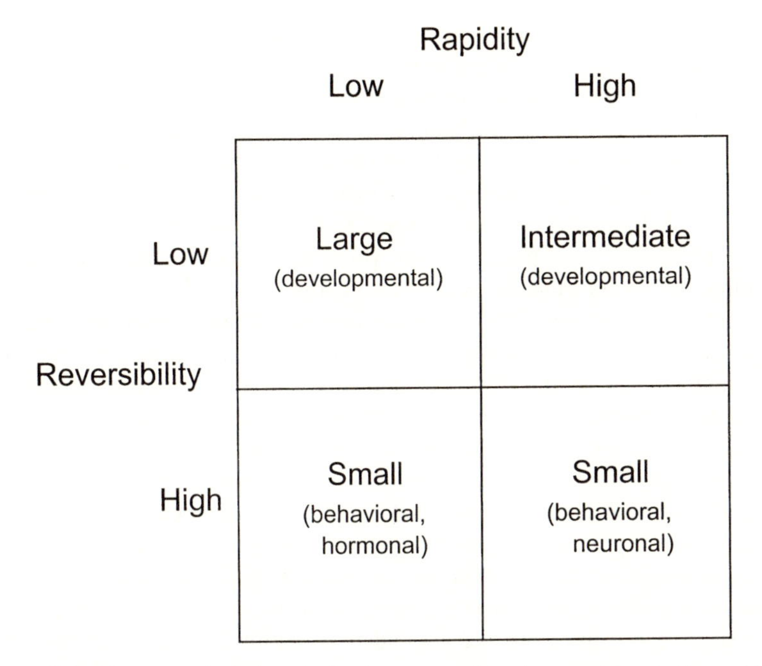

Figure 12 5. The amount of phenotypic change possible given variation in the rapidity, reversibility with which phenotypic plasticity occurs.

when changes occur rapidly. Such changes are likely to be primarily neuronal in origin and execution.

The fifth element of phenotypic plasticity, *competence*, refers not to relative male mating success, as in "making the best of a bad job," but rather to the ability of a phenotype to respond appropriately at a particular time. Greater competence describes a wider window for change, whereas lesser competence describes a narrower window for change. Clearly, neuronally mediated behavioral flexibility affords the greatest competency in this sense, hormonally influenced behavior is next, and intermediate and large developmental changes present increasingly narrow windows for phenotypic change. As figs. 12.4 and 12.5 show, our scheme distinguishes developmental from behavioral plasticity based on the rapidity, reversibility, amount, and therefore competency with which phenotypic plasticity in alternative mating tactics occurs. Categories A and B coincide with phenotypic plasticity influencing developmental patterns, that is, those that are likely to evolve in coarse-grained environments, which become fine grained due to the existence of environmental cues. Categories C and D coincide with phenotypic plasticity influencing behavioral phenotypes, that is, those that evolve in fine-grained environments in which circumstances influencing mating success change quickly and can reverse within individual lifetimes. By identifying the pattern of phenotypic plasticity as belonging to categories A–D, the possible evolutionary context in which this phenotypic plasticity in alternative mating tactics has evolved may be inferred. We recommend that this approach be used primarily as a means for gaining experimental insight. Identification of the developmental mechanisms as well as the genetic architectures likely to underlie phenotypic plasticity for alternative mating strategies will require more detailed hypothesis formation and testing. Moreover, rigorous investigation of the evolutionary context in which these characteristics have arisen requires further information as well (see below).

Moore's (1991) hypothesis for endocrine mediation of alternative mating strategies dovetails nicely with the reaction norm approach for describing variation in male mating behavior and morphology. According to this hypothesis, hormonal mediation of alternative mating strategies can occur at distinct periods of an individual's lifetime with different effects. Hormones secreted early in development mediate *organizational effects*. These hormones cause individuals to proceed down a developmental pathway that leads to the expression of one or another adult phenotype (fig. 12.6). Obvious examples of this process are the primary sexes in most species that arise as a result of hormonal organization of the development of primary reproductive organs. However, secondary variation in either sex may also occur if organizational hormones also produce morphologically distinct types of males, as appears to occur in horned beetles (*Onthophagus*; Emlen 1996) or in side-blotched lizards (*Uta stansburiana*; Sinervo et al. 2000). Clearly, the genes underlying these hormonal effects on development could lie in either the male or the maternal genome.

Hormones secreted later in development mediate *activational effects*. These hormones cause individuals to prepare for reproductive activities that do not lead to permanent changes in phenotype. Instead, these effects are associated with temporary changes in behavior, pigmentation, or physiological processes immediately associated with mate acquisition or parental care (fig. 12.6). Obvious examples include the development of aggressive behaviors in territorial males in many lizard species (Hews 2000) that are often associated with temporary changes in energy metabolism and can also involve changes in skin color.

There is some debate over whether Moore's (1991) framework is weakened by the observation that both organizational and activational effects can influence alternative mating strategies in the same species. For example, in side-blotched lizards (*Uta stansburiana*), three different male morphs, orange, blue, and yellow, coexist in the same population. There is empirical evidence that the three male morphs represent genetically distinct alternatives (Sinervo 2001b; Sinervo and Lively 1996). Orange-throated males are "ultradominant" and defend breeding territories containing females. Orange usurpers out-compete blue-throated males, who are primarily mate guarders. However, orange males are vulnerable to the sneaker strategy of yellow-

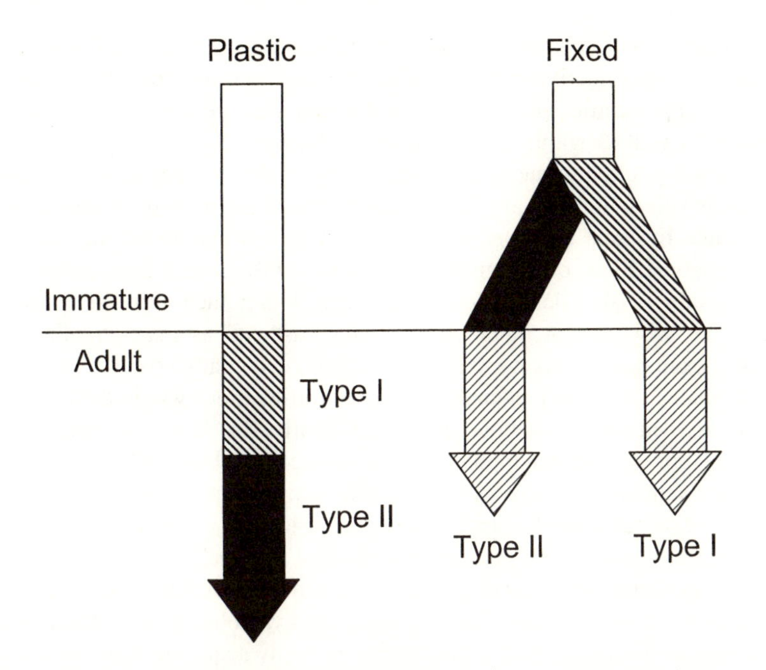

Figure 12.6. Moore's (1991) model for alternative mating strategies arising from organizational and activational hormonal effects; shaded areas indicate stages of development in which hormone levels are predicted to be similar or different between morphs as well as when hormone differences are expected to be important in establishing different adult phenotypes; the single arrow on the left represents the simplest case, in which hormonal differences do not affect morphology during development, but can cause changes in adult phenotype or behavior; the bifurcating arrow on the right represents fixed morphs which differentiate in phenotype before sexual maturation is complete and do not change in phenotype as adults; hormone differences between morphs are manifested during development but not during adulthood.

throated males that mimic females. In turn, yellow males are outcompeted by the mate-guarding strategy of blue males. Thus, the relative fitnesses of the three morphs are intransitive and frequency dependent.

As genetically determined alternatives, differences among males appear to have arisen as a result of the organizational effects of hormones associated with males of each type. However, some blue males (heterozygotes for yellow and blue alleles) also appear to have the ability to switch from the guarder to the sneaker phenotype in response to changes in female availability (Sinervo 2001b). According to the organization/activation hypothesis, this change in behavior and morphology must be caused by the activational effects of hormones associated with changes in environmental cues predicting mate availability perceived by males who are genetically programmed to express the blue male strategy. Do such observations invalidate this theoretical framework? We think not. Instead, we suggest that such variation in the morphology and behavior of male side-blotched lizards, as well as the hormonal effects underlying them, reflect the fact that environmental grain in mating opportunities perceived by males can vary on two temporal scales, among as well as within individual male lifetimes.

Evidently, circumstances exist that allow males to specialize as orange, blue, or yellow individuals. These lizards breed on isolated patches of rock separated by large expanses of grassland. Females can breed only on these rock islands. Thus, orange males who successfully defend favored locations are certain to also defend several females within their territories, and by doing so will prevent other orange males, as well as mate-guarding blue males, from mating. As we explained earlier, such conditions are expected to favor males who obtain a small fraction of fertilizations within the territories of orange males, as yellow males apparently do. When variance arises among yellow males attempting to mate as sneakers, blue males, who defend a single female successfully, may invade populations because their mating success exceeds that of orange males unable to secure territories, as well as that of yellow males that are unable to mate at all. Such mating opportunities appear to represent a coarse-grained environment that favors different male morphs from year to year. Yet heterozygous (*yb*) blue males appear to behave opportunistically like yellow males, not only acquiring mates within the territories of deceased orange males, but also adjusting their pigmentation to appear more yellow in color (Sinervo 2001a,b). This observation suggests that sufficient variance in the mating success of yellow males may arise, that is detectable within a particular breeding season, and which allows blue males to opportunistically switch their behavior and morphology. If anything, Moore's argument becomes more convincing in light of these results.

Alternative mating strategies insensitive to environmental variation are best suited for environments with clearly defined mating opportunities, which do not change appreciably within the lifetimes of individual males. In contrast, within-season variation in the availability of mating opportunities and perceptible cues about the mating environment permit plastic responses. Sinervo (2001b) also reports individual variations in the behavioral repertoires of males that are responsible for variation in mating success within each of the three morphs, a response likely to arise from fine-grained mating opportunities favoring the evolution of behavioral polyphenism.

Genetic Architectures Underlying Alternative Mating Strategies

Three fundamental patterns of phenotypic expression exist for alternative mating strategies. These patterns include Mendelian strategies, developmental strategies, and behavioral strategies. Each pattern of expression depends, at the most proximate level, on hormonal and neurological factors that regulate the timing and degree of gene expression (Moore 1991; Goodson and Bass 2000). The nature of each regulatory mechanism depends, in turn, on its underlying mode of inheritance, direct or indirect maternal. Ultimately, these genetic architectures depend on the circumstances in which mating opportunities arise; that is, on the intensity of selection favoring distinct reproductive morphologies, as well as on the predictability or unpredictability of mating opportunities relative to individual life span (Bradshaw 1965; Levins 1968; Lively 1986; Moran 1992).

Mendelian Strategies

Alternative mating strategies controlled by a few loci of major effect, which segregate among males in populations according to Mendelian rules, are well documented in diverse animal taxa. Examples include bulb mites (Radwan 1995); damselflies (Tsubaki et al. 1997); fig wasps (Hamilton 1979; Cook et al. 1997); marine isopods (Shuster 1992); poeciliid fish (Kallman 1983; Zimmerer and Kallman 1989; Ryan et al. 1992; Morris et al. 1992); side-blotched lizards (Sinervo and Lively 1996); and ruffs (Lank et al. 1995, 1999). In each of these cases, specific combinations of alleles in males produce behaviorally and morphologically distinct male phenotypes. Although these genetic results seem unambiguous, distinguishing the segregation of direct acting genes in sons from indirect acting genes in their mothers from matings based on male phenotypes can be difficult, because of the expected genetic correlation between mothers and sons, which is 0.5 with random mating. As a further complication, inbreeding, either in nature or in controlled laboratory crosses, inflates the conditional probability that a male homozygote will have a mother of the same genotype. Epistasis between genes in the maternal genome and genes in the sons may appear "additive" in the sons, if the maternal genotype is fixed or nearly so, and vice versa. Once we admit the possibility of maternal effects or maternal-son interactions, approximate Mendelian segregation may not be an adequate method of inference for genes segregating among males. However, until these possibilities are more fully explored, allelic difference among males at loci of major effect remains the simplest explanation for phenotypic differences among males in these species.

Mendelian strategies are expected to arise when sexual selection favors specialized male mating phenotypes, *and* when the relative mating success of each phenotype is unpredictable within male lifetimes (Levins 1968; chapter 11). A morph is, by chance, well or poorly suited for securing mates in a given environment, and its relative fitness, as well as its relative population frequency, soars or plummets accordingly. In such circumstances, genes of major effect are expected to exclude genetic architectures that allow a phenotypic response to environmental cues predicting mating success. When such cues are lacking, this ability is unnecessary. When trait heritabilities approach unity, Mendelian strategies may cycle rapidly over brief intervals (Shuster 1989b; Shuster and Wade 1991b; Sinervo and Lively 1996; Sinervo, 2001b). Over longer durations, different morphs are expected to persist in the population when average fitnesses are equal across strategies (Darwin 1871; Shuster and Wade 1991a; Ryan et al. 1992; Sinervo and Lively 1996; Lank et al. 1995).

Three discrete male morphotypes coexist in the marine isopod *Paracerceis sculpta* (fig. 11.8). As we have explained, phenotypic differences among males appear to be controlled by a major gene (*Ams* = Alternative mating strategy), whose inheritance is Mendelian and whose alleles exhibit directional dominance ($Ams^{\beta} > Ams^{\gamma} > Ams^{\alpha}$). This gene is not itself responsible for *all* of the phenotypic differences that exist among males. Instead, the different *Ams* alleles interact with alleles at other loci, switching on developmental cascades that lead to discontinuous adult phenotypes. For example, interactions between *Ams* and a second autosomal locus (*Tfr* = Transformer) produce deviations in family sex ratio by causing genetic males to mature as females and vice versa (Shuster and Wade 1991a; Shuster and Sassaman 1997). The sum of all of these interactions throughout the genome determines individual phenotype, which in turn determines how well or poorly each individual reproduces.

At breeding sites, mating success among the male morphs varies with the number of females, as well as with the number and type of other males (Shuster 1989a). Thus, the fitness of an *Ams-Tfr* allele combination depends on the sex ratio and on the relative frequencies of each male morph (Shuster et al. 2001). As explained in chapter 11, the average mating success of all males of each type can be calculated by dividing the number of females mated by each male type, by the total number of males of each type, mating as well as nonmating. Thus, the average number of mates for all α-, β-, and γ-males equals (average ±95% CI, 1.52 ± 0.16, 1.25 ± 0.86, and 1.37 ± 0.45), respectively. It is clear that the average fitnesses of α-, β-, and γ-males are equal (fig. 11.8).

We have already described male polymorphism in side-blotched lizards (*Uta stansburiana*). The three male morphs in this species also represent genetically distinct alternatives (Sinervo and Lively 1996). Also, the relative fitnesses of each morph are *frequency dependent*, that is, the fitness of each morph is highest when it is rare. Thus, these lizards provide a colorful, living example of the "rock-paper-scissors" game, familiar to children and game theoreticians alike (Sinervo 2001a,b; Maynard Smith 1982). Female color morphs of side-blotched lizards also exist and occur as two density- and frequency-dependent strategies. Orange-throated females lay large clutches of small progeny, which are favored at low density. Yellow-throated females, on the other hand, lay small clutches of large progeny, which are favored at high population density. This female "game" interacts with the male rock-paper-scissors "game." Observed oscillations in morph frequencies occur every five years for the male game and every two years for the female game. In simulations, only a genetic model consisting of one locus with three alleles (*o*,*b*,*y*) is capable of driving the rapid cycles observed in nature for male and female morph frequencies. As an additional nuance, the *o* allele must be dominant to the *y* allele in females, indicating that allelic expression at this locus differs between the sexes (Zamudio and Sinervo 2000; Sinervo and Zamudio 2001). Sex differences in gene expression, by definition, involve epistatic gene action.

Developmental Strategies

Several "developmental strategies" are possible regarding male alternative mating phenotypes. Maternally acting genes or maternal physiological condition may affect the expression of the son mating phenotype. In this case, maternal condition may be a predictor of the adult phenotype and, hence, mating success of her sons. Mothers in poor condition might "make the best of a bad batch" and maximize brood fitness by producing β-males instead of α-males. Viewed entirely from the perspective of sons, it might appear as though the average fitness of β-males was lower than that of α-males and, thus, that β-males were making the "best of a bad job." However, from the perspective of mothers, mothers able to alter the male phenotypic composition of their family in response to maternal condition would have a higher fitness than mothers unable to adjust son phenotype in response to condition.

Alternatively, male alternative mating phenotypes could result from the plastic response of the male genome to the environment. Such distinct developmental trajectories, which do *not* segregate in a Mendelian manner, are well documented among animals with discontinuous male phenotypes. Examples include earwigs (Tomkins 1999), horned beetles (Emlen 1996), amphipods (Borowsky 1985; Clark 1997), freshwater prawns (Kuris et al. 1987; Ra'anan and Sagi 1985), salmonids (Gross 1985; Hutchings and Myers 1994), coral reef fish (Warner 1984), midshipmen fish (Goodson and Bass 2000), and, once again, side-blotched lizards (Sinervo 2001a; Sinervo and Lively 1996). Developmental strategies in males are expected to arise when sexual selection favors specialized mating phenotypes, *and* when environmental cues detectable by males predict the type of mating opportunities likely to become available (Levins 1968; Lively 1986; chapter 11). Genetic architectures sensitive to environmental cues allow males to tune their mating phenotypes in response to changing environments. Such architectures are expected to exclude major genes not permissive of phenotypic plasticity.

Although they fail to segregate according to Mendelian rules, threshold models of quantitative inheritance or maternal effect models can often explain the appearance of discrete, developmental phenotypes within a popula-

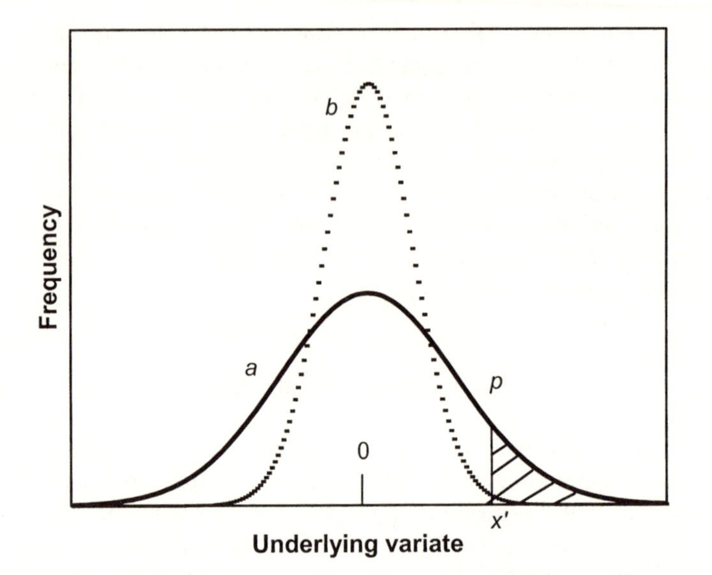

Figure 12.7. The distribution of individuals expressing and not expressing a threshold trait. Curve *a* represents the distribution of the underlying heritable character, x, for the entire population; the average genotype, $\bar{x} = 0$; x' represents the threshold value of x; the hatched area, p, represents the proportion of individuals exhibiting the character. Curve *b* represents the distribution of individuals with the mean genotype in the population ($x_a = 0$) that are expected to express the character or not; note that if curve *b* is shifted so that x_a is less than 0, the probability that individuals will express the character is decreased (redrawn from Dempster and Lerner 1950).

tion. As with most complex characters, continuous genetic variation is expected, on theoretical grounds, to underlie threshold traits. A threshold of "liability" within this distribution also exists, which makes trait expression discontinuous. Individuals with genotypes below the threshold express a default phenotype, whereas individuals with genotypes above the threshold express a modified phenotype (fig. 12.7).

Because the expression of threshold traits is not absolute, each male genotype may have its own probability of trait expression, which depends upon trait heritability, threshold position, and the environment (Dempster and Lerner 1950; Roff 1996). For this reason, environmental influences on threshold characters can be viewed in two ways. When the environment is constant or reasonably so, as might exist over a period of maturation, trait expression appears as described above. Genotypes above the threshold usually express the trait, trait expression becomes increasingly unlikely for genotypes below the threshold, and the population appears dimorphic. Alternative mating strategies involving distinct developmental trajectories are well described by this hypothesis.

When the environment changes over shorter time scales, few or no ge-

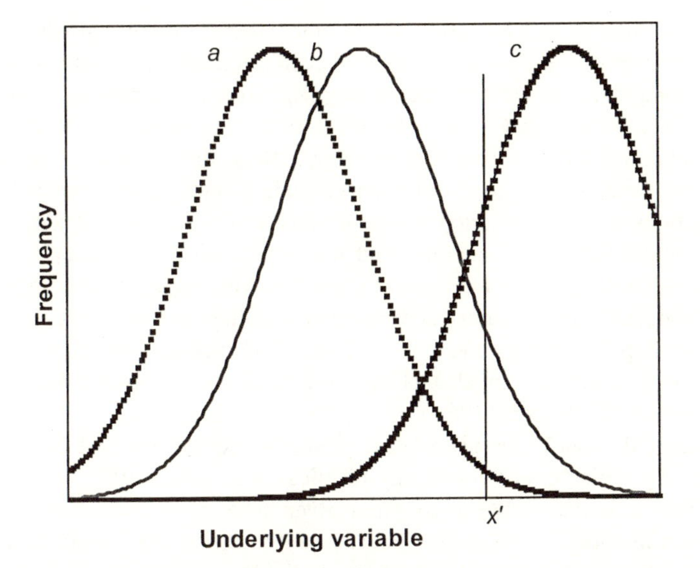

Figure 12.8. The expression of a threshold character in a variable environment. Solid vertical line at *x′* represents the threshold of trait expression; all individuals to the right of the line express the character, all individuals to the left of the line do not. Curve *a* represents the distribution of character expression when the environmental cue stimulating the expression of the character is weak; *b* represents the distribution of expression when the cue is of usual intensity; *c* represents the distribution of expression when the cue is strong.

notypes may express the trait at one environmental extreme, whereas at the other extreme all or nearly all genotypes will become modified (fig. 12.8). Although the probability of trait expression remains constant for each genotype, depending on the intensity of the environmental "cue" at any time, few, some, or all individuals in the population may express the trait. Alternative mating strategies involving behavioral polyphenism are well described by this hypothesis.

Threshold models may also explain age-dependent mating strategies, although, contrary to current models of this phenomenon, a threshold view predicts that few males will perform both "young" and "old" mating strategies within their lifetimes. Instead, quantitative genetic variation is expected to predispose males to mate as satellites when young, *or* as territorial males when old, with frequency-dependent selection maintaining the position of the threshold within the distribution of male maturation rates.

In many species, the environmental cue to which males respond appears to be based on feedback from their own growth rate (Fairbairn 1994; Fairbairn and Roff 1990; Emlen 1996; Hutchings and Myers 1994). In some species, males unable to reach some threshold size within a given duration tend to mature early as satellites, whereas males who cross this threshold continue to grow and mature later as territorials. In other species, rapidly growing males become satellites and slower growers become territorial. Genetic variation controlling male growth rate, like most quantitative traits, appears to be normally distributed (Cade 1981; Hedrick 1988; Fairbairn and Roff 1990; Hutchings and Myers 1994; Emlen 1996). Thus, the position of the body size threshold within the distribution of male growth rates determines the proportions of the population likely to consist of satellites and territorials. The relative success of satellites and territorials, in turn, appears to determine where the average male growth rate lies with respect to the body size threshold. Although circumstances favoring satellites or territorials can influence the population frequencies of each morph (Hutchings and Myers 1994), directional selection can also change the position of the threshold itself (Fairbairn 1994; Emlen 1996).

Threshold models for phenotypic plasticity require that individuals are genetically variable, *not* genetically identical (e.g., Gross 1996; see chapter 11). This is a reasonable assumption given what is known about genetic variation in natural populations (Falconer 1989; Avise 1994; Roff 1996). In the case of growth rate polymorphisms, individuals either commit to an accelerated developmental trajectory or not, depending on the position of their genotype relative to the population threshold. Thus, only *part* of the population, not the *entire* population, must respond to an environmental cue for male polymorphism to appear (e.g., Gross 1996; Gross and Repka 1998). Threshold inheritance explains much about variation in male phenotype within and among populations. As the frequency of circumstances favoring one or another male phenotype changes, the proportion of a population likely to respond to environmental cues is also expected to change. Interpopulational variation in the proportion of males exhibiting each phenotype is expected to exist, and it does (Travis and Woodward 1979; Hutchings and Myers 1994; Vuturo and Shuster 2000). Such variation is *not* expected if males are genetically monomorphic vis à vis their ability to express one or another mating phenotype.

Males in most species of the genus *Onthophagus* (Coleoptera: Scarabaeidae) bear impressive horns (fig. 12.9). In all *Onthophagus* species, these structures are present only on large males and are used in contests for access to females. Horns vary in size, shape, and location on individual beetles, and a steplike relationship exists between horn length and body size in adult males. Small individuals possess rudimentary horns or lack them entirely, whereas large individuals possess well-developed weaponry. Although smaller males lack effective combat structures, these males are sexually mature, highly mobile, and steal matings within the burrows of horned males. Horn size correlates with success in burrow defense, but small body size correlates with success in stealing matings. Thus, males that are intermediate in either of these characteristics are unlikely to mate at all.

Douglas Emlen (1996) and his colleagues (Moczek and Emlen 2000; Em-

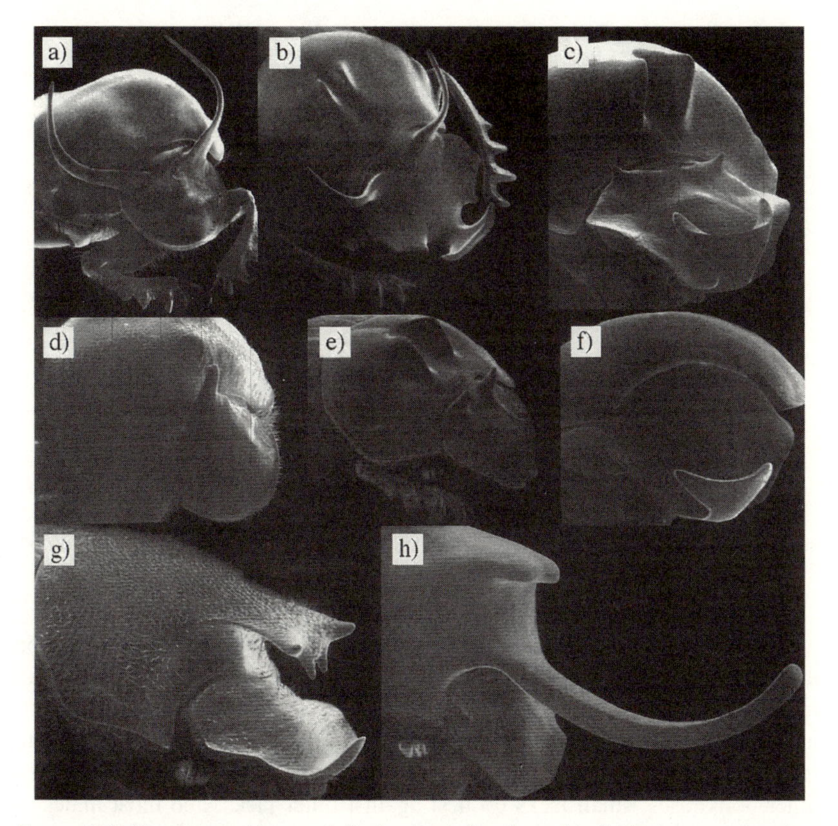

Figure 12.9. The enlarged horns of *Onthophagus* beetles; all individuals are male unless otherwise indicated. (a) *O. taurus*; (b) *O. clypeatus*; (c) *O. praecellens*; (d) *O. nuchicornis*; (e) *O. clypeatus*, female; (f) *O. sharpi*; (g) *O. hecate* (photo by D. Emlen).

len and Nijhout 1999) have used controlled breeding designs and diet-manipulation experiments to explore the genetic foundations of intraspecific variation in male horn length in several *Onthophagus* species. Within beetle families, morph frequencies do not segregate in a Mendelian manner. However, Emlen found that he could produce males with unusually large and unusually small horns by selecting in opposite directions on male horn length. This approach changed horn length, as well as the size within each lineage at which males developed horns (fig. 12.10). These results do not just confirm that horn length is heritable. They also demonstrate that a heritable "threshold" exists, which influences how males respond to their feeding history and thus to their own growth rate.

Similarly, two male morphs coexist in southern populations of Coho salmon (*Oncorhynchus kisutch*). The fry of this species leave their natal streams and mature in the Pacfic Ocean. Larger, *hooknose* males return to spawn after three or more years. Smaller males or *jacks* mature early and

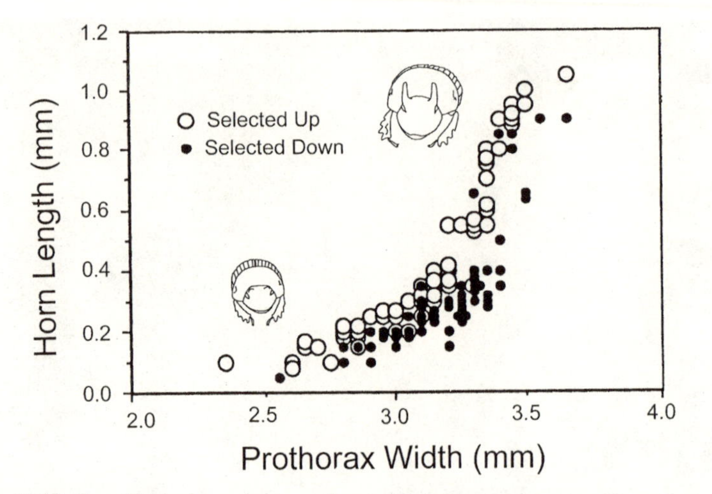

Figure 12.10. The effect of seven generations of selection on horn length, on horn size, and on the position of the body size threshold at which males respond to growth rates made variable by reduced food (redrawn from Emlen 1996).

return after only two years. Each hooknose defends a *redd*, a gravel nest on the stream bottom where females spawn. Hooknoses usually fertilize the ova in their redd. However, jacks lurking nearby may steal fertilizations by darting in and ejaculating when spawning begins. The male closest to the spawning female is most likely to sire young. Large hooknoses do best against other hooknoses, small jacks do best against other jacks, and most males of intermediate size are excluded from mating altogether.

Because jacks must be rare to mate successfully, the relative fitness of hooknoses and jacks appears to be frequency dependent. Gross (1985, 1991a,b) demonstrated this relationship by considering the simultaneous influences of survival probability, reproductive tenure, and frequency-dependent spawning efficiency for males of each type. To standardize the units of these diverse fitness measures, Gross first estimated the ratio of each jack to hooknose fitness estimate, and then calculated the product of the three ratios. His result, 0.95, while not bounded by confidence limits, suggested that hooknoses and jacks experience approximately equal fitness. Such conditions are necessary and sufficient to maintain a genetic polymorphism in male phenotype by frequency-dependent selection.

The mating system of Atlantic salmon, *Salmo charr*, resembles that of its Pacific relatives (Montgomery et al. 1987), although *anadromous males*, the equivalents of Coho hooknoses, mature in 4–7 years, whereas *parr*, the equivalents of jacks, remain in natal streams and mature in just 1–4 years. Quantitative genetic variation underlies maturation rate among parr and anadromous males in this and in other salmonid species (Hutchings and Myers 1994). Hutchings and Myers took these results a step further and generated multidimensional fitness surfaces for each male morph from existing data on age-specific survival, fertilization, and growth rates. The intersection of these surfaces predicted the equilibrium frequencies of each morph in different populations. When mapped onto a threshold inheritance model for male growth rate, these surfaces showed that stabilizing selection can maintain the position of growth rate thresholds within the normal distribution of parr maturation ages. The frequencies of anadromous males and parr were also predicted to vary, depending on the relative fitnesses of the male morphs among salmon populations. Indeed they did, closely matching observed frequencies within and among natural populations, and demonstrating that salmon provide excellent examples of how threshold traits controlling developmental plasticity can maintain alternative mating strategies by frequency-dependent selection.

Behavioral Strategies

Behavioral strategies are expected to evolve when environmental changes influencing mating success occur often within individual lifetimes, *as well as* when the circumstances in which successful matings occur are highly variable (Levins 1968; chapter 11). Thus, mobile, iteroparous species are most likely to exhibit behavioral polyphenism in the context of mating (Alcock 1995; Schlichting and Pigliucci 1998). Examples of such variation in mating behavior include dungflies (Parker 1970a,b), scorpionflies (Thornhill 1981), frogs (Perrill et al. 1978; Howard 1978, 1984), toads (Sullivan 1983), salamanders (Verrell 1982), songbirds (Kempenaers et al. 2001), ruffs (Lank et al. 1995), rodents (Stockley et al. 1996), ungulates (Clutton-Brock et al. 1992; Hogg 1984; Hogg and Forbes 1997), felids (Packer and Pusey 1987), and primates (Bulger 1993; Brereton 1994). In each of these cases, males, and often *females* as well, rapidly change their behaviors to allow them to exploit mating opportunities as they arise.

The underlying genetic architectures responsible for such variability are not well understood, either theoretically or experimentally, but they appear to be similar to those described above for developmental strategies (Gross 1985; Hutchings and Myers 1994; Emlen 1996). That is, genetic variation underlying quantitative traits is expected to influence the likelihood that individuals will express a particular mating behavior. In a given situation, individuals with phenotypes below the liability threshold express one set of mating behaviors, whereas individuals with phenotypes above this threshold express another behavioral set.

It is known that individuals vary genetically in their sensitivity to crowding and to circulating hormone levels, both factors associated with behavioral liability. Heritable traits likely to influence mating behavior include sensitivity to pheromones, density of competitors, and observations of conspecific matings. In the presence of a strong environmental cue, most individuals are expected to adjust their phenotype. Weaker cues, may induce few or no individuals to adjust their phenotype. This behavioral threshold hypothesis, like the developmental threshold hypothesis, predicts differential responsiveness among individuals to the same environmental stimuli. Thus, the same female distributions that induce some males to assume satellite behavior are expected to cause other males to persist as territorial males, as is widely observed.

Ruffs (*Philomachus pugnax*) are delicate sandpipers known to inhabit marshy regions in northern Europe and Asia. Ruffs are named for the mane of feathers adult males display on their shoulders, necks and heads. Breeding ruffs cluster on mating courts called leks, and in most populations males exhibit two distinct color morphs. About 85% of males consist of territorial residents, bearing darker plumage, who defend mating courts against other court residents. Nonterritorial, satellite males, bearing lighter plumage, make up the remainder of male populations. Satellites are recruited onto residents' courts, where pairs of males form contentious, temporary breeding alliances.

David Lank and his colleagues (Lank et al. 1995, 1999) have shown that plumage patterns are heritable and that a single Mendelian locus, or perhaps by a chromosomal inversion that segregates within families, is responsible for differences between resident and satellite males. Females do not exhibit plumage variation like males, although females treated with testosterone implants express male characteristics, indicating that, once again, the expression of genetic factors influencing alternative mating strategies varies between the sexes.

Males do not change plumage color during their lifetime, but the mating tactics of residents and satellites can be variable. When a female arrives on a mating territory, both males court her and the resident often drives the satellite away. However, if a neighboring resident challenges the courting resident, also a frequent event, a satellite may return to the court and mate with the female while the residents fight. Females visiting mating courts routinely mate with both males, particularly on smaller leks where residents are more tolerant of satellites. The combined displays of residents and satellites on small leks may be more attractive than larger leks containing only residents. Co-occupation appears to result in higher mating success for the participants, although such magnanimity may not lead to equal fertilization success among court partners. Among passerine birds, it is common for species with high levels of multiple paternity to have longer sperm. Ruff sperm are the longest known for any shorebird, about 125 μm (Briske et al. 1997). Despite temporal variation in ruff behavior on leks, the average fitnesses of resident and satellite males appear to be equal.

Sexual dimorphism in group-nesting, monogamous birds intrigued Darwin (1874). He reasoned that sexual selection could operate in such species if females in superior condition nested earlier with the most attractive males.

The enhanced fecundity of their mates could thus favor the evolution of elaborate male characters. An alternative hypothesis exists, now that DNA fingerprinting has shown males and females in a large number of "socially monogamous" species to engage in EPCs or extra-pair copulations (Kempenaers et al. 1999, 2001; chapter 9). Elaborate males are indeed favored in these species, but not by the fecundity of one, physically superior female. Instead, if *multiple* females sneak matings with the most attractive males, and then rear these offspring with cuckolded males, variance in offspring numbers among males will increase, and sexual dimorphism could readily evolve. The intensity of this sexual selection is stronger than the fecundity selection proposed by Darwin.

"Socially monogamous" animals also engage in EPCs, and in surprising numbers. In these usually nondimorphic species, both sexes routinely seek multiple mates, but particular individuals do not appear to be favored. Paternity in most nests is mixed, and mixed maternity may also exist if females lay eggs in the nests of other females. If EPCs and egg dumping increase the variance in offspring numbers among males, sexual dimorphism could evolve. However, if individuals of both sexes tend to mate more than once, the sex difference in the variance in offspring numbers may not increase and could diminish. If so, the species will remain monomorphic.

The Source and Intensity of Sexual Selection

As we discussed in previous chapters, the intensity of sexual selection on males, I_{mates}, arises from two primary sources: (1) the distribution of females in space, and (2) the distribution of sexual receptivity among females in time. The magnitude of I_{mates} is negligible when females are spatially dispersed and synchronous in their receptivity. However, extremely large values of I_{mates} can arise when females are clumped in space and are asynchronous in their sexual receptivity (chapters 2, 3, 8, and 9). As we have explained, large values of I_{mates} need not arise only from the physical positions of receptive females in space and time. Differential mating success can also occur when female mating preferences lead females to change their positions in order to mate with particular males. This latter situation causes the functional distribution of females in space to become extremely clumped around preferred males. Moreover, since females with similar mate preferences must mate with a preferred male in sequence, similar female preferences cause the functional distribution of females in time to become extremely asynchronous as well (chapters 8 and 9). In each of these situations, I_{mates} can be augmented by female copying or sperm competition, as well as depressed by multiple mating or multiple reproductive episodes by females (chapters 5 and 6). Yet even when these latter factors are considered, positive values of

ΔI always indicate that some males have succeeded in mating with more than one female. When this happens, some males will not mate at all.

Intermediate values of ΔI arise under at least four sets of conditions. These conditions include circumstances in which receptive females (1) are dispersed in space and mate asynchronously, (2) are moderately aggregated in space and mate with moderate synchrony, (3) are highly aggregated in space and mate with moderately synchrony, and (4) are moderately aggregated in space and mate asynchronously (chapter 8). Again, in each of these situations, I_{mates} can be augmented by female copying and sperm competition, or depressed by multiple mating and multiple breeding episodes. Although ΔI resulting from these circumstances may be smaller than that which occurs when I_{mates} is extreme (see above), positive values of ΔI indicate that some males have mated and others have not. When ΔI is positive, variance in mating success exists among males and alternative mating strategies are expected to evolve (chapter 9).

The Expected Forms of Alternative Mating Strategies

We suggest that the degree to which some males mate and others do not can influence the potential for alternative mating strategies to evolve. According to our reasoning, polymorphisms in male mating behavior are expected to appear whenever ΔI, the sex difference in the opportunity for selection, is nonzero. This result is consistent with the observations of Gadgil (1972), as well as of most other researchers in alternative mating strategies (reviews in Shuster and Wade 1991a; Sinervo 2001a,b; Sinervo and Lively 1996; Gross 1996). As stated in chapters 8 and 9, the circumstances that cause ΔI to become large and positive depend on the distribution of receptive females in space and time, as well as on the details of female life history. We have argued that male responses to female distributions are predictable. Thus, since male alternative mating strategies represent possible male responses to female distributions, we expect the pattern of these characters to be predictable as well.

Depending on the spatial and temporal distribution of females and female life history, males are expected to (1) defend individual females, (2) defend groups of females, (3) defend resources required by females, (4) attempt to mate with many females in sequence, or attempt some combination of these mating behaviors (chapter 9). If variance in male mating success arises due to differential male abilities to perform these behaviors, the context in which an alternative male mating strategy may be successful can vary too. As we explained above, the four general contexts in which sexual selection occurs lead to four major categories of alternative mating strategies, although each context can generate more than one pattern of strategies (table 12.1).

The patterns within each context are as follows: (1) If most males in the population tend to *guard individual females*, males able to usurp guarded females from their potential mates, or otherwise make male mate guarding ineffective, are expected to invade the guarding male population. Within this general context of mate acquisition, (a) large size and/or aggressiveness may enhance usurpation or harassment ability, (b) small size may allow males to insinuate themselves between guarding males and their mates, or (c) surreptitious courting of already paired females may allow males to secure extra-pair matings. (2) If most males tend to *guard groups of females*, males who sneak into groups to steal matings from peripheral females might be favored. To accomplish this, satellite males may (a) adopt stealthy behaviors and/or diminutive morphologies, (b) become inconspicuous by mimicking the behavior and morphology of females in harems, or (c) focus their mating and parental effort on a single female. (3) If most males tend to *guard resources required by females*, the same alternative mating strategies described in (2a–c) may arise since females are expected to cluster around resources. (4) If most males attempt to *mate with large numbers of females in sequence*, as might occur on leks, monogynist males (e.g., [2c]) are likely to invade such populations. In any of these categories, if females mate more than once or tend to release masses of unfertilized ova, sperm competition can increase the variance in fertilization success and favor sperm morphologies or male secretions that depart from the population average (chapter 8).

As shown in table 12.1, when the specific evolutionary contexts are removed, the patterns of alternative mating strategies condense to four major categories, (1) usurper, (2) sneaker (including insinuator, cuckholder, satellite, subordinate, and female mimic), (3) monogynist, and (4) sperm competitor. Note that within each context more than one of these categories may exist, depending on the degree to which mating success among conventional males is variable. Also note that individuals may display more than one category of alternative strategy within their individual behavioral repertoires. We will consider these differences in more detail below.

The Observed Forms of Alternative Mating Strategies

A large body of literature exists describing variation in male mating behavior (table 12.2). Observed patterns of male polymorphism conform to the four general categories we have described (usurper, sneaker, monogynist, sperm competitor). Again, as Darwin (1871) and others (Gadgil 1972; Dawkins 1980; Maynard Smith 1982; Gross 1996) have noted, the greater the intensity of sexual selection, that is, the greater the variance in mating success among males, the more likely alternative male strategies will be observed. This could explain the lack of described cases of alternative mating strategies in organisms in which male gametes are dispersed on the wind, in the water column, by the use of pollinators, or in which simultaneous spawn-

Table 12.2
Some examples of alternative mating strategies among animal species.

Taxon	*Reference*
Platyhelminthes	Michiels and Newman 1998
Acanthocephala	Abele and Gilchrist 1977; Poulin and Morand 2000
Nematoda	Ainsworth 1990
Annelida	Sella 1990; Vuturo and Shuster 2000
Mollusca	Norman et al. 1999; Hanlon 1998, Buresch et al. 2001
Arthropoda	
Merostomata	Cohen and Brockmann 1983; Brockmann 1996; Brockmann and Penn 1991; Brockmann et al. 1994, 2000
Chelicerata	Christiansen 1984; Lubin 1986; Zeh and Zeh 1988; Whitehouse 1991; Clark and Uetz 1992, 1993
Acari	Atyeo 1984; Cowan 1984; Baker et al. 1987; Klompen et al. 1987; Radwan 1993, 1995, 2001
Insecta	Parker 1970a,b, 1974; reviews in Alcock 1979; Hamilton 1979; Thornhill 1980, 1981; Cade 1981; Eberhard 1982; O'Neill and Evans 1983; reviews in Thornhill and Alcock 1983; Day 1984; Rubenstein 1980, 1984; Goldsmith 1985; Hayashi 1985; Greenfield and Shelly 1985; Shelly and Greenfield 1985, 1989; Alcock and Houston 1987; Beani and Turillazzi 1988; Crespi 1988; Hedrick 1988; Loiselle and Francoeur 1988; Waltz and Wolf 1984, 1988; Arnquist 1989; O'Neill et al. 1989; Danforth and Neff 1992; Eberhard and Gutierrez 1991; Field and Keller 1993; Krupa and Sih 1993; Wolf and Waltz 1993; Joseph 1994; Tsuji et al. 1994; Kawano 1995; Andersen 1996; Belovsky et al. 1996; Emlen 1996, 1997; Emlen and Nijhout 1999; Kukuk 1996; Lauer 1996; Laur et al. 1996; Danforth and Desjardins 1999; Nomakuchi and Higashi 1996; Higashi and Nomakuchi 1997; Cook et al. 1997; Greeff 1996, 1998; Greeff and Ferguson 1999; Herre et al. 1997; Tsubaki et al. 1997; Heinze et al. 1998; Sauer et al. 1998; Dunn et al. 1999; Takamura 1999; Tomkins 1999; Tomkins and Simmons 2000; Preston-Mafham 2001
Crustacea	Haq 1972; Bocquet and Veville 1973; Ra'anan and Sagi 1985; Borowsky 1985; Kuris et al. 1997; Shuster 1987, 1989b; Shuster and Wade 1991a,b; Shuster and Sassaman 1997; Shuster et al. 2001; Stancyk and Moreira 1988; Gotto 1990; Waddy and Aiken

Table 12.2
(*continued*)

Taxon	*Reference*
	1990; Lindberg and Lockhart 1993; Ahl and Laufer 1996; Clark 1997; Koga et al. 1993; Koga 1998; Marioghae and Ayinla 1999
Vertebrata	
Pisces	Warner et al. 1975; Robertson and Warner 1978; Warner and Robertson 1978; Warner 1984; Warner et al. 1995b; Gross 1985, 1991; Dominey 1980; Chernitskij 1986; Petersen 1987; Petersen et al. 2001; Hutchings and Myers 1988, 1994; Travis and Woodward 1989; Kallman 1983; Castillo and Brull 1989; De Jonge and Videler 1989; Sigurjonsdottir and Gunnarsson 1989; Taylor 1989; Zimmerer and Kallman 1989; van den Berghe et al. 1989; Goldschmidt et al. 1992; Morris et al. 1992; Ryan et al. 1990, 1992; Godin 1995; Brantley et al. 1993; Bass 1993; Bass et al. 1996; Erbelding-Denk et al. 1994; Santos et al. 1996; Foote et al. 1997; Pilastro et al. 1997; Pilastro and Bisazza 1999; Blanchfield and Ridgway 1999; Sunobe and Nakazono 1999; Goodson and Bass 2000; Jirotkul 2000; Jones et al. 2001a,b; Jones and Hutchings 2001; Bass and Grober 2001
Amphibia	Perrill et al. 1978; Howard 1978, 1984; Wells 1979; Verrell 1982; Ryan 1983b; Sullivan 1983, 1992; Krupa 1989; Woolbright et al. 1990; Fukuyama 1991; Haddad 1991; Pfennig 1992; Lucas and Howard 1995; Marler et al. 1999
Reptilia	Hover 1985; Duvall et al. 1985; Moore 1991; Hews et al. 1994; Wikelski et al. 1996; Sinervo 2001a,b; Sinervo et al. 2000; Sinervo and Lively 1996; Alonzo and Sinervo 2001; Shine et al. 2001
Aves	van Rhijn 1973; Lank and Smith 1987; Lank et al. 1995, 1999; Gowaty 1985; Westneat et al. 1987; Westneat and Sherman 1997; Brodsky 1988; Morrill and Robertson 1990; Negro et al. 1992; Grant et al. 1995; Hugie and Lank 1997; Lanctot et al. 1998; Bachman and Widemo 1999; P. O. Dunn et al. 1999; D. W. Dunn et al. 1999; Pitcher and Stutchbury 2000; Johnson et al. 2000; Kempenaers et al. 2001
Mammalia	Clutton-Brock et al. 1982; Sandell 1983; reviews in Austad 1984; Gren 1984; Waltz and Wolf 1984; Hogg 1984; Gosling 1986, 1991; Packer and Pusey 1987; Johnson 1989; Carranza et al. 1990; Kovacs

Table 12.2
(*continued*)

Taxon	*Reference*
	1990; Thirgood 1990; Bulger 1993; Koprowski 1993; Brereton 1994; Komers et al. 1994; Stockley et al. 1996; Hogg and Forbes 1997; Soltis et al. 1997; Coltman et al. 1999; Linklater et al. 1999; Thornhill and Palmer 2000; Gemmell et al. 2001

ing occurs (e.g., within most Plantae, as well as within most Porifera, Cnidaria, Echinodermata, and marine Annelida; see also chapter 4). In such species, sperm limitation is possible (Levitan and Petersen 1995) and considerable variance in offspring numbers can exist within and among members of both sexes. In such cases, while I_{mates} and $I_{clutches}$ may become large, the sex difference in the opportunity for selection, ΔI, is likely to be negligible. As we have explained, alternative strategies in either sex are unlikely to become conspicuous unless ΔI is nonzero (see below for discussion of situations in which ΔI is near zero).

The expected lack of alternative strategies also appears to hold for most hermophroditic taxa, including the Ctenophora, Gnathostomulida, Gasterotricha, most Platyhelminthes, and many sessile Mollusca, Crustacea, and Lophophorata). As explained in chapters 1 and 4, reciprocal gamete transfer limits variance in both male and female mate numbers and therefore limits the degree to which alternative mating strategies may evolve. However, individuals who specialize as males *do* exist in at least two hermaphroditic annelid species (*Ophryotrocha*, Sella 1996; *Ophiodromus*, Vuturo and Shuster 1999), as well as in at least one hermaphroditic flatworm species (*Pseudoceros bifurcus* Prudhoe, Michiels and Newman 1998). Males also coexist with hermaphrodites in androdioecious plants and clam shrimp (Mayer and Charlesworth 1991; Weeks et al. 1999). These cases suggest that reciprocal gamete transfer among hermaphrodites need not always occur and in some cases may be impossible. In *P. bifurcus*, for example, individuals engage in nonreciprocal gamete transfer using hypodermic insemination. Similar patterns of mating behavior are likely to be found in other hermaphroditic taxa with hypodermic insemination, including the Gnathostomulida, Turbellaria, and rhynchobdellid leeches (Brusca and Brusca 1990; Michiels and Newman 1998). The invasion of hermaphroditic populations by individuals who specialize as males is likely to proceed rapidly toward gonochorism unless circumstances favoring hermaphroditism (e.g., low effective population size, Ghiselin 1969; negative covariance between male and female size fecundity relationships, Charnov 1979, 1982) persist. Thus, populations with persistent hermaphrodites provide clues to factors that prevent extreme variance in male mating success (see below).

In taxa in which separate sexes already exist and in which mobility allows the distribution of females to become aggregated in space as well as dispersed in time, alternative male mating strategies have proliferated. Particular diversity has arisen within the cephalopod Mollusca, the Arthropoda, and throughout the Chordata (table 12.2). The reproductive biology and mating behavior of other gonochoristic, non-gamete-dispersing, mobile taxa such as the Rotifera, the Kinorhyncha, the Chaetognatha, the Nematoda, the Acanthocephala, and the schistosomatid and didymozoid Trematoda, are poorly known (Brusca and Brusca 1990). However, sexually dimorphic species in these groups are likely to exhibit patterns of alternative mating strategies similar to those described above (Abele and Gilchrist 1977; Ainsworth 1990; Poulin 1997; Poulin and Morand 2000).

Unlike sexually dimorphic species, alternative mating strategies in monomorphic species are likely to be quite subtle. Moreover, mating strategy variation is expected only in species in which females produce more than one brood of offspring, in which members of both sexes mate more than once, and in which variance in offspring numbers is approximately equal among males and females (chapter 4). In such cases, the sexes are likely to be monomorphic because the sex difference in offspring numbers, ΔI, is small. However, similar variance in offspring numbers among males as well as among females means that alternative mating strategies can invade both sexes. Recent evidence of widespread extra-pair copulations and egg dumping in monomorphic and ostensibly "monogamous" birds (see chapter 5) may provide examples of alternative mating strategies existing in both sexes. Since this terminology is already in use to describe such activities, we invite researchers to measure the variance in offspring numbers among males and females, due to differences in mate numbers, as well as to differences in clutch numbers among members of both sexes. We predict that I_{clutches} and I_{mates} in such species will be approximately equal in magnitude.

Reconciling Observed and Expected Forms of Male Mating Behavior

A major conclusion of Levins' (1968) approach is that under weak natural selection the ability of individuals to adjust to environmental changes will be common, whereas genetic polymorphisms will be relatively rare. This conclusion has been substantiated by others, as previously described in this chapter (Slatkin 1978, 1979; Via and Lande 1985; Lively 1986; Moran 1992). We have argued that under sexual selection genetic polymorphism in male mating behavior will be common, perhaps even more common than phenotypic plasticity, because the strength of sexual selection is often large (see also Sinervo 2001a). If genetic polymorphisms are expected to be common under sexual selection, what explains the apparent preponderance of so-called "conditional" male mating strategies (Gross 1996; Lank et al. 1999)?

There are three possible answers. First, until now, any evidence that fitness differences exist among different male phenotypes has been interpreted as confirmation that some males in the population "make the best of a bad job" (review in Gross 1996). According to this hypothesis, all males in the population are genetically identical in their ability to shift between conventional and alternative mating tactics. With this assumption made, there is little to motivate even the most enthusiastic graduate student, much less the grant-starved principal investigator, to begin the hard work of searching for genetic differences among males. Thus, the "rarity" of genetic polymorphisms in male mating behavior may be more apparent than real.

A second explanation for the apparent preponderance of conditional mating strategies arises from the assumption that phenotypic plasticity *must* be widespread (Bradshaw 1965; Levins 1968; Slatkin 1979; Dawkins 1980; Thornhill and Alcock 1983; Gross 1996). With this theoretical tradition well established, it is easy to suggest that *any* difference in behavior or morphology associated with environmental change might represent phenotypic plasticity. This assertion is often made in the absence of studies on selection or the nature of the genetic system responsible for producing the character, particularly when applied to mating behavior. An organism that does well in one environment and poorly in another may display distinct behaviors in each location, but these differences need not have been shaped by selection. Thus, apparent phenotypic plasticity may actually illustrate phenotypic constraint rather than adaptation by natural selection (Via and Lande 1985; Schlichting and Pigliucci 1998).

A third explanation for the apparent preponderance of conditional mating strategies is consistent with our predictions in the sections above. Under intense sexual selection, concave fitness sets are likely, and if environmental circumstances influencing mating success change over brief time scales, behavioral polyphenism is expected to evolve. Thus, conditional polymorphisms in male mating behavior are likely to exist. However, as we describe below, the genetic basis for behavioral polyphenism as well as the relative fitnesses of individuals exhibiting it, are quite distinct from that which is now presumed to exist for condition-dependent mating tactics (Dawkins 1980; Eberhard 1982; Gross and Repka 1998).

We will address these three possible explanations in the following section of this chapter. We will begin by describing current models that describe genetic architectures underlying Mendelian polymorphisms and phenotypic plasticity. We will generate predictions about how to distinguish these mechanisms observationally and how they can be disentangled experimentally. We will next examine the evidence that genetic polymorphisms in male mating behavior are relatively rare, and we will examine the types of data that will allow evolved variation in male mating behavior to be distinguished from phenotypic constraint. Males may appear to behave in different ways in different environments, but there is good reason to conclude that such variation need not represent adaptation. We will then examine the ways in which gene × environment interactions can generate particular patterns of phenotypic plasticity in alternative mating behavior. We will lastly examine the observed patterns of phenotypic plasticity in alternative mating strategies in nature to determine how well our predictions are met. This last section will reveal gaps in the existing data that may be productively explored. We predict that as more researchers investigate the inheritance of alternative mating strategies, condition-dependent polymorphisms will cease to appear ubiquitous. This process is already under way (see above).

The Evolution of Condition Dependence

The reaction norm approach to phenotypic plasticity focuses on how plasticity is expressed more than on how it has evolved. Yet practitioners of this method are unequivocal in their view that plasticity itself is a complex trait whose expression is *shaped by selection* (review in Schlichting and Pigliucci 1998). By this definition, phenotypic plasticity exists within a population because it has a *heritable* basis, that is, the tendency to exhibit phenotypic plasticity is *genetically variable* among individuals. If this were not so, it would be impossible to measure the heritability of phenotypic plasticity (Falconer 1989). More importantly, without an underlying genetic basis for phenotypic plasticity, invasion and persistence of individual abilities to respond to changing environments would be impossible, regardless of the strength of selection (Via and Lande 1985).

Phenotypic plasticity is expected to displace genetic polymorphism if environments become fine grained, or if mutants arise that anticipate environmental change and adjust their development accordingly (see above). Under these circumstances, individuals with generalized phenotypes, or individuals with flexible phenotypes, will experience higher fitness than more specialized, less flexible individuals. Yet after this invasion occurs, circumstances are likely to remain dynamic. Refinement of phenotypic plasticity after invasion can occur if selection favors modifiers that improve individual abilities to respond to environmental changes. Such modifiers are likely to vary in frequency and effect within the population, and must be inheritable for their effects to be manifested each generation.

In species in which phenotypic plasticity has evolved, individuals will neither be exposed to identical degrees of environmental variation, nor be equally flexible in their phenotypic expression of characters. Such variation is due to interactions between alleles responsible for plasticity, and alleles coding for other characters. Populations exhibiting apparent phenotypic plasticity are expected to vary in their average response to a particular environment, and individuals within populations are likely to be variable in their liability to respond to environmental changes (Roff 1986, 1992, 1996). With these factors in mind, it is unlikely that there are "fixed, conditional strategies" without genetic variation as argued by some (e.g., Dawkins 1980; Eberhard 1982; Gross 1996; Gross and Repka 1998).

Researchers since Darwin have argued that for male polymorphisms of any kind to perist, the fitnesses of the different morphs must be equal. We suggest that the fraction of the male population exhibiting an alternative mating tactic will depend on the average success of males performing the conventional male mating tactic. The cues males may use to detect mating opportunities and adopt an alternative mating tactics such as sneaking may be the number of females that have accumulated in the harems of successful males. As we have shown, when some territorial males mate with more than one female, other territorial males will obtain no mates at all. The average male who attempted to steal mates under such circumstances would likely experience greater mating success than the average male attempting to guard them. Males in such populations clearly exhibit a conditional mating strategy in which they may shift between guarding and sneaking as circumstances demand. However, when averaged across all males employing a particular tactic at a particular time, neither tactic is doing more poorly than the other.

As we have shown, the rate at which males are likely to shift from one mating tactic to another will depend on the variance in the success of males performing each tactic at any time. Some males may be capable of shifting between tactics faster than other males. Some will be slower and less successful in acquiring mates. When most males are capable of acquiring mates, little variance in male mating success exists, and observed variation in male mating behavior is expected to be small. However, when some fraction of the male population gains multiple mates, other males *must* be excluded from mating, even if only over brief intervals of time. Such differential male mating success will favor males who adopt alternative mating tactics because the average mating success of such males can exceed the success of the average conventional male during that time interval.

As explained above, as soon as males begin to employ alternative mating tactics, the variance in mating success will rapidly erode because the number of mating males increases. The greater the success of males employing the alternative tactic, the faster this erosion will take place and the lower the frequency of males able to exploit the temporal mating niche can be before mating opportunities disappear. This rapid change in mating success among males employing conventional and alternative mating *tactics* will be more difficult to measure than the similar changes that occur among alternative mating strategies, in say, the marine isopod, *Paracerceis sculpta*. Yet there is no reason to suggest that the circumstances allowing the invasion and persistence of multiple mating strategies over longer time scales should be any different from the circumstances that permit them to invade and persist over shorter durations. The only difference is the developmental mechanism influencing the speed with which the trait changes frequency within the population.

Alternative Mating Strategies and Precocial Male Maturation

Precocial male maturation refers to the tendency among males in a population, to achieve sexual maturity *before* acquiring secondary sexual characteristics. That is, young males develop testes before they acquire male structures typically used in combat with other males, or in displays to attract females. Precocial male maturation is known to occur in a variety of species, most of which are noteworthy because young males are so distinct in their appearance compared to fully mature males. As explained in chapter 10, Darwin (1874) was skeptical of sexual selection's ability to produce such characteristics as permanent elements of male populations. However, numerous examples of precocial male maturation appear to exist in a diversity of taxa (e.g., Hutchings and Myers 1988, 1994; Taylor 1989; Taborsky 1994; Blanchfield and Ridgway 1999, in addition to those noted by Darwin himself [1874]; chapter 10).

There is little doubt that, if variance in mate numbers exists among adult males in a population, males who reach sexual maturity at an early age might acquire mates as sneakers (see above). We have shown that this tendency can persist as a genetic polymorphism because the circumstances in which territorial males and precocially mature males may experience equal fitness are likely to be quite common. However, examples appear to exist in which males acquire testes at preadult size, successfully mate, and survive to later acquire the weapons or ornaments of older males. Are such examples real or was Darwin correct in his original assessment?

Two possible explanations for such observations exist. First, precocial male maturation plus later acquisition of secondary sexual characters may exist in situations in which there is little cost to this combined life history, *and* the gains obtained through sneaking are rare and intermittent. It is easy to imagine the fitness increment gained by males who mature early, mate only occasionally, but later mature to mate often, compared to males who delay maturation and fail to obtain *any* matings as an early maturing male. However, this hypothetical gain is possible only if there is very little cost to adult fitness incurred by males who mature at a young age. Given the ease with which genetic polymorphism for early male maturation can evidently invade a territorial male population (chapter 11), the cost of precocial maturation plus later ornamentation would have to be extremely low to allow such mutants to compete against specialists. To Darwin (1874), this cost seemed prohibitive.

A second possible explanation for the appearance of early maturing males is that the advantage of precocial maturation comes not from the *combination* of fitness gained as a young male with fitness as an older male, but instead arises solely from the fitness gained as a young male. Some males may survive to mate as young males *and* as older males, but the frequency with which this occurs in nature may actually be small. This explanation is consistent with the hypothesis that the appearance of early male maturation represents part of a genetic polymorphism, in which a fraction of the male population matures early and the remaining males mature later (Hutchings and Myers 1988, 1994). This second explanation could easily be mistaken for the first if a fraction of the male population is reared under laboratory or protected field conditions.

In current literature on precocial male maturation, it is assumed that all males in the population are able to mature early and survive to reach an older, more competitive age. However, the degree to which all males in the population have this capacity has only been examined in detail in a few species, and in each of these cases, a certain fraction of the male population appears to consist of precocially maturing specialists (review in Hutchings and Myers 1994). In a related experiment, Lively et al. (2000) showed that in the presence of extremely strong and prolonged cues, only 22% of barnacles expected to transform to a bent morph, actually transformed. While other mechanisms may exist to explain these results, such observations draw into serious question the notion that precocially mature males routinely mate when old. Indeed, Darwin (1874) may have been correct, and clearly more detailed investigations are needed.

Alternative Mating Strategies and Sex Ratio

Fisher (1930, 1958) proposed a mechanism by which equal frequencies of males and females might persist in natural populations. He argued that since every offspring has one mother and one father, the average number of offspring per individual in each sex will depend on the frequency of each sex in the population. Thus, if $N_{\text{offspring}}$ equals the total number of offspring produced per generation, and N_{males} and N_{females} equal the number of males and females in the population, respectively, the average number of offspring per male, M, that is, the average fitness of males, will equal

$$M = N_{\text{offspring}}/N_{\text{males}}. \qquad [12.1]$$

Similarly, the average number of offspring per female, F, that is, the average fitness of females, will equal

$$F = N_{\text{offspring}}/N_{\text{females}}. \qquad [12.2]$$

If N_{males} equals $N_{females}$, the sex ratio, R, equals 1. Thus, M and F are equivalent.

A sex ratio bias will cause the average number of offspring per individual to be greater among members of the minority sex. Thus, any allele that causes females to produce an excess of the minority sex among their progeny will increase in frequency and thereby tend to equalize the sex ratio (Futuyma 1998; Ridley 1995). Although the population genetic mechanism by which sex ratio evolution occurs is usually explained in these terms, Fisher (1930; Edwards 1998) himself stated his sex ratio principle in terms of "investment" in male and female offspring. In particular, Fisher suggested that members of the majority sex were "cheaper" in terms of energetic expense to female parents. This view coincides well with Trivers' (1972) parental investment hypotheses, because it suggests that females manipulate their family sex ratios by differential investment in offspring of a particular sex. This view is also apparently responsible for the hypothesis that females in good physical condition may tend to bias their family sex ratios in favor of males, since maternal effects on offspring condition may enhance the ability of sons to compete for (Trivers and Willard 1973; Clutton-Brock et al. 1984; Charnov and Bull 1989).

The evolution of maternal sex ratio biasing requires a positive covariance across environments between the sex ratio bias toward sons and the mating success of sons. This covariance alone, however, is not a sufficient condition for the evolution of maternal sex ratio biasing. Unequal primary sex ratios awaken the sleeping giant of Fisherian sex ratio selection, which acts relentlessly to favor a balanced sex ratio (i.e., $R = 1$). Thus, the strength of selection in favor of maternal sex ratio biasing must be sufficient to outweigh the opposing Fisherian selection. The interaction of these two forces is asymmetrical, however, and as we have shown elsewhere total selection favors a female-biased sex ratio.

To see this, consider the relationship between primary sex ratio and sexual selection, where H equals (R/p_s). Female-biased sex ratios, where $R > 1$, increase H and, thus, increase the strength of sexual selection. In contrast, male-biased sex ratios, where $R < 1$, decrease H and weaken sexual selection. Fisherian sex ratio selection is egalitarian, opposing biases in either direction with equal force, proportional to the degree of bias. As a result, biasing broods toward females in a poor environment is more strongly favored than the same degree of bias toward males in a good environment. Consequently, strong sexual selection will result more frequently in female-biased than in male-biased primary sex ratios. Our argument is concordant with the primary sex ratio biases toward females observed in many strongly sexually selected species (e.g., Freedberg and Wade, in press, and below).

The foundation of our argument regarding maternal sex ratio biasing is the increased strength of sexual selection that attends a female-biased primary sex ratio. This component of our argument is the opposite of that made in almost every mating system analysis conducted over the past three decades (see, for example, Emlen and Oring 1977; Thornhill and Alcock 1983; Reynolds 1996; Ahnesjö et al. 2001). Current theory, based on the operational sex ratio (R_O = [1/R]), predicts that sexual selection is strongest when males are more abundant than females and weaker when females are in excess. While such conditions may appear to apply when sex ratios are locally biased (reviews in Reynolds 1996; Ahnesjö et al. 2001), as we have explained in chapter 3, the OSR, R_O, is only one component of the sex difference in the variance in fitness and it modifies $I_{females}$ rather than I_{mates} (see eq. [1.24a]) With very strong sexual selection, we showed (chapter 2) that V_{mates} is approximately equal to ($H - 1$). Thus, sexual selection through the variance in mate numbers among males outweighs and is enhanced by the effects of a biased sex ratio on the total opportunity for selection on males (chapter 11). Moreover, as we showed in chapter 3, instantaneous estimates of the breeding sex ratio tend to *overestimate* the influence of sex ratio on total selection, exaggerating their actual effect.

In spite of the positive relationship between the OSR and one component of the total opportunity for selection on males, sexual selection, proportional to H when strong, predominates. As a result, a positive covariance across environments between a sex ratio bias toward sons and the mating success of sons will lead more often to female sex ratio biases, which increases the value of H. In sex-role-reversed species, the OSR is presumed to have become female biased. However, in such species, we predict the opposite, namely, that primary sex ratios will be male biased, although a mechanism by which this could occur is not clear. It is possible that males in role-reversed species preferentially rear sons when they find themselves in poor environments or that mothers in poor condition produce more sons via maternal condition-dependent meiotic drive.

Many studies of population sex ratio in other species document female-biased sex ratios and sex ratio fluctuations in nature (Bull 1983; Conover and Van Voorhees 1990; Basolo 1994; Godfray and Werren 1996; Komdeur et al. 1997; Whiteman 1997; Atlan et al. 1997; Hurst and Pomiankowski 1998; Wilkinson et al. 1998; Freedberg and Wade 2002). These deviations are usually attributed to greater preadult mortality among males than females, and are therefore assumed to represent a form of Fisherian sex ratio adjustment in which males are produced in quantities necessary to compensate for sexual differences in viability (Fisher 1930; but see Rigaud and Juchault 1993; Shuster and Sassaman 1997; Shuster et al. 2001).

In *P. sculpta*, Shuster et al. (2001) showed that a primary sex determination mechanism capable of generating a 50:50 sex ratio can be modified by interactions with autosomal sex ratio distorters, and that these factors can generate female-biased population sex ratios and sex ratio fluctuations without requiring differential male mortality. They also showed that interactions between *Ams-Tfr* and primary sex determination factors can routinely produce fluctuating sex ratios in the natural *P. sculpta* population, and that such fluctuations can result from differential mating success among the three male morphotypes in this species. The *Ams-Tfr* model indicates that sexual selection and sex ratio distortion are genetically interrelated phenomena in this species. This result provides a compelling context for the co-occurrence of sexual selection and sex ratio distortion in a diversity of animal taxa (Komdeur et al. 1997; Whiteman 1997; Atlan et al. 1997; Hurst and Pomian kowski 1998; Wilkinson et al. 1998).

Alternative Mating Strategies and Sex Change

Species in which individuals change sex during their lifetimes provide an interesting combination of these last two issues. In protrandrous sequential hermaphrodites, individuals mature first as males and later transform in to females (Charnov 1979). In protogynous sequential hermaphrodites, individuals mature first as females and later transform into males (Warner 1984). In androdioecy, certain individuals mature as males while others mature as hermaphrodites (Mayer and Charlesworth 1991; Weeks et al. 1999; Vuturo and Shuster 2000). In gynodioecy, certain individuals mature as females and other individuals mature as females (Mayer and Charlesworth 1991; McCauley and Brock 1998). As has been explored in detail by these authors and others, these life history changes by individuals produce populational changes in sex ratio. Without exception, the framework within which the theoretical basis of sex ratio equalization has been considered involves family selection. We suggest that in some if not all of these species there may exist genetic polymorphisms for the ability to change sex or remain unchanged. A simple method for investigating this possibility is to examine the frequency of the population that exists as a single sex. If this fraction represents an alternative mating strategy, then, as explained above, their frequency may approximate the fraction of the hermaphroditic population that is unsuccessful in performing male function (in the case of androdioecy) or female function (in the case of gynodioecy). The increasing sophistication of molecular methods for examining parentage could allow detailed tests of this hypothesis in each of these mating systems.

Classification of Alternative Mating Strategies

Existing classification schemes for alternative mating strategies are typological. They rely wholly or in part on the degree to which mating phenotypes are "genetically fixed" or "condition dependent" (table 12.2; Austad 1984; Gross 1996). This approach can be useful, as we have explained above, because such patterns are expected to arise from the temporal grain of conditions in which mating opportunities occur, relative to the life span of individuals. Indeed, the resulting phenotypes are expected to reflect specific ways in which selection shapes behavior and life history. However, as we have also shown, the source of selection favoring of alternative mating behaviors ultimately arises from the spatial and temporal distributions of females and the details of female life history. Because these characters define the competitive arena in which alternative mating strategies evolve, they are likely to contain more useful phylogenetic information than typological descriptions of relative plasticity in mating phenotypes.

Thus, a more instructive evolutionary scheme for classifying alternative mating strategies can proceed as we have outlined in the previous chapters. By first identifying the spatial and temporal distributions of females, the opportunity for sexual selection can be identified (chapters 1–3). This information provides the most reliable information for investigating where alternative mating strategies evolve and why. By considering the life history and environmental circumstances in which individual species exist, the degree to which I_{mates} is eroded by I_{clutch} may be determined. This permits identification of possible female alternative mating strategies as well as circumstances that could permit sexual conflict. Next, by identifying the mating system expected to evolve under such circumstances, it is possible to identify which forms of alternative mating strategies correlate with each set of circumstances. We have suggested some possibilities (chapter 9), but more rigorous phylogenetic analyses (Harvey and Pagel 1991) will be necessary to test our initial suggestions. The general forms alternative mating strategies assume are summarized in table 12.1. When combined with information on life history and environmental circumstances influencing reproduction, the degree to which behavioral polyphenism, developmental conversion, or genetic polymorphism may be explicitly tested, and the degree to which these forms correlate with particular circumstances in existing mating systems may be systematically explored.

This approach is admittedly laborious. However, it is preferable to post hoc classifications of mating systems based on observations of male behavior, in which information on the distribution and abundance of mates inevitably fits the preconceptions of the researchers. Quantification of the opportunity for selection, either from measurable distributions of mates, or from actual distributions of progeny among adults, provides the only concrete method for identifying how and when sexual selection is a significant evolutionary force. As we have shown, when sexual selection occurs, alternative mating strategies are likely to evolve.

Future Directions

We maintain that no special rules are necessary to explain the expression, inheritance, dynamics, or persistence of alternative mating strategies in natural populations. Alternative mating strategies, like other strategies discussed in an evolutionary context, can be investigated following well-established principles from population genetics and evolutionary game theory (Maynard Smith 1982; Crow 1986; Falconer 1989; Lucas and Howard 1995; Roff and Simmons 1997; Flaxman 2000). Such studies are likely to bear early fruit. Alternative mating strategies clearly evolve under intense sexual selection. Thus, changes in morph frequency resulting from differential mating success can be readily observed over short periods of time (Shuster 1989a; Shuster et al. 2001; Sinervo and Lively 1996; Sinervo 2001b).

Other research on alternative mating strategies, beginning at almost any level, can also yield a rich harvest. Analyses of variation in mating behavior within and among species could identify cross-taxonomic patterns in mating strategy evolution (Cook et al. 1997). Within-species analyses of mating success during satellite invasions could determine whether the average success of satellites routinely exceeds that of territorial males (Fu et al. 2001). Analyses of mating success over time could determine whether average mating success calculated for all males within each phenotypic class is equivalent among all male classes (Shuster and Wade 1991a; Sinervo and Lively 1996). Investigations of the neurological and hormonal mechanisms controlling the expression of alternative mating strategies could reveal the ways in which alternative mating phenotypes arise from ancestral states (Goodson and Bass 2000).

A world of information awaits discovery amid genetic architectures underlying alternative mating strategies. Experimental methods already exist for exploring the inheritance and dynamics of Mendelian as well as threshold traits (Crow 1986; Falconer 1989; Roff 1996). In the hands of a few pioneers, these well-honed tools have cleared paths toward a general understanding of how alternative mating strategies arise and persist in nature. Such research has and will continue to influence one of the most rapidly expanding areas of evolutionary biology, the study of phenotypic plasticity and evolution in changing environments (Levins 1968; Schlichting and Pigliucci 1998). In any field, the articulation of initial research questions requires courage and insight (Gadgil 1972; Dawkins 1980; Dominey 1984; Austad 1984; Gross 1996). With the power of the quantitative methods we have identified focused on these issues, a general theory of mating systems and alternative mating strategies is close at hand.

Chapter Summary

The conventional wisdom with respect to alternative mating strategies is that most variation in male mating behavior represents "conditional" polymorphism. To determine the degree to which this approach may be appropriate, we review existing theory concerning the evolution of phenotypic plasticity. We first clarify Levins' (1968) terminology and approach as it relates to alternative strategies, specifically, environmental circumstances favoring genetic architectures leading to Mendelian inheritance or phenotypic flexibility in mating behavior. We then use a reaction norm approach to illustrate the forms alternative mating strategies may assume, including genetic polymorphism, developmental conversion, or behavioral polyphenism. Using established population genetic methods and quantitative genetic hypotheses for the evolution of phenotypic plasticity, we propose an alternative to the hypothesis of status-dependent selection (SDS) for condition dependence in male mating behavior. We show not only that genetic polymorphism underlying alternative mating strategies is more common than previously assumed, but also that the conditions in which it may evolve are widespread.

We have argued that male responses to female distributions are predictable. Thus, if male alternative mating strategies represent male responses to female distributions, we expect the pattern of these characters to be predictable as well. We review the expected forms of alternative mating strategies and show that these patterns condense to four major categories: (1) usurper, (2) sneaker (including insinuator, cuckholder, satellite, subordinate, and female mimic), (3) monogynist, and (4) sperm competitor. We argue that the greater the intensity of sexual selection, the more likely alternative male strategies will be observed. This could explain the lack of described cases of alternative mating strategies in organisms in which male gametes are dispersed on the wind, in water, by the use of pollinators, in which simultaneous spawning occurs, and in hermaphrodites. In taxa with separate sexes and in which mobility allows the distribution of females to become aggregated in spacc as well as dispersed in time, alternative male mating strategies have proliferated, particularly within the Arthropoda and Vertebrata.

We suggest that the ubiquity of conditional strategies may be more apparent than real. With the assumption of genetic monomorphism among males exhibiting alternative mating strategies, there is little to stimulate genetic investigations. We also argue that despite its frequent observation as such, not all variation in male mating behavior can legitimately be identified as adaptive. Conditional polymorphisms clearly exist. However, we suggest that their genetic architectures are most easily understood as threshold traits, whose underlying genetic variation and means of expression differ greatly from those now presumed to exist for condition-dependent mating tactics.

Our perspective explains much about alternative mating strategies involving differences in behavior, precocial male maturation, arising from sex ratio biases, and involving sex change. We suggest that the circumstances that allow behavioral polymorphisms to arise and persist are no different from those that allow multiple mating strategies to arise over ecological or evolutionary time. The appearance of early male maturation within a male population may in fact represent a genetic polymorphism, in which a fraction of the male population matures early and the remaining males mature later. We suggest that sexual selection itself creates functional sex ratio biases. When such biases occur, particularly when nonrandom mating causes the bias, modifiers acting directly on the primary expression of sex can equalize the sex ratio more rapidly than is possible by family selection, favoring genetic polymorphism for the ability to change sex or remain unchanged. We conclude this chapter by proposing a new, nontypological scheme for classifying alternative mating strategies based on the theoretical framework outline in this book.

References

Aars, J., H. P. Andreassen, and R. A. Ims. 1995. Root voles: Litter sex-ratio variation in fragmented habitat. *J. Anim. Ecol.* 64:459–472.

Abele, L. G., and S. Gilchrist. 1977. Homosexual rape and sexual selection in acanthocephalan worms. *Science* 197:81–83.

Agrawal, A. F. 2001. Sexual selection and the maintenance of sexual reproduction. *Nature* 411:692–695.

Agrawal A. F., E. D. Brodie, and J. Brown. 2001a. Parent-offspring coadaptation and the dual genetic control of maternal care. *Science* 292:1710–1712.

Agrawal, A. F., E. D. Brodie III, and M. J. Wade. 2001b. On indirect genetic effects in structured populations. *Am. Nat.* 158:308–323.

Agren, J., and L. Ericson. 1996. Population structure and morph-specific fitness differences in tristylous *Lythrum salicaria*. *Evolution* 50:126–139.

Ahl, J.S.B., and H. Laufer. 1996. The pubertal molt in Crustacea revisited. *Invertebr. Reprod. Dev.* 30:177–180.

Ahnesjö, I. 1995. Temperature affects male and female potential reproductive rates differently in the sex-role reversed pipefish, *Syngnathus typhle*. *Behav. Ecol.* 6:229–233.

Ahnesjö, I., C. Kvarnemo, and S. Merilaita. 2001. Using potential reproductive rates to predict mating competition among individuals qualified to mate. *Behav. Ecol.* 12:397–401.

Ainsworth, R. 1990. Male dimorphism in two new species of nematode (Pharyngodonidae: Oxyurida) from New Zealand lizards. *J. Parasit.* 76:812–822.

Alatalo R. V., J. Kotiaho, J. Mappes, and S. Parri. 1998. Mate choice for offspring performance: major benefits or minor costs? *Proc. R. Soc. London, Ser. B* 265:2297–2301.

Alcock, J. 1975. *Animal Behavior: An Evolutionary Approach*, Sinauer Associates, Sunderland, MA.

Alcock, J. 1979. The evolution of intraspecific diversity in reproductive strategies in some bees and wasps. In M. S. Blum and N. A. Blum (eds.), *Sexual Selection and Reproductive Competition in Insects*, Academic Press, New York, pp. 381–402.

Alcock, J. 1995. *Animal Behavior: An Evolutionary Approach*, 5th Ed., Sinauer Associates, Sunderland, MA.

Alcock, J., and T. F. Houston. 1987. Resource defense and alternative mating tactics in the banksia bee, *Hylaeus alcyoneus* (Erichson). *Ethology* 76:177–188.

Alexander, R. D., and G. Borgia. 1979. On the origin and basis of the male female phenomenon. In M. S. Blum and N. A. Blum (eds.), *Sexual Selection and Reproductive Competition in Insects*, Academic Press, New York, pp. 417–440.

Alonzo, H., and Sinervo, B. 2001. Mate choice games, context-dependent good genes, and genetic cycles in the side-blotched lizard, *Uta stansburiana*. *Behav. Ecol. Sociobiol.* 49:176–186.

Altenberg, L., and M. W. Feldman. 1987. Selection, generalized transmission, and the evolution of modifier genes. I. The reduction principle. *Genetics* 117:559–572.

Altmann J. 1990. Primate males go where the females are. *Anim. Behav*. 39:193–195.

Altmann, S. A., S. S. Wagner, and S. Lenington. 1977. Two models for the evolution of polygyny. *Behav. Ecol. Sociobiol*. 2:397–410.

Amat, J. A., R. M. Fraga, and G. M. Arroyo. 1999. Brood desertion and polygamous breeding in the Kentish Plover *Charadrius alexandrinus*. *Ibis* 141:596–607.

Amos, W., J. Worthington Wilmer, and H. Kokko. 2001. Do female grey seals select genetically diverse mates? *Anim. Behav*. 62:157–164.

Andersen, N. M. 1996. Ecological phylogenetics of mating systems and sexual dimorphism in water striders (Heteroptera: Gerridae). *Vie milieu* 46:103–114.

Andersson M. 1994. *Sexual Selection*, Princeton University Press, Princeton, NJ.

Andersson M., and Y. Iwasa. 1996. Sexual selection. *Trends Ecol. Evol*. 11:53–58.

Andrade, M.C.B. 1996. Sexual selection for male sacrifice in the Australian redback spider. *Science* 271:70–72.

Andrade, M.C.B. 1998. Female hunger can explain variation in cannibalistic behavior despite male sacrifice in redback spiders. *Behav. Ecol*. 9:33–42.

Arak, A. 1983. Mate choice. *Nature* 306:261.

Araujo, R., M. A. Ramos, and R. Molinet. 1999. Growth pattern and dynamics of a southern peripheral population of *Pisidium aminicum* (Mueller 1774) (Bivalvia: Sphaeriidae) in Spain. *Malacologia* 41:119–137.

Armendáriz, L. C. 2000. Population dynamics of *Stylaria lacustris* (Linnaeus 1767) (Oligochaeta: Naididae) in Los Talas, Argentina. *Hydrobiologia* 438:217–226.

Arnaud, F., and R. Bamber. 1987. The biology of the Pycnogonida. In J.H.S. Blaxter and A. J. Southward (eds.), *Advances in Marine Biology*, Academic Press, New York, Vol. 24.

Arnold, S. J. 1994. Bateman's principles and the measurement of sexual selection in plants and animals. *Am. Nat*. 144:S126–S149.

Arnold, S. J., and D. Duvall. 1994. Animal mating systems: A synthesis based on selection theory. *Am. Nat*. 143:317–348.

Arnold, S. J., and M. J. Wade. 1984a. On the measurement of selection in natural and laboratory populations: Theory. *Evolution* 38:709–719.

Arnold, S. J., and M. J. Wade. 1984b. On the measurement of selection in natural and laboratory populations: Applications. *Evolution* 38:720–734.

Arnquist, G. 1989. Multiple matings in a water strider: Mutual benefits or intersexual conflict. *Anim. Behav*. 38:749–756.

Atlan, A., H. Mercot, C. Landre, and C. Monchamp-Moreau. 1997. The sex ratio trait in *Drosophila simulans*: Geographical distribution of distortion and resistance. *Evolution* 51:1886–1895.

Atwood, J. L., V. L. Fitz, and J. E. Bamesberger. 1991. Temporal patterns of singing activity at leks of the white-bellied emerald. *Wilson Bull*. 103:373–386.

Atyeo, W. T. 1984. A new genus of feather mites and a new expression of male polymorphism (Analgoidea: Avenzoariidae). *J. Kans. Entomol. Soc*. 57:437–455.

Austad, S. N. 1984. A classification of alternative reproductive behaviors, and methods for field testing ESS models. *Am. Zool*. 24:309–320.

Avise, J. C. 1994. *Molecular Markers, Natural History and Evolution*, Chapman and Hall, New York.

Bachman, G., and F. Widemo. 1999. Relationships between body composition, body size and alternative reproductive tactics in a lekking sandpiper, the Ruff (*Philomachus pugnax*). *Funct. Ecol.* 13:411–416.

Backwell, P.R.Y., M. D. Jennions, J. H. Christy, and U. Schober. 1995. Pillar building in the fiddler crab, *Uca beebi*—evidence for a condition dependent ornament. *Behav. Ecol. Sociobiol.* 36:185–192.

Backwell, P.R.Y., J. H. Christy, S. R. Telford, M. D. Jennions, and N. I. Passmore. 2000. Dishonest signalling in a fiddler crab. *Proc. R. Soc. London, Ser. B* 267:719–724.

Badyaev, A. V., and G. E. Hill. 2000. Evolution of sexual dichromatism:contribution of carotenoid- versus melanin-based coloration. *Biol. J. Linn. Soc.* 69:153–172.

Bain, B., 1992. Pycnogonid higher classification and a revision of the genus *Austropallene* (Family Callipallenidae). Ph.D. diss., City University of New York.

Baker, E. W., D. W. Roubik, and M. Delfinado-Baker. 1987. The developmental stages and dimorphic males of *Chaetodactylus panamensis*, n. sp. (Acari: Chaetodactylidae) associated with solitary bee (Apoidea: Anthophoridae). *Int. J. Acarol.* 13:65–73.

Baker, M. B., M. Shachak, and S. Brand. 1998. Settling behavior of the desert isopod *Hemilepistus reaumuri* in response to variation in soil moisture and other environmental cues. *Isr. J. Zool.* 44:345–354.

Baker, R. R., and M. A. Belis 1995. *Human Sperm Competition*, Chapman and Hall, London.

Balmford, A., I. L. Jones, and A.L.R. Thomas. 1994. How to compensate for costly sexually selected tails—the origin of sexually dimorphic wings in long-tailed birds. *Evolution* 48:1062–1070.

Balshine-Earn, S., and D.J.D. Earn. 1998. On the evolutionary pathway of parental care in mouth-brooding cichlid fishes. *Proc. R. Soc. London, Ser. B* 265:2217–2222.

Barber, I., and S. A. Arnott. 2000. Split-clutch IVF: A technique to examine indirect fitness consequences of mate preferences in sticklebacks. *Behaviour* 137:1129–1140.

Baroiller, J.-F., I. Nakayama, F. Foresti, and D. Chourrout. 1996. Sex determination studies in two species of teleost fish, *Oreochromis niloticus* and *Leporinus elongatus*. *Zool. Stud.* 35:279–285.

Bartels-Hardege, H. D., and E. Zeeck. 1990. Reproductive behaviour of *Nereis diversicolor* (Annelida: Polychaeta). *Mar. Biol.* 106:409–412.

Baskin, C. C., and J. M. Baskin. 1998. *Seed Dormancy and Germination: Ecology, Biogeography and Evolution*, Academic Press, San Diego, CA.

Basolo, A. L. 1990a. Female preference predates the evolution of the sword in the swordtail fish. *Science* 250:808–810.

Basolo, A. L. 1990b. Female preference for male sword length in the green swordtail, *Xiphophorus helleri* (Pisces: Poeciliidae). *Anim. Behav.* 40:332–338.

Basolo, A. L. 1994. The dynamic of Fisherian sex-ratio evolution: Theoretical and experimental investigations. *Am. Nat.* 144:473–490.

Basolo, A. L. 1995. Phylogenetic evidence for the role of preexisting bias in sexual selection. *Proc. R. Soc. London, Ser. B* 50:365–375.

Basolo, A. L., and J. A. Endler. 1995. Sensory biases and the evolution of sensory systems. *Trends Ecol. Evol.* 10:489–489.

Bass, A. H. 1993. From brains to behaviour: Hormonal cascades and alternative mating tactics in teleost fishes. *Rev. Fish Biol. Fish.* 3:181–186.

Bass, A. H., and M. S. Grober. 2001. Social and neural modulation of sexual plasticity in teleost fish. *Brain Behav. Evol.* 57:293–300.

Bass, A. H., B. J. Horvath, and E. B. Brothers. 1996. Nonsequential developmental trajectories lead to dimorphic vocal circuitry for males with alternative reproductive tactics. *J. Neurobiol.* 30:493–504.

Bateman, A. J. 1948. Intra-sexual selection in *Drosophila*. *Heredity* 2:349–368.

Beani, L., and S. Turillazzi. 1988. Alternative mating tactics in males of *Polistes dominulus* (Hymenoptera: Vespidae). *Behav. Ecol. Sociobiol.* 22:257–264.

Beck, C. W. 1998. Mode of fertilization and parental care in anurans. *Anim. Behav.* 55:439–449.

Begun, D. J., P. Whitley, B. L. Todd, H. M. Waldrip-Dail, and A. G. Clark. 2000. Molecular population genetics of male accessory gland proteins in *Drosophila*. *Genetics* 156:1879–1888.

Belk, D. 1991. Anostracan mating behavior: A case of scramble-competition polygyny. In R. T. Bauer and J. W. Marten (eds.), *Crustacean Sexual Biology*, Columbia University Press, New York, pp. 111–125.

Bell, C. H. 1994. A review of diapause in stored-product insects. *J. Stored Prod. Res.* 30:99–120.

Bella, J. L., R. K. Butlin, C. Ferris, and G. M. Hewitt. 1992. Asymmetrical homogamy and unequal sex-ratio from reciprocal mating-order crosses between *Chorthippus parallelus* subspecies. *Heredity* 68:345–352.

Belovsky, G. E., J. B. Slade, and J. M. Chase. 1996. Mating strategies based on foraging ability: An experiment with grasshoppers. *Behav. Ecol.* 7:438–444.

Berglund, A., G. Rosenqvist, and I. Svensson. 1986. Mate choice, fecundity, and sexual dimorphism in two pipefish species (Sygnathidae). *Behav. Ecol. Sociobiol.* 19:301–307.

Berglund, A., G. Rosenqvist, and I. Svensson. 1989. Reproductive success of females limited by males in two pipefish species. *Am. Nat.* 133:506–516.

Birkhead, T. R. 2000. *Promiscuity: An Evolutionary History of Sperm Competition and Sexual Conflict*, Harvard University Press, Cambridge, MA.

Birkhead, T. R., and F. Fletcher. 1995. Depletion determines sperm numbers in male zebra finches. *Anim. Behav.* 49:451–456.

Birkhead T. R., and A. P. Møller. 1993a. Sexual selection and the temporal separation of reproductive events: Sperm storage data from reptiles, birds and mammals. *Biol. J. Linn. Soc.* 50:295–311.

Birkhead, T. R., and A. P. Møller 1993b. Female control of paternity. *Trends Ecol. Evol.* 8:100–104.

Birkhead, T. R., and A. P. Møller (eds.). 1998. *Sperm Competition and Sexual Selection*, Academic Press, New York.

Blanchfield, P. J., and M. S. Ridgway. 1999. The cost of peripheral males in a brook trout mating system. *Anim. Behav.* 57:537–544.

Blumstein, D. T., and W. Arnold. 1998. Ecology and social behavior of golden marmots (*Marmota caudata aurea*). *J. Mammal.* 79:873–886.

Boake, C.R.B. 1986. A method for testing adaptive hypotheses of mate choice. *Am. Nat.* 127:654–666.

Boake, C.R.B. 1991. The coevolution of sending and receiving sexual signals: Perspectives from quantitative genetics and sensory physiology. *Trends Ecol. Evol.* 6:225–227.

Boake, C.R.B. 1994. Outlining the issues. In C.R.B. Boake (ed.), *Quantitative Genetic Studies of Behavioral Evolution*, University of Chicago Press, Chicago.

Bocquet, C., and M. Veuille. 1973. Le polymorphisme des variants sexuels des males chez *Jaera (albifrons) ischiosetosa* Forsman (Isopoda: Asellota). *Arch. Zool. Exp. Gen.* 114:111–128.

Borgia, G. 1979. Sexual selection and the evolution of mating systems. In M. S. Blum and N. A. Blum (eds.), *Sexual Selection and Reproductive Competition in Insects*, Academic Press, New York, pp. 19–80.

Borgia, G. 1986. Sexual selection in bowerbirds. *Sci. Am.* 254:92–100.

Borgia, G., and S. W. Coleman. 2000. Co-option of male courtship signals from aggressive display in bowerbirds. *Proc. R. Soc. London, Ser. B* 267:1735–1740.

Borgia, G., and D. C. Presgraves. 1998. Coevolution of elaborated male display traits in the spotted bowerbird: An experimental test of the threat reduction hypothesis. *Anim. Behav.* 56:1121–1128.

Borowsky, B. 1985. Differences in reproductive behavior between two male morphs of the amphipod crustacean, *Jassa falcata* Montagu. *Physiol. Zool.* 58:497–502.

Bourdon, R., and A. J. Bruce. 1983. On *Probynia*, a new genus of bopyrid (Isopoda Epicaridea) parasitic on pontoniine shrimps from the Great Barrier Reef, Australia. *Crustaceana* 44:310–316.

Bradbury, J. W. 1981. The evolution of leks. In R. D. Alexander and D. W. Tinkle (eds.), *Natural Selection and Social Behavior*, Chiron Press, New York, pp. 138–169.

Bradbury, J. W. 1985. Contrasts between insects and vertebrates in the evolution of male display, female choice and lek mating. In B. Holldobler and M. Lindauer (eds.), *Fortschritte der Zoologie: Experimental Behavioral Biology*, Fisher Verlag, Stuttgart, Germany, pp. 273–288.

Bradbury, J. W., and R. M. Gibson. 1983. Leks and mate choice. In P. Bateson (ed.), *Mate Choice*, Cambridge University Press, Cambridge, U.K., pp. 109–138.

Bradbury, J. W., and S. L. Vehrencamp 1977. Social organization and foraging in emballonurid bats. III. Mating systems. *Behav. Ecol .Sociobiol.* 2:1–17.

Bradbury, J. W., and S. L. Vehrencamp. 2000. Economic models of animal communication. *Anim. Behav.* 59:259–268.

Bradbury, J. W., S. L. Vehrencamp, and R. Gibson. 1985. Leks and the unanimity of female choice. In P. J. Greenwood, P. H. Harvey, and M. Slatkin (eds.), *Evolution: Essays in Honor of John Maynard Smith*, Cambridge University Press, Cambridge, U.K., pp. 301–314.

Bradshaw, A. D. 1965. Evolutionary significance of phenotypic plasticity in plants. *Adv. Genet.* 13:115–155.

Brantley, R. K., J. C. Wingfield, and A. H. Bass. 1993. Sex steroid levels in *Porichthys notatus*, a fish with alternative reproductive tactics, and a review of the hormonal bases for male dimorphism among teleost fishes. *Horm. Behav.* 27:332–347.

Breden, F. 1987. The effect of post-metamorphic dispersal on the population genetic-structure of the Fowler toad, *Bufo woodhousei fowleri*. *Copeia* 2:386–395.

Breden, F. J., and M. J. Wade. 1991. "Runaway" social evolution: Reinforcing selection for inbreeding and altruism. *J. Theor. Biol.* 153:323–337.

Brereton, A. R. 1994. Return-benefit spite hypothesis: An explanation for sexual interference in stumptail macaques (*Macaca arctoides*). *Primates* 35:123–136.

Briske, J. V., R. Montgomerie, and T. R. Birkhead. 1997. The evolution of sperm size in birds. *Evolution* 51:937–945.

Brockmann, H. J. 1996. Satellite male groups in horseshoe crabs, *Limulus polyphemus*. *Ethology* 102:1–21.

Brockmann, H. J., T. Colson, and W. Potts. 1994. Sperm competition in horseshoe crabs (*Limulus polyphemus*). *Behav. Ecol. Sociobiol.* 35:153–160.

Brockmann, H. J., A. Grafen, and R. Dawkins. 1979. Evolutionary stable nesting strategy in a digger wasp. *J. Theor. Biol.* 77:473–496.

Brockmann, H. J., C. Nguyen, and W. Potts. 2000. Paternity in horseshoe crabs when spawning in multiple-male groups. *Anim. Behav.* 60:837–849.

Brockmann, H. J., and D. Penn. 1991. Conditional mating strategies in *Limulus polyphemus*. Abstract, 22nd Int. Ethology Conf., Otani Univ., Kyoto, Japan, p. O10.

Brodsky, L. 1988. Mating tactics of male rock ptarmigan, *Labopus mutus*: A conditional mating strategy. *Anim. Behav.* 36:335–342.

Brooks, R., and D. J. Kemp. 2001. Can older males deliver the good genes? *Trends Ecol. Evol.* 16:308–313.

Brotherton, P.N.M., J. M. Pemberton, P. E. Komers, and G. Malarky. 1997. Genetic and behavioural evidence of monogamy in a mammal, Kirk's dik-dik (*Madoqua kirkii*). *Proc. R. Soc. London, Ser. B* 264:675–681.

Brown, J. L. 1997. A theory of mate choice based on heterozygosity. *Behav. Ecol.* 8:60–65.

Bruce, A. J., and K. Baba. 1973. *Spongiocaris*, a new genus of Stenopodidean shrimp from New Zealand and South African waters, with a description of two new species (Decapoda Nutantia, Stenopodidea). *Crustaceana* 25:153–170.

Bruce, N. L. 1994. The Cassidininae Hansen, 1905 (Crustacea: Isopoda: Sphaeromatidae) of Australia. *J. Nat. Hist.* 28:1077–1173.

Brusca, R. C. 1980. *Common Intertidal Invertebrates of the Gulf of California*, 2nd Ed., University of Arizona Press, AZ.

Brusca, R. C., and G. J. Brusca, 1990. *Invertebrates*, Sinauer Associates, Sunderland, MA.

Bucher, T. L., M. U. Ryan, and G. A. Bartholomew. 1982. Oxygen consumption during resting, calling, and nest building in the frog *Physalaemus pustulosus*. *Physiol. Zool.* 55:10–22.

Bulger, J. B. 1993. Dominance rank and access to estrous females in male savanna baboons. *Behaviour* 127:67–104.

Bull, C. M. 2000. Monogamy in lizards. *Behav. Process.* 51:7–20.

Bull, J. J. 1983. *Evolution of Sex Determining Mechanisms*, Benjamin Cummings, Menlo Park, CA.

Bull, J. J., and E. L. Charnov. 1989. Enigmatic reptilian sex ratios. *Evolution* 43:1561–1566.

Burda, H., R. L. Honeycutt, S. Begall, L. Gruetjen, and O. Scharff. 2000. Are naked and common mole-rats eusocial and if so, why? *Behav. Ecol. Sociobiol.* 47:293–30.

Buresch, K. M., R. T. Hanlon, M. R. Maxwell, and S. Ring. 2001. Microsatellite DNA markers indicate a high frequency of multiple paternity within individual field-collected egg capsules of the squid *Loligo pealeii*. *Mar. Ecol. Prog. Ser.* 210:161–165.

Burley, N. T. 1988. Wild zebra finches have band-color preferences. *Anim. Behav.* 36:1235–1237.

Burley, N. T., and R. Symanski. 1998. "A taste for the beautiful:" Latent aesthetic mate preferences for white crests in two species of Australian finches. *Am. Nat.* 152:792–802.

Butchart, S.H.M. 2000. Population structure and breeding system of the sex-role reversed, polyandrous Bronze-winged Jacana, *Metopidius indicus. Ibis* 142:93–102.

Butchart, S.H.M., N. Seddon, and J.M.M. Ekstrom. 1999. Polyandry and competition for territories in bronzewinged jacanas. *J. Anim. Ecol.* 68:928–939.

Cade, W. 1981. Alternative male strategies: Genetic differences in crickets. *Science* 212:563–564.

Caldwell, R. L. 1986. Assessment strategies in stomatopods. *Bull. Mar. Sci.* 41:135–150.

Campbell, B. G. (ed). 1972. *Sexual Selection and the Descent of Man, 1871–1971*, Aldine, Chicago, IL.

Cannon, R. W., and F. L. Knopf. 1981. Lek numbers as a trend index to prairie grouse populations. *J. Wildlife Manag.* 45:776–778.

Carranza, J., F. Alvarez, and T. Redondo. 1990. Territoriality as a mating strategy in red deer. *Anim. Behav.* 40:79–88.

Cassinello, J., and M. Gomendio. 1996. Adaptive variation in litter size and sex ratio at birth in a sexually dimorphic ungulate. *Proc. R. Soc. London, Ser. B* 263:1461–1466.

Castillo, O. G., and G. O. Brull. 1989. *Ageneiosus magoi*, una nueva especie de bagre Ageneiosido (Teleostei, Siluriformes) para Venezuela y algunas notas sobre su historia natural. *Acta. Biol. Venez.* 12:72–87.

Cezilly, F., M. Preault, F. Dubois, B. Faivre, and B. Patris. 2000. Pair-bonding in birds and the active role of females: A critical review of the empirical evidence. *Behav. Process.* 51:83–92.

Chapman T., and L. Partridge. 1996. Sexual conflict as fuel for evolution. *Nature* 381:189–190.

Charnov, E. L. 1979. Natural selection and sex change in pandalid shrimp: Test of a life history theory. *Am. Nat.* 113:715–734.

Charnov, E. L. 1982. *The Theory of Sex Allocation*, Princeton University Press, Princeton, NJ.

Charnov, E. L. 1989. Evolution of the breeding sex-ratio under partial sex change. *Evolution* 43:1559–1561.

Charnov, E. L. 1997. Trade-off-invariant rules for evolutionarily stable life histories. *Nature* 387:393–394.

Charnov, E. L., and D. Berrigan. 1991. Dimensionless numbers and the assembly rules for life histories. *Philos. Trans. R. Soc. London, Ser. B* 332:41–48.

Charnov, E. L., and J. J. Bull. 1989. The primary sex ratio under environmental sex determination. *J. Theor. Biol.* 139:431–436.

Chernitskij, A. G. 1986. Ehkologo-fiziologicheskie prichiny polimorfizma proizvoditelej atlanticheskogo lososya *Salmo salar* L. (Salmonidae). *Vopr. Ikhtiol. Moscow* 26:524–526.

Cheverud, J. M., and A. J. Moore. 1994. Quantitative genetics and the role of the environment provided by relatives in the evolution of behavior. In C.R.B. Boake, (ed.), *Quantitative Genetic Studies of Behavioral Evolution*, University of Chicago Press, Chicago, pp. 67–100.

Choe, J. C., and B. J. Crespi (eds). 1997. *The Evolution of Mating Systems in Insects and Arachnids*, Cambridge University Press, Cambridge, MA.

Christiansen, T. 1984. Alternative reproductive tactics in spiders. *Am. Zool.* 24:321–332.

Christy, J. H. 1983. Female choice in the resource defense mating system of the sand fiddler crab, *Uca pugilator. Behav. Ecol. Sociobiol.* 12:160–180.

Christy, J. H. 1995. Mimicry, mate choice, and the sensory trap hypothesis. *Am. Nat.* 146:171–181.

Civetta, A., and A. G. Clark. 2000. Correlated effects of sperm competition and postmating female mortality (trade-offs between sperm competitive ability and female longevity). *Proc. Natl. Acad. Sci. USA* 97:13162–13165.

Clark, A. G., M. Aguade, T. Prout, L. G. Harshman, and C. H. Langley. 1995. Variation in sperm displacement and its association with accessory-gland protein loci in *Drosophila melanogaster*. *Genetics* 139:189–201.

Clark, A. G., and D. J. Begun. 1998. Female genotypes affect sperm displacement in *Drosophila*, *Genetics* 149:1487–1493.

Clark, A. G., D. J. Begun, and T. Prout. 1999. Female × male interactions in *Drosophila* sperm competition. *Science* 283:217–220.

Clark, D. L., and G. W. Uetz. 1992. Morph-independent mate selection in a dimorphic jumping spider: Demonstration of movement bias in female choice using video-controlled courtship behaviour. *Anim. Behav*. 43:247–254.

Clark, D. L., and G. W. Uetz. 1993. Signal efficacy and the evolution of male dimorphism in the jumping spider, *Maevia inclemens*. *Proc. Natl. Acad. Sci. USA* 90:11954–11957.

Clark, R. A. 1997. Dimorphic males display alternative reproductive strategies in the marine amphipod *Jassa marmorata* Holmes (Corophioidea: Ischyroceridae). *Ethology* 103:531–553.

Clutton-Brock, T. H. 1982. The functions of antlers. *Behaviour* 79:108–125.

Clutton-Brock, T. H. (ed). 1988. *Reproductive Success: Studies of Individual Variation in Contrasting Breeding Systems*, University of Chicago Press, Chicago.

Clutton-Brock, T. H. 1989. Mammalian mating systems. *Proc. R. Soc. London, Ser. B* 236:339–372.

Clutton-Brock, T. H. 1991a. *The Evolution of Parental Care*, Princeton University Press, Princeton, NJ.

Clutton-Brock, T. H. 1991b. Lifetime data and the measurement of selection. *Evolution* 45:454.

Clutton-Brock, T. H., S. D. Albon, and F. E. Guinness. 1984. Maternal dominance, breeding success and birth sex ratios in red deer. *Nature* 308:358–360.

Clutton-Brock, T. H., F. E. Guiness, and S. O. Albon. 1982. *Red Deer: The Behavior and Ecology of Two Sexes*, University of Chicago Press, Chicago, IL.

Clutton-Brock, T. H., and G. A. Parker. 1992. Potential reproductive rates and the operation of sexual selection. *Q. Rev. Biol.* 67:437–456.

Clutton-Brock, T. H., and G. A. Parker. 1995. Sexual coercion in animal societies. *Anim. Behav*. 49:1345–1365.

Clutton-Brock, T. H., O. F. Price, S. D. Albon, and P. A. Jewell. 1992. Early development and population fluctuations in Soay sheep. *J. Anim. Ecol.* 61:381–396.

Clutton-Brock, T. H., and A.C.J. Vincent. 1991. Sexual selection and the potential reproductive rates of males and females. *Nature* 351:58–60.

Cochran, P. A., A. P. Kinziger, and W. J. Poly. 1999. Predation on horsehair worms (phylum Nematomorpha). *J. Freshw. Ecol.* 14:211–218.

Cohen, J. A., and H. J. Brockmann. 1983. Breeding activity and mate selection in the horseshoe crab, *Limulus polyphemus*. *Bull. Mar. Sci.* 33:274–281.

Coltman, D. W., R. D. Bancroft, A. Robertson, A. J. Smith, T. H. Clutton-Brock, and M. J. Pemberton. 1999a. Male reproductive success in a promiscuous mammal: Behavioural estimates compared with genetic paternity. *Mol. Ecol.* 8:1199–1120.

Coltman, D. W., J. A. Smith, D. R. Bancroft, J. Pilkington, A.D.C. MacColl, T. H. Clutton-Brock, and J. M. Pemberton. 1999b. Density-dependent variation in lifetime breeding success and natural and sexual selection in soay rams. *Am. Nat.* 154:730–746.

Colwell, R. K. 1981. Group selection is implicated in the evolution of female-biased sex ratios. *Nature* 290:401–404.

Comes, H. B., and H. J. Abbott. 1998. The relative importance of historical events and gene flow on the population structure of a Mediterranean ragwort, *Senecio gallicus* (Asteraceae). *Evolution* 52:355–367.

Conner, J. 1988. Field measurements of natural and sexual selection in the fungus beetle, *Bolitotherus cornutus*. *Evolution* 42:735–749.

Conover, D. O., and D. A. van Voorhees. 1990. Evolution of a balanced sex ratio by frequency-dependent selection in a fish. *Science* 250:1556–1557.

Cooch, E., D. Lank, R. Robertson, and F. Cooke. 1997. Effects of parental age and environmental change on offspring sex ratio in a precocial bird. *J. Anim. Ecol.* 66:189–202.

Cook, J. M., S. G. Compton, E. A. Herre, and S. A. West. 1997. Alternative mating tactics and extreme male dimorphism in fig wasps. *Proc. R. Soc. London, Ser. B* 264:747–751.

Cooper, S.J.B., C. M. Bull, and M. G. Gardner. 1997. Characterization of microsatellite loci from the socially monogamous lizard *Tiliqua rugosa* using a PCR-based isolation technique. *Mol. Ecol.* 6:793–795.

Cowan, D. P. 1984. Life history and male dimorphism in the mite *Kennethiella trisetosa* (Acarina: Winterschmidtiidae), and its symbiotic relationship with the wasp *Ancistrocerus antilope* (Hymenoptera: Eumenidae). *Ann. Entomol. Soc. Am.* 77: 725–732.

Crespi, B. J. 1988. Alternative male mating tactics in a thrips: Effects of sex ratio variation and body size. *Am. Midl. Nat.* 119:83–92.

Crow, J. F. 1958. Some possibilities for measuring selection intensities in man. *Human Biol.* 30:1–13.

Crow, J. F. 1962. Population genetics: Selection. In W. J. Burdette (ed.), *Methodology in Human Genetics*, Holden-Day, San Francisco, pp. 53–75.

Crow, J. F. 1986. *Basic Concepts in Population, Quantitative and Evolutionary Genetics*, W. H. Freeman, New York.

Curtsinger, J. W. 1991. Sperm competition and the evolution of multiple mating. *Am. Nat.* 138:93–102.

Dacks, J., and A. J. Roger. 1999. The first sexual lineage and the relevance of facultative sex. *J. Mol. Evol.* 48:779–783.

Dahan, M., and Y. Benayahu. 1997. Reproduction of *Dendronephthya hemprichi* (Cnidaria: Octocorallia) year round spawning in a azoxanthellate soft coral. *Mar. Biol.* 129:573–579.

Daly, M., and M. Wilson 1992. The man who mistook his wife for a chattel. In J.

Barkow and L. Cosmides (eds.), *The Adapted Mind*, Oxford University Press, Oxford, U.K.

Danforth, B. N., and C. A. Desjardins. 1999. Male dimorphism in *Perdita portalis* (Hymenoptera, Andrenidae) has arisen from preexisting allometric patterns. *Insectes Soc.* 46:18–28.

Danforth, B. N., and J. L. Neff. 1992. Male polymorphism and polyethism in *Perdita texana* (Hymenoptera:Andrenidae). *Ann. Entomol. Soc. Am.* 85:616–626.

Darling, F. 1937. *A Herd of Red Deer.* Doubleday, New York.

Darwin, C. R. 1859. *On the Origin of Species*, facsimile of the first edition 1964, Harvard University Press, Cambridge, MA.

Darwin, C. 1868. *The Variation of Animals and Plants Under Domestication*, J. Murray, London.

Darwin, C. R. 1871. *The Descent of Man and Selection in Relation to Sex*, Appleton, New York.

Darwin, C. R. 1874. *The Descent of Man and Selection in Relation to Sex*, 2nd Ed., Rand, McNally and Co., New York.

Darwin, C. 1882. A preliminary notice: "On the modification of a race of Syrian street dogs by means of sexual selection." In P. H. Barrett (ed.), *The Collected Papers of Charles Darwin*, University of Chicago Press, Chicago, IL, Vol. 2, pp. 278–280.

Davies, N. B. 1983. Polyandry, cloaca-pecking and sperm competition in dunnocks. *Nature* 302:334–336.

Davies, N. B. 1985. Cooperation and conflict among dunnocks, *Prunella modularis*, in a variable mating system. *Anim. Behav.* 33:628–648.

Davies, N. B. 1989. Sexual conflict and the polygamy threshold. *Anim. Behav.* 38:226–234.

Davies, N. B. 1991. Mating systems. In J. R. Krebs and N. B. Davies (eds.), *Behavioral Ecology: An Evolutionary Approach*, 3rd Ed., Blackwell Scientific Publications, Oxford, U.K.

Davies, N. B., and T. R. Halliday. 1979. Competitive mate searching in male common toads, *Bufo bufo*. *Anim. Behav.* 27:1253–1267.

Davies, N. B., and I. R. Hartley. 1996. Food patchiness, territory overlap and social systems: An experiment with dunnocks *Prunella modularis*. *J. Anim. Ecol.* 65: 837–846.

Dawkins, R. 1980. Good strategy or evolutionary stable strategy? In G. W. Barlow and J. Silverberg (eds.), *Sociobiology: Beyond Nature/Nurture?*, Westview, Boulder, CO, pp. 331–367.

Dawkins, R., and T. R. Carlisle. 1976. Parental investment, mate desertion, and a fallacy. *Nature* 262:131–133.

Day, M. C. 1984. Male polymorphism in some Old World species of *Cryptocheilus* Panzer (Hymenoptera: Pompilidae). *Zool. J. Linn. Soc.* 80:83–101.

De Jonge, J., and J. J. Videler. 1989. Differences between the reproductive biologies of *Tripterygion tripteronotus* and *T. delaisi* (Pisces, Perciformes, Tripterygiidae): The adaptive significance of an alternative mating strategy and a red instead of a yellow nuptial colour. *Mar. Biol.* 100:431–437.

Dearborn, D. C. 1998. Interspecific territoriality by a rufous-tailed hummingbird (*Amazilia tzacatl*): Effects of intruder size and resource value. *Biotropica* 30:306–313.

Delehanty, D. J., R. C. Fleischer, M. A. Colwell, and L. W. Oring. 1998. Sex-role reversal and the absence of extra-pair fertilization in Wilson's phalaropes. *Anim. Behav.* 55:995–1002.

Delph, L. F., L. F. Galloway, and M. L. Stanton. 1996. Sexual dimorphism in flower size. *Am. Nat.* 148:299–320.

Delph, L. F., M. H. Johannsson, and A. G. Stephenson. 1997. How environmental factors affect pollen performance:ecological and evolutionary perspectives. *Ecology* 78:1632–1639.

Delph, L. F., C. Weinig, and K. Sullivan. 1998. Why fast-growing pollen tubes give rise to vigorous progeny: The test of a new mechanism. *Proc. R. Soc. London, Ser. B* 265:935–939.

Dempster, E. R., and I. M. Lerner. 1950. Heritability of threshold characters. *Genetics* 35:212–236.

Demski, L. S. 1987. Diversity in reproductive patterns and behavior in teleost fishes. In D. Crews (ed.), *Psychobiology of Reproductive Behavior: An Evolutionary Perspective*, Prentice-Hall, Englewood Cliffs, NJ, pp. 1–27.

Denno, R. F. 1995. The evolution of dispersal polymorphism in insects:the influence of habitats, host plants and mates. *Res. Popul. Ecol.* 36:127–135.

Denno, R. F., and Roderick, G. K. 1992. Density-related dispersal in planthoppers: Effects of interspecific crowding. *Ecology* 73:1323–1334.

Desrosiers, G. A. Caron, M. Olivier, and G. Miron. 1994. Life history of the polychaee *Nereis virens* (Sars) in an intertidal flat of the Lower St. Laurence Estuary. *Oceanol. Acta* 17:683–695.

Deutsch, J. C. 1994. Lekking by default: Female habitat preferences and male strategies in Uganda kob. *J. Anim. Ecol.* 63:101–115.

Diesel, R. 1989. Structure and function of the reproductive system of the symbiotic spider crab *Inachus phalangium* (Decapoda:Majidae): Observations on sperm transfer, sperm storage, and spawning. *J. Crust. Biol.* 9:266–277.

Dingle, H. 1991. Evolutionary genetics of animal migration. *Am. Zool.* 31:253–264.

Dinsmore, D. 1985. Demographic structure and opportunity for selection of Halfmoon Township, Pennsylvania: 1850–1900. *Human Biol.* 57:335–352.

Ditchkoff, S. S., R. L. Lochmiller, R. E. Masters, S. R. Hoofer, and R. A. Van Den Bussche. 2001. Major-histocompatibility-complex-associated variation in secondary sexual traits of white-tailed deer (*Odocoileus virginianus*): Evidence for good-genes advertisement. *Evolution* 55:616–625.

Dominey, W. 1981. Maintenance of female mimicry as a reproductive strategy in Bluegill Sunfish (*Lepomis macrochirus*). *Environ. Biol. Fish.* 6:59–64.

Dominey, W. J. 1980. Female mimicry in male bluegill sunfish—a genetic polymorphism? *Nature* 294:546–548.

Dominey, W. J. 1984. Alternative mating tactics and evolutionarily stable strategies. *Am. Zool.* 24:385–396.

Doty, G. V., and A. M. Welch. 2001. Advertisement call duration indicates good genes for offspring feeding rate in gray tree frogs (*Hyla versicolor*). *Behav. Ecol. Sociobiol.* 49:150–156.

Downes, J. A. 1970. The feeding and mating behaviour of the specialized Empidinae (Diptera): Observations on four species of *Rhamphomyia* in the arctic and a general discussion. *Can. Entomol.* 102:769–791.

Downhower, J. F., and K. B. Armitage. 1971. The yellow-bellied marmot and the evolution of polygamy. *Am. Nat.* 105:355–370.

Downhower, J. F., and L. Brown, L. 1979. Seasonal changes in the social structure of a mottled sculpin (*Cottus bairdi*) population. *Anim. Behav.* 27:451–458.

Downhower, J. F., L. S. Blumer, and L. Brown. 1987. Opportunity for selection: An appropriate measure for evaluating variation in the potential for selection? *Evolution* 41:1395–1400.

Drickamer, L. C., P. A. Gowaty, and C. M. Holmes. 2000. Free female mate choice in house mice affects reproductive success and offspring viability and performance. *Anim. Behav.* 59:371–378.

Drickamer, L. C., S. H. Vessey, and E. M. Jacob. 2002. *Animal Behavior: Mechanisms, Ecology, Evolution*, 5th Ed., McGraw-Hill, Boston.

Drnevich, J. M. 2002. Assessing postcopulatory sexual selection in simple versus complex competitive environments. *Anim. Behav.*, in review.

Droney, D. C. 1992. Sexual selection in a lekking Hawaiian *Drosophila*: The roles of male competition and female choice in male mating success. *Anim. Behav.* 44: 1007–1020.

Dugatkin, L. A. 1996. Interface between culturally based preferences and genetic preferences: Female mate choice in *Poecilia reticulata*. *Proc. Natl. Acad. Sci. USA* 93:2770–2773.

Dugatkin, L. A. 1998. Genes, copying, and female mate choice: Shifting thresholds. *Behav. Ecol.* 9:323–327.

Dugatkin, L. A., and J.-G. J. Godin. 1993. Female mate copying in the guppy (*Poecilia reticulata*): Age-dependent effects. *Behav. Ecol.* 4:289–292.

Dugatkin, L. A., and M. Kirkpatrick. 1994. Sexual selection and the evolutionary effects of copying mate choice. *Behav. Ecol. Sociobiol.* 34:443–449.

Dugatkin, L. A., and H. K. Reeve (eds). 1998. *Game Theory and Animal Behavior*, Oxford University Press, Oxford, U.K.

Duncan, T. K. 1983. Sexual dimorphism and reproductive behavior in *Almyracuma proximoculi* (Crustacea:Cumacea): The effect of habitat. *Biol. Bull.* 165:370–378.

Dunn, D. W., C. S. Crean, C. L. Wilson, and A. S. Gilburn. 1999. Male choice, willingness to mate and body size in seaweed flies (Diptera: Coelopidae). *Anim. Behav.* 57:847–853.

Dunn, P. O., and A. Cockburn. 1999. Extrapair mate choice and honest signaling in cooperatively breeding superb fairy-wrens. *Evolution* 53:938–946.

Dunn, P. O., A. D. Afton, M. L. Gloutney, and R. T. Alisauskas. 1999. Forced copulation results in few extrapair fertilizations in Ross's and lesser snow geese. *Anim. Behav.* 57:1071–1081.

Dunn, P. O., L. A. Whittingham, and T. E. Pitcher. 2001. Mating systems, sperm competition, and the evolution of sexual dimorphism in birds. *Evolution* 55:161–175.

Duvall, D., M. B. King, and K. J. Gutzwiller. 1985. Behavioral ecology and ethology of the prairie rattlesnake. *Natl. Geogr. Res.* 1985:80–111.

Duvall, D., G. W. Schuett, and S. J. Arnold. 1995. Ecology and evolution of snake mating systems. In R. A. Siegel and J. T. Collins (eds.), *Snakes: Ecology and Behavior*, McGraw-Hill, New York.

Dybas, H. S. 1978. Polymorphism in featherwing beetles, with a revision of the genus *Ptinellodes* (Coleoptera: Ptiliidae). *Ann. Entomol. Soc. Am.* 71:695–714.

Eberhard, W. G. 1979. The function of horns in *Podichnus agenor* (Dynastinae) and other beetles. In M. S. Blum and N. A. Blum (eds.), *Sexual Selection and Reproductive Competition in Insects*, Academic Press, New York, pp. 231–258.

Eberhard, W. G. 1982. Beetle horn dimorphism: Making the best of a bad lot. *Am. Nat.* 119:420–426.

Eberhard, W. G. 1996. *Female Control: Sexual Selection by Cryptic Female Choice*, Princeton University Press, Princeton, NJ.

Eberhard, W. G., and E. E. Gutierrez. 1991. Male dimorphisms in beetles and earwigs and the question of developmental constraints. *Evolution* 45:18–28.

Eckert, C. G., and S.C.H. Barrett. 1992. Stochastic loss of style morphs from populations of tristylous *Lythum salicaria* and *Decodon vericillatus* (Lythraceae). *Evolution* 46:1014–1029.

Edwards, A.W.F. 1998. Natural selection and the sex ratio: Fisher's sources. *Am. Nat.* 151:564–569.

Eens, M., and R. Pinxten. 1995. Inter-sexual conflicts over copulations in the European starling: Evidence for the female mate guarding hypothesis. *Behav. Ecol. Sociobiol.* 36:71–81.

Eens, M., and R. Pinxten. 2000. Sex role reversal in vertebrates: Behavioural and endocrinological accounts. *Behav. Process.* 51:135–147.

Eggert, A. K., and S. K. Sakaluk. 1995. Female-coerced monogamy in burying beetles. *Behav. Ecol. Sociobiol.* 37:147–153.

Ellegren, H., L. Gustafsson, and B. C. Sheldon. 1996. Sex ratio adjustment in relation to paternal attractiveness in a wild bird population. *Proc. Natl. Acad. Sci. USA* 93:11723–11728.

Embry, S. J., and S. M. Shuster. 2001. The frequency of multiple insemination in a natural population of marine isopods. *J. Ariz. Nevad. Acad. Sci.* 36:8.

Emerson, S. B. 2000. Vertebrate secondary sexual characteristics—Physiological mechanisms and evolutionary patterns. *Am. Nat.* 156:84–91.

Emlen, D. J. 1996. Artificial selection on horn length–body size allometry in the horned beetle, *Onthophagus acuminatus* (Coleoptera: Scarabaeidae). *Evolution* 50:1219–1230.

Emlen, D. J. 1997. Alternative reproductive tactics and male-dimorphism in the horned beetle *Onthophagus acuminatus* (Coleoptera: Scarabaeidae). *Behav. Ecol. Sociobiol.* 41:335–341.

Emlen, D. J. 2000. Integrating development with evolution: A case study with beetle horns. *Bioscience* 50:403–418.

Emlen, D. J., and H. F. Nijhout. 1999. Hormonal control of male horn length dimorphism in the dung beetle *Onthophagus taurus* (Coleoptera: Scarabaeidae). *J. Insect Physiol.* 45:45–53.

Emlen, S. T., and L. W. Oring. 1977. Ecology, sexual selection, and the evolution of mating systems. *Science* 197:215–223.

Endler, J. 1986. *Natural Selection in the Wild*, Princeton University Press, Princeton, NJ.

Endler, J. A., and A. L. Basolo. 1998. Sensory ecology, receiver biases and sexual selection. *Trends Ecol. Evol.* 13:415–420.

Endler, J. A., and M. Théry. 1996. Interacting effects of lek placement, display behavior, ambient light, and color patterns in three neotropical forest-dwelling birds. *Am. Nat.* 148:421–452.

Erbelding-Denk, C., J. H. Schroeder, M. Schartl, I. Nanda, M. Schmid, and J. T. Epplen. 1994. Male polymorphism in *Limia perugiae* (Pisces: Poeciliidae). *Behav. Genet.* 24:95–101.

Evans, M. R., and B. J. Hatchwell. 1992. An experimental study of male adornment in the scarlet-tufted malachite sunbird: I. The role of pectoral tufts in territorial defense. *Behav. Ecol. Sociobiol.* 29:413–420.

Evans, M. R., and A.L.R. Thomas. 1997. Testing the functional significance of tail streamers. *Proc. R. Soc. London, Ser. B* 264:211–217.

Fairbairn, D. J. 1993. The costs of loading associated with mate-carrying in the water strider, *Aquarius remigis*. *Behav. Ecol.* 4:224–231.

Fairbairn, D. J. 1994. Wing dimorphism and the migratory syndrome:correlated traits for migratory tendency in wing dimorphic insects. *Res. Popul. Ecol.* 36:157–163.

Fairbairn, D. J., and Roff, D. A. 1990. Genetic correlations among traits determining migratory tendency in the sand cricket, *Gryllus firmus*. *Evolution* 44:1787–1795.

Fairbairn, D. J., and A. E. Wilby. 2001. Inequality of opportunity: measuring the potential for sexual selection. *Evol. Ecol. Res.* 3:667–686.

Falconer, D. S. 1989. *Introduction to Quantitative Genetics*, 2nd Ed., Longman Scientific and Technical, New York.

Falconer, D. S., and T.F.C. Mackay. 1996. *Introduction to Quantitative Genetics*, 4th Ed., Longman Scientific Publications/J. Wiley and Sons, New York.

Ferguson, I. M., and D. J. Fairbairn. 2001. Is selection ready when opportunity knocks? *Evol. Ecol. Res.* 3:199–207.

Field, S. A., and M. A. Keller. 1993. Alternative mating tactics and female mimicry as post-copulatory mate-guarding behaviour in the parasitic wasp, *Cotesia rubecula*. *Anim. Behav.* 46:1183–1189.

Finke, O. M. 1986. Lifetime reproductive success and the opportunity for selection in a nonterritorial damselfly (Odonata: Coenagrionidae). *Evolution* 40:791–803.

Fischer, K., and K. Fiedler. 2000. Sex-related differences in reaction norms in the butterfly *Lycaena tityrus* (Lepidoptera: Lycaenidae). *Oikos* 90:372–380.

Fischer, K., and K. Fiedler. 2001. Dimorphic growth patterns and sex-specific reaction norms in the butterfly *Lycaena hippothoe sumadiensis*. *J. Evol. Biol.* 14:210–218.

Fisher, R. A. 1928. The possible modification of the response of the wild type to recurrent mutations. *Am. Nat.* 62:115–126.

Fisher, R. A. 1930. *The Genetical Theory of Natural Selection*, Clarendon Press, Oxford, U.K.

Fisher, R. A. 1958. *The Genetical Theory of Natural Selection*, 2nd Ed., Dover Press, New York.

Fitzsimmons, N. N. 1998. Single paternity of clutches and sperm storage in the promiscuous green turtle (*Chelonia mydas*). *Mol. Ecol.* 7:575–584.

Flaxman, S. M. 2000. The evolutionary stability of mixed strategies. *Trends Ecol. Evol.* 15:482.

Fleming, I. A., and M. R. Gross. 1994. Breeding competition in a Pacific salmon (Coho: *Oncorhynchus kisutch*): Measures of natural and sexual selection. *Evolution* 48:637–657.

Foote, C. J., G. S. Brown, and C. C. Wood. 1997. Spawning success of males using alternative mating tactics in sockeye salmon, *Oncorhynchus nerka*. *Can. J. Fish. Aquat. Sci.* 54:1785–1795.

Forsyth, A., and J. Alcock. 1990. Female mimicry and resource defense polygyny by males in a tropical rove beetle, *Leistotrophus versicolor* (Coleoptera: Staphylinidae). *Behav. Ecol. Sociobiol.* 26:325–330.

Franck, P., N. Koeniger, G. Lahner, R. M. Crewe, and M. Solignac. 2000. Evolution of extreme polyandry: An estimate of mating frequency in two African honeybee subspecies, *Apis mellifera monticola* and *A. m. scutellata. Insectes Soc.* 47:364–370.

Frank, S. A., and M. Slatkin. 1990. Evolution in a variable environment. *Am. Nat.* 136:244–260.

Freedberg, S., and M. J. Wade, 2001. Cultural inheritance as a mechanism for population sex ratio bias. *Evolution* 55:1049–1055.

Freedberg, S., and M. J. Wade. 2002. Sexual selection results in female-biased sex ratios under environmental sex determination. *Evolution*, in press.

Fu, P., B. D. Neff, and M. R. Gross. 2001. Tactic-specific success in sperm competition. *Proc. R. Soc. London, Ser. B* 268:1105–1112.

Fukuyama, K. 1991. Spawning behavior and male mating tactics of a foam-nesting treefrog, *Rhacophorus schlegelii. Anim. Behav.* 42:193–200.

Funch, P., and R. M. Kristensen. 1995. Cycliophora is a new phylum with affinities to Entoprocta and Ectoprocta. *Nature* 378:711–714.

Futuyma, D. J. 1998. *Evolutionary Biology*, 3rd Ed., Sinauer, Sunderland, MA.

Gadgil, M. 1972. Male dimorphism as a consequence of sexual selection. *Am. Nat.* 106:574–58.

Geist, V. 1966. The evolution of horn-like organs. *Behaviour* 27:175–214.

Geist, V., and M. Bayer. 1988. Sexual dimorphism in the Cervidae and its relation to habitat. *J. Zool.* 214:45–53.

Gemmell, J. N., M. T. Burg, L. I. Boyd, and W. Amos. 2001. Low reproductive success in territorial male Antarctic fur seals (*Arctocephalus gazella*) suggests the existence of alternative mating strategies. *Mol. Ecol.* 10:451–460.

Ghani, N., and N. M. Tirmizi. 1993. Occurrence of *Bopyrella saronae* Bourdon and Bruce, 1979 (Isopoda, Epicaridea) in Karachi waters, northern Arabian Sea. *Crustaceana* 65:117–120.

Ghiselin, M. 1969. *The Triumph of the Darwinian Method*, University of California Press, Berkeley, CA.

Gilliard, E. T. 1962. On the breeding behavior of the cock-of-the-rock (Aves, *Rupicola rupicola*). *Bull. Am. Mus. Nat. Hist.* 124:31–68.

Godfray, H.C.J., and J. H. Werren. 1996. Recent developments in sex ratio studies. *Trends Ecol. Evol.* 11:59–63.

Godin, J.-G. J., 1995. Predation risk and alternative mating tactics in male Trinidadian guppies (*Poecilia reticulata*). *Oecologia* 103:224–229.

Goldizen, A. W., J. C. Buchan, D. A. Putland, A. R. Goldizen, and E. A. Krebs. 2000. Patterns of mate-sharing in a population of Tasmanian Native Hens, *Gallinula mortierii. Ibis* 142:40–47.

Goldschmidt, T., S. A. Foster, and P. Sevenster. 1992. Inter-nest distance and sneaking in the three-spined stickleback. *Anim. Behav.* 44:793–795.

Goldsmith, S. K. 1985. Male dimorphism in *Dendrobias mandibularis* Audinet-Serville (Coleoptera: Cerambycidae). *J. Kans. Entomol. Soc.* 58:534–538.

Gomendio, M., and E.R.S. Roldan. 1993. Mechanisms of sperm competition: Linking physiology and behavioral ecology. *Trends Ecol. Evol.* 8:95–100.

Gomulkiewicz, R., J. N. Thompson, R. D. Holt, S. L. Nuismer, and M. E. Hochberg. 2000. Hot spots, cold spots, and the geographic mosaic theory of coevolution. *Am. Nat.* 156:156–174.

Goodnight, C. J. 1991. Intermixing ability in 2-species communities of Tribolium flour beetles. *Am. Nat.* 138:342–354.

Goodnight, C. J. 2000. Heritability at the ecosystem level. *Proc. Natl. Acad. Sci. USA* 97:9365–9366.

Goodnight, C. J., and D. M. Craig. 1996. Effect of coexistence on competitive outcome in *Tribolium castaneum* and *Tribolium confusum*. *Evolution* 50:1241–1250.

Goodson, J. L., and A. H. Bass. 2000. Forebrain peptides modulate sexually polymorphic vocal circuitry. *Nature* 403:769.

Goodwillie, C. 1999. Wind pollination and reproductive assurance in *Linanthus parviflorus* (Polemoniaceae), a self-incompatible annual. *Am. J. Bot.* 86:948–954.

Gosling, L. M. 1986. The evolution of mating strategies in male antelopes. In D. I. Rubenstein and R. W. Wrangham (eds.), *Ecological Aspects of Social Evolution*, Princeton University Press, Princeton, NJ, pp. 244–281.

Gosling, L. M. 1991. The alternative mating strategies of male topi, *Damaliscus lunatus*. *Appl. Anim. Behav. Sci.* 29:107–119.

Gotto, R. V. 1990. Some aspects of host- and mate-finding in the symbiotic copepods of invertebrates. Proceedings of the 4th International Conference on Copepoda, Karuizawa, Japan, 16–20 September 1990. *Bull. Plankt. Soc. Jpn.* 1990:19–23.

Gould, S. J., and R. C. Lewontin. 1979. The spandrels of San Marco and the Panglossian paradigm: A critique of the adaptationist programme. *Proc. R. Soc. London, Ser. B* 205:581–598.

Gowaty, P. A. 1985. Multiple parentage and apparent monogamy in birds. In P. A. Gowaty and D. W. Mock (eds.), *Avian Monogamy*, American Ornithologists' Union, Washington, DC.

Gowaty, P. A. 1997. Sexual dialectics, sexual selection and variation in reproductive behavior. In P. A. Gowaty (ed.), *Feminism and Evolutionary Biology: Boundaries, Intersections and Frontiers*, Chapman and Hall, New York.

Gowaty, P. A., and N. Buschhaus. 1998. Ultimate causation of aggressive and forced copulation in birds: Female resistance, the CODE hypothesis, and social monogamy. *Am. Zool.* 38:207–225.

Grafe, T. U. 1999. A function of synchronous chorusing and a novel female preference shift in an anuran. *Proc. R. Soc. London, Ser. B* 266:2331–2336.

Grafen, A. 1987. Measuring sexual selection: Why bother? In J. W. Bradbury and M. B. Andersson (eds.), *Sexual Selection: Testing the Alternatives*, John Wiley and Sons, New York, pp. 221–233.

Grafen, A. 1988. On the used of data on lifetime reproductive success. In T. H. Clutton-Brock (ed.), *Reproductive Success*, University of Chicago Press, Chicago, pp. 454–471.

Grafen, A. 1990. Biological signals as handicaps. *J. Theor. Biol.* 144:517–546.

Grafen, A. 1992. Behavioral ecology—of mice and the MHC. *Nature* 360:530.

Grant, J.W.A., M. J. Bryant, and C. E. Soos. 1995. Operational sex ratio, mediated by synchrony of female arrival, alters the variance of male mating success in Japanese medaka. *Anim. Behav.* 49:367–375.

Grant, V. 1995. Sexual selection in plants—Pros and cons. *Proc. Natl. Acad. Sci. USA* 92:1247–1250.

Greeff, J. M. 1996. Alternative mating strategies, partial sibmating and split sex ratios in haplodiploid species. *J. Evol. Biol.* 9:855–869.

Greeff, J. M. 1998. Local mate competition, sperm usage and alternative mating strategies. *Evol. Ecol.* 12:627–628.

Greeff, J. M., and J. W. Ferguson. 1999. Mating ecology of the nonpollinating fig wasps of *Ficus ingens*. *Anim. Behav.* 57:215–222.

Green, D. M., C. W. Zeyl, and T. F. Sharbel. 1993. The evolution of hypervariable sex and supernumerary (B) chromosomes in the relict New Zealand frog, *Leiopelma hochstetteri*. *J. Evol. Biol.* 6:417–441.

Greene, E. 1999. Phenotypic variation in larval development and evolution: Polymorphism, polyphenism, and developmental reaction norms. In M. Wake and B. Hall (eds.), *The Origin and Evolution of Larval Forms*, Academic Press, New York, pp. 379–410.

Greenfield, M. D., and T. E. Shelly. 1985. Alternative mating strategies in a desert grasshopper: Evidence of density-dependence. *Anim. Behav.* 33:1192–1210.

Gren, G. A. 1984. Alternative mating strategies in the Mongolian gerbil. *Behaviour* 91:229–244.

Grether, G. F. 1996. Sexual selection and survival selection on wing coloration and body size in the rubyspot damselfly, *Hetaerina americana*. *Evolution* 50:1939–1948.

Gross, M. R. 1982. Sneakers, satellites and parentals: Polymorphic mating strategies in North American sunfishes. *Z. Tierpsychol.* 60:1–16.

Gross, M. R. 1985. Disruptive selection for alternative life histories in salmon. *Nature* 313:47–48.

Gross, M. R. 1991a. Salmon breeding-behavior and life-history evolution in changing environments. *Ecology* 72:1180–1186.

Gross, M. R. 1991b. Evolution of alternative reproductive strategies: Frequency dependent sexual selection in male bluegill sunfish. *Philos. Trans. R. Soc. London, Ser. B* 332:59–66.

Gross, M. R. 1996. Alternative reproductive strategies and tactics: Diversity within sexes. *Trends Ecol. Evol.* 11:92–97.

Gross, M. R., and E. L. Charnov. 1980. Alternative male life histories in bluegill sunfish. *Proc. Natl. Acad. Sci. USA* 77:6937– 6948.

Gross, M. R., and J. Repka. 1995. The evolutionarily stable strategy under individual condition and tactic frequency. *J. Theor. Biol.* 176:27–31.

Gross, M. R., and J. Repka. 1998. Game theory and inheritance of the conditional strategy. In L. A. Dugatkin and H. K. Reeve (eds.), *Game Theory and Animal Behavior*, Oxford University Press, Oxford, U.K.

Gross, M. R., and R. Shine, 1981. Parental care and mode of fertilization in ectothermic vertebrates. *Evolution* 35:775–793.

Guewuch, W. T., and R. A. Croker. 1973. Microfauna of northern New England marine sand: I. The biology of *Mancocuma stellifera* Zimmer, 1943 (Crustacea: Cumacea). *Can. J. Zool.* 51:1011–1020.

Gutterman, Y. 1994. Strategies of seed dispersal and germination in plants inhabiting deserts. *Bot. Rev.* 60:373–425.

Gwynne, D. T. 1984. Male mating effort, confidence of paternity and insect sperm competition. In R. L. Smith (ed.), *Sperm Competition and the Evolution of Animal Mating Systems*, Academic Press, San Diego, CA.

Gwynne, D. T. 1997. The evolution of edible "sperm sacs" and other forms of courtship feeding in crickets, katydids and their kin (Orthoptera: Ensifera). In J. C. Choe and B. J. Crespi (eds.), *The Evolution of Mating Systems in Insects and Arachnids*, Cambridge University Press, Cambridge, MA.

Haddad, C.F.B. 1991. Satellite behavior in the neotropical treefrog *Hyla minuta*. *J. Herpetol.* 25:226–229.

Hall, D. W., M. Kirkpatrick, and B. West. 2000. Run-away sexual selection when female preferences are directly selected. *Evolution* 54:1862–1869.

Halliday, T., and S. J. Arnold. 1987. Multiple mating by females—a perspective from quantitative genetics. *Anim. Behav.* 35:939–941.

Hamilton, W. D. 1979. Wingless and fighting males in fig wasps and other insects. In M. S. Blum and N. A. Blum (eds.), *Sexual Selection and Reproductive Competition in Insects*, Academic Press, New York, pp. 167–222.

Hamilton, W. D., and M. Zuk. 1982. Heritable true fitness and bright sires: A role for parasites. *Science* 218:384–387.

Hanlon, R. T. 1998. Mating systems and sexual selection in the squid *Loligo*: How might commercial fishing on spawning squids affect them? *Calcofi Rep.* 39:92–100.

Haq, S. M. 1972. Breeding of *Euterpina acutifrons*, a harpacticoid copepod, with special reference to dimorphic males. *Mar. Biol.* 15:221–235.

Hardege, J. D., H. D. Bartels-Hardege, Y. Yang, B. L. Wu, M. Y. Zhu, and E. Zeeck. 1994. Environmental control of reproduction in *Perinereis nuntia* var. *brevicirrus*. *J. Mar. Biol. Assoc. UK.* 74:903–918.

Harrison, K. 1984. The morphology of the sphaeromatid brood pouch (Crustacea: Isopoda: Sphaeromatidae). *Zool. J. Linn. Soc.* 82:363–407.

Hartl, D. L., and A. G. Clark. 1989. *Principles of Population Genetics*, 2nd Ed., Sinauer Associates, Sunderland, MA.

Harvey, P. H., and M. D. Pagel. 1991. *The Comparative Method in Evolutionary Biology*, Oxford University Press, Oxford, U.K.

Hayashi, K. 1985. Alternative mating strategies in the water strider *Gerris elongatus* (Heteroptera, Gerridae). *Behav. Ecol. Sociobiol.* 16:301–306.

Hayashi, K.-I., and Y. Ogawa. 1987. *Spongicola levigata* sp. nov., a new shrimp associated with a hexactinellid sponge from the East China Sea (Decapoda, Stenopodidae). *Zool. Sci.* 4:367–373.

Hayes, T. B. 1997. Hormonal mechanisms as potential constraints on evolution: Examples from the Anura. *Am. Zool.* 37:482–490.

Hazel, W. N., R. Smock, and M. D. Johnson. 1990. A polygenic model for the evolution and maintenance of conditional strategies. *Proc. R. Soc. London, Ser. B* 242:181–187.

Hed, H. 1984. Opportunity for selection during the 17th–19th centuries in the diocese of Linkoping as estimated with Crow's index in a population of clergymen's wives. *Human Hered.* 34:378–387.

Hed, H.M.E. 1986. Selection opportunities in seven Swedish 19th century populations. *Human Biol.* 58:919–931.

Hedrick, A. V. 1988. Female choice and the heritability of attractive male traits:an empirical study. *Am. Nat.* 128:267–276.

Heifetz, Y., U. Tram, and M. F. Wolfner. 2001. Male contributions to egg production:

The role of accessory gland products and sperm in *Drosophila melanogaster*. *Proc. R. Soc. London, Ser. B* 268:175–180.

Heinz, J., B. Hoelldobler, and K. Yamaguchi. 1998. Male competition in *Cardiocondyla* ants. *Behav. Ecol. Sociobiol.* 42:239–246.

Herre, E. A., S. A. West, J. M. Cook, S. G. Compton, and F. Kjellberg. 1997. Fig-associated wasps: Pollinators and parasites, sex-ratio adjustment and male polymorphism, population structure and its consequences. In J. C. Choe and B. J. Crespi (eds.), *The Evolution of Mating Systems in Insects and Arachnids*, Cambridge University Press, Cambridge, U.K., pp. 226–239.

Herrera, C. M. 2000. Measuring the effects of pollinators and herbivores: Evidence for non-additivity in a perennial herb. *Ecology* 81:2170–2176.

Hews, D. K. 1993. Food resources affect female distribution and male mating opportunities in the iguanian lizard *Uta palmeri*. *Anim. Behav.* 46:279–291.

Hews, D. K., R. Knapp, and M. C. Moore. 1994. Early exposure to androgens affects adult expression of alternative mating types in tree lizards. *Horm. Behav.* 28:96–115.

Higashi, K., and S. Nomakuchi. 1997. Alternative mating tactics and aggressive male interactions in *Mnais nawai* Yamamoto (Zygoptera: Calopterygidae). *Odonatologica* 26:159–169.

Hoeg, J. T. 1990. "Akentrogonid" host invasion and an entirely new type of life cycle in the rhizocephalan parasite *Clistosaccus paguri* (Thecostraca:Cirripedia). *J. Crust. Biol.* 10:37–52.

Hoeg, J. T. 1995. The biology and life cycle of the Rhizocephala (Cirripedia). *J. Mar. Biol. Assoc. UK.* 75:517–550.

Hoekstra, H. E., J. M. Hoekstra, D. Berrigan, S. N. Vignieri, A. Hoang, C. E. Hill, P. Beerli, and J. G. Kingsolver. 2001. Strength and tempo of directional selection in the wild. *Proc. Natl. Acad. Sci. USA* 98:9157–9160.

Hogg, J. T. 1984. Mating in bighorn sheep: Multiple creative male strategies. *Science* 225:526–529.

Hogg, J. T., and S. H. Forbes. 1997. Mating in bighorn sheep: Frequent male reproduction via a high-risk "unconventional" tactic. *Behav. Ecol. Sociobiol.* 41: 33–48.

Höglund, J., and R. V. Alatalo. 1995. *Leks*, Princeton University Press, Princeton, NJ.

Höglund, J., F. Widemo, W. J. Sutherland, and H. Nordenfors. 1998. Ruffs, *Philomachus pugnax*, and distribution models: Can leks be regarded as patches? *Oikos* 82:370–376.

Hoikkala, A., and K. Y. Kaneshiro. 1997. Variation in male wing song characters in *Drosophila plantibia* (Hawaiian picture-winged Drosophila group). *J. Insect Behav.* 10:425–436.

Holland, B., and W. R. Rice. 1998. Perspective: Chase-away sexual selection: Antagonistic seduction versus resistance. *Evolution* 52:1–7.

Holmes, S. J. 1904. Remarks on the sexes of the sphaeromids with a description of a new species of *Dynamene*. *Proc. Cal. Acad. Sci.* 3:295–306.

Hosken, D. J. 1999. Sperm displacement in yellow dung flies: A role for females. *Trends Ecol. Evol.* 14:251–252.

Houde, A. E. 1992. Sex-linked heritability of a sexually selected character in a natural population of *Poecilia reticulata* (Pisces: Poeciliidae) (guppies). *Heredity* 69:229–235.

Houde, A. E., and J. A. Endler. 1990. Correlated evolution of female mating preferences and male color patterns in the guppy, *Poecilia reticulata. Science* 248:1405–1408.

Hover, E. L. 1985. Differences in aggressive behavior between two throat color morphs in a lizard, *Urosaurus ornatus. Copeia* 1985:933–940.

Howard, D. J. 1999. Conspecific sperm and pollen precedence and speciation. *Ann. Rev. Ecol. Syst.* 30:109–132.

Howard, D. J., and P. G. Gregory. 1993. Post-insemination signaling systems and reinforcement. *Philos. Trans. R. Soc. London, Ser. B* 340:231–236.

Howard, R. D. 1978. The evolution of mating strategies in bullfrogs, *Rana catesbeiana. Evolution* 32:850–871.

Howard, R. D. 1984. Alternative mating behaviors of young male bullfrogs. *Am. Zool.* 24:397–406.

Howie, D.I.D. 1982. The reproductive biology of the lugworm, *Arenicola marina. Fortschr. Zool.* 29:247–263.

Hughes, L., B. Siew Woon Chang, D. Wagner, and N. E. Pierce. 2000. Effects of mating history on ejaculate size, fecundity, longevity, and copulation duration in the ant-tended lycaenid butterfly, *Jalmenus evagoras. Behav. Ecol. Sociobiol.* 47:119–128.

Hugie, D. M., and D. B. Lank. 1997. The resident's dilemma: A female choice model for the evolution of alternative mating strategies in lekking male ruffs (*Philomachus pugnax*). *Behav. Ecol.* 8:218–225.

Hurley, D. E., and P. K. Jansen. 1977. The marine fauna of New Zealand: Family Sphaeromatidae (Crustacea:Isopoda:Flabellifera). *New Zealand Oceanogr. Inst. Mem.* 63:1–95.

Hurst, L. D., and A. Pomiankowski. 1998. Sexual selection: The eyes have it. *Nature* 391:223.

Hutchings, J. A., and R. A. Myers. 1988. Mating success of alternative maturation phenotypes in male Atlantic salmon, *Salmo salar. Oecologia* 75:169–174.

Hutchings, J. A., and R. A. Myers. 1994. The evolution of alternative mating strategies in variable environments. *Evol. Ecol.* 8:256–268.

Hutchings, J. A., T. D. Bishop, and C. R. McGregor-Shaw. 1999. Spawning behaviour of Atlantic cod, *Gadus morhua*: Evidence of mate competition and mate choice in a broadcast spawner. *Can. J. Fish. Aquat. Sci.* 56:97–104.

Ichikawa, N. 1995. Male counterstrategy against infanticide of the female giant water bug *Lethocerus deyrollei* (Hemiptera: Belostomatidae). *J. Insect Behav.* 8:181–18.

Iribarne, O. 1996. Habitat structure, population abundance and the opportunity for selection on body weight in the amphipod *Eogammarus oclairi. Mar. Biol.* 127:143–150.

Jennions, M. D., and M. Petrie. 2000. Why do females mate multiply? A review of the genetic benefits. *Biol. Rev. Camb. Philos. Soc.* 75:21–64.

Jernigan, R. W., D. C. Culver, and D. W. Fong. 1994. The dual role of selection and evolutionary history as reflected in genetic correlations. *Evolution* 48:587–596.

Jia, F.-Y., and M. D. Greenfield. 1997. When are good genes good? Variable outcomes of female choice in wax moths. *Proc. R. Soc. London, Ser. B* 264:1057–1063.

Jia, F. Y., M. D. Greenfield, and R. D. Collins. 2000. Genetic variance of sexually selected traits in waxmoths: Maintenance by genotype × environment interaction. *Evolution* 54:953–967.

Jirotkul, M. 2000. Male trait distribution determined alternative mating tactics in guppies. *J. Fish Biol.* 56:1427–1434.

Johnsen, A., V. Andersen, C. Sunding, and J. T. Lifjeld. 2000. Female bluethroats enhance offspring immunocompetence through extra-pair copulations. *Nature* 406:296–299.

Johnson, C. N. 1989. Social interactions and reproductive tactics in red-necked wallabies (*Macropus rufogriseus banksianus*). *J. Zool.* 217:267–280.

Johnson, K., R. Thornhill, J. D. Ligon, and M. Zuk. 1993. The direction of mothers' and daughters' preferences and the heritability of male ornaments in red jungle fowl (*Gallus gallus*). *Behav. Ecol.* 4:254–259.

Johnson, K., E. DuVal, M. Kielt, and C. Hughes. 2000. Male mating strategies and the mating system of great-tailed grackles. *Behav. Ecol.* 11:132–141.

Johnson, N. L., and F. C. Leone. 1964. *Statistics and Experimental Design*, John Wiley and Sons, New York.

Johnstone, R. A. 2000. Models of reproductive skew: A review and synthesis. *Ethology* 106:5–26.

Jones, A. G., and J. C. Avise. 1997. Polygynandry in the dusky pipefish *Syngnathus floridae* revealed by microsatellite DNA markers. *Evolution* 51:1611–1622.

Jones, A. G., D. Walker, C. Kvarnemo, K. Lindstroem, and J. C. Avise. 2001a. How cuckoldry can decrease the opportunity for sexual selection: Data and theory from a genetic parentage analysis of the sand goby, *Pomatoschistus minutus*. *Proc. Natl. Acad. Sci. USA* 98:9151–9156.

Jones, A. G., D. Walker, K. Lindstroem, C. Kvarnemo, and C. J. Avise. 2001b. Surprising similarity of sneaking rates and genetic mating patterns in two populations of sand goby experiencing disparate sexual selection regimes. *Mol. Ecol.* 1:461–469.

Jones, J. S. 1987. The heritability of fitness: Bad news for "good genes"? *Trends Ecol. Evol.* 2:35–38.

Jones, K. W., and L. Singh. 1985. Snakes and the evolution of sex chromosomes. *Trends Genet.* 1:55–61.

Jones, M. W., and J. A. Hutchings. 2001. The influence of male parr body size and mate competition on fertilization success and effective population size in Atlantic salmon. *Heredity* 86:675–684.

Jones, T. M. 2001. A potential cost of monandry in the lekking sandfly *Lutzomyia longipalpis*. *J. Insect Behav.* 14:385–400.

Jones, T. M., A. Balmford, and R. J. Quinnell. 2000. Adaptive female choice for middle-aged mates in a lekking sandfly. *Proc. R. Soc. London, Ser. B* 267:681–686.

Jones, T. M., R. J. Quinnell, and A. Balmford. 1998. Fisherian flies: Benefits of female choice in a lekking sandfly. *Proc. R. Soc. London, Ser. B* 265:1651–1657.

Jormalainen, V. 1998. Precopulatory mate guarding in crustaceans: Male competitive strategy and intersexual conflict. *Q. Rev. Biol.* 73:275–304.

Jormalainen, V., and S. Merilaita. 1993. Female resistance and precopulatory guarding in the isopod *Idotea baltica* (Pallas). *Behaviour* 125:219–231.

Jormalainen, V., and S. Merilaita. 1995. Female resistance and duration of mate-guarding in three aquatic peracarids (Crustacea). *Behav. Ecol. Sociobiol.* 36:43–48.

Jormalainen, V., S. Merilaita, and J. Riihimaeki. 2001. Costs of intersexual conflict in the isopod *Idotea baltica*. *J. Evol. Biol.* 14:763–772.

Jormalainen, V., and S. M. Shuster. 1997. Microhabitat segregation and cannibalism in an endangered freshwater isopod, *Thermosphaeroma thermophilum*. *Oecologia* 111:271–279.

Jormalainen, V., and S. M. Shuster. 1999. Female reproductive cycles and sexual conflict over precopulatory mate-guarding in *Thermosphaeroma* isopods. *Ethology* 105:233–246.

Jormalainen, V., S. M. Shuster, and H. Wildey. 1999. Reproductive anatomy, sexual conflict and paternity in *Thermosphaeroma thermophilum*. *Mar. Freshw. Behav. Physiol.* 32:39–56.

Jormalainen, V., and J. Tuomi. 1989. Sexual differences in habitat selection and activity of the colour polymorphic isopod, *Idotea baltica*. *Anim. Behav.* 38:576–585.

Joseph, K. J. 1994. Sexual dimorphism and intra-sex variations in the elephant dung-beetle, *Heliocopris dominus* (Coprinae: Scarabaeidae). *Entomon* 19:165–168.

Jouventin, P., B. Lequette, and F. S. Dobson. 1999. Age-related mate choice in the wandering albatross. *Anim. Behav.* 57:1099–1106.

Kaitala, A., and C. Wiklund. 1995. Female mate choice and mating costs in the polyandrous butterfly *Pieris napi* (Lepidoptera:Pieridae). *J. Insect Behav.* 8:355–363.

Kallman, K. D. 1983. The sex-determining mechanism of the poeciliid fish, *Xiphophorus montezumae* Jordan and Snyder and the genetic control of the sexual maturation process and adult size. *Copeia* 3:755–769.

Kawano, K. 1995. Horn and wing allometry and male dimorphism in giant rhinoceros beetles (Coleoptera: Scarabaeidae) of tropical Asia and America. *Ann. Entomol. Soc. Am.* 88:92–99.

Kempenaers, B., B. Congdon, P. Boag, and R. J. Robertson. 1999. Extrapair paternity and egg hatchability in tree swallows: Evidence for the genetic compatibility hypothesis? *Behav. Ecol.* 10:304–311.

Kempenaers, B., S. Everding, C. Bishop, P. Boag, and R. J. Robertson. 2001. Extra-pair paternity and the reproductive role of male floaters in the tree swallow (*Tachycineta bicolor*). *Behav. Ecol. Sociobiol.* 49:251–259.

Kempenaers, B., G. R. Verheyen, and A. Dhondt. 1997. Extrapair paternity in the blue tit (*Parus caeruleus*): Female choice, male characteristics, and offspring quality. *Behav. Ecol.* 8:481–492.

Kessel, E. L. 1955. Mating activities of balloon flies. *Syst. Zool.* 4:97–104.

Ketterson, E. D., and V. Nolan. 1994. Male parental behavior in birds. *Ann. Rev. Ecol. System.* 25:601–628.

Ketterson, E. D., P. G. Parker, S. A. Raouf, V. Nolan, Jr., C. Ziengenfus, and C. R. Chandler. 1997. The relative impact of extra-pair fertilizations on variation in male and female reproductive success in dark-eyed juncos (*Junco hyemalis*). *Ornithol. Monog.* 1997:81–101.

Keyser, A. J., and G. E. Hill. 2000. Structurally based plumage coloration is an honest signal of quality in male blue grosbeaks. *Behav. Ecol.* 11:202–209.

Kiester, A. R. 1979. Conspecifics as cues: A mechanism for habitat selection in the Panamanian grass anole (*Anolis auratus*). *Behav. Ecol. Sociobiol.* 5:323–330.

Kiflawi, M. 2000. Adaptive gamete allocation when fertilization is external and sperm competition is absent: Optimization models and evaluation using coral reef fish. *Evol. Ecol. Res.* 2:1045–1066.

King, P. E., 1973. *Pycnogonids*, Hutchinson and Co., London/St. Martin's Press, New York.

Kirkendall, L. R., and N. C. Stenseth. 1985. On defining "breeding once." *Am. Nat.* 125:189–204.

Kirkpatrick, M. 1982. Sexual selection and the evolution of female choice. *Evolution* 36:1–12.

Kirkpatrick, M. 1985. Evolution of female choice and male parental investment in polygynous species: The demise of the "sexy son." *Am. Nat.* 125:788–810.

Kirkpatrick, M. 1996. Good genes and direct selection in the evolution of mating preferences. *Evolution* 50:2125–2140.

Kirkpatrick, M., and N. H. Barton. 1997. The strength of indirect selection on female mating preferences. *Proc. Natl. Acad. Sci. USA* 94:1282–1286.

Klompen, J.S.H., F. S. Lukoschus, and B. M. O'Connor. 1987. Ontogeny, life history and sex ratio evolution in *Ensliniella kostylevi* (Acari: Winterschmidtiidae). *J. Zool.* 213:591–607.

Knowles, L. L., and T. A. Markow. 2001. Sexually antagonistic coevolution of a postmating-prezygotic reproductive character in desert *Drosophila. Proc. Natl. Acad. Sci. USA* 98:8692–8696.

Knowlton, N., and S. R. Greenwell. 1984. Male sperm competition avoidance mechanisms: The influence of female interests. In R. L. Smith (ed.), *Sperm Competition and the Evolution of Animal Mating Systems*, Academic Press, San Diego, CA.

Kodric-Brown, A., and J. L. Brown. 1984. Truth in advertising: The kinds of traits favored by sexual selection. *Am. Nat.* 124:309–323.

Koenig, W. D., and S. S. Albano. 1986. On the measurement of sexual selection. *Am. Nat.* 127:403–409.

Koga, T. 1998. Reproductive success and two modes of mating in the sand-bubbler crab, *Scopimera globosa. J. Exp. Mar. Biol. Ecol.* 229:197–207.

Koga, T., Y. Henmi, and M. Murai. 1993. Sperm competition and the assurance of underground copulation in the sand-bubbler crab, *Scopiemera globosa* (Brachyura: Ocypodidae). *J. Crust. Biol.* 13:134–137.

Kokko, H. 1999. Cuckoldry and the stability of biparental care. *Ecol. Lett.* 2:247–255.

Kokko, H. 2001. Fisherian and "good genes" benefits of mate choice: How (not) to distinguish between them. *Ecol. Lett.* 4:322–326.

Komdeur, J. 1996. Facultative sex ratio bias in the offspring of Seychelles warblers. *Proc. R. Soc. London, Ser. B* 263:661–666.

Komdeur, J., S. Daan, J. Tinbergen, and C. Mateman. 1997. Extreme adaptive modification in sex ratio of the Seychelles Warbler's eggs. *Nature* 385:522–525.

Komers, P. E. 1996. Obligate monogamy without paternal care in Kirk's dikdik. *Anim. Behav.* 51:131–140.

Komers, P. E., and P.N.M. Brotherton. 1997. Female space use is the best predictor of monogamy in mammals. *Proc. R. Soc. London, Ser. B* 264:1261–1270.

Komers, P. E., F. Messier, and C. C. Gates. 1994. Plasticity of reproductive behaviour in wood bison bulls: On risks and opportunities. *Ethol. Ecol. Evol.* 6:481–495.

Koprowski, J. L. 1993. Behavioral tactics, dominance, and copulatory success among male fox squirrels. *Ethol. Ecol. Evol.* 5:169–176.

Kotiaho, J. S., L. W. Simmons, and J. L. Tomkins. 2001. Towards a resolution of the lek paradox. *Nature* 410:684–686.

Kovacs, K. M. 1990. Mating strategies in male hooded seals (*Cystophora cristata*)? *Can. J. Zool.* 68:2499–2502.

Kraak, S.B.M., and E.M.A. De Looze. 1993. A new hypothesis on the evolution of sex determination in vertebrates. Big females ZW, big males XY. *Neth. J. Zool.* 43:260–273.

Krebs, J. R., and N. B. Davies. 1987. *An Introduction to Behavioral Ecology*, Blackwell Scientific, New York.

Krebs, J. R., and R. Dawkins. 1984. Animal signals: Mindreading and manipulation. In J. R. Krebs and N. B. Davies (eds.), *Behavioural Ecology. An Evolutionary Approach*, Blackwell Scientific, Oxford, U.K., pp. 380–402.

Krebs, R. A. 1987. The mating behavior of *Papilio glaucus* (Papilionidae). *J. Res. Lepidop*. 26:27–31.

Kruger, A., and M. H. Schleyer. 1998. Sexual resproduction in the coral *Pocillopora verrucosa* (Cnidaria: Scleractinia) in KwaZulu-Natal, South Africa. *Mar. Biol.* 132:703–710.

Kruijt, J. P., and G. J. deVos. 1988. Individual in reproductive success in male black grouse, *Tetrao tetrix* L. In T. H. Clutton-Brock (ed.), *Reproductive Success: Studies of Individual Variation in Contrasting Breeding Systems*, University of Chicago Press, Chicago, IL.

Krupa, J. J., 1989. Alternative mating tactics in the Great Plains toad. *Anim. Behav.* 37:1035–1043.

Krupa, J. J., and A. Sih 1993. Experimental studies on water strider mating dynamics: Spatial variation in density and sex ratio. *Behav. Ecol. Sociobiol.* 33:107–120.

Kruuk, L.E.B., T. H. Clutton-Brock, and F. E. Guinness. 1999. Early determinants of lifetime reproductive success differ between the sexes in red deer. *Proc. R. Soc. London, Ser. B* 266:1655–1661.

Kuenzler, R., and T.C.M. Bakker. Pectoral fins and paternal quality in sticklebacks. *Proc. R. Soc. London, Ser. B* 267:999–1004.

Kukuk, P. F. 1996. Male dimorphism in *Lasioglossum* (*Chilalictus*) *hemichalceum*: The role of larval nutrition. *J. Kans. Entomol. Soc.* 69:147–157.

Kuris, A. M., Z. Ra'anan, A. Sagi, and D. Cohen. 1987. Morphotypic differentiation of male Malaysian Giant prawns, *Macrobrachium rosenbergii. J. Crust. Biol.* 7:219–237.

Kvarnemo, C. 1996. Temperature affects operational sex ratio and intensity of male-male competition: An experimental study of sand gobies, *Pomatoschistus minutus. Behav. Ecol.* 7:208–212.

Kvarnemo, C. 1997. Food affects the potential reproductive rates of sand goby females but not of males. *Behav. Ecol.* 8:605–611.

Kvarnemo, C., and I. Ahnesjö. 1996. The dynamics of operational sex ratios and competition for mates. *Trends Ecol. Evol.* 11:404–408.

Kvarnemo, C., and E. Forsgren. 2000. The influence of potential reproductive rate and variation in mate quality on male and female choosiness in the sand goby, *Pomatoschistus minutus. Behav. Ecol. Sociobiol.* 48:378–384.

Lanctot, R. B., P. J. Weatherhead, B. Kempenaers, and K. T. Scribner. 1998. Male traits, mating tactics and reproductive success in the buff-breasted sandpiper, *Tryngites subruficollis. Anim. Behav*. 56:419–432.

Lande, R. 1981. Models of speciation by sexual selection on polygenic traits. *Proc. Natl. Acad. Sci. USA*. 78:3721–3725.

Lande, R. 2000. Quantitative genetics and phenotypic evolution. In R. S. Singh and

C. B. Krimbas (eds.), *Evolutionary Genetics from Molecules to Morphology*, Cambridge Univesity Press, Cambridge, U.K., pp. 335–350.

Lande, R., and S. J. Arnold. 1983. The measurement of selection on correlated characters. *Evolution* 37:1210–1226.

Lande, R., D. W. Schemske, and S. T. Schultz. 1994. High inbreeding depression, selective interference among loci, and the threshold selfing rate for purging recessive lethal mutations. *Evolution* 48:965–978.

Lank, D. B., M. Coupe, and K. E. Wynne-Edwards. 1999. Testosterone-induced male traits in female ruffs (*Philomachus pugnax*): Autosomal inheritance and gender differentiation. *Proc. R. Soc. London, Ser. B* 266:2323–2330.

Lank, D. B., and C. M. Smith. 1987. Conditional lekking in ruff (*Philomachus pugnax*). *Behav. Ecol. Sociobiol.* 20:137–145.

Lank, D. B., C. M. Smith, O. Hanotte, T. Burke, and F. Cooke. 1995. Genetic polymorphism for alternative mating behaviour in lekking male ruff. *Nature* 378:59–62.

Lauer, M. J. 1996. Effect of sperm depletion and starvation on the mating behavior of the water strider, *Aquarius remigis*. *Behav. Ecol. Sociobiol.* 38:89–96.

Lauer, M. J., A. Sih, and J. J. Krupa. 1996. Male density, female density and intersexual conflict in a stream-dwelling insect. *Anim. Behav.* 52:929–939.

Lawrence, S. E. 1992. Sexual cannibalism in the praying mantid, *Mantis religiosa*: A field study. *Anim. Behav.* 43:569–583.

Lederhouse, R. C. 1981. The effect of female mating frequency on egg fertility in the Black Swallowtail, *Papilio polyxenes asterius* (Papilionidae). *J. Lepidop. Soc.* 35:266–277.

Legrand, R. S., and D. H. Morse. 2000. Factors driving extreme sexual size dimorphism of a sit-and-wait predator under low density. *Biol. J. Linn. Soc.* 71:643–664.

Lesna, I., and M. W. Sabelis. 1999. Diet-dependent female choice for males with "good genes" in a soil predatory mite. *Nature* 401:581–584.

Levins, R. 1968. *Evolution in Changing Environments*, Princeton University Press, Princeton, NJ.

Levitan, D. R., and C. W. Petersen. 1995. Sperm limitation in the sea. *Trends. Ecol. Evol.* 10:228–231.

Lewis, S. M., and S. N. Austad. 1994. Sexual selection in flour beetles—the relationship between sperm precedence and male olfactory attractiveness. *Behavioral Ecol.* 5:219–224.

Lewis, S. M., and E. Jutkiewicz. 1998. Sperm precedence and sperm storage in multiply mated red flour beetles. *Behav. Ecol. Sociobiol.* 43:365–369.

Ligon, D. J. 1999. *The Evolution of Avian Mating Systems*, Oxford University Press, Oxford.

Liker, A., and T. Szekely. 1997. Aggression among female lapwings, *Vanellus vanellus*. *Anim. Behav.* 54:797–802.

Lill, A. 1974a. Sexual behavior of the lek forming white-bearded manakin (*Manacus manacus trinitatis* Hartert.) *Z. Tierpsych.* 36:1–36.

Lill, A. 1974b. Social organization and space utilizaton in the lek-forming white-bearded manakin, *M. manacus trinitatis* Hartert. *Z. Tierpsych.* 36:513–530.

Lindberg, W. J., and F. D. Lockhart. 1993. Depth-stratified population structure of geryonid crabs in the eastern Gulf of Mexico. *J. Crust. Biol.* 13:713–722.

Linklater, W. L., E. Z. Cameron, E. O. Minot, and K. J. Stafford. 1999. Stallion harassment and the mating system of horses. *Anim. Behav.* 58:295–306.

Linsenmair, K. E., and C. Linsenmair. 1971. Paarbildung and Paarzumsammenhalt bei der monogamen Wüstenassel, *Hemilepistus reamuri* (Crustacea, Isopods, Oniscoida). *Z. Tierpsychol.* 29:134–155.

Lively, C. M. 1986. Canalization versus developmental conversion in a spatially variable environment. *Am. Nat.* 128:561–572.

Lively, C. M., W. N. Hazel, M. J. Schellenberger, and K. S. Michelson. 2000. Predator-induced defense: Variation for inducibility in an intertidal barnacle. *Ecology* 81:1240–1247.

Lloyd, D. G. 1984. Variation strategies of plants in heterogeneous environments. *Biol. J. Linn. Soc.* 21:357–385.

Lloyd, J. E. 1966. Studies on the flash communication system in *Photinus* fireflies. *University of Michigan Museum of Zoology, Misc. Publications* 130:1–95.

Lloyd, J. E. 1975. Aggressive mimicry in *Photuris*: Signal repertoires by femmes fatales. *Science* 187:452–453.

Lloyd, J. E. 1979. Sexual selection in luminescent beetles. In M. S. Blum and N. A. Blum (eds.), *Sexual Selection and Reproductive Competition in Insects*, Academic Press, New York, pp. 293–342.

Lloyd, M. 1967. Mean crowding. *J. Anim. Ecol.* 36:1030.

Loiselle, R., and A. Francoeur. 1988. Regression du dimorphisme sexuel dans le genre *Formicoxenus* et polymorphisme compare des sexes dans la famille des Formicidae (Hymenoptera). *Nat. Can.* 115:367–378.

Lubin, Y. D. 1986. Courtship and alternative mating tactics in a social spider. *J. Arachnol.* 14:239–257.

Lubjuhn, T., W. Winkel, J. T. Epplen, and J. Bruen. 2000. Reproductive success of monogamous and polygynous pied flycatchers. *Behav. Ecol. Sociobiol.* 48:12–17.

Lucas, J., and R. D. Howard. 1995. On alternative reproductive tactics in anurans: Dynamic games with density and frequency dependence. *Am. Nat.* 146:365–397.

Luetzen, J., H. Sakamoto, A. Taguchi, and T. Takahashi. 2001. Reproduction, dwarf males, sperm dimorphism, and life cycle in the commensal bivalve *Peregrinamor ohshimai* Shoji (Heterodonta: Galeommatoidea: Montacutidae). *Malacologia* 43: 313–325.

Luis, J., A. Carmona, J. Delgado, F. A. Cervantes, and R. Cardenas. 2000. Parental behavior of the volcano mouse, *Neotomodon alstoni* (Rodentia: Muridae), in captivity. *J. Mammal.* 81:600–605.

Lung, O., U. Tram, C. M. Finnerty, M. A. Eipper-Mains, J. M. Kalb, and M. F. Wolfner. 2002. The *Drosophila melanogaster* seminal fluid Acp62F is a protease inhibitor that is toxic upon ectopic expression. *Genetics* 160:211–224.

Lynch, M., and B. Walsh. 1998. *Genetics and Analysis of Quatitative Traits*. Sinauer Associates, Sunderland, MA.

Lynch, M., L. Latta, J. Hicks, and M. Giorgianni. 1998. Mutation, selection, and the maintenance of life-history variation in a natural population. *Evolution* 52:727–733.

Lyons, E. E., N. M. Waser, M. V. Price, J. Antonovics, and A. F. Motten. 1989. Sources of variation in plant reproductive success and implications for concepts of sexual selection. *Am. Nat.* 134:409–433.

Lytle, D. A. 2001. Disturbance regimes and life-history evolution. *Am. Nat.* 157:525–536.

Machado, G., and R. L. G. Raimundo. 2001. Parental investment and the evolution of subsocial behaviour in harvestmen (Arachnida Opiliones). *Ethol. Ecol. Evol.* 13: 133–150.

Marchetti, K. 1993. Dark habitats and bright birds illustrate the role of the environment in species divergence. *Nature* 362:149–152.

Marden, J. H. 1987. In pursuit of females: Following and contest behavior by males of a Namib Desert tenebrionid beetle, *Physadesmia globosa*. *Ethology* 75:15–24.

Marioghae, I. E., and O. A. Ayinla. 1999. The reproductive biology and culture of *Macrobrachium vollenhovenii* (Herklots, 1857) and *Macrobrachium macrobrachion* (Herklots 1851) in Nigeria. *Tech. Paper Niger. Inst. Oceanogr. Mar. Res., NIOMR, Port Harcourt (Nigeria)*, no. 100.

Markham, J. C. 1979. Epicaridean isopods of Bermuda. *Bull. Mar. Sci.* 29:522–529.

Markham, J. C. 1992. The Isopoda Bopyridae of the eastern Pacific—Missing or just hiding? *Proc. San Diego So. Nat. Hist.* 17:1–5.

Marks, J. S., J. L. Dickinson, and J. Haydock. 2002. Serial polyandry and alloparenting in long-eared owls. *Condor* 104:202–204.

Marler, C. A., S. K. Boyd, and W. Wilczynski. 1999. Forebrain arginine vasotocin correlates of alternative mating strategies in cricket frogs. *Horm. Behav.* 36: 53–61.

Martens, J. 1993. Further cases of paternal care in Opiliones (Arachnida). *Trop. Zool.* 6:97–107.

Marzluff, J. M., and R. P. Balda. 1992. *The Pinon Jay: Behavioral Ecology of a Colonial and Cooperative Corvid*, Poyser, London.

Masonjones, H. D., and S. M. Lewis. 2000. Differences in potential reproductive rates of male and female seahorses related to courtship roles. *Anim. Behav.* 59:1–20.

Mauck, R. A., T. A. Waite, and P. G. Parker. 1995. Monogamy in Leach's storm-petrel: DNA-fingerprinting evidence. *Auk* 112:473–482.

Mayer, S. S., and D. Charlesworth. 1991. Cryptic dioecy in flowering plants. *Trends Ecol. Evol.* 6:320–325.

Maynard Smith, J. 1977. Parental investment: A prospective analysis. *Anim. Behav.* 25:1–9.

Maynard Smith, J. 1981. Will a sexual population evolve to an ESS? *Am. Nat.* 117:1015–1018.

Maynard Smith, J. 1982. *Evolution and the Theory of Games*, Cambridge University Press, Cambridge, U.K.

Maynard Smith, J., and G. R. Price 1973. The logic of animal conflict. *Nature* 246:15–18.

McCauley, D. E. 1981. Application of the Kence-Bryant model of mating behavior to a natural population of soldier beetles. *Am. Nat.* 117:400–402.

McCauley, D. E. 1982. The behavioural components of sexual selection in the Milkweed Beetle, *Tetraopes tetraophthalmus*. *Anim. Behav.* 30:23–28.

McCauley, D. E., and M. T. Brock. 1998. Frequency-dependent fitness in *Silene vulgaris*, a gynodioecious plant. *Evolution* 52:30–36.

McCauley, D. E., and M. J. Wade. 1978. Female choice and the mating structure of a natural population of the soldier beetle, *Chauliognathus pennsylvanicus*. *Evolution* 32:771–775.

McCauley, D. E., M. J. Wade, F. J. Breden, and M. Wohltman. 1988. Spatial and temporal variation in group relatedness: Evidence from the imported willow leaf beetle. *Evolution* 42:184–192.

McCoy, E. E., A. G. Jones, and J. C. Avise. 2001. The genetic mating system and tests for cuckoldry in a pipefish speices in which males fertilize eggs and brood offspring externally. *Mol. Ecol.* 10:1793–1800.

McCurdy, D. G., J. S. Boates, and M. R. Forbes. 2000. Reproductive synchrony in the intertidal amphipod, *Corophium volutator*. *Oikos* 88:301–309.

McHugh, D. 1993. A comparative study of reproduction and development in the polychaete family Terebellidae. *Biol. Bull.* 185:153–167.

McKenzie, W. D., D. Crews, K. D. Kallman, D. Policansky, and J. J. Sohn. 1983. Age, weight and the genetics of sexual maturation in the platyfish, *Xiphophorus maculatus*. *Copeia* 1983:770–774.

McKone, M. J., C. P. Lund, and J. M. O'Brien. 1998. Reproductive biology of two dominant prairie grasses (*Andropogon gerardii* and *Sorghastrum nutans*, Poaceae): Male-biased sex allocation in wind-pollinated plants? *Am. J. Bot.* 85:776–783.

Merilae, J., B. C. Sheldon, and H. Ellegren. 1998. Quantitative genetics of sexual size dimorphism in the collared flycatcher, *Ficedula albicollis*. *Genome* 41:870–876.

Mesnick, S. L. 1997. Sexual alliances: Evidence and evolutionary implications. In P. A. Gowaty (ed.), *Feminism and Evolutionary Biology: Boundaries, Intersections and Frontiers*, Chapman and Hall, New York.

Messina, F. J., and J.A.A. Renwick. 1985. Dispersal polymorphism of *Callosobruchus maculatus* (Coleoptera:Bruchidae): Variation among populations in response to crowding. *Ann. Entomol. Soc. Am.* 78:201–206.

Michiels, N. K., and L. J. Newman. 1998. Sex and violence in hermaphrodites. *Nature* 391:647.

Moczek, A. P., and D. J. Emlen 2000. Male horn dimorphism in the scarab beetle, *Onthophagus taurus*: Do alternative reproductive tactics favour alternative phenotypes? *Anim. Behav.* 59:459–466.

Møller, A. P. 1997. Immune defence, extra-pair paternity, and sexual selection in birds. *Proc. R. Soc. London, Ser. B* 264:561–566.

Møller, A. P. 2000. Male parental care, female reproductive success, and extrapair paternity. *Behav. Ecol.* 11:161–168.

Møller, A. P., and R.V. Alatalo 1999. Good-genes effects in sexual selection. *Proc. R. Soc. London, Ser. B* 266:85–91.

Møller, A. P., and P. Ninni. 1998. Sperm competition and sexual selection: A meta-analysis of paternity studies of birds. *Behav. Ecol. Sociobiol.* 43:345–358.

Møller, A. P., and R. Thornhill. 1998. Developmental stability and sexual selection: A meta-analysis. *Am. Nat.* 151:174–192.

Montgomery, L. M., G. E. Goslow, Jr., and K. B. Staley. 1987. Alternative mating behaviors in male Atlantic Salmon (*Salmo salar*), with special reference to mature male parr. In W. J. Matthews and D. C. Heins (eds.), *Community and Evolutionary Ecology of North American Stream Fishes*, University of Oklahoma Press, Norman, OK.

Moore, M. C. 1991. Application of organization-activation theory to alternative male reproductive strategies: A review. *Horm. Behav.* 25:154–179.

Mora, G. 1990. Paternal care in a neotropical harvestman, *Zygopachylus albomarginis* (Arachnida, Opiliones: Gonyleptidae). *Anim. Behav.* 39:582–593.

Moran, N. A. 1992. Evolutionary maintenance of alternative phenotypes. *Am. Nat.* 139:971–989.

Morgan, M. 1994. Models of sexual selection in hermaphrodites, especially plants. *Am. Nat.* 144:S100–S125.

Morgan, M. T., and D. J. Schoen. 1997. Selection on reproductive characters: Floral morphology in *Asclepias syriaca. Heredity* 79:433–441.

Morgan, S. G., and J. H. Christy. 1994. Plasticity, constraint and optimality in reproductive timing. *Ecology* 75:2185–2203.

Morin, P. J. 1987. Predation, breeding asynchrony, and the outcome of competition among treefrog tadpoles. *Ecology* 68:675–683.

Morrill, S. B., and R. J. Robertson. 1990. Occurrence of extra-pair copulation in tree swallow (*Tachycineta bicolor*). *Behav. Ecol. Sociobiol.* 26:291–296.

Morris, M. R., P. Batra, and M. J. Ryan. 1992. Male-male competition and access to females in the swordtail, *Xiphophorus nigrensis. Copeia* 1992:980–986.

Morris, M. R., L. Gass, and M. J. Ryan. 1995. Assessment and individual recognition of opponents in the pygmy swordtails, *Xiphophorus nigrensis* and *X. multilineatus. Behav. Ecol. Sociobiol.* 37:303–310.

Morton, B. 1985. The population dynamics, reproductive strategy and life history tactics of *Musculium lacustre* (Bivalvia: Pisidiidae) in Hong Kong. *J. Zool.* 207:581–603.

Moynihan, M. 1982. Why is lying about intentions rare during some kinds of contests? *J. Theor. Biol.* 97:7–12.

Murie, J. O. 1995. Mating behavior of Columbian ground squirrels. I. Multiple mating by females and multiple paternity. *Can. J. Zool.* 73:1819–1826.

Murphy, C. G. 1994. Chorus tenure of male barking treefrogs, *Hyla gratiosa. Anim. Behav*. 48:763–777.

Murphy, C. G. 1998. Interaction-independent sexual selection and the mechanisms of sexual selection. *Evolution* 52:8–18.

Myrberg, A. A., Jr., W. L. Montgomery, and L. Fishelson. 1988. The reproductive behavior of *Acanthurus nigrofuscus* (Forskal) and other surgeonfishes (Fam. Acanthuridae) off Eilat, Israel (Gulf of Aquaba, Red Sea). *Ethology* 79:31–61.

Nakano, S. 1985. Effect of interspecific mating on female fitness in two closely related ladybirds (*Henosepilachna*). *Kontyu* 53:112–119.

Negro, J. J., J. A. Donazar, and F. Hiraldo. 1992. Copulatory behaviour in a colony of lesser kestrels: Sperm competition and mixed reproductive strategies. *Anim. Behav.* 43:921–930.

Neubaum, D. M., and M. F. Wolfner. 1999. Mated *Drosophila melanogaster* females require a seminal fluid protein, Acp36DE, to store sperm efficiently. *Genetics* 153:845–857.

Nievergelt, C. M., L. J. Digby, U. Ramakrishnan, and D. S. Woodruff. 2000. Genetic analysis of group composition and breeding system in a wild common marmoset (*Callithrix jacchus*) population. *Int. J. Primatol.* 21:1–20.

Nijhout, H. F., and S. M. Paulsen. 1997. Developmental models and polygenic characters. *Am. Nat.* 149:394–405.

Nilsson, S. O., and G. E. Nilsson. 2000. Free choice by female sticklebacks: Lack of preference for male dominance traits. *Can. J. Zool.* 78:1251–1258.

Nomakuchi, S., and K. Higashi. 1996. Competitive habitat utilization in the damselfly, *Mnais nawai* (Zygoptera: Calopterygidae) coexisting with a related species, *Mnais pruinosa*. *Res. Popul. Ecol.* 38:41–50.

Norman, M. D., J. Finn, and T. Tregenza. 1999. Female impersonation as an alternative reproductive strategy in giant cuttlefish. *Proc. R. Soc. London, Ser. B* 266: 1347–1349.

Nowicki, S., D. Hasselquist, S. Bench, and S. Peters. 2000. Nestling growth and song repertoire of sire in great reed warblers: Evidence for song learning as an indicator mechanism in mate choice. *Proc. R. Soc. London, Ser. B* 267:2419–2424.

O'Donald, P. 1962. The theory of sexual selection. *Heredity* 17:237–246.

Oestlund, S., and I. Ahnesjö. 1998. Female fifteen-spined sticklebacks prefer better fathers. *Anim. Behav.* 56:1177–1183.

Oestlund-Nilsson, S. 2001. Fifteen-spined stickleback (*Spinachia spinachia*) females prefer males with more secretional threads in their nests: An honest-condition display by males. *Behav. Ecol. Sociobiol.* 50:263–269.

Okuda, N. 1999. Sex roles are not always reversed when the potential reproductive rate is higher in females. *Am. Nat.* 153:540–548.

Olff, H., D. M. Pegtel, J. M. Vangroenendael, and J. P. Bakker. 1994. Germination strategies during grassland succession. *J. Ecol.* 82:69–77.

Olson, V. A., and I.P.F. Owens 1998. Costly sexual signals: Are carotenoids rare, risky or required? *Trends Ecol. Evol.* 13:510–514.

Olsson, M., A. Gullberg, H. Tegelström, T. Madsen, and R. Shine. 1994a. Scientific correspondence ("Promiscuous" matings enhance maternal fitness in lizards). *Nature* 369:528.

Olsson, M., T. Madsen, and R. Shine. 1997. Is sperm really so cheap? Costs of reproduction in male adders, *Vipera berus*. *Proc. R. Soc. London, Ser. B* 264:455–459.

Olsson, M., T. Madsen, R. Shine, A. Gullberg, and H. Tegelström. 1994b. Rewards of promiscuity. *Nature* 372:230.

Olsson, M., R. Shine, A. Gullberg, T. Madsen, and H. Tegelström. 1996. Female lizards control paternity of their offspring by selective use of sperm. *Nature* 383:585.

O'Neill, K. M., and H. E. Evans. 1983. Body size and alternative mating tactics in the beewolf *Philanthus zebratus* (Hymenoptera; Sphecidae). *Biol. J. Linn. Soc. London* 20:175–184.

O'Neill, K. M., H. E. Evans, and R. P. O'Neill. 1989. Phenotypic correlates of mating success in the sand wasp, *Bembecinus quinquespinosus* (Hymenoptera: Sphecidae). *Can. J. Zool.* 67:2557–2568.

Orians, G. 1969. On the evolution of mating systems in birds and mammals. *Am. Nat.* 103:589–603.

Oring, L. W., J. M. Reed, and J.A.R. Alberico. 1994. Mate acquisition tactics in polyandrous spotted sandpipers (*Actitis macularia*): The role of age and experience. *Behav. Ecol.* 5:9–16.

Osorio-Beristain, M., and H. Drummond. 1999. Non-aggressive mate guarding by the blue-footed booby: A balance of female and male control. *Behav. Ecol. Sociobiol.* 43:307–315.

Ostfeld, R. S. 1987. On the distinction between female defense and resource defense polygyny. *Oikos* 48:238–240.

Packer, C. 1986. The ecology and sociality in felids. In D. I. Rubenstein and R. W.

Wrangham (eds.), *Ecological Aspects of Social Evolution*, Princeton University Press, Princeton, NJ.

Packer, C., and A. C. Pusey. 1987. Intrasexual cooperation and the sex ratio in African lions. *Am. Nat.* 130:636–642.

Parker, G. A. 1970a. Sperm competition and its evolutionary consequences in the insects. *Biol. Rev. Camb. Phil. Soc.* 45:525–567.

Parker, G. A. 1970b. The reproductive behavior and the nature of sexual selection in *Scatophaga stercoraria*. II. The fertilization rate and the spatial and temporal relationships of each around the site of mating and oviposition. *J. Anim. Ecol.* 39:205–228.

Parker, G. A. 1974. Courtship persistence and female quarding as male time investment strategies. *Behaviour* 48:157–184.

Parker, G. A. 1978a. The evolution of competitive mate searching. *Ann. Rev. Entomol.* 23:173–196.

Parker, G. A. 1978b. Searching for mates. In J. R. Krebs and N. B. Davies (eds.), *Behavioral Ecology: An Evolutionary Approach*, Blackwell Scientific, New York, pp. 214–244.

Parker, G. A. 1998. Sperm competition and the evolution of ejaculates: Towards a theory base. In T. R. Birkhead and A. P. Møller (eds.), *Sperm Competition and Sexual Selection*, Academic Press, New York, pp. 3–54.

Parker, G. A., R. R. Baker, and V.G.F. Smith. 1972. The origin and evolution of gamete dimorphism and the male-female phenomenon. *J. Theor. Biol.* 36:529–553.

Parker, G. A., and L. W. Simmons 1996. Parental investment and the control of sexual selection: Predicting the direction of sexual competition. *Proc. R. Soc. London, Ser. B* 263:315–321.

Partridge, L. 1994. Genetic and nongenetic approaches to questions about sexual selection. In C.R.B. Boake (ed.), *Quantitative Genetic Studies of Behavioral Evolution*, University of Chicago Press, Chicago, pp. 126–141.

Partridge, L., and L. D. Hurst. 1998. Sex and conflict. *Science* 281:2003–2008.

Pascual, M. 2000. Dwarf males in the puelche oyster (*Ostrea puelchana*, D'orb.): Differential mortality or selective settlement? *J. Shellfish Res.* 19:815–820.

Penn, D. J., and W. K. Potts. 1999. The evolution of mating preferences and major histocompatibility complex genes. *Am. Nat.* 153:145–164.

Perrill, S. A., H. C. Gerhardt, and R. Daniel. 1978. Sexual parasitism in the green tree frog, *Hyla cinerea*. *Science* 200:1179–1180.

Petersen, C. W. 1987. Reproductive behaviour and gender allocation in *Serranus fasciatus*, a hermaphroditic reef fish. *Anim. Behav.* 35:1602–1614.

Petersen, C. W., R. R. Warner, D. Y. Shapiro, and A. Marconato. 2001. Components of fertilization success in the bluehead wrasse, *Thalassoma bifasciatum*. *Behav. Ecol.* 12:237–245.

Petersen, M. E. 1999. Reproduction and development in Cirratulidae (Annelida: Polychaeta). *Hydrobiologia* 402:107–128.

Petrie, M. 1994. Improved growth and survival of offspring of peacocks with more elaborate trains. *Nature* 371:598–599.

Pfennig, D. W. 1992. Polyphenism in spadefoot toad tadpoles as a locally-adjusted evolutionarily stable strategy. *Evolution* 46:1408–1420.

Pica-Mattoccia, L., R. Moroni, L.A.T. Tchuente, V. R. Southgate, and D. Cioli. 2000. Changes of mate occur in *Schistosoma mansoni*. *Parasitology* 120:495–500.

Pilastro, A., and A. Bisazza. 1999. Insemination efficiency of two alternative male mating tactics in the guppy (*Poecilia reticulata*). *Proc. R. Soc. London, Ser. B* 266:1887–1891.

Pilastro, A., E. Giacomello, and A. Bisazza. 1997. Sexual selection for small size in male mosquitofish (*Gambusia holbrooki*). *Proc. R. Soc. London, Ser. B* 264:1125–1129.

Pitcher, T. E., and B. J. Stutchbury. 2000. Extraterritorial forays and male parental care in hooded warblers. *Anim. Behav.* 59:1261–1269.

Pitnick, S. 1996. Investment in testes and the cost of making long sperm in *Drosophila*. *Am. Nat.* 148:57–80.

Pitnick, S., T. A. Markow, and G. S. Spicer. 1999. Evolution of multiple kinds of female sperm-storage organs in *Drosophila*. *Evolution* 53:1804–1822.

Porchet, M., and P. Olive. 1988. Aspects physiologiques des stratégies de reproduction chez les annelides Polychetes. 1. Semelparity et iteroparite. *Bull. Soc. Zool. France* 113:245–250.

Poulin, R. 1997. Covariation of sexual size dimorphism and adult sex ratio in parasitic nematodes. *Biol. J. Linn. Soc.* 62:567–580.

Poulin, R., and S. Morand. 2000. Testes size, body size and male-male competition in acanthocephalan parasites. *J. Zool.* 250:551–558.

Preston-Mafham. K. 2001. Resource defence mating system in two flies from Sulawesi: *Gymnonerius fuscus* Wiedemann and *Teleostylinus* sp. near *duplicatus* Wiedemann (Diptera: Neriidae). *J. Nat. Hist.* 35:149–156.

Price, P. W. 1997. *Insect Ecology*, 3rd Ed., J. Wiley and Sons, New York.

Provine, W. B. 1971. *The Origins of Theoretical Population Genetics*, University of Chicago Press, Chicago, IL.

Pruett-Jones, S. 1992. Independent versus nonindependent mate choice: Do females copy each other? *Am. Nat.* 140:1000–1009.

Pruett-Jones, S., and M. Pruett-Jones. 1994. Sexual competition and courtship disruptions: Why do male bowerbirds destroy each other's bowers? *Anim. Behav.* 47:607–620.

Quintana-Murci, L., S. Jamain, and M. Fellous. 2001. The origin and evolution of mammalian sex chromosomes. *C. R. Acad. Sci., Ser. III. Sci. Vie/Life Sci.* 324:1–11.

Qvarnstroem, A., and E. Forsgren. 1998. Should females prefer dominant males? *Trends Ecol. Evol.*13:498–501.

Ra'anan, Z., and A. Sagi. 1985. Alternative mating strategies in male morphotypes of the freshwater prawn *Machrobrachium rosenbergii* (DeMan). *Biol. Bull.* 169:592–601.

Radwan, J. 1993. The adaptive significance of male polymorphism in the acarid mite, *Caloglyphus berlesei*. *Behav. Ecol. Sociobiol.* 33:201–208.

Radwan, J. 1995. Male morph determination in two species of acarid mites. *Heredity* 74:669–673.

Radwan, J. 1997. Sperm precedence in the bulb mite, *Rhizoglyphus robini:* context-dependent variation. *Ethol. Ecol. Evol.* 9:373–383.

Radwan, J. 2001. Male morph determination in *Rhizoglyphus echinopus* (Acaridae). *Exp. Appl. Acarol.* 25:143–149.

Raikov, I. B. 1995. Meiosis in protists: Recent advances and persisting problems. *Eur. J. Protistol.* 31:1–7.

Rajanikumari, J., C. R. Srikumari, and T. V. Rao. 1985. Variability of selection opportunities with changing socio-cultural environments. *Human Hered.* 35:218–222.

Reavis, R. H., and G. W. Barlow. 1998. Why is the coral-reef fish *Valenciennea strigata* (Gobiidae) monogamous? *Behav. Ecol. Sociobiol.* 43:229–237.

Reid, M. L. 1999. Monogamy in the bark beetle *Ips latidens*: Ecological correlates of an unusual mating system. *Ecol. Entomol.* 24:89–94.

Relya, R. A. 2002. Costs of plasticity. *Am. Nat.* 159:272–282.

Repka, J., and M. R. Gross. 1995. The evolutionarily stable strategy under individual condition and tactic frequency. *J. Theor. Biol.* 176:27–31.

Reynolds, J. D. 1987. Mating system and nesting biology of red-necked phalarope *Phalaropus lobatus*: What constrains polyandry? *Ibis* 129:225–242.

Reynolds, J. D. 1996. Animal breeding systems. *Trends Ecol. Evol.* 11:68–72.

Reynolds, J. D., and Szekely, T. 1997. The evolution of parental care in shorebirds: Life histories, ecology, and sexual selection. *Behav. Ecol.* 8:126–134.

Reznick, D. 1983. The structure of guppy life histories: The tradeoff between growth and reproduction. *Ecology* 64:862–873.

Reznick, D., and H. Bryga. 1996. Life-history evolution in guppies (*Poecilia reticulata*: Poeciliidae). 5. Genetic basis parallelism in life histories. *Am. Nat.* 147: 339–359.

Reznick, D. N., M. J. Butler, F. H. Rodd, and P. Ross. 1996. Life-history evolution in guppies (*Poecilia reticulata*). 6. Differential mortality as a mechanism for natural selection. *Evolution* 50:1651–1660.

Rice, W. R. 1996. Sexually antagonistic male adaptation triggered by experimental arrest of female evolution. *Nature* 381:232–234.

Ridley, M. 1978. Paternal care. *Anim. Behav.* 26:904–932.

Ridley, M. 1983. *The Explanation of Organic Diversity: The Comparative Method and Adaptations for Mating*, Oxford University Press, Oxford, U.K.

Ridley, M. 1995. *Evolution*, Blackwell Scientific, Boston.

Rigaud, T., and P. Juchault. 1993. Conflict between feminizing sex ratio distorters and an autosomal masculinizing gene in the terrestrial isopod, *Armadillidium vulgare* Latr. *Genetics* 133:247–252.

Ritchie, M. G., D. Sunter, and L. R. Hockham. 1998. Behavioral components of sex role reversal in the tettigoniid bushcricket, E*phippiger ephippiger. J. Insect Behav.* 11:481–491.

Roberts, L. S., and J. Janovy, Jr. 2000. *Foundations of Parasitology*, 6th Ed. McGraw-Hill, Boston.

Robertson, A. (ed.) 1980. *Selection Experiments in Laboratory and Domestic Animals*, Commonwealth Agricultural Bureau, Slough, U.K.

Robertson, D. R., and R. R. Warner. 1978. Sexual patterns in the labroid fishes of the Western Caribbean. 2. The Parrotfishes (Scaridae.) *Smithson. Contrib. Zool.* 255:1–26.

Robinson, J. V., and R. Allgeyer 1996. Covariation in life-history traits, demographics and behaviour in ischnuran damselflies: The evolution of monandry. *Biol. J. Linn. Soc.* 58:85–98.

Robinson, T., N. Johnson, and M. J. Wade. 1994. Postcopulatory, prezygotic isolation: Intraspecific and interspecific sperm precedence in flour beetles, *Tribolium* spp. *Heredity* 74:155–159.

Rodriguez, A., P. Drake, and A. M. Arias. 1992. First records of *Paracerceis sculpta*

(Holmes 1904) and *Paradella dianae* (Menzies 1962) (Isopoda, Sphaeromatidae) at the Atlantic coast of Europe. *Crustaceana* 63:94–97.

Roeder, G., and K. E. Linsenmair. 1999. The mating system in the subsocial desert woodlouse, *Hemilepistus elongatus* (Isopoda Oniscidea): A model of an evolutionary step towards monogamy in the genus *Hemilepistus* sensu stricto? *Ethol. Ecol. Evol.* 11:349–369.

Roff, D. A. 1986. The evolution of wing dimorphism in insects. *Evolution* 40:1009–1020.

Roff, D. A. 1992. *The Evolution of Life Histories*, Chapman and Hall, New York.

Roff, D. A. 1994. Habitat preference and the evolution of wing dimorphism in insects. *Am. Nat.* 144:772–798.

Roff, D. A. 1995. Why is there so much genetic variation for wing dimorphism? *Res. Popul. Ecol.* 36:145–150.

Roff, D. A. 1996. The evolution of threshold traits in animals. *Q. Rev. Biol.* 71:3–35.

Roff, D. A. 1998a. Evolution of threshold traits: The balance between directional selection, drift and mutation. *Heredity* 80:25–32.

Roff, D. A. 1998b. The maintenance of phenotypic and genetic variation in threshold traits by frequency-dependent selection. *J. Evol. Biol.* 11:513–529.

Roff, D. A., and A. M. Simmons. 1997. The quantitative genetics of wing dimorphism under laboratory and "field" conditions in the cricket *Gryllus pennsylvanicus. Heredity* 78:235–240.

Roff, D. A., G. Stirling, and D. J. Fairbairn. 1998. The evolution of threshold traits: Quantitative genetic analysis of physiological and life history correlates of wing dimorphism in the sand cricket. *Evolution* 51:1910–1919.

Rohwer, S., and F. C. Rohwer. 1978. Status signalling in Harris sparrows: Experimental deceptions achieved. *Anim. Behav.* 26:1012–1022.

Rose, R. W., C. M. Nevison, and A. F. Dixson. 1997. Testes weight, body weight and mating systems in marsupials and monotremes. *J. Zool.* 243:523–531.

Rosenqvist, G. 1993. Sex role reversal in a pipefish. Behavioural ecology of fishes, Erice, Sicily (Italy), Oct. 1991. In F. A. Huntingford and P. Torricelli (eds.), *Mar. Behav. Physiol.* 23:219–230.

Roulin, A., T. W. Jungi, H. Pfister, and C. Dijkstra. 2000. Female barn owls (*Tyto alba*) advertise good genes. *Proc. R. Soc. London, Ser. B* 267:937–941.

Rowe L., G. Arnqvist, A. Sih, and J. J. Krupa. 1994. Sexual conflict and the evolutionary ecology of mating patterns: Water striders as a model system. *Trends Ecol Evol* 9:289–93.

Rubenstein, D. I. 1980. On the evolution of alternative mating strategies. In J.E.R. Staddon (ed.), *Limits to Action: The Allocation of Individual Behavior*, Academic Press, New York, pp. 65–100.

Rubenstein, D. I. 1984. Resource acquisition and alternative male mating strategies in water striders. *Am. Zool.* 24:345–354.

Ruzzante, D. E., D. C. Hamilton, D. L. Kramer, and J.W.A. Grant. 1996. Scaling of the variance and the quantification of resource monopolization. *Behav. Ecol.* 7:199–207.

Ryan, M. J. 1983a. Sexual selection and communication in a neotropical frog, *Physalemus pustulosus. Evolution* 37:261–272.

Ryan, M. J. 1983b. *The Tungara Frog*, University of Chicago Press, Chicago, IL.

Ryan, M. J. 1988. Energy, calling, and selection. *Am. Zool.* 28:885–898.

Ryan, M. J. 1998. Sexual selection, receiver biases, and the evolution of sex differences. *Science* 281:1999–2003.

Ryan, M. J., D. K. Hews and W. E. Wagner, Jr. 1990. Sexual selection on alleles that determine body size in the swordtail, *Xiphophorus nigrensis. Behav. Ecol. Sociobiol.* 26:224–226.

Ryan, M. J., C. M. Pease, and M. R. Morris. 1992. A genetic polymorphism in the swordtail, *Xiphophorus nigrensis*: Testing the prediction of equal fitnesses. *Am. Nat.* 139:21–31.

Ryan, M. J., and M. D. Tuttle. 1981. Bat predation and the evolution of frog vocalizations in the neotropics. *Science* 214:677–678.

Ryan, M. J., M. D. Tuttle, and A. S. Rand. 1982. Bat predation and sexual advertisement in a neotropical anuran. *Am. Nat.* 119:136–139.

Ryan, M. R., L. W. Burger, Jr., D. P. Jones, and A. P. Wywialowski. 1998. Breeding ecology of greater prairie-chickens (*Tympanuchus cupido*) in relation to prairie landscape configuration. *Am. Midl. Nat.* 140:111–121.

Sagar, P. M. 1985. Breeding of the bellbird on the Poor Knights Islands, New Zealand. *N. Z. J. Zool.* 12:643–648.

Saino, N., A. M. Bolzern, and A. P. Møller. 1997. Immunocompetence, ornamentation, and viability of male barn swallows (*Hirundo rustica*). *Proc. Natl. Acad. Sci. USA* 94:549–552.

Saito, T., and K. Konishi. 1999. Direct development in the sponge-associated deep-sea shrimp *Spongicola japonica* (Decapoda: Spongicolidae). *J. Crust. Biol.* 19:46–52.

Sandell, M. 1983. Alternative mating strategies of male stoats. *Acta Zool. Fenn.* 74:173–174.

Santos, R. S., S. J. Hawkins, and R.D.M. Nash. 1996. Reproductive phenology of the Azorean rock pool blenny, a fish with alternative mating tactics. *J. Fish. Biol.* 48:842–858.

Sauer, K. P., T. Lubjuhn, J. Sindern, H. Kullmann, J. Kurtz, C. Epplen, and J. T. Epplen. 1998. Mating system and sexual selection in the scorpionfly, *Panorpa vulgaris* (Mecoptera: Panorpidae). *Naturwissenschaften* 85:219–228.

Schaffer, W. M. 1974. Optimal reproductive effort in fluctuating enviroments. *Am. Nat.* 108:783–790.

Scheib, J. E., S. W. Gangestad, and R. Thornhill. 1999. Facial attractiveness, symmetry and cues of good genes. *Proc. R. Soc. London, Ser. B* 266:1913–1917.

Scheiner, S. M. 1993. Genetics and evolution of phenotypic plasticity. *Annu. Rev. Ecol. Syst.* 24:35–68.

Scheiner, S. M., and D. Berrigan. 1998. The genetics of phenotypic plasticity. VII. The cost of plasticity in *Daphnia pulex. Evolution* 52:368–378.

Schlichting, C. D., and M. Pigliucci. 1998. *Phenotypic Evolution: A Reaction Norm Perspective*, Sinauer Associates, Sunderland, MA.

Schroeder, M. A., and G. C. White. 1993. Dispersion of greater prairie chicken nests in relation to lek location: Evaluation of the hot-spot hypothesis of lek evolution. *Behav. Ecol.* 4:266–270.

Schwagmeyer, P. L. 1986. Effects of multiple mating on reproduction in female thirteen-lined ground squirrels. *Anim. Behav.* 34:297–299.

Schwagmeyer, P. L., R.C.S. Clair, J. D. Moodie, T. C. Lamey, G. D. Schnell, and M. N. Moodie. 1999. Species differences in male parental care in birds: A reexamination of correlates with paternity. *Auk* 116:487–503.

Schwagmeyer, P. L., and D. W. Foltz. 1990. Factors affecting the outcome of sperm competition in thirteen-lined ground squirrels. *Anim. Behav*. 39:434–443.

Schwagmeyer, P. L., and S. J. Woontner. 1985. Mating competition in an asocial ground squirrel, *Spermophilus tridecemlineatus*. *Behav. Ecol. Sociobiol*. 17:291–296.

Scott, M. P. 1990. Brood guarding and the evolution of male parental care in burying beetles. *Behav. Ecol. Sociobiol*. 26:31–39.

Searcy, W. A. 1979. Male characteristics and pairing success in red-winged blackbirds. *Auk* 96:353–363.

Seeger, E. 1996. Christmas crabs. *Wildlife Conserv*. 99:28–35.

Sella, G., 1990. Sex allocation in the simultaneously hermaphroditic polychaete worm, *Ophryotrocha diadema*. *Ecology* 71:27–32.

Sever, Z., and H. Mendelssohn. 1985. Copulation as a possible mechanism to maintain monogamy in porcupines, *Hystrix indica*. *Anim. Behav*. 36:1541–1542.

Shapiro, D. Y., A. Marconato, and T. Yoshikawa. 1994. Sperm economy in a coral reef fish, *Thalassoma bifasciatum*. *Ecology* 75:1334–1344.

Shelly, T. E., and M. D. Greenfield. 1985. Alternative mating strategies in a desert grasshopper: A transitional analysis. *Anim. Behav*. 33:1211–1222.

Shelly, T. E., and M. D. Greenfield. 1989. Satellites and transients: Ecological constraints on alternative mating tactics in male grasshoppers. *Behaviour* 109:200–219.

Shelly, T. E., and T. S. Whittier 1997. In J. C. Choe and B. J. Crespi (eds.), *The Evolution of Mating Systems in Insects and Arachnids*, Cambridge University Press, Cambridge, U.K.

Shine, R., B. Phillips, H. Waye, M. Lemaster, and R. T. Mason. 2001. Animal behaviour: Benefits of female mimicry in snakes. *Nature* 414:267.

Shuster, S. M. 1981a. Life history characteristics of *Thermosphaeroma thermophilum*, the Socorro isopod (Crustacea: Peracarida). *Biol. Bull*. 161:291–302.

Shuster, S. M. 1981b. Sexual selection in the Socorro Isopod, *Thermosphaeroma thermophilum* (Cole and Bane)(Crustacea: Peracarida). *Anim. Behav*. 29:698–707.

Shuster, S. M. 1986. The reproductive biology of *Paracerceis sculpta* (Crustacea: Isopoda), Ph.D. University of California, Berkeley, CA.

Shuster, S. M. 1987. Alternative reproductive behaviors: Three discrete male morphs in *Paracerceis sculpta*, an intertidal isopod from the northern Gulf of California. *J. Crust. Biol*. 7:318–327.

Shuster, S. M. 1989a. Female sexual receptivity associated with molting and differences in copulatory behavior among the three male morphs in *Paracerceis sculpta*, (Crustacea: Isopoda). *Biol. Bull*. 177:331–337.

Shuster, S. M. 1989b. Male alternative reproductive behaviors in a marine isopod crustacean (*Paracerceis sculpta*): The use of genetic markers to measure differences in fertilization success among α-, β- and γ-males. *Evolution* 34:1683–1698.

Shuster, S. M. 1990. Courtship and female mate selection in a semelparous isopod crustacean (*Paracerceis sculpta*). *Anim. Behav*. 40:390–399.

Shuster, S. M. 1991a. Changes in female anatomy associated with the reproductive molt in *Paracerceis sculpta* (Holmes), a semelparous isopod crustacean. *J. Zool*. 225:365–379.

Shuster, S. M. 1991b. The ecology of breeding females and the evolution of polygyny in *Paracerceis sculpta*, a marine isopod crustacean. In R. Bauer and J. Martin (eds.), *Crustacean Sexual Biology*, Columbia University Press, New York, pp. 91–110.

Shuster, S. M. 1992. The reproductive behaviour of α-, β- and γ-males in *Paracerceis sculpta*, a marine isopod crustacean. *Behaviour* 121:231–258.

Shuster, S. M., J.O.W. Ballard, G. Zinser, C. Sassaman, and P. Keim. 2001.The influence of genetic and extrachromosomal factors on population sex ratio in *Paracerceis sculpta*. In R. C. Brusca and B. Kensley (eds.), *Isopod Systematics and Evolution, Crustacean Issues*, Balkema Press, Amsterdam, Vol. 13.

Shuster, S. M., and R. L. Caldwell. 1989. Male defense of the breeding cavity and factors affecting the persistence of breeding pairs in the stomatopod, *Gonodactylus bredini* (Crustacea: Hoplocarida). *Ethology* 82:192–207.

Shuster, S. M., and E. E. Guthrie. 1999. The effects of temperature and food availability on adult body length in natural and laboratory populations of *Paracerceis sculpta* (Holmes), a Gulf of California isopod. *J. Exp. Mar. Biol. Ecol.* 233:269–284.

Shuster, S. M., and L. Levy. 1999. Sex-linked inheritance of a cuticular pigmentation marker in a marine isopod, *Paracerceis sculpta*. *J. Hered.* 90:304–307.

Shuster, S. M., and C. Sassaman. 1997. Genetic interaction between male mating strategy and sex ratio a marine isopod. *Nature* 388:373–376.

Shuster, S. M., and M. J. Wade. 1991a. Equal mating success among male reproductive strategies in a marine isopod. *Nature* 350:606–610.

Shuster, S. M., and M. J. Wade. 1991b. Female copying and sexual selection in a marine isopod crustacean. *Anim. Behav.* 42:1071–1078.

Shuster, S. M., and M. J. Wade. 1997. Hiring selection. In P. A. Gowaty (ed.), *Feminism and Evolutionary Biology*, Chapman and Hall, New York, pp. 153–183.

Sigurjonsdottir, H., and K. Gunnarsson. 1989. Alternative mating tactics of arctic charr, *Salvelinus alpinus*, in Thingvallavatn, Iceland. *Environ. Biol. Fish.* 26:159–176.

Simmons, L. W. 1995. Relative parental expenditure, potential reproductive rates, and the control of sexual selection in katydids. *Am. Nat.* 145:797–808.

Sinervo, B. 2001a. Selection in local neighborhoods, graininess of social environments, and the ecology of alternative strategies. In L. A. Dugatkin (ed.), *Model Systems in Behavioral Ecology*, Princeton University Press, Princeton, NJ, pp. 191–226.

Sinervo, B. 2001b. Runaway social games, genetic cycles driven by alternative male and female strategies, and the origin of morphs. *Genetica* 112:417–434.

Sinervo, B., C. Bleay, and C. Adamopoulou. 2001. Social causes of correlational selection and the resolution of a heritable throat color polymorphism in a lizard. *Evolution* 55:2040–2052.

Sinervo, B., and C. M. Lively. 1996. The rock-paper-scissors game and the evolution of alternative male strategies. *Nature* 380:40–243.

Sinervo, B., D. B. Miles, W. A. Frankino, M. Klukowski, and D. F. DeNardo. 2000. Testosterone, endurance, and Darwinian Fitness: Natural and sexual selection on the physiological bases of alternative male behaviors in side-blotched lizards. *Horm. Behav.* 38:222–233.

Sinervo, B., and K. Zamudio, 2001. The evolution of alternative reproductive strategies: Fitness differential, heritability and genetic correlations between the sexes. *J. Hered.* 92:198–205.

Siva-Jothy, M. T. 1987. The structure and function of the female sperm-storage organs in libellulid dragonflies. *J. Insect Physiol.* 33:559–567.

Siva-Jothy, M. T., and F. Skarstein. 1998. Towards a functional understanding of "good genes." *Ecol. Lett.* 1:178–185.

Sivinski, J. 1980. Sperm in competition. *Fla. Entomol.* 63:99–111.

Sivinski, J. M., and E. Petersson. 1997. Mate choice and species isolation in swarming insects. In J. C. Choe and B. J. Crespi (eds.), *The Evolution of Mating Systems in Insects and Arachnids*, Cambridge University Press, Cambridge, U.K., pp. 294–209.

Slagsvold, T., and G. Saetre. 1991. Evolution of plumage color in male pied flycatchers (*Ficedula hypoleuca*): Evidence for female mimicry. *Evolution* 45:910–917.

Slatkin, M. 1973. Gene flow and selection in a cline. *Genetics* 75:733–756.

Slatkin, M. 1978. On the equilibration of fitnesses by natural selection. *Am. Nat.* 112:845–859.

Slatkin, M. 1979. The evolutionary response to frequency and density dependent interactions. *Am. Nat.* 14:384–398.

Slatkin, M. 1984. Ecological causes of sexual dimorphism. *Evolution* 38:622–630.

Slatkin, M. 1987. Gene flow and the geographical structure of natural populations. *Science* 236:787–792.

Smith, R. L. 1979. Repeated copulation and sperm precedence: Paternity assurance for a mal brooding water bug. *Science* 205:1029–1031.

Smith, R. L. (ed.). 1984. *Sperm Competition and the Evolution of Animal Mating Systems*, Academic Press, San Diego, CA.

Smith, R. L. 1997. Evolution of paternal care in the giant water bugs (Heteroptera: Belostomatidae). In J. C. Choe and B. J. Crespi (eds.), *The Evolution of Social Behavior in Insects and Arachnids*, Cambridge University Press, Cambridge, U.K., pp. 116–149.

Snedden, W. A., and M. D. Greenfield. 1998. Females prefer leading males: Relative call timing and sexual selection in katydid choruses. *Anim. Behav.* 56:1091–1098.

Snook, R. R. 1997. Is the production of multiple sperm types adaptive? *Evolution* 57:797–808.

Snow, A. A. 1994. Postpollination selection and male fitness in plants. *Am. Nat.* 144:S69–S83.

Snow, A. A., and P. O. Lewis. 1993. Reproductive traits and male-fertility in plants—Empirical approaches. *Annu. Rev. Ecol. Syst.* 24:331–351.

Sokal, R. R., and J. F. Rohlf. 1981. *Biometry*, 2nd Ed., W. H. Freeman, San Francisco.

Sokal, R. R., and J. F. Rohlf. 1995. *Biometry*, 3rd Ed., W. H. Freeman, San Francisco.

Soltis, J., F. Mitsunaga, K. Shimizu, M. Nozaki, Y. Yanagihara, X. Domingo-Roura, and O. Takenaka. 1997. Sexual selection in Japanese macaques II: Female mate choice and male-male competition. *Anim. Behav.* 54:737–746.

Soukup, S. S., and C. F. Thompson. 1998. Social mating system and reproductive success in house wrens. *Behav. Ecol.* 9:43–48.

Souroukis, K., and W. H. Cade. 1993. Reproductive competition and selection on male traits at varying sex ratios in the field cricket, *Gryllus pennsylvanicus. Behaviour* 126:45–62.

Southgate, V. R., D. Rollinson, G. C. Ross, and R. J. Knowles. 1982. Mating behaviour in mixed infections of *Schistosoma haematobium* and *S. intercalatum. J. Nat. Hist.* 16:491–496.

Sparkes, T. C. 1996. Effects of predation risk on population variation in adult size in a stream-dwelling isopod. *Oecologia* 106:85–92.

Sparkes T. C., D. P. Keogh, and R. A. Pary. 1996. Energetic costs of mate guarding behavior in male stream-dwelling isopods. *Oecologia* 106:166–171.

Spieth, H. T. 1981. *Drosophila heteroneura* and *Drosophila silvestris*: Head shapes, behavior and evolution. *Evolution* 35:921–930.

Spieth, H. T. 1984. Courtship behaviors of the Hawaiian picture-winged *Drosophila. Publ. Entom. Univ. Calif.* 103:1–94.

Stancyk, S. E., and G. S. Moreira. 1988. Inheritance of male dimorphism in Brazilian populations of *Euterpina acutifrons* (Dana) (Copepoda: Harpacticoida). *J. Exp. Mar. Biol. Ecol.* 120:125–144.

Stanton, M. L. 1994. Male-male competition during pollination in plant populations. *Am. Nat.* 144:S40–S68.

Stanton, M. L., B. A. Roy, and D. A. Thiede. 2000. Evolution in stressful environments. I. Phenotypic variability, phenotypic selection, and response to selection in five distinct environmental stresses. *Evolution* 54:93–111.

Stanton, M. L., A. A. Snow, and S. N. Handel. 1986. Floral evolution: Attractiveness to pollinators increases male fitness. *Science* 232:1625–1627.

Stearns, S. C. 1976. Life history tactics: A review of the ideas. *Q. Rev. Biol.* 51:3–47.

Stearns, S. C. 1977. The evolution of life history traits:a critique of the theory and a review of the data. *Annu. Rev. Ecol. Syst.* 8:145–171.

Stearns, S. C. 1989. The evolutionary significance of phenotypic plasticity. *BioScience* 39:436–445.

Steger, R., and R. L. Caldwell. 1983. Intraspecific deception by bluffing: A defense strategy of newly molted stomatopods (Arthropoda: Crustacea). *Science* 217:558–560.

Stockley, P., J. B. Searle, D. W. Macdonald, and C. S. Jones. 1996. Correlates of reproductive success within alternative mating tactics of the common shrew. *Behav. Ecol.* 3:334–340.

Stoner, G., and F. Breden. 1988. Phenotypic differentiation in female preference related to geographic variation in male predation risk in the Trinidad guppy (*Poecilia reticulata*). *Behav. Ecol. Sociobiol.* 22:285–291.

Sullivan, B. K. 1983. Sexual selection in the great plains toad (*Bufo cognatus*). *Behaviour* 84:258–264.

Sullivan, B. K. 1992. Sexual selection and calling behavior in the American toad (*Bufo americanus*). *Copeia* 1:1–7.

Sunobe, T., and A. Nakazono. 1999. Alternative mating tactics in the gobiid fish, *Eviota prasina. Ichthyol. Res.* 46:212–215.

Sutherland, W. J. 1985. Measures of sexual selection. *Oxford Surv. Evol. Biol.* 2:90–101.

Svensson, B. C. 1997. Swarming behavior, sexual dimorphism, and female reproductive status in the sex role-reversed dance fly species *Rhamphomyia marginata. J. Insect Behav.* 10:783–804.

Svensson, B. C., and J.-A. Nilsson. 1996. Mate quality affects offspring sex ratio in blue tits. *Proc. R. Soc. London, Ser. B* 263:357–361.

Swanson, W. J., A. G. Clark, H. M. Waldrip-Dail, M. F. Wolfner, and C. F. Aquadro.

2001. Evolutionary EST analysis identifies rapidly evolving male reproductive proteins in *Drosophila. Proc. Natl. Acad. Sci. USA* 98:7375–7379.

Swenson, R. O. 1997. Sex-role reversal in the tidewater goby, *Eucyclogobius newberryi. Environ. Biol. Fish.* 50:27–40.

Sword, G. A. 1999. Density-dependent warning coloration. *Nature* 397:217.

Taborsky, M. 1994. Sneakers, satellites, and helpers: Parasitic and cooperative behavior in fish reproduction. *Adv. Stud. Anim. Behav.* 23:1–100.

Takamura, K. 1999. Wing length and asymmetry of male *Tokunagayusurika akamusi* chironomid midges using alternative mating tactics. *Behav. Ecol.* 10:498–503.

Tallamy, D. W. 2001. Evolution of exclusive paternal care in arthropods. *Annu. Rev. Entomol.* 46:139–165.

Tallarovic, S. K., J. M. Melville, and P. H. Brownell 2000. Courtship and mating in the Giant Hairy Desert Scorpion, Hadrurus arizonensis (Scorpionida, Iuridae). *J. Insect Behav.* 13:827–838.

Tanaka, K., and M. Aoki. 1999. Spatial distribution patterns of the sponge-dwelling gnathiid isopod *Elaphognathia cornigera* (Nunomura) on an intertidal rocky shore of the Isu Peninsula, southern Japan. *Crust. Res.* 28:160–167.

Tanaka, K., and M. Aoki. 2000. Seasonal traits of reproduction in a gnathiid isopod *Elaphognathia cornigera* (Nunomura 1992). *Zool. Sci.* 17:467–475.

Taylor, E. B. 1989. Precocial male maturation in laboratory-reared populations of chinook salmon, *Oncorhynchus tshawytscha. Can. J. Zool.* 67:1665–1669.

Taylor, E. L., D. Blache, D. Groth, J. D. Wetherall, and G. B. Martin. 2000. Genetic evidence for mixed parentage in nests of the emu (*Dromaius novaehollandiae*). *Behav. Ecol. Sociobiol.* 47:359–364.

Thirgood, S. J. 1990. Alternative mating strategies and reproductive success in fallow deer. *Behaviour* 116:1–10.

Thompson, J. N. 1999. Specific hypotheses on the geographic mosaic of coevolution. *Am. Nat.* 153:S1–S14.

Thomson, D.A., and K. A. Muench. 1977. Influence of tides and waves on the spawning behavior of the Gulf of California grunion, *Leuresthes sardina* (Jenkins and Evermann). *Bull. S. Calif. Acad. Sci.* 75:198–203.

Thornhill, R. 1976. Reproductive behavior of the lovebug, *Plecia nearctica* (Diptera: Bibionidae). *Ann. Entomol. Soc. Am.* 69:843–847.

Thornhill, R. 1980. Rape in *Panorpa* scorpionflies and a general rape hypothesis. *Anim. Behav.* 28:52–59.

Thornhill, R. 1981. *Panorpa* (Mecoptera: Panorpidae) scorpionflies: Systems for understanding resource-defense polygyny and alternative male reproductive efforts. *Annu. Rev. Ecol. Syst.* 12:355–386.

Thornhill, R., and J. Alcock 1983. *The Evolution of Insect Mating Systems*, Harvard University Press, Cambridge, MA.

Thornhill, R., and C. T. Palmer. 2000. *A Natural History of Rape: Biological Bases of Sexual Coercion*, MIT Press, Cambridge, MA.

Tomkins, J. L. 1999. Environmental and genetic determinants of the male forceps length dimorphism in the European earwig, *Forficula auricularia* L. *Behav. Ecol. Sociobiol.* 47:1–8.

Tomkins, J. L., and L. W. Simmons. 2000. Sperm competition games played by dimorphic male beetles: Fertilization gains with equal mating access. *Proc. R. Soc. London, Ser. B* 267:1547–1553.

Tram, U., and M. F. Wolfner. 1998. Seminal fluid regulation of female sexual attractiveness in *Drosophila melanogaster. Proc. Natl. Acad. Sci. USA* 95:4051–4054.

Travis, J. 1983. Variation in development patterns of larval anurans in temporary ponds. I. Persistent variation in a *Hyla granulosa* population. *Evolution* 37:496–512.

Travis, J., and B. D. Woodward. 1989. Social context and courtship flexibility in male sailfin mollies, *Poecilia latipinna* (Pisces: Poeciliidae). *Anim. Behav.* 38:1001–1012.

Travis, S. E., C. N. Slobodchikoff, and P. Keim. 1997. DNA fingerprinting reveals low genetic diversity in Gunnison's prairie dog (*Cynomys gunnisoni*). *J. Mammal.* 78:725–732.

Tregenza, T., and N. Wedell. 2000. Genetic compatibility, mate choice and patterns of parentage: Invited Review. *Mol. Ecol.* 9:1013–1027.

Trillmich, F.G.H., and K.G.K. Trillmich. 1984. The mating systems of pinnipeds and marine iguanas: Convergent evolution of polygyny. *Biol. J. Linn. Soc.* 21:209–216.

Trivers, R. L. 1972. Parental investment and sexual selection. In B. Campbell (ed.), *Sexual Selection and the Descent of Man*, Aldine, Chicago, pp. 136–179.

Trivers, R. L. 1985. *Social Evolution*, Benjamin/Cummings, Menlo Park, CA.

Trivers, R. L., and D. Willard. 1973. Natural selection of parental ability to vary the sex ratio of offspring. *Science* 179:90–92.

Tschinkel, W. R. 1993. Crowding, maternal age, age at pupation, and life-history of *Zophobas atratus* (Coleoptera: Tenebrionidae). *Ann. Entomol. Soc. Am.* 86:278–297.

Tsubaki Y., R. E. Hooper, and M. T. Siva-Jothy. 1997. Differences in adult and reproductive lifespan in the two male forms of *Mnais pruinosa costalis* (Selys) (Odonata: Calopterygidae). *Res. Popul. Ecol.* 39:149–155.

Tsuji, N., K. Yamauchi, and N. Yamamura. 1994. A mathematical model for wing dimorphism in male *Cardiocondyla* ants. *J. Ethol.* 12:19–24.

Upton, N.P.D. 1987. Asynchronous male and female life cycles in the sexually dimorphic, harem-forming isopod *Paragnathia formica* (Crustacea: Isopoda). *J. Zool.* 212:677–690.

van den Berghe, E. P., F. Wernerus, and R. R. Warner. 1989. Female choice and the mating cost of peripheral males. *Anim. Behav.* 38:875–884.

van Duivenboden, Y. A., A. W. Pieneman, and A. ter Maat. 1985. Multiple mating suppresses fecundity in the hermaphrodite freshwater snail, *Lymnaea stagnalis*: A laboratory study. *Anim. Behav.* 53:1184–1191.

van Rhijn, J. G. 1973. Behavioral dimorphism in male ruffs, *Philomachus pugnax* (L.). *Behaviour* 47:153–229.

Vehrencamp, S. L. 2000. Evolutionary routes to joint-female nesting in birds. *Behav. Ecol.* 11:334–344.

Verner, J. 1964. The evolution of polygamy in the long-billed marsh wren. *Evolution* 18:252–261.

Verner, J., and M. F. Willson. 1966. The influence of habitats on mating systems of North American passerine birds. *Ecology* 47:143–147.

Verrell, P. 1982. The sexual behaviour of the red-spotted newt, *Notophthalmus viridescens* (Amphibia: Urodela: Salamandridae). *Anim. Behav.* 30:1224–1236.

Verrell, P. 1997. Courtship behaviour of the Ouachita dusky salamander, *Desmognathus brimleyorum*, and a comparison with other desmognathine salamanders. *J. Zool.* 243:21–27.

Via, S., R. Gomulkiewicz, G. De Jong, S. M. Scheiner, C. D. Schlichting, and P. H. Van Tienderen. 1995. Adaptive phenotypic plasticity: Consensus and controversy. *Trends Ecol. Evol.* 10:212–217.

Via, S., and R. Lande. 1985. Genotype-environment interactions and the evolution of phenotypic plasticity. *Evolution* 39:505–522.

Vincent, A., I. Ahnesjö, and A. Berglund. 1994. Operational sex ratios and behavioural sex differences in a pipefish population. *Behav. Ecol. Sociol.* 34:435–442.

Vincent, A.C.J. 1994. Operational sex ratios in seahorses. *Behaviour* 128:153–166.

Vincent, A.C.J. 1995. A role for daily greetings in maintaining seahorse pair bonds. *Anim. Behav.* 49:258–260.

Vincent, A.C.J., and L. M. Sadler. 1995. Faithful pair bonds in wild seahorses, *Hippocampus whitei. Anim. Behav.* 50:1557–1569.

Vogel, S. 1988. *Life's Devices: The Physical World of Animals and Plants*, Princeton University Press, Princeton, NJ.

Vollrath, F. 1998. Dwarf males. *Trends Ecol. Evol.* 13:159–163.

Vuturo, S., and S. Shuster. 2000. Evidence of simultaneous hermaphroditism with a protandrous phase in the commensal polychaete, *Ophiodromus pugettensis. J. Ariz. Nev. Acad. Sci.* 35:7.

Waage, J. K. 1979. Dual function of the damselfly penis: Sperm removal and transfer. *Science* 203:916–918.

Waage, J. K. 1997. Parental investment—minding the kids or keeping control? In P.A. Gowaty (ed.), *Feminism and Evolutionary Biology: Boundaries, Intersections and Frontiers*, Chapman and Hall, New York, pp. 527–553.

Wachtmeister, C. 2001. Display in monogamous pairs: A review of empirical data and evolutionary explanations. *Anim. Behav.* 61:861–868.

Waddy, S. L., and D. E. Aiken. 1990. Intermolt insemination, an alternative mating strategy for the American lobster (*Homarus americanus*). *Can. J. Fish. Aquat. Sci.* 47:2402–2406.

Wade, M. J. 1979. Sexual selection and variance in reproductive success. *Am. Nat.* 114:742–764.

Wade, M. J. 1987. Measuring sexual selection. In J. W. Bradbury and M. B. Andersson (eds.), *Sexual Selection: Testing the Alternatives*, John Wiley and Sons, New York, pp. 197–207.

Wade, M. J. 1992. Sewall Wright: Gene interaction in the shifting balance theory. *Oxford Surv. Evol. Biol.* 6:35–62.

Wade, M. J. 1995. The ecology of sexual selection: Mean crowding of females and resource-defence polygyny. *Evol. Ecol.* 9:118–124.

Wade, M. J. 1998. The evolutionary genetics of maternal effects. In T. Mousseau and C. Fox (eds.), *Maternal Effects as Adaptations*, Oxford University Press, Oxford, pp. 5–21.

Wade, M. J. 2001. Epistasis, complex traits, and rates of evolution. *Genetica* 112:59–69.

Wade, M. J. 2002. A gene's eye view of epistasis, selection, and speciation. *J. Evol. Biology*, in press.

Wade, M. J. 2003. Community genetics and species interactions: a commentary. *Ecology*, in press.

Wade, M. J., and S. J. Arnold. 1980. The intensity of sexual selection in relation to male sexual behaviour, female choice, and pserm precedence. *Anim. Behav.* 28:446–461.

Wade, M. J., and S. Kalisz. 1990. The causes of natural selection. *Evolution* 44:1947–1955.

Wade, M. J., H. Patterson, N. W. Chang, and N. A. Johnson. 1994. Postcopulatory, prezygotic isolation in flour beetles. *Heredity* 72:163–167.

Wade, M. J., and S. J. Pruett-Jones. 1990. Female copying increases the variance in male mating success. *Proc. Natl. Acad. Sci. USA* 87:5749–5753.

Wade, M. J., and S. M. Shuster. 2002. Sexual selection favors female-biased sex ratios: The balance between Fisherian sex-ratio selection and sexual selection. *Am. Nat.*, in press.

Wagner, A. 1996. Does evolutionary plasticity evolve? *Evolution* 50:1008–1023.

Wagner, G. P., and L. Altenberg. 1996. Complex adaptations and the evolution of evolvability. *Evolution* 50:967–976.

Walker, W. F. 1980. Sperm utilization strategies in nonsocial insects. *Am. Nat.* 115:780–799.

Wallace, A. R. 1867. *Westminster Review*, July.

Wallace, A. R. 1868. *Journal of Travel*, Vol. 1, A. Murray (ed.), n.p.

Waltz, E. C., and L. L. Wolf. 1984. By Jove!! Why do alternative mating tactics assume so many different forms? *Am. Zool.* 24:333–344.

Waltz, E. C., and L. L. Wolf. 1988. Alternative mating tactics in male white-faced dragonflies (*Leucorrhinia intacta*): Plasticity of tactical options and consequences for reproductive success. *Evol. Ecol.* 2:205–231.

Wang, H., G. C. Paesen, P. A. Nuttall, and A. G. Barbour. 1998. Male ticks help their mates to feed. *Nature* 391:753–754.

Warner, R. R. 1984. Deferred reproduction as a response to sexual selection in a coral reef fish:a test of the life historical consequences. *Evolution* 38:148–162.

Warner, R. R., and D. R. Robertson. 1978. Sexual patterns in the labroid fishes of the western Caribbean. 1. The wrasses (Labridae). *Smithson. Contrib. Zool.* 254:1–27.

Warner, R. R., D. R. Robertson and E. G. Leigh. 1975. Sex change and sexual selection. *Science* 190:633–638.

Warner R. R., D. Y. Sharpio, A. Marcanato, and C. W. Petersen. 1995a. Sexual conflict: Males with highest mating success convey the lowest fertilization benefits to females. *Proc. R. Soc. London, Ser. B* 262:135–139.

Warner, R. R., F. Wernerus, P. Lejeune, and E. van den Berghe. 1995b. Dynamics of female choice for parental care in a fish species where care is facultative. *Behav. Ecol.* 6:73–81.

Waser, N. M., and M. V. Price. 1983. Pollinator behavior and natural selection for flower color in *Delphinium nelsonii. Nature* 302:422–424.

Weatherhead, P. J., K. W. Dufour, S. C. Lougheed, and C. G. Eckert. 1999. A test of the good-genes-as-heterozygosity hypothesis using red-winged blackbirds. *Behav. Ecol.* 10:619–625.

Weatherhead, P. J., and R. J. Robertson. 1979. Offspring quality and the polygyny threshold: "The sexy son hypothesis." *Am. Nat.* 113:201–208.

Webster, M. S., H. C. Chuang-Dobbs, and R. T. Holmes. 2001. Microsatellite identification of extrapair sires in a socially monogamous warbler. *Behav. Ecol.* 12:439–446.

Webster, M. S., S. Pruett-Jones, D. F. Westneat, and S. J. Arnold. 1995. Measuring the effects of pairing success, extra-pair copulations and mate quality on the opportunity for sexual selection. *Evolution* 49:1147–1157.

Wedell, N., and P. A. Cook. 1999. Strategic sperm allocation in the Small White butterfly *Pieris rapae* (Lepidoptera: Pieridae). *Funct. Ecol.* 13:85–93.

Weeks, S. C., V. Marcus, and B. R. Crosser. 1999. Inbreeding depression in a self-compatible, androdioecious crustacean, *Eulimnadia texana. Evolution* 53:472–483.

Wells, K. D. 1979. Reproductive behavior and male mating success in a neotropical toad, *Bufo typhonius. Biotropica* 11:301–307.

West-Eberhard, M. J. 1989. Phenotypic plasticity and the origins of diversity. *Annu. Rev. Ecol. Syst.* 20:249–278.

Westneat, D. F., and T. R. Birkhead. 1998. Alternative hypotheses linking the immune system and mate choice for good genes. *Proc. R. Soc. London, Ser. B* 265:1065–1073.

Westneat, D. F., P. C. Frederick, and R. H. Wiley. 1987. The use of genetic markers to estimate the frequency of alternative reproductive tactics. *Behav. Ecol. Sociobiol.* 21:35–45.

Westneat, D. F., L. A. McGraw, J. M. Fraterrigo, T. R. Birkhead, and F. Fletcher. 1998. Patterns of courtship behavior and ejaculate characteristics in male red-winged blackbirds. *Behav. Ecol. Sociobiol.* 43:161–171.

Westneat, D. F., and R. C. Sargent. 1996. Sex and parenting: The effects of sexual conflict and parentage on parental strategies. *Trends Ecol Evol.* 11:87–91.

Westneat, D. F., and P. W. Sherman. 1997. Density and extra-pair fertilizations in birds: A comparative analysis. *Behav. Ecol. Sociobiol.* 41:205–215.

Westneat, D. F., A. Walters, T. M. McCarthy, M. I. Hatch, and W. K. Hein. 2000. Alternative mechanisms of nonindependent mate choice. *Anim. Behav*. 59:467–476.

Whitehouse, M.E.A. 1991. To mate or fight? Male-male competition and alternative mating strategies in *Argyrodes antipodiana* (Theridiidae, Araneae). *Behav. Process.* 23:163–172.

Whiteman, H. H. 1997. Maintenance of polymorphism promoted by sex-specific fitness payoffs. *Evolution* 51:2039–2044.

Whitlock, M. C., and M. J. Wade. 1995. Speciation: Founder events and their effects on X-linked and autosomal genes. *Am. Nat.* 145:676–685.

Whittingham, L. A., P. D. Taylor, and R. J. Robertson. 1992. Confidence of paternity and male parental care. *Am. Nat.* 139:1115–1125.

Wickler, W., and U. Seibt. 1981. Monogamy in Crustacea and man. *Z. Tierpsychol.* 57:215–234.

Widemo, F., and I.P.F. Owens. 1995. Lek size, male mating skew and the evolution of lekking. *Nature* 373:148–151.

Wikelski, M., C. Carbone, and F. Trillmich. 1996. Lekking in marine iguanas: Female grouping and male reproductive strategies. *Anim. Behav*. 52:581–596.

Wiklund, C., A. Kaitala, and N. Wedell. 1998. Decoupling of reproductive rates and parental expenditure in a polyandrous butterfly. *Behav. Ecol.* 9:20–25.

Wilkinson, G. S., D. C. Presgraves, and L. Crymes. 1998. Male eyespan in stalk-eyed flies indicates genetic quality by meiotic drive suppression. *Nature* 391:276–279.

Williams, G. C. 1966. *Adaptation and Natural Selection*, Princeton University Press, Princeton, NJ.

Williams, G. C. 1975. *Sex and Evolution*, Princeton University Press, Princeton, NJ.

Willson, M. F. 1990. Sexual selection in plants and animals. *Trends Ecol. Evol.* 5:210–214.

Willson, M. F. 1994. Sexual selection in plants: Perspective and overview. *Am. Nat.* 144:S13–S39.

Willson, M. F. 1997. Variation in salmonid life histories: Patterns and perspectives. *USDA Res. Paper PNWW-RP-498*, pp. 1–50.

Willson, M. F., and N. Burley. 1983. *Mate Choice in Plants*, Princeton University Press, Princeton, NJ.

Wilson, E. O. 1971. *The Insect Societies*, Harvard University Press, Cambridge, MA.

Wilson, E. O. 1975. *Sociobiology: The New Synthesis*, Harvard University Press, Cambridge, MA.

Winn, A. A. 1996. The contributions of programmed developmental change and phenotypic plasticity to within-individual variation in leaf traits in *Dicerandra linearifolia*. *J. Evol. Biol.* 9:737–752.

Winslow, J. T., N. Hastings, C. S. Carter, C. R. Harbaugh, and T. R. Insel. 1993. A role for central vasopressin in pair bonding in monogamous prairie voles. *Nature* 365:545–548.

Wittenberger, J. A. 1981. *Animal Social Behavior*, Duxbury Press, Boston, MA.

Wolf, J. B. 2000. Gene interactions from maternal effects. *Evolution* 54:1882–1898.

Wolf, J. B., and E. D. Brodie. 1998. The coadaptation of parental and offspring characters. *Evolution* 52:299–308.

Wolf, J. B., E. D. Brodie, and A. J. Moore. 1999. The role of maternal and paternal effects in the evolution of parental quality by sexual selection. *J. Evol. Biol.* 12:1157–1167.

Wolf, J. B., E. D. Brodie III, and M. J. Wade. 2002. Genotype-environment interaction and evolution when the environment contains genes. In T. DeWitt and S. Scheiner (eds.), *Phenotypic Plasticity. Functional and Conceptual Approaches*, Oxford University Press, Oxford, in press.

Wolf, J. B., and M. J. Wade. 2001. On the assignment of fitness to parents and offspring: Whose fitness is it and when does it matter? *J. Evol. Ecol.* 14:347–358.

Wolf, J. B., M. J. Wade, and E. D. Brodie III. 2002. Evolution when the environment contains genes. In T. DeWitt and S. Scheiner (eds.), *Phenotypic Plasticity. Functional and Conceptual Approaches*, Oxford University Press, Oxford, in press.

Wolf, L. L., and E. C. Waltz. 1993. Alternative mating tactics in male white-faced dragonflies: Experimental evidence for a behavioural assessment ESS. *Anim. Behav.* 46:325–334.

Woolbright, L. L., E. J. Greene, and G. C. Rapp. 1990. Density dependent mate searching strategies in male wood frogs. *Anim. Behav.* 40:135–142.

Wright, S. 1929a. Fisher's theory of dominance. *Am. Nat.* 63:274–279.

Wright, S. 1929b. The evolution of dominance. *Am. Nat.* 63:556–561.

Wright, S. 1969. *Evolution and the Genetics of Populations*, University of Chicago Press, Chicago, Vol. 2.

Zahavi, A. 1975. Mate selection: A selection for a handicap. *J. Theor. Biol.* 53:205–214.

Zahavi, A. 1977. The cost of honesty. *J. Theor. Biol.* 67:603–605.

Zahavi, A., and A. Zahavi. 1997. *The Handicap Principle: A Missing Piece of Darwin's Puzzle*, Oxford University Press, Oxford, U.K.

Zamudio, K. R., and B. Sinervo. 2000. Polygyny, mate-guarding, and posthumous fertilization as alternative male mating strategies. *Proc. Natl. Acad. Sci. USA* 97:14427–14432.

Zamudio, K. R., and B. Sinervo. In press. Ecological and social contexts for the evolution of alternative mating strategies. In S. Fox, T. Baird, and K. McCoy (eds.), *Lizard Mating Systems*, John's Hopkins University Press, Baltimore, MD.

Zeh, D. W., and J. A. Zeh. 1988. Condition-dependent sex ornaments and field tests of sexual selection theory. *Am. Nat.* 132:454–459.

Zeh, J. A., and D. W. Zeh. 1994. Last-male sperm precedence breaks down when females mate with three males. *Proc. R. Soc. London, Ser. B* 257:287–292.

Zera, A. J., and R. F. Denno. 1997. Physiology and ecology of dispersal polymorphism in insects. *Annu. Rev. Entomol.* 42:207–230.

Zimmerer, E. J., and K. D. Kallman. 1989. Genetic basis for alternative reproductive tactics in the pygmy swordtail, *Xiphorphorus nigrensis. Evolution* 43:1298–1307.

Zuk, M., R. Thornhill, J. D. Ligon, K. Johnson, S. Austad, S. H. Ligon, N. W. Thornhill, and C. Costin. 1990. The role of male ornaments and courtship behavior in female mate choice of red jungle fowl. *Am. Nat.* 136:459–473.

Author Index

Word Index

Taxonomic Index